DBA 实战手记

薛晓刚 王明杰 宋希 李欢 唐政 冯元泊 唐永金 张博 著

机械工业出版社
CHINA MACHINE PRESS

本书是一本指导DBA进行数据库开发和运维的实用手册，本书共9章，包括漫谈数据库、如何提升数据库性能、如何运维好数据库、如何进行数据库设计、如何做好数据库之间的数据同步、认识HTAP技术、认识数据库的功能原理、认识数据库中的数学（逻辑与算法），以及DBA的日常：数据库管理及开发的最佳实践。另有附录DBA杂谈。

本书内容是资深DBA多年实践经验的归纳总结，涵盖DBA日常工作中的主要方面，通过剖析众多的真实案例，并解读原理、分享技巧、提供思路，内容具备较强的启发性和指导性。

本书的受众包括但不限于想要了解如何提升数据库性能的应用开发人员、DBA、业务架构师、IT架构师、数据架构师、数据分析师、系统架构师、解决方案架构师和企业高级管理人员。

图书在版编目（CIP）数据

DBA实战手记 / 薛晓刚等著. —北京：机械工业出版社，2024.6
ISBN 978-7-111-75766-5

Ⅰ.①D… Ⅱ.①薛… Ⅲ.①关系数据库系统 Ⅳ.①TP311.132.3

中国国家版本馆CIP数据核字（2024）第092538号

机械工业出版社（北京市百万庄大街22号　邮政编码100037）
策划编辑：王　斌　　　　　责任编辑：王　斌　解　芳
责任校对：杨　霞　陈　越　责任印制：常天培
固安县铭成印刷有限公司印刷
2024年7月第1版第1次印刷
184mm×240mm・30.75印张・722千字
标准书号：ISBN 978-7-111-75766-5
定价：169.00元

电话服务　　　　　　　　　网络服务
客服电话：010-88361066　　机　工　官　网：www.cmpbook.com
　　　　　010-88379833　　机　工　官　博：weibo.com/cmp1952
　　　　　010-68326294　　金　书　网：www.golden-book.com
封底无防伪标均为盗版　　　机工教育服务网：www.cmpedu.com

序言 1

晓刚和几位作者的新书（样稿）摆在案前，翻阅纵览，心有所感，两句话跃然纸上：于变化处见不变，于平淡处见真知。

我从 2009 年开始编写一个系列的作品，命名为《Oracle DBA 手记》；2024 年，晓刚的书命名为《DBA 实战手记》，这两者的差异，足见一个时代的变迁。

上一代 DBA 所处的时代是 Oracle 的世界，一家独大，所向无敌；而这一代 DBA 所处的则是商业数据库、开源数据库百花齐放、各展所长的新时代。

时代在改变，然而 DBA 不变，DBA 维护生产环境安全、稳定、高效运行的使命不变。在乱花渐欲迷人眼的数据库新时代，安稳不动如山岳的 DBA 精神则更加宝贵。

《DBA 实战手记》来自几位作者的生产实践总结，其中包含了 Oracle、MySQL、PostgreSQL、Elasticsearch、Redis 等不同数据库的知识分享，这充分表明：多样性数据库应用已经成为企业的现状。

在变革的浪潮中，如何避免焦虑与不安，从容应对新时代的挑战，我认为本书提供了一个出色的范例。

禅宗有个故事说，有人问大珠慧海禅师是怎么用功的，他答道："饥来吃饭困来眠。"对方说："大家都是这样的啊！那他们都跟你一样用功吗？"大珠禅师说："不同。他吃饭时不肯吃饭，百种需索；睡时不肯睡，千般计较。"

于 DBA 而言，千般计较都不如饥来吃饭困来眠。我们只要将过往的经验和方法抽象总结，平心静气地将其转移到各种新数据库上来——无外乎安装部署、SQL 优化、故障诊断……，一一经历、分析研究、记录总结，如是再三，则专业可成。

《DBA 实战手记》开宗明义，于篇首即指出 SQL 的重要性，更是在行文中通过对于索引知识的介绍、执行计划的分析、智能索引技术的解读，逐步强化这一对于性能具备关键影响的核心要素，作者更是在附录中给出了一份 SQL 开发规范。这些关于 SQL 的认知和经验，其实是具备普适价值的，DBA 们通过无数血泪教训总结的经验，可以推及到任何数据库中，而 SQL 能力将是 DBA 知识技能的"硬通货"。

本书通过大量实践应用案例，将数据库应用中的种种日常娓娓道来，于平淡处见真知，也一定能让读者在潜移默化中了解不同数据库的应用精髓。我认为，理解了各种数据库的原理知识，在生产中，就是如此见招拆招、履险如夷。

最后，我衷心祝贺《DBA 实战手记》的顺利出版！当下，能够经年累月地安心著述实属不易，而作者分享的经验更是宝贵。期待读者朋友们能够从书中获取对自己有价值的内容，开卷有益，共同进步。

云和恩墨（北京）信息技术有限公司创始人、鲲鹏 MVP　盖国强

2024 年 5 月 31 日

序言 2

在数据库运维领域，每年都有很多技术书籍出版，不过这些书籍大多数都是一些知识的总结，想要把这些知识运用到实际工作中，还存在一定的鸿沟。DBA 们都知道，对于工作中实战案例的抽象与总结以及 DBA 运维经验的总结才是最为难得的技术资料。这些资料必须出自于身经百战的资深 DBA 之手，他们自己在长期的工作中积累了丰富的案例，结合数据库的知识总结出发现问题、分析问题和解决问题的方法。不仅如此，这些案例与经验不能简单地呈现给读者，还需要具有知识分享精神的有心人认真记录下来，并经过提炼和总结才能形成体系化的知识，传授给需要学习这方面知识的人。

幸运的是，薛晓刚、王明杰、宋希等人恰好是这样一群善于总结、乐于分享的有心人，于是就有了《DBA 实战手记》这本书的诞生。这本书是作者对十多年工作的一次全面的总结，从一些细小的案例入手，经过对知识的体系化的梳理，最终上升到数据库理论层面，让读者能够从浅显的案例入手，从深处理解数据库原理性的知识。更为可贵的是，本书覆盖的数据库知识十分全面，不仅仅限于某种数据库的原理与某些具体的技术，在数据库运维、数据架构设计、系统优化、数据同步、高可用与容灾备份等都有涉猎。

技术人员写书比较容易写得干涩，不过本书的读者不必担心，本书的可读性很好，我拿到样稿后原本只是想随便翻翻，不过看过几页后就被吸引住了。能用一种十分愉快的心情去阅读一本技术书籍是十分难得的。我觉得这本书不仅对从事数据库工作或者想要踏上 DBA 职业之路的朋友有所帮助，应用架构师、系统架构师也会从中受益良多。

白鳝（徐戟）

南京基石数据技术有限责任公司技术总监

南瑞子衿技术团队首席架构师

2024 年 6 月

序言 3

在现代信息技术高速发展的今天,数据库作为信息系统的核心组件,其重要性不言而喻。《DBA 实战手记》凝聚了多位资深 DBA(数据库管理员)多年来的经验和智慧,全面涵盖了数据库管理、运维、设计及优化的方方面面,是一本关于数据库管理和运维的实战手册,能够为广大从事数据库工作的专业人员提供宝贵的指导和借鉴。

作为本书的主要作者,薛晓刚先生及其团队在数据库领域深耕多年,通过实践积累了丰富的经验。他们将这些经验通过本书加以总结,不仅分享了如何提升数据库性能、优化 SQL 语句、进行数据库同步等技术,还详细介绍了数据库设计和 HTAP(混合事务/分析处理)技术的应用。本书通过真实案例的剖析,揭示了许多数据库工作中的常见问题及解决方案,具有极高的实用价值和指导意义。

本书内容涵盖了数据库的多个方面,从基础概念、性能优化、故障处理、数据库设计到数据同步,每一章都以丰富的案例和详实的技术细节为读者提供了深入浅出的讲解。特别是对于那些希望在数据库领域深入发展的技术人员来说,本书不仅是一本参考手册,更是指导实践的宝贵资源。

在飞速发展的 AI 时代,数据的重要性愈加凸显。随着人工智能和机器学习技术的飞速发展,数据库作为数据存储和管理的核心,其性能和稳定性直接影响到智能应用的效果。《DBA 实战手记》不仅总结了传统数据库管理和运维的最佳实践,还展望了未来数据库技术的发展方向。书中对于 HTAP 技术的介绍和实践,更是为中小企业如何在降低成本的同时满足业务需求提供了宝贵的经验。

可以说,《DBA 实战手记》是一本不可多得的好书,无论是对于初学者还是有经验的 DBA,都能从中受益良多。它不仅帮助读者提升数据库管理和运维的能力,还提供了许多关于职业发展和定位的思考与建议。相信本书的出版,将为广大数据库从业人员带来巨大的帮助和启发,也为企业的信息化和智能化建设提供坚实的技术支持。

期待广大数据领域的从业者,尤其是 DBA 们能够翻开这本凝聚了作者多年心血和智慧的《DBA 实战手记》,从中汲取营养,在充满无尽可能性的人工智能时代从容应对更多机遇与挑战。

贺仁龙
上海流程智造科技创新研究院有限公司 董事、总经理
IEEE Shanghai Section 副主席
上海人工智能与社会发展研究会副会长
2024 年 6 月 12 日

前言

为什么要写这本书

数据库是一个独立的技术领域，几乎所有的后端开发都要和数据库（关系数据库和非关系数据库）打交道，不读写数据库的后端应用程序几乎不存在。而应用程序中 50% 的代码甚至极端情况下 80% 的代码都是 SQL。所以与其说应用程序开发不如说是基于数据库的应用开发。

作为本书的第一作者，笔者一开始并没有从事真正意义上的数据库工作，而是从其他工作岗位转到数据库岗位的。在经过前辈和名师指导后才慢慢了解了数据库。后来还成为数据库的认证讲师，培训过的学员来自各行各业，有近千人之多。

在不断深入了解数据库的过程中，笔者发现很多研发问题归根结底都是数据库的问题，而且发现并解决这些数据库的相关问题，可以帮助和指导研发人员更好地开展工作，于是就全身心地投入到了数据库的应用、运维、设计、架构、治理等相关工作中。在多年的工作和授课中，笔者还发现，即便是有多年工作经验的 DBA 也仍然对数据库技术有很多误解和错误的认知，很多基于数据库技术的问题和教训反反复复地出现，而且这些问题在全国范围内也具有很强的共性。

因此，笔者决定撰写一本介绍数据库管理与运维、数据架构以及数据库开发技巧的书籍来为广大 DBA 解惑，让开发人员知道如何使用数据库，让 DBA 知道如何指导开发人员更好地进行应用的设计。希望开发人员和 DBA 都明白数据才是架构的中心。具体来说，笔者希望这本书起到以下作用。

分享经验：笔者和笔者带领的团队在实践中积累了大量的经验和实践方法，这对于企业数据库的开发、运维、设计、架构十分有价值。我们通过本书把这些经验和方法加以总结，供广大 DBA 和开发人员参考，帮助他们更好地发挥出数据库应有的性能，解决或避免大部分在应用开发层面出现的问题和故障，让读者少走一些弯路。

DBA 重新定位：希望通过这本书，让广大 DBA 明确自己的职业规划和定位，认识想运维好数据库就要从幕后走到台前，不是单纯地去应对需求，而是指导开发人员编写好的 SQL，一方面提升工作效率，另一方面提升自身价值。从运维"小工"成长为更具价值的架构师。

降低企业 IT 成本：本书分享的经验、实施方法论和案例，可以为广大企业提供非常有价值的借鉴。希望通过这本书,帮助广大企业降低数据库的故障率,提升系统的稳定性。通过选择合适的数据架构,避免因为数据库导致的架构复杂而产生的高昂的 IT 成本。

开发人员能力提升:希望通过本书给开发岗位的读者一些思路,不再做"CRUD 工程师",而是成为可以集设计和开发于一身、从事企业智能化应用建设和咨询的专家,更好地帮助企业实现数字化转型和智能化应用的建设。

构建生态圈:我们希望通过这本书可以广结朋友,无论是 DBA、开发者,还是从事企业数字化建设的朋友。希望广大读者朋友可以通过这本书加入到 DBA 交流群中(扫描封面勒口二维码加入),能够相互学习、相互启发、共同进步。

本书的内容

本书内容具体介绍如下。

第 1 章　漫谈数据库。介绍了数据库的概念和数据库的分类,通过本章,读者将了解到主流的数据库种类以及数据库技术的发展趋势,包括大数据、区块链这些数据库技术栈和 DBA 与数据库的关系。

第 2 章　如何提升数据库性能。介绍数据库性能优化技术,并对不同类型的索引进行了分类和介绍,还介绍了从优化需求和设计两个方面来提升数据库性能的具体案例。

第 3 章　如何运维好数据库。介绍数据库的主要故障类型和现象,列举了十几种各式各样的数据库故障案例及故障原因分析。

第 4 章　如何进行数据库设计。介绍了数据库选型、数据库架构设计以及数据库分开与合并的优缺点。

第 5 章　如何做好数据库之间的数据同步。介绍了数据库同步的各种场景和技术栈。其中有同构数据库的同步,也有异构数据库之间的同步。尤其是 OGG 的微服务版的实践,包括官方文档中没有全面覆盖(或者疏漏)的知识点。

第 6 章　认识 HTAP 技术。介绍了 HTAP 技术架构的原理以及对中小企业的意义,结合国内大多数企业现状,HTAP 技术是去除 Hadoop 技术栈的最佳实践,可以使企业在满足业务的前提下降低成本。

第 7 章　认识数据库的功能原理。介绍了数据库优化器和执行器的一些底层原理。有助于读者了解数据库的运行机制。

第 8 章　认识数据库中的数学(逻辑与算法)。介绍了数据库优化中涉及的一些逻辑和算法的知识。从逻辑和算法上切入,让读者明白优化的核心是逻辑和算法。

第 9 章　DBA 的日常:数据库管理及开发的最佳实践。集中介绍了若干笔者在工作中总结积累的数据库管理及应用开发的实践案例,示范性和启发性很强,通过这些案例可以解决读者在数据库开发、运维、构建上的许多问题。

除以上内容之外,本书还有附录 1 篇:DBA 杂谈,分享了包括 DBA 的职业规划以及笔者总结的 SQL 开发规范等的六个话题。本书另将部分限于篇幅没有在正文中体现的内容以在线的扩展阅读文档加以呈现,主要内容包括介绍各类型数据库的应用场景、若干提升数据库性能的实践案例。读者可以通过扫描二维码访问笔者的微信公众号阅读这部分内容。

本书的价值创新

笔者认为，本书分享的众多实践案例是最具价值的内容。这些案例具备很强的启发性，例如：从对 SQL 进行优化、提升数据库性能的案例中可以看出，即使不了解业务，凭借基本的逻辑和算法也可以对数据库进行优化，提升数据库性能。对于数据库设计的案例，可以引导读者认真思考应该怎么做数据库的架构设计，因为良好的架构设计所获收益远大于对 SQL 本身优化获得的收益。在一些案例中强调了 DBA 应当认真考虑如何控制需求、引导需求方提出合理的需求，而不是一味执行，因为不合理的需求或者是无意义的需求带来的问题是很多的，技术人员是可以指导业务方开展工作的。对于数据库故障的案例则侧重强调了分析解决问题的思路，如果读者按照本书给出的思路去管理、运维、设计数据库，那么由应用开发层面出现的问题，基本是可以得到控制的。

本书适合的读者人群

本书适合对数据库开发和运维有一定兴趣和基础的读者，包括但不限于以下人群：
- 后端基于数据库的 Java 开发人员。
- 使用 SQL 语言从事数据分析的分析师。
- 从事数据库运维的 DBA。
- 企业研发主管、企业运维主管。

致谢

在完成本书的过程中，我获得了许多支持和帮助。首先要感谢我的家人，他们在整个写作过程中一直给予我勇气、支持和鼓励。在写作期间，我家中发生了很多事情，但家人们仍然全力支持我去完成书稿。没有他们的支持和帮助，我不可能完成这本书。

我还要感谢本书的其他合作者，他们是：王明杰博士，现担任欧冶云商物联网技术首席师，负责撰写本书物联网场景和时序数据库的相关内容；宋希，现担任欧冶云商资深算法工程师，负责撰写本书的动态规划相关内容；李欢，来自字节跳动，作为资深的移动开发者，他撰写了移动端数据库的相关内容。唐政、冯元泊、唐永金三位同事帮助归纳整理了我的案例、文章和想法。感谢张博参与书中的配图制作，还有朱盈易后期帮助笔者解决了一些配图问题。几位合作者贡献良多，也在此表示感谢！

我还要感谢本书的策划编辑王斌，他为我们提供了很多有益的建议，他的专业知识和经验对我们来说是宝贵的。

由于笔者水平有限，书中难免出现错误和不足之处，敬请广大读者批评指正。

薛晓刚

2024 年 6 月 24 日

目录

序言 1
序言 2
序言 3
前言

第 1 章　漫谈数据库　　　　　　　　　　　　　　　　　　　　　　1
　1.1　什么是数据库　　　　　　　　　　　　　　　　　　　　　　　2
　1.2　数据库发展史　　　　　　　　　　　　　　　　　　　　　　　2
　　1.2.1　国外数据库的历史　　　　　　　　　　　　　　　　　　　2
　　1.2.2　国内数据库的历史　　　　　　　　　　　　　　　　　　　4
　1.3　数据库的主要分类　　　　　　　　　　　　　　　　　　　　　6
　　1.3.1　关系数据库：传统交易数据库　　　　　　　　　　　　　　6
　　1.3.2　键值数据库：基于 KV 键值对的内存数据库　　　　　　　　7
　　1.3.3　列式数据库：列式存储的分析型数据库　　　　　　　　　　9
　　1.3.4　文档型数据库：松散型数据结构的数据库　　　　　　　　 10
　　1.3.5　图数据库：用于社交图谱的数据库　　　　　　　　　　　 11
　　1.3.6　时序数据库：适用于物联网场景的数据库　　　　　　　　 12
　　1.3.7　搜索引擎数据库：全文索引的数据库　　　　　　　　　　 14
　　1.3.8　多模数据库：具有多种数据库模式的融合数据库　　　　　 17
　　1.3.9　最容易忽视的数据库——移动端数据库　　　　　　　　　 19
　1.4　数据库应用的发展趋势　　　　　　　　　　　　　　　　　　 22
　　1.4.1　国外数据库应用发展趋势　　　　　　　　　　　　　　　 22
　　1.4.2　国内数据库应用发展趋势　　　　　　　　　　　　　　　 24
　1.5　数据库与新兴数字技术　　　　　　　　　　　　　　　　　　 30
　　1.5.1　数据库与大数据技术　　　　　　　　　　　　　　　　　 30
　　1.5.2　数据库的延伸：区块链、物联网　　　　　　　　　　　　 32
　1.6　与数据库长相厮守的 DBA　　　　　　　　　　　　　　　　　 33

1.6.1　DBA 的定义与内涵　33
1.6.2　DBA 的工作职责　34
1.6.3　DBA 需要具备的能力　35
1.6.4　DBA 的价值　35
1.6.5　DBA 是可以做一辈子的职业　36

第 2 章　如何提升数据库性能　38

2.1　通过索引提升性能　39
2.1.1　索引的概念及原理　39
2.1.2　索引的种类　44
2.1.3　规避索引使用的误区　68
2.1.4　通过索引实现海量数据中的高效查询　70
2.1.5　自动化索引：数据库自治的趋势　71

2.2　通过 SQL 优化提升性能　78
2.2.1　SQL 优化实现高速执行任务　79
2.2.2　慎用分页有效提升性能　86
2.2.3　从认知上杜绝低效 SQL　89

2.3　避免数据库对象设计失误　91
2.3.1　避免不必要的多表关联导致的低效查询　91
2.3.2　避免动态计算结果没有单独存储导致的低效查询　93
2.3.3　避免没有明确需求查询条件导致的低效查询　97
2.3.4　避免过度分表　98

2.4　从识别需求的合理性提升性能　101
2.4.1　拒绝无效需求　101
2.4.2　正确理解开发需求　104
2.4.3　拒绝不合理的需求　106
2.4.4　引导业务改善需求　107

2.5　减少 IO 操作提升数据库性能　108
2.5.1　正确认识数据库的性能　108
2.5.2　减少 IO 交互——批量写入数据（MySQL）　110
2.5.3　减少 IO 交互——批量写入数据（Oracle）　122
2.5.4　减少 IO 交互——批量写入数据（PostgreSQL）　123
2.5.5　精简架构　127

第 3 章　如何运维好数据库　130

3.1　运维好数据库的关键：处理故障　131
3.1.1　常见的数据库故障类型　131

  3.1.2 数据库故障的危害 132
3.2 分析处理数据库故障的关键点 133
  3.2.1 分析数据库故障的主要原因——SQL 133
  3.2.2 分析 SQL 的主要问题——处理速度慢 134
  3.2.3 分析数据库参数对数据库的影响 134
  3.2.4 分析硬件短板带来的问题——IO 吞吐能力与 CPU 计算能力 135
3.3 数据库故障处理的典型案例分析 138
  3.3.1 一次 Elasticsearch 误删除的故障分析 139
  3.3.2 一次配置文件丢失导致的 MySQL 数据库故障分析 141
  3.3.3 一次 In-Memory 丢失引起的故障分析 146
  3.3.4 一次疑似分区查询异常的故障分析 150
  3.3.5 一次数据库归档导致的故障分析 159
  3.3.6 一次数据库 binlog 写入失败的故障分析 163
  3.3.7 一次两表关联导致的故障分析 165
  3.3.8 数据库连接数与连接复用不当的故障分析 166
  3.3.9 一次数据库 CPU 使用率 100% 的故障分析 168
  3.3.10 一次数据库索引不当引起的故障分析 169
  3.3.11 一次数据库主从延迟过大的故障分析 172
  3.3.12 一次数据库主从不一致的故障分析 174
  3.3.13 一次 Redis 数据库无法启动的故障分析 182
  3.3.14 一次 MySQL 数据库数据类型不恰当导致的故障 183
  3.3.15 一次数据库全表查询的优化 187

第 4 章 如何进行数据库设计 197

4.1 数据库都有哪些架构 198
  4.1.1 集中式架构 198
  4.1.2 分布式架构 200
  4.1.3 数据库内部的体系架构 201
  4.1.4 "烟囱"式的数据库架构 201
  4.1.5 独立业务线的数据库架构 202
4.2 根据实际场景选择数据库架构 203
4.3 五个维度谈数据库选型 203
  4.3.1 从业务场景特征维度 204
  4.3.2 从数据规模大小维度 204
  4.3.3 从用户自身开发团队能力维度 205
  4.3.4 从用户自身运维团队能力维度 205
  4.3.5 从公司管理能力维度 206

4.4 数据库拆分的利与弊   206
4.4.1 数据库拆分的背景   207
4.4.2 数据库拆分的三大问题：一致性、数据关联、数据同步   208
4.4.3 分表带来的问题：一致性、聚合、排序、扩缩容   208
4.5 如何看待数据库的合并   210
4.5.1 为何要做数据库合并   210
4.5.2 数据库合并的意义：降成本、提升稳定性   211
4.5.3 数据库合并带来的问题：鸡蛋放在一个篮子里   211
4.5.4 数据库合并的前提：高质量 SQL、硬件的进步、稳定的基础环境   212
4.6 CAP 理论与分布式数据库   212
4.6.1 CAP 理论概述   213
4.6.2 CAP 理论的延展   213
4.6.3 分库分表不是分布式数据库   214
4.7 如何看待数据库与中间件   215
4.7.1 数据库与中间件：上下游的"难兄难弟"   215
4.7.2 中间件内存溢出的原因：源头还是数据库   215
4.7.3 适当减少数据库与中间件的数量：避免不必要的故障节点   216

第 5 章 如何做好数据库之间的数据同步   217
5.1 数据同步的作用   218
5.1.1 实现数据传输   218
5.1.2 实现数据汇聚   218
5.1.3 实现数据迁移   219
5.2 数据库同步的分类   219
5.2.1 同构数据库的同步   219
5.2.2 异构数据库的同步   219
5.2.3 数据库到消息队列的同步   219
5.2.4 数据库到 Hadoop 的数据同步   219
5.3 同构数据库数据同步的示例   220
5.3.1 使用 dblink 实现 Oracle 数据库之间的数据同步   220
5.3.2 使用物化视图实现远程 Oracle 数据库之间的数据同步   226
5.3.3 使用插件实现 MySQL 数据库之间的数据同步   231
5.3.4 利用数据同步实现 MySQL 的版本升级   234
5.3.5 采用多源复制的功能实现 MySQL 数据库之间的数据同步   239
5.3.6 采用主从模式实现 PostgreSQL 主从数据同步   242
5.4 异构数据库同步的实例——基于 OGG   248
5.4.1 CDC 简介   248

5.4.2　OGG 概述　249
　　5.4.3　OGG 微服务版的安装部署　249
　　5.4.4　使用 OGG for Oracle 实现 Oracle 数据库之间的数据同步　251
　　5.4.5　使用 OGG for MySQL 实现 MySQL 与异构数据库的同步　279
　　5.4.6　使用 OGG for BigData 实现大数据组件与异构数据库的数据同步　284

第 6 章　认识 HTAP 技术　294

6.1　HTAP 的概念及其价值　295
　　6.1.1　HTAP：混合事物/分析处理　295
　　6.1.2　HTAP 的价值　295
6.2　HTAP 的几种实现方式　297
　　6.2.1　垂直方向的实现：以 Oracle 数据库为例　297
　　6.2.2　水平方向的实现：以 MySQL、TiDB 数据库为例　298
　　6.2.3　其他类型数据库的实现方式　298

第 7 章　认识数据库的功能原理　300

7.1　优化器——基于统计学原理　301
7.2　数据库的查询——火山模型　316
7.3　数据库 AI——向量化　319
7.4　编译执行　323

第 8 章　认识数据库中的数学（逻辑与算法）　326

8.1　数据库中的典型逻辑与算法　327
　　8.1.1　从斐波那契数列到数据分析　327
　　8.1.2　增加数据维度处理——减少关联　332
　　8.1.3　多表关联算法——一对多与多对多　336
　　8.1.4　排除处理算法——笛卡儿积　341
　　8.1.5　函数算法——对 SQL 的影响　348
8.2　动态规划法在数据库中的应用　351
　　8.2.1　动态规划原理 1：爬楼梯　351
　　8.2.2　动态规划原理 2：背包问题　355
　　8.2.3　动态规划的五部曲　361
8.3　数据库开发中的逻辑思维　362
　　8.3.1　元数据的空与非空　362
　　8.3.2　优化器处理极值和极限：加速查询速度　369
　　8.3.3　并发处理热点的逻辑：避免锁　371
　　8.3.4　减库存的逻辑：从设计出发防止超卖　373

第 9 章　DBA 的日常：数据库管理及开发的最佳实践　375

9.1　七个针对数据库特性的最佳实践　376
9.1.1　Oracle 的 DML 重定向　376
9.1.2　Oracle 的资源隔离　381
9.1.3　MySQL 的延迟复制——挽救误删除　389
9.1.4　PostgreSQL 的延迟复制　395
9.1.5　数据库对于磁盘 IO 吞吐的要求　402

9.2　面向执行器和优化器的最佳实践　403
9.2.1　主流数据库的事务异常处理机制对比（MySQL、PostgreSQL、Oracle）　403
9.2.2　MySQL 和达梦两个数据库优化器的对比　406
9.2.3　SQL 语句的解析过程分析　407
9.2.4　优化器的多表关联　411
9.2.5　SQL 使用 in 进行子查询　412
9.2.6　数据归档和数据迁移　419

9.3　数据库的复制（克隆）与高可用受控切换的最佳实践　424
9.3.1　Oracle 的 PDB 数据库克隆　425
9.3.2　MySQL 克隆的关键点　428
9.3.3　MySQL 的 MGR 架构以及受控切换　431
9.3.4　PostgreSQL 的高可用切换　434
9.3.5　Redis 的高可用切换　440

9.4　三个 SQL 编写的最佳实践　445
9.4.1　关于 MyBatis 开发框架使用绑定变量的实践　445
9.4.2　使用 exists 的 SQL 语句的改写　451
9.4.3　设计上出发减少 SQL 的标量子查询　458

9.5　时序数据库使用的最佳实践　463
9.5.1　时序数据库的数据库表设计　463
9.5.2　时序数据库的数据分析　466

附录　DBA 杂谈　468

第 1 章

漫谈数据库

本章引领读者重温与广大 DBA 日夜相伴的数据库的概念、发展史、分类,介绍数据库应用的发展趋势,以及数据库和大数据、区块链、物联网等的关系。

本章内容
什么是数据库
数据库发展史
数据库的几种主要分类
数据库应用的发展趋势
数据库与新兴数字技术
与数据库长相厮守的 DBA

1.1 什么是数据库

如果问什么是数据库，大家的认知差别很大，比如有人会认为 Oracle 和 MySQL 是数据库，而 Redis、Elasticsearch 这种算中间件。其实有这样的认识主要是因为对数据库定义的认知仍然停留在存储结构化数据的关系数据库阶段，而没有意识到存储半结构化数据的 NoSQL 数据库也是数据库。

就如百度百科中提到：数据库是"按照数据结构来组织、存储和管理数据的仓库"，是一个长期存储在计算机内的、有组织的、可共享的、统一管理的大量数据的集合。这就是数据库的定义。即存储数据库的仓库就是数据库。

数据结构是多种多样的，所以数据库也是多种多样的。可共享是指并发的访问，可以是有锁的也可以是没有锁，这是数据库区别于电子表格的一个不同点，否则 Excel 也可以被称为数据库了。

Excel 的确具备了结构化数据库中表的样子，故 Excel 也可以通过 ODBC 导入到 Access、SQL Sever、MySQL、Oracle 等关系数据库中，而这些关系数据库也可以将结果集导出成为 Excel。

有 Excel 就能解决大多数问题，为什么还要使用数据库进而创建一个信息化系统呢？这是因为 Excel 不能解决多人同时操作数据，也不能处理海量数据的存储。

1.2 数据库发展史

数据库起源于 20 世纪 60 年代，到目前为止，在数据库领域中表现出色且被广泛使用的大部分产品，都来自那些起步较早、经验丰富的美国老牌数据库厂商。

1.2.1 国外数据库的历史

云和恩墨公司编制了纸质版的数据库发展年鉴，本书特意参照绘制了数据库发展时间轴，可以对数据库的演进历史有一个更好的认识。如图 1-1 所示，数据库从出现到今天经历了三个阶段，分别是 1960—1970 年前关系数据库阶段、1970—2008 年关系数据库阶段和 2008 年至今的后关系数据库阶段。

在前关系数据库这个阶段，美国加州大学伯克利分校一个名为 Ingres 研究项目诞生了，全称为 Interactive Graphics and Retrieval System，是最早实现的数据库系统之一。Ingres 的源代码使用 BSD 许可证分发，在其基础上衍生了很多商业数据库软件，是历史上最有影响的计算机研究项目之一。

PostgreSQL 数据库就是源自 Ingres 的对象关系数据库管理系统。1985 年 Stonebraker 启动 Post-Ingres 计划；1994 年，来自香港的研究生 Andrew Yu 和 Jolly Chen 为该项目增加 SQL 能力，推出 Postgres95 大获成功；1996 年正式更名为 PostgreSQL，这就是 PostgreSQL 的历史变迁。

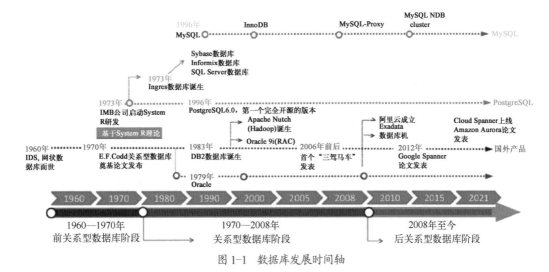

图 1-1 数据库发展时间轴

1970 年，E.F.Codd 博士在 IBM 发表了题为《A Relational Model of Data for Large Shared Data Banks》的论文，并由此创建了数据库的关系模型。1973 年，Ingres 研究项目诞生，在同一时期，IBM 还启动了 System R 项目。1977 年，System R 原型进入客户实践；1983 年，IBM 发布了 Database2(DB2)for MVS，作为 IBM 公司开发的关系数据库管理系统，DB2 正式诞生。

就在 IBM 的 System R 原型进入客户实践的同时，Larry Ellison、Bob Miner 和 Ed Oates 于 1977 年创建 SDL（Software Development Laboratories），后更名为 Oracle，其公司中文翻译为甲骨文。作为甲骨文公司的一款关系数据库管理系统。在 1998 年的 v8.15 版本中加入"i"（Oracle 8i）代表互联网；2001 年 6 月，在 Oracle Open World 大会上，Oracle 发布了 Oracle 9i，其诸多新特性中，最重要的就是 Real Application Clusters(RAC)。在 2003 年推出 Oracle 10g（g 代表网格）；在 2013 年推出 Oracle 12c（c 代表云）；现以年度命名版本，最新发布的版本为 21c。笔者在编写本书的时候 Oracle 官方已经发布了 23c 的开发版。也是目前数据库领域霸主地位的数据库产品。

随后创立于 1980 年的 Relational Database Systems 公司推出一款关系数据库管理系统 Informix，含义为 Information on UNIX，1996 年，Relational Database Systems 公司收购了 Stonebraker 创立的 Illustra 公司。在 2001 年 IBM 收购了 Informix 数据库。

美国 Sybase 公司研制了一种关系数据库系统，该数据库与公司同名，叫作 Sybase。是一种典型的 UNIX 或 Windows NT 平台上客户机/服务器环境下的大型数据库系统。

以上数据库都是商用数据库，下面介绍一下同一时期的开源数据库。

MySQL 是目前最流行的开源关系数据库管理系统，由瑞典 MySQL AB 公司开发并于 1995 年推出。MySQL AB 公司于 2008 年被 Sun 公司收购。到 2009 年，Oracle 又收购了 Sun 公司。现在，MySQl 是 Oracle 旗下产品，其 License（许可证）分为免费的社区版与收费的标准版、企业版等。借助互联网，MySQL 得以流行。几乎所有的互联网公司都在使用 MySQL。

"MySQL 之父"Monty 在 Sun 公司被甲骨文收购以后，担心 MySQL 未来的开源问题，于是创立了一个数据库——MariaDB。MariaDB 是 MySQL 的一个分支，由社区开发，有商业支持，旨在 GNUGPL 下保持免费和开源。MariaDB 与 MySQL 高度兼容，包括 API 和命令行，使之能轻松成为 MySQL 的替代品，MariaDB 的名字来源于 Monty 的女儿的名字。

以上介绍的都是关系数据库的历史，相比较而言非关系数据的历史较短。

MongoDB 是一种面向文档的数据库管理系统，用 C++等语言撰写而成，以解决应用程序开发中的大量现实问题。MongoDB 由 MongoDB Inc（当时是 10gen 团队）于 2007 年 10 月开发，2009 年 2 月首度推出，现以服务器端公共许可（SSPL）分发。

Neo4j 是由 Java 实现的开源图数据库。自 2003 年开始研发，直到 2007 年正式发布第一版。Neo4j 是一个高性能的 NoSQL 数据库，有一个嵌入式的、基于磁盘的、具备完全的事务特性的 Java 持久化引擎，将结构化数据存储在网络（即图）上而非表中，具有成熟数据库的所有特性。

Redis 是一个使用 ANSI C 编写的开源、支持网络、基于内存、分布式、可选持久性的键值数据库。诞生于 2009 年，在 2013 年 5 月之前，其开发由 VMware 赞助；2013 年 5 月至 2015 年 6 月期间，其开发由 Pivotal 赞助；从 2015 年 6 月开始，Redis 的开发由 Redis Labs 赞助。

Snowflake 是一个云原生的数据仓库。公司成立于 2012 年 7 月，提供基于云的数据存储和分析服务，通常称为"数据仓库即服务"。它允许企业用户使用基于云的硬件和软件来存储和分析数据。

2003—2006 年，Google 三年内陆续发表了 *Google File System*（缩写为 GFS）、*MapReduce* 和 *BigTable* 三篇技术论文，成为后来云计算发展的重要基石也是大数据技术的理论依据。

SQLite 是一款轻型的、遵守 ACID 且实现大多数 SQL 标准的嵌入式数据库。SQLite 不是一个 C/S 结构的数据库引擎，而是被集成在 App 中。它是 D.Richard Hipp 创建的公有领域项目，是部署最广泛的数据库引擎，可以说是全世界装机最多的数据库，因为人们现在的手机中几乎都有 SQLite 的存在。

1.2.2　国内数据库的历史

1978 年，萨师煊等学者最早引入"信息"一词作为我国高等学校经济管理类的专业名称，创建了经济信息管理系。这是我国高等学校中第一个以信息技术在经济管理领域中的应用为特色的系科，萨师煊是第一任系主任。1982 年，萨师煊起草了国内第一个本科计算机专业"数据库系统概论"课程的教学大纲。萨师煊老师将"数据库"带给了进入大学的年轻学子。国产数据库的发展历程如图 1-2 所示。

1992 年，华中理工大学（后改名为华中科技大学）达梦数据库与多媒体技术研究所成立。达梦的拼音缩写是 DM，英文则是 Database Management 的缩写。1996 年，第二代达梦数据库"DM2"问世。这是我国最早自主研发的数据库。达梦数据库的体系结构比较像 Oracle，兼容 Oracle 数据库，当然其也兼容其他数据库。

人大金仓数据库最早起源于中国数据库学科发源地——中国人民大学信息学院。萨师煊和王珊老师在 1999 年创建人大金仓公司，并基于自研技术推出数据库产品 KingbaseES。

2020 年，TDSQL 团队成立；2024 年，TDSQL 上线，其以 MySQL 为基础研发，主要支撑腾讯计费交易系统。2020 年，腾讯云的 TDSQL、TBase、CynosDB 三个产品线统一为"企业级分布式数据库 TDSQL"，涵盖分布式、分析型、云原生等多引擎融合的完整数据库产品体系。

图 1-2 国产数据库发展图鉴

GBase 是天津南大通用数据技术股份有限公司推出的自主品牌数据库系列产品。南大通用成立于 2005 年。其产品主要包括事务型数据库 GBase 8s（2007 年）、分布式分析型数据库 GBase 8a（2010 年）、多模多态分布式数据库 GBase 8c（2021 年）。GBase 8s 最新版本是基于共享。

2008 年天津神舟通用数据技术有限公司推出了自己的数据库产品，OSCAR 数据库。后改名为神通数据库管理系统。

同在 2008 年，亚信科技打造的一款分布式关系数据库，名为 AntDB。AntDB 数据库采用原生分布式架构，融合事务处理和在线分析。

OceanBase 是一款自主研发的企业级原生分布式数据库，在阳振坤博士的带领下于 2010 年立项，2019 和 2020 年连续两次打破 TPC-C 世界纪录，创造 7.07 亿 tpmC 的成绩，2021 年以 1526 万 QphH@30,000GB 成绩登顶 TPC-H 榜单。

SequoiaDB 是由前北美 IBM 的 DB2 分布式内核团队成员于 2011 年归国后自研的一款原生分布式数据库。于 2013 年发布巨杉数据库 v1.0；2020 年发布 V5.0，提出融合多模数据湖、实时数据湖及数仓能力，形成"湖仓一体"数据基础设施。

GaussDB 最早源于华为在 2007 年开始研发的内存数据库，2010 年进行重构并向通用关系数据库转变，2013 年 Gauss OLTP 正式上线，2020 年华为云进行了品牌和业务上的全面战略升级，如今的 GaussDB 是一个产品系列。

GoldenDB 是由中兴通讯推出的一款金融级交易型分布式数据库，经过多年的研发积累和在金融行业的锤炼，于 2014 年正式问世。2019—2020 年，GoldenDB 先后在中信银行信用卡核心业务系统及总行账务核心业务系统成功投产。

TiDB 是 PingCAP 公司自主研发的企业级开源分布式数据库，具备水平扩缩容、金融级高可用、实时 HTAP 兼容 MySQL 协议、云原生等特点。2020 年 9 月，《TiDB: A Raft-based HTAP Database》论文入选 VLDB 2020。

PolarDB 是阿里巴巴自主研发的云原生关系数据库，于 2017 年 9 月首次发布。自 2020 年起，PolarDB 数据库连续两年入选 Gartner 全球数据库领导者象限。同年，PolarDB 获中国电子学会科学技术奖——科技进步一等奖。

openGauss 是一款开源关系数据库管理系统，2020 年 6 月 30 日正式开源，采用木兰宽松许可证 v2 发行。openGauss 内核深度融合华为在数据库领域多年的经验，结合企业级场景需求，持续构建竞争力特性。

1.3 数据库的主要分类

数据库发展到今天已经不再是刚刚诞生时的模样了，从刚开始的数据库到如今针对不同场景的专用数据库，数据库已经是一个大家庭了。

1.3.1 关系数据库：传统交易数据库

最早出现的关系数据库一诞生就受到了青睐，而且在数据库领域经久不衰，时至今日依然是数据库领域占绝对主导地位的品类。关系数据库简称是 RDBMS，代表产品有甲骨文公司的 Oracle 数据库和 MySQL 数据库、IBM 公司的 DB2 数据库、微软公司的 SQL Server 数据库和诞生于加州大学伯克利分校的 PostgreSQL 数据库。关系数据库也是大家涉及最多的数据库种类，也是所有数据库中体系最复杂的数据库。NoSQL 数据库的体系和关系数据库的体系比起来，是小巫见大巫。关系数据库最重要的是 ACID 四个属性。

ACID，是指数据库管理系统在写入或更新资料的过程中，为保证事务（Transaction）是正确可靠的，所必须具备的四个特性：原子性（Atomicity，或称不可分割性）、一致性（Consistency）、隔离性（Isolation，又称独立性）、持久性（Durability）。例如，一次转账将 A 账户扣款 10 元，给 B 账户增加 10 元。这个操作在一个事务中完成，要么一起成功，要么一起失败。如今有的 NoSQL 也支持了简单的事务，如 Redis 和 MongoDB。不过关系数据库的强一致性还是 NoSQL 所不能达到的。

所以一般来说，关系数据库占据主导的都涉及交易场景（联机事务处理过程，On-Line Transaction Processing，OLTP），如金融、支付、电商、物流、医疗、保险等。

OLTP 数据库通常来说是企业的生命线。对于一个依靠计算机系统进行业务的公司来说，毫不夸张地说，选择了一种 OLTP 数据库，就相当于将公司的资产和生命交付在了这个数据库上。美国的双子大楼在"911"事件中倒塌，当时纽约银行和德意志银行的数据中心因此毁于一旦。6 个月后，纽约银行破产清盘，而德意志银行因为在几十公里外做了数据备份，得以存活下来。

关系数据库是最重要也是最复杂的数据库，本书中的核心以及主要篇幅都是围绕着关系数据库而写的。

1.3.2　键值数据库：基于 KV 键值对的内存数据库

键值数据库又称为键值存储，是 NoSQL 数据库的一种。与关系数据库使用实体关系模型来管理数据不同，键值数据库使用一个主键 Key 以及主键对应内容的 Value 来管理数据。常见的键值数据库的代表是 Redis 和 Memcached，而这其中又以 Redis 更为大家所熟知。

在键值数据库中，每条数据都有一个由简单、无重复字符串组成的 Key，以及该 Key 对应的任意类型的 Value。Value 的内容从简单的数值、字符串，到诸如列表、集合、JSON 等多元素组合类型，甚至 BLOB、向量等可以任意指定，能够满足不同的业务需要。

由于键值数据库采用一对一的键值对来存储数据，这种简化的数据的关系也让数据读写操作更加直接，效率更高，因此键值数据库常使用 DRAM 内存来加速其性能。这类数据库通常会使用一个哈希表，表中包含每一个 Key 以及指向其对应 Value 的指针。通过哈希表的查询和更新，可以高效地实现数据读写，并利用 DRAM 内存的优势有效地减少磁盘读写的次数，从而达到优于关系型存储的读写性能。

Redis 作为一种基于 DRAM 内存的键值存储，它提供了灵活的数据结构支持，比如 String、List、Set、Sorted Set、Hash、Bitmap、Streams 等，也可以通过模组（Module）支持 JSON、概率化数据（Probabilistic）等，满足不同场景下的业务需求，如图 1-3 所示。

在实际使用中，根据业务场景，Redis 键值存储可以作为缓存来使用（关系数据库作为持久化层，Redis 作为缓存层），也可以作为 NoSQL 主数据库来使用。这两种使用模式，Redis

的读写操作都是相同的，区别主要是前者在 Redis 之外提供持久化层，而后者由 Redis 自身提供持久化并支持基于键值对的查询（搜索+聚合）。

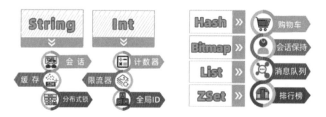

图 1-3　Redis 数据类型与应用场景

在缓存模式下，Redis 会定期存储部分（或全部）关系数据库的内容，并设置过期机制定期自动清理缓存。读操作的业务到来时，会先查询所需的数据是否存在于 Redis 数据库中，如果存在（缓存命中）则读取并返回，如果不存在（缓存缺失）则查询数据库获取内容、存入 Redis 并返回。这种模式可以大量降低对关系数据库的读操作次数，避免关系数据库过载导致服务缓慢或中断。但是这种模式需要维护好从关系数据库到 Redis 的数据同步以及 Redis 的数据定期清理，同时还要避免大量缓存击穿、缓存雪崩带来的关系数据库性能下降甚至是不可用的问题。

为了实现从关系数据库到 Redis 的数据同步，Redis 还提供了一套企业级的应用——Redis Data Integration（RDI），用户只需要配置系统间连接、映射关系、触发条件等，并以 Job 的方式将预先设置需要加载的相关数据自动载入 Redis，Redis 就可以响应关系数据库的写入操作并将这些需要加载的数据自动同步到 Redis 的数据结构中，业务只需要关注从 Redis 读取数据即可，而无须从关系数据库手动将数据导入 Redis，也不用在缓存缺失时再去查询关系数据库获取数据。目前，RDI 已经在 Redis 企业版中提供，并支持主流的关系数据库。具体请参见 https://redis.io/docs/latest/integrate/redis-data-integration/。

Redis 通过搜索模组（RediSearch Module）支持对在 Redis 中存储的 Hash 或 JSON 类型的键值对进行查询（搜索、聚合）操作。用户可以指定搜索条件（文本、数值、GPS 坐标，以及向量），在 Redis 中查询所有满足条件的键值对，并可指定返回的 ResultSet 的内容（如只返回 UserID 和 UserLoginTime 两项）。Redis 搜索进一步拓展了键值数据库的操作，并支持将一部分关系数据库的查询操作在 Redis 中直接完成，从而满足更快的查询速度、更高的吞吐量。

与缓存模式相比，主数据库模式可以进一步降低系统时延，达到高吞吐量、低时延的读写操作要求。在这种模式下，每个读写操作都应尽量简单，尽量降低同一读/写操作所需的数据量，并避免同一操作对不同数据的依赖（避免数据 Join 等）。

以下列举一些常见的、将 Redis 作为主数据库使用的业务场景。

- 大语言模型、生成式 AI 所需的向量存储及查询。
- 人工智能领域中使用的在线特征值存储（Online Feature Store）。
- 在线游戏的账号及令牌存储。

- 实时交易系统的交易信号计算。

Redis 支持 Snapshot（RDB）以及 Append-only File（AOF）两种数据落盘的方式，用户可以根据需求选择。需要注意的是，社区版 Redis 在开启数据落盘时会导致性能有所下降，用户有时不得不在数据落盘和访问速度之间做出取舍。与社区版 Redis 不同的是，Redis 企业版优化了数据落盘的机制和效率，在保证数据写入磁盘的同时依然保持了 Redis 的高性能特性。Redis 企业版的内容参见 https://redis.io/redis-enterprise/technology/durable-redis/。

1.3.3 列式数据库：列式存储的分析型数据库

列式数据库是按照数据的字段（列）为单元维度存储的，这与传统关系数据库以行为单元维度存储的有较大区别。其区别主要就是因为应用场景的不同。代表产品有 Cassandra、ClickHouse、GreenPlum 等。

一般来说，由于关系数据库遵循 ACID 的特性，所以关系数据库处理数据都是以行为单位的。比如在电商网站上下一个订单，这个订单一定包含买家信息、卖家信息、商品信息、商品价格、交易时间等。这种产生的订单数据，称为一条订单记录，这是以一行的形式存储的。在数据库中称之为"行存"。如图 1-4 所示理想的数据库订单表的存储形式，代表了 6 条订单数据。

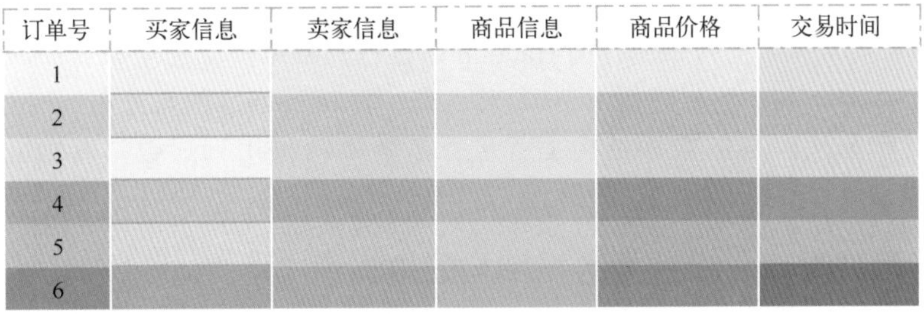

图 1-4 理想的数据库订单表存储形式

真实情况其实和理想情况大不同。因为想象的模型是人们基于 Excel 的认知的，而数据库中是以行为单位存储的。数据库的行存不会整整齐齐对齐，实际数据库中数据的存储更像图 1-5 所示的数据库订单表的存储形式。

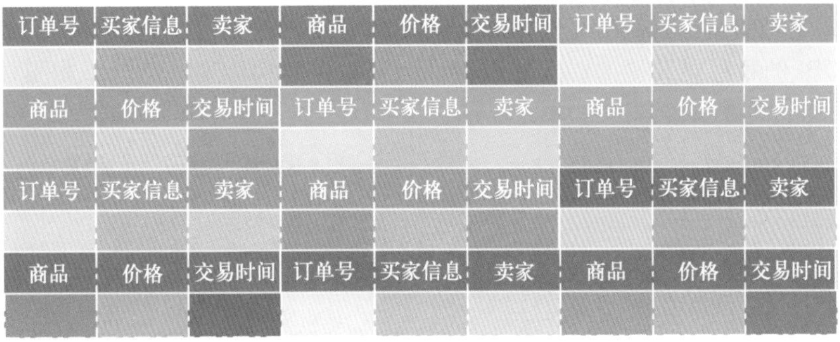

图 1-5 实际的数据库订单表存储形式

对比两张图可以发现，在图 1-5 中，计算 5 条订单的数据需要把其他的列也读取一遍（尽管运算不需要但也要读取，因为行是最基本的单位，不能只读取一列），这些列也要参与运算过程。举例来说，如果一个表有 1000 行数据，每行 10 个字段，在行式存储的数据库中，对一列进行运算，需要读取 1000×10=10000 个字段。如果行数和列数都增加，那么每次分析都是行与列的几何乘积。

在这种模型下，虽然传统关系数据库的行存也能作为分析，但是无用功太多，运算效率不如列存数据库的运算效率高。一旦读取的数据多了，调动的 CPU 和 IO 资源就非常多，会影响到 OLTP 数据库的写入。所以读写分离的架构就是为了缓解这种情况。

但如果是列式存储的数据库，即如图 1-4 所示，计算 5 条订单的数据只要对第 5 列（价格）求和就行，不需要其他列参与运算。同样，对一个有 1000 行数据、每行 10 个字段的表的某列进行计算，列式存储的数据库只对统计的列进行运算，其他的列即使再多也不会参与到运算中来，大大提升了运算效率。

其实一般用户进行数据分析的时候通常是对某列进行统计，也就是俗称的聚合。在这里也给这类分析做一个定义，即少量列的聚合称之为分析。

由于分析定义为列运算，那么分析的场景就比较适合列式数据库。如果是纯列式的数据库，其运算效果会完全不一样。在数据库领域有单独的列式数据库，也有在关系数据库上实现的列式数据库。

作为对比，从十几年前开始流行的 Hadoop 是在行式数据库下的全表查询，效率并不高。当然 Hive 的效率也不高，所以才用 Impala 来完成，而 Impala 靠的是分布式内存提升效率。但是列式数据库是从底层的数据结构上就进行了解决，采用列式数据库则完全是不需要 Hadoop 技术栈来进行数据分析的。

1.3.4 文档型数据库：松散型数据结构的数据库

随着各种业务模式的拓展，又出现了一种新的数据库类型——文档型数据库。以 MongoDB 为代表的文档型数据库使用越来越广泛。这种松散型的数据结构特别受到欢迎。

在文档型数据库中，文档是处理信息的基本单位，一个文档相当于关系数据库中的一条记录。该类型的数据模型是版本化的文档，半结构化的文档以特定的格式存储，比如 JSON。文档型数据库可以看作键值数据库的升级版，允许嵌套键值，在处理网页等复杂数据时其查询效率更高。常见的文档型数据库有 MongoDB。

MongoDB 的特点是其松散的数据结构，这种数据结构称为文档型。如图 1-6 所示，MongoDB 写入了 3 条数据。

```
> db.xuexiaogang.insert({id:1,name:"xuexiaogang",sex:"男",city:"上海"});
WriteResult({ "nInserted" : 1 })
> db.xuexiaogang.insert({id:2,name:"张三",sex:"男",QQ:"12345678"});
WriteResult({ "nInserted" : 1 })
> db.xuexiaogang.insert({id:3,name:"Jack",Tel:"130003131111"});
WriteResult({ "nInserted" : 1 })
```

图 1-6 MongoDB 的数据写入

每条数据固有 ID 和 NAME 两个关键字,犹如在关系数据库中的数据列。不过其余的关键字或者列(SEX、CITY、QQ、TEL),并不是每行数据都必有。这种特殊的数据类型使得增加一个新字段是很快速的操作,并且不会影响到已有的数据。另外,当业务数据发生变化时,也不需要 DBA 修改表结构。所以 MongoDB 在游戏场景中存储游戏用户信息、装备等优势明显。

1.3.5 图数据库:用于社交图谱的数据库

图数据库是以点、边为基础存储单元,以高效存储、查询图数据为设计目标的数据管理系统。图形结构的数据库同其他行列以及刚性结构的 SQL 数据库不同,它使用灵活的图形模型并且能够扩展到多个服务器上。图数据库把数据间的关联作为数据的一部分进行存储,关联上可添加标签、方向以及属性,这使得其在关系查询上相比其他类型数据库有巨大性能优势。所以在社交图谱、知识图谱和企业关联等场景优势显著,最具有代表性的图数据库就是 Neo4J。

用图数据库 Neo4J 展示社交图谱的场景如图 1-7 所示。图中展示了笔者通过线上和线下结识的部分朋友,主要是展现图数据库在社交图谱的场景应用。

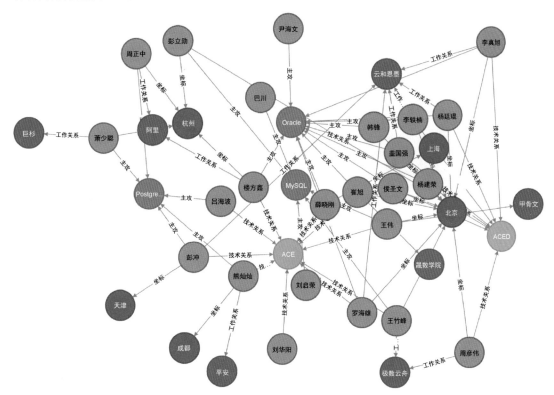

图 1-7 图数据库的社交展示

1.3.6 时序数据库：适用于物联网场景的数据库

这几年工业互联网和物联网的发展使得特定于这个场景的数据库也走进了大家的视野。时序数据库是目前最适用于物联网场景的数据库类型。时序数据库（Time Series Database）正如其名，是专门用于采集、处理和存储时序数据的数据库类型。时序数据是按照时间顺序排列的数据，通常包含多个字段，如时间戳、数值和标签等。时间戳是其中最重要的字段，它记录了数据的时间信息，使得按照时间顺序进行排序和查询成为可能。

图1-8展示了时序数据的典型字段构成。

设备ID	时间戳	采集数据			标签	
Device ID	Timestamp	Current	Voltage	Phase	Location	Type
XXG1	2022-05-15 12:30:45	4.6	100	0.24	baoshan	1
XXG2	2022-09-18 09:15:20	3.6	130	0.64	minhang	2
XXG3	2022-06-03 18:45:10	5.2	110	0.42	putuo	1
XXG4	2022-08-07 08:10:35	3.8	120	0.31	xuhui	2
XXG5	2022-07-29 16:55:55	4.1	105	0.27	changning	1
XXG6	2022-04-12 22:40:15	3.9	125	0.58	jingan	2
XXG7	2022-03-02 06:25:30	4.8	115	0.39	hongkou	1
XXG8	2022-11-11 14:05:50	3.5	135	0.72	yangpu	2
XXG9	2022-10-05 19:50:05	4.3	100	0.21	huangpu	1
XXG10	2022-12-25 03:55:40	4	130	0.47	pudong	2

图1-8 时序数据的典型字段构成

时序数据通常随时间推移而变化，因此具有时间维度属性。时间维度有助于更好地分析和理解数据，例如，可以通过时间维度的变化来识别数据中的趋势和周期性变化。

时序数据库能够有效地处理庞大且统一的数据，其独特属性意味着可以在存储空间和性能方面提供比通用数据库更加显著的改进。比如基于统一性的专门的压缩算法可以提供优于常规压缩算法的数据处理效率；对重复或过于陈旧的数据，可以定期删除以节省空间；特殊的数据库索引还可以提高查询性能。

时序数据库在物联网环境中扮演着关键的角色。物联网涉及大量设备和传感器，它们不断产生时序数据，记录了设备状态、环境参数和各种事件的发生时间。这些数据对于实时监测、故障诊断、预测分析和决策制定至关重要。时序数据库能够高效地存储和处理这些数据，提供快速的查询和分析能力，使用户能够及时获取有价值的信息。

此外，时序数据库还在工业监控、能源管理、交通运输、物流和供应链等领域得到广泛应用。工业设备的运行状态监测、能源消耗的分析、车辆位置跟踪和货物运输的时效性管理都离不开时序数据的存储和分析。时序数据库提供了高性能、可扩展和可靠的解决方案，满足这些领域对大规模时序数据处理的需求。

随着物联网技术（如通信技术和定位技术）在工业、农业、环境、交通、物流和安防等基础领域的广泛应用，接入的传感器和其他硬件设备数量不断增加，采集的时序数据量成指数级增长。预计这些数据将占据世界数据总量的 90% 以上。

尽管关系数据库可以存储时序数据，但由于缺乏针对时序数据的优化，处理大量时序数据时常常力不从心。传统的关系数据库主要应用于 OLTP 场景，一般指高可用的在线系统，如电子商务平台和银行交易系统，以处理增、删、改为主要任务，写入操作主要是执行数据更新。然而，时序数据通常是一系列关联时间戳的连续度量值，其写入操作主要是执行数据插入。

此外，关系数据库通常采用 B-Tree 结构来实现快速索引和数据查询，减少磁盘寻道时间并避免内存与磁盘间的频繁数据交换。然而，对于主要进行插入操作的时序数据库而言，业界主流趋向于采用基于磁盘的新型 LSM（Log-Structure Merge）树结构。LSM 树通过将最新的数据操作存储在内存队列中，定时或当数据量达到一定阈值时，批量将数据刷回磁盘，降低磁盘 I/O 成本，提升写入性能。

时序数据库的发展历史可以追溯到 20 世纪 90 年代。当时，由于对硬件性能指标的监控需求增加，早期版本的时序数据库应运而生，其中以 RRDTool 为代表。环形数据库（Round Robin Database，RRD）通过循环使用固定大小的存储空间，实现了高性能的时间序列数据记录和图形化渲染。这种设计既有效地存储了监控数据，又限制了存储空间的使用。

随着物联网场景的兴起，时序数据量迅速增长，过去十年间时序数据库也迎来了快速发展。自 2014 年起，DB-Engines 在其统计中将时序数据库作为一个独立目录进行分类。如图 1-9 所示的 DB-Engines 细分领域排名展示了不同数据库类别的流行度趋势，时间跨度从 2013 年 1 月到 2023 年 5 月。从图 1-9 可以清晰地看出，时序数据库（Time Series DBMS）在流行程度排名中位居第二，仅次于图数据库（Graph DBMS），并且随着时间的推移，其流行程度成逐渐增长的趋势。

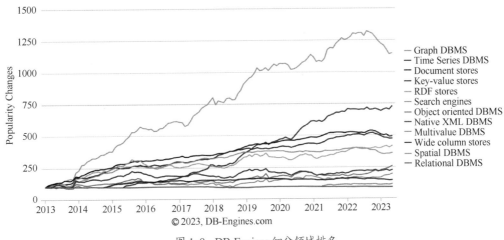

图 1-9　DB-Engines 细分领域排名

从 2011 年开始，出现了基于分布式存储的时序数据库，如 OpenTSDB。相较于早期的时序数据库，这些数据库在时间维度上进行了查询性能和存储空间的优化，性能提升明显。它们利用分布式存储和索引技术，能够高效处理大规模时序数据，并支持并行查询和水平扩展。然而，部署和维护这些分布式时序数据库的复杂性逐渐显露，对非专业用户而言，使用和维护的难度较高。

在这一背景下，垂直型时序数据库（如 InfluxDB）崭露头角，逐渐成为市场的主流选择。垂直型时序数据库专注于时序数据的存储和处理，采用针对时间特征的高效数据压缩算法，提供卓越的数据处理能力，同时具备部署简单和开发友好等特点。InfluxDB 引入了时间结构反转索引（Time-Structured Inverted Index，TSI）作为索引结构，能够快速定位和检索时序数据。此外，InfluxDB 还提供了专用查询语言 InfluxQL，使用户能够轻松执行复杂的时序数据查询和聚合操作。

2017 年问世的 TimescaleDB 也是备受关注的开源时序数据库。TimescaleDB 是一个用于处理时间序列数据的开源时序数据库，它采用插件形式存在，可以随着 PostgreSQL 的版本升级而升级。

时序数据库领域在互联网企业中得到了广泛的关注和发展，包括阿里、Facebook、亚马逊都纷纷推出各自的时序数据库产品。2015 年，Facebook 推出了 Gorilla，是专为处理大规模时序数据而设计的开源存储引擎；在 2018 年，亚马逊 AWS 推出了托管的时序数据库服务 Amazon TimeStream。同年，阿里巴巴推出了 AliTSDB，这是基于开源 OpenTSDB 优化的时序数据库解决方案。此外，阿里巴巴还提供了托管的云服务 Aliyun TSDB，以满足不同规模的时序数据存储和分析需求。

随着物联网、工业监控和金融分析等领域对时序数据处理需求的不断增长，时序数据库在数据处理方面扮演着越来越重要的角色。总的来说，时序数据库在物联网场景下的主要作用如下。

- 实时存储：时序数据库能快速接收和存储物联网设备产生的实时数据。
- 高速查询：针对时间相关查询模式进行了优化，支持快速的数据检索和查询。
- 数据分析和可视化：与分析工具相结合，实现实时监测和分析，发现趋势和异常事件。
- 实时监控和告警：用于监控设备状态和指标，触发告警和通知。
- 存储和传输优化：提供高效的数据存储和压缩技术，与其他组件集成，优化数据传输和通信效率。

通过以上功能，时序数据库能够为物联网应用提供强大的数据管理和分析能力。它支持实时监控、智能决策、资源优化和预测维护等关键目标，帮助实现高效的物联网系统和应用。

1.3.7　搜索引擎数据库：全文索引的数据库

关系数据库有定义明确的表和列，而且要求对于充当谓词的列建立索引，再通过列去检索行。比如查询一个银行卡号获得银行卡所有人的姓名和联系方式等。现实生活中有一种需求就是不确定列查询。请看下面的几句诗：

借问酒家何处有，牧童遥指杏花村。出自唐·杜牧《清明》

牧童骑黄牛，歌声振林樾。出自清·袁枚《所见》

牧童敲火牛砺角，谁复著手为摩挲。出自唐·韩愈《石鼓歌》

这几句诗都是有"牧童"关键字的，可以使用"牧童"作为一个自定义的列，而它的几个出处（诗词原文）作为一个 URL 或一个 Value。如果说关系数据库的银行卡号是这个牧童的话，那么银行卡所有人的姓名和联系方式等就是它的出处（诗词原文）。这就是全文搜索数据库，也叫作搜索引擎数据库。

其应用场景主要有类似百度的搜索。其搜索原理是在数据库中利用倒排索引，代表产品是 Elasticsearch。Elasticsearch 作为搜索引擎，它的检索和关系数据库完全不一样。Elasticsearch 以文档中的字、词为关键字（如上面讲的"牧童"），对这些关键字自动建立索引。每次新增数据就会对新加入的数据进行分词拆解，拆解到不同的维度中去，以便后续按照分词进行检索。通常用的分词有汉语分词和拼音分词。之所以这样是因为英文句子有空格，而中文没有空格。"北京大学"可以是"北、京、大、学""北京、大学""北、京大学"等，计算机不知道这些规则，要使用者设置分词去告诉它。

此外，搜索引擎数据库还有一个特点就是同义词分析。

比如查找"李寻欢"，那么结果应该如下：古龙武侠小说《多情剑客无情剑》及其衍生作品中的第一男主角。曾是朝廷殿试第三名「探花」，故人称「小李探花」，而后厌倦功名，弃官归隐；百晓生所著「兵器谱」上排名第三：「小李飞刀」。

同样如果搜索"小李飞刀"也应该能搜索到"李寻欢"。

搜索"六如公子"也应该能查到李寻欢。描述如下：古龙三公子之一，男，《小李飞刀》男主角李寻欢因为贪酒如命、嫉恶如仇、爱友如己、挥金如土、出刀如飞、视死如归，被称为六如公子。

前者李寻欢和后者小李飞刀、六如公子做成了同义词。

Elasticsearch 作为数据库，需要配套的操作数据库的客户端。Elasticsearch 的客户端一般是采用 Kibana 工具，Kibana 的控制台中开发工具页面如图 1-10 所示。

图 1-10　Kibana 控制台中开发工具页面

其风格类似于 Redis+MongoDB 的结合。当然查询也是不一样的，用 GET 命令在 Kibana 控制台获取数据如图 1-11 所示。

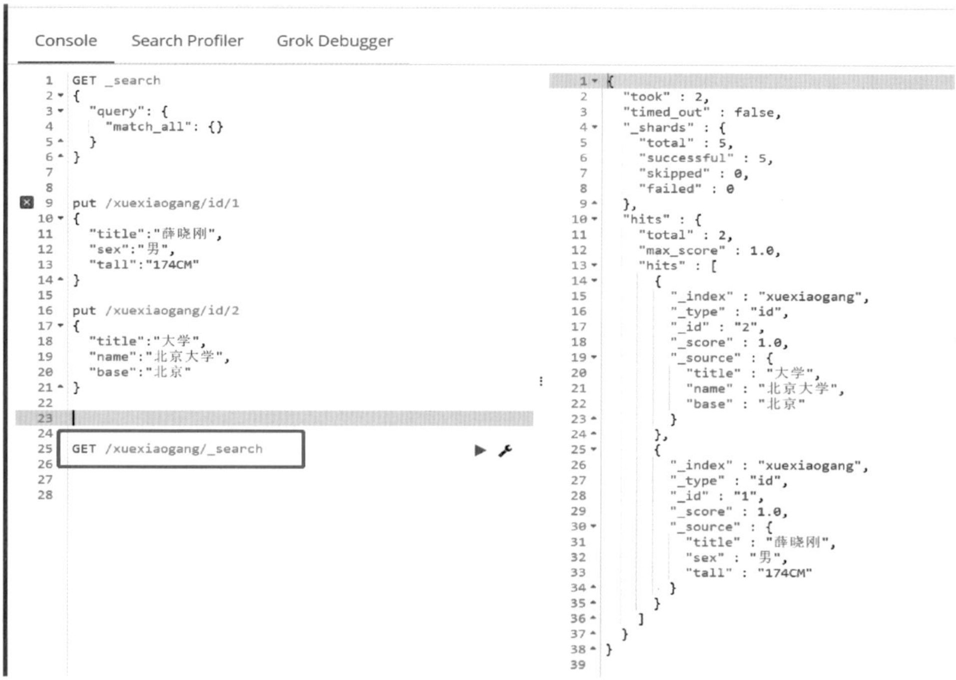

图 1-11　在 Kibana 控制台获取数据

如果没有 Kibana 组件或者没有安装该组件，Elasticsearch 的查询功能也是仍然可以使用的。如图 1-12 所示，在 SHELL 下也可以执行查询。

```
[es@dev-dbteam-143107 ~]$ curl -XGET http://10.60.143.107:9200/xuexiaogang/_search
{"took":3,"timed_out":false,"_shards":{"total":5,"successful":5,"skipped":0,"failed":0},"hits"
:{"total":2,"max_score":1.0,"hits":[{"_index":"xuexiaogang","_type":"id","_id":"2","_score":1.0,"_source":{
  "title":"大学",
  "name":"北京大学",
  "base":"北京"
}
},{"_index":"xuexiaogang","_type":"id","_id":"1","_score":1.0,"_source":{
  "title":"薛晓刚",
  "sex":"男",
  "tall":"174CM"
}
```

图 1-12　Elasticsearch 的命令行查询

除此之外，Elasticsearch 从 v6.0 开始支持 SQL。Elasticsearch 中的 SQL，是官方自带的。在 Elasticsearch 数据库的服务器上执行 elasticsearch-sql-cli（elasticsearch-sql-cli 这个程序是小写）服务器 IP:9200（默认端口）就可以进入 Elasticsearch 进行查询，其语法类似于 MySQL，如图 1-13 所示，Elasticsearch 已经支持 SQL 查询。

```
                         asticElasticE
                       ElasticE  sticEla
              sticEl  ticEl              Elast
           lasti  Elasti                   tic
           cEl         ast                 icE
          icE          as                  cEl
          icE          as                  cEl
          icEla       las                   El
        sticElasticElast                  icElas
         las          last              ticElast
         El           asti              asti  stic
         El          asticEla          Elas      icE
         El        Elas  cElasticE    ticEl       cE
         Ela       ticEl       ticElasti         cE
          las     astic              last      icE
           sticElas               asti        stic
             icEl              sticElasticElast
             icE                sticE   ticEla
             icE                 sti    cEla
             icEl                sti     Ela
             cEl                 sti     cEl
              Ela               astic   ticE
               asti             ElasticElasti
                 ticElasti   lasticElas
                    ElasticElast

                           SQL
                          6.8.6

sql> show tables;
     name         |    type
-----------------+---------------
xuexiaogang      |BASE TABLE

sql> select * from xuexiaogang;
     sex          |    tall         |    title
-----------------+-----------------+----------------
男                |174CM            |薛晓刚

sql> select * from xuexiaogang;
     base         |    name         |    sex          |    tall         |    title
-----------------+-----------------+-----------------+-----------------+----------------
北京              |北京大学          |null             |null             |大学
null              |null             |男                |174CM            |薛晓刚

sql>
```

图 1-13　Elasticsearch 中的 SQL 客户端查询

1.3.8　多模数据库：具有多种数据库模式的融合数据库

多模数据库是指在原本各自数据库领域的产品为了融合更多的业务场景需求，做出了兼容其他数据库特性的功能。使得原有数据库适用场景不再是单一场景，通过对其他数据库特性非深度的替换，使得数据库可以扩展两个甚至更多的场景。

Oracle、MySQL 和 PostgreSQL 这些数据库都是关系数据库，但是它们又不仅仅是关系数据库，它们也拥有或兼容非关系数据库的功能。

1. 关系数据库兼容文档型数据库特性

以 Oracle 对文档数据库的支持举例，如图 1-14 所示，Oracle 支持文档型数据库的半结构化字段类型——JSON。本书使用的版本是 Oracle 19c，即将发布的 Oracle 23c 有更为强大的支持。

```
SQL> create table jtab (id int,name varchar2(20),info CLOB CONSTRAINT ensure_json CHECK (info IS JSON));

表已创建。
```

图 1-14　Oracle 的 JSON 列创建方法

如图 1-15 所示，写入 JSON 类型的数据格式。

```
SQL> insert into jtab values (1,'xuexiaogang','{"tel":"13712345678","QQ":"123456","Base":"shanghai"}');

已创建 1 行。

SQL> insert into jtab values (2,'xuexiaogang2','{"tel":"13012345678","QQ":"123456","地点":"北京"}');

已创建 1 行。

SQL> insert into jtab values (3,'xuexiaogang3','{"tel":"13012345678","SEX":"男","职业":"数据库"}');

已创建 1 行。

SQL> commit;

提交完成。
```

图 1-15　Oracle 写入 JSON 列的方法

对 JSON 列进行检索，读取 JSON 类型的数据格式，如图 1-16 所示，既可以全列展示，也可以选择其中一个关键字进行展示。

```
SQL> select * from jtab;

        ID NAME           INFO
---------- -------------- --------------------------------------------------
         1 xuexiaogang    {"tel":"13712345678","QQ":"123456","Base":"shanghai"}
         2 xuexiaogang2   {"tel":"13012345678","QQ":"123456","地点":"北京"}
         3 xuexiaogang3   {"tel":"13012345678","SEX":"男","职业":"数据库"}

SQL> select id,name,t.info.tel from jtab t;

        ID NAME           TEL
---------- -------------- --------------------------------------------------
         1 xuexiaogang    13712345678
         2 xuexiaogang2   13012345678
         3 xuexiaogang3   13012345678
```

图 1-16　Oracle 读取 JSON 列的方法

2. 关系数据库兼容 KV 型数据库的特性

以 MySQL 对 KV 数据库的支持举例，MySQL 有一个 KV 的插件-memcache。该插件是模拟 memcache 的样子将 KV 场景集成到 MySQL 数据库当中。具体安装实施步骤详见附件。如图 1-17 所示，使用可以持久化的 MySQL 的 innodb 表作为数据存储。

```
mysql> select * from x.xxg;
+----+----------+----------------------------+----------------+
| id | username | password                   | email          |
+----+----------+----------------------------+----------------+
|  1 | a        | aaaaaaaaaaaaaaaaaaaaaaaaaa | a@domain.com   |
|  2 | b        | bbbbbbbbbbbbbbbbbbbb       | b@domain.com   |
|  3 | c        | cccccccccccccccc           | c@domain.com   |
|  4 | d        | dddddddddddddddddddd       | d@domain.com   |
+----+----------+----------------------------+----------------+
4 rows in set (0.00 sec)
```

图 1-17　MySQL 的 innodb 表数据样本

MySQL 的 memcache 引擎模拟的是 memcache 数据库，所以访问端口也是 memcache 的端口-11211。memcache 和 Redis 在持久化上有区别，不支持持久化。但是 MySQL 的 memcache 引擎则具备了持久化的功能。

使用 telnet 访问数据库的 11211 端口，如图 1-18 所示，通过 MySQL 的 mencache 插件访问 MySQL 的 innodb 表。输入：get a，得到了 aaaaaaaaaaaaaaaaaaaaaaaaaa | a@domain.com 的值和在数据库中 xxg 表中 name=a 的数据是一致的。

```
[root@base ~]# telnet 10.60.143.134 11211
Trying 10.60.143.134...
Connected to 10.60.143.134.
Escape character is '^]'.
get a
VALUE a 0 41
aaaaaaaaaaaaaaaaaaaaaaaaaa|a@domain.com
END
```

图 1-18　MySQL 的 mencache 插件访问 innodb 引擎

mencache 插件的意义如下。
1）实现了 KV 数据的持久化。
2）不必引入 Redis，实现了多模态的数据库。
3）比起一般的 Select 主键的查找省去了解析的过程。
4）可以保障与其他表的一致性。
5）可以和其他表进行关联查询。

3．关系数据库兼容图数据库的特性

以 Oracle 对图数据库的支持举例，在 Oracle 数据库中用 SQL，Oracle 兼容图数据库特性的功能如图 1-19 所示，展现出和图数据库一样的效果。

以上内容总结了部分不同关系数据库在 NoSQL 上的兼容场景，可以看出关系数据库已经在不同程度上实现了多模态。

1.3.9　最容易忽视的数据库——移动端数据库

要是问世界上装机最多的数据库是哪种？首先一定不是 Oracle，尽管它是数据库中的"霸主"。其次看看是不是最流行的开源数据库 MySQL 和最先进的开源数据库 PostgreSQL？都不是，答案是移动端数据库，这种数据库现在几乎每台手机中都有，所以它可以说是全世界装机量最高的数据库了。

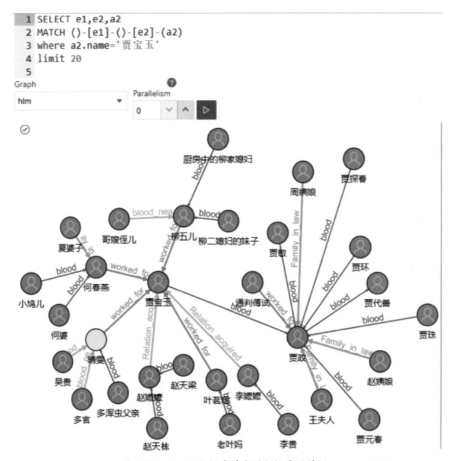

图 1-19　Oracle 数据库兼容图数据库的特性

1．移动端数据库的特点

在移动端这类嵌入式系统平台上，SQLite 是一种轻量级、易于使用的关系数据库，广泛应用于移动应用的数据存储，Android 和 iOS 的数据库都基于 SQLite 3 版本，作为一个开源的数据库引擎，它可以将数据存储在设备的本地文件系统中，而无须搭建服务器或进行复杂的配置，其主要特点如下。

- 轻量级：SQLite 具有较小的二进制文件大小，对设备资源的占用较低，尤其适合移动设备。
- 无服务器：SQLite 不需要服务器支持，数据存储在设备本地，有利于降低开发和维护成本。
- 支持 SQL 语言：SQLite 支持大部分的 SQL 语言标准，可直接用于执行查询和操作数据。
- 事务支持：SQLite 支持事务处理，确保数据的一致性和完整性。
- 安全：SQLite 可以通过对数据库文件进行加密来保证用户的数据安全。

2．移动端数据库与服务端数据库区别

Android 和 iOS 端数据库与服务端数据库在各项数据库指标上存在以下区别。

- 数据库类型：Android 和 iOS 端的数据库类型为 SQLite 3，而服务端数据库常用的数据库类型有 MySQL、PostgreSQL、MongoDB 等。服务端数据库通常更加适合处理大量数据的存储、管理和查询。
- 数据库规模：服务端数据库通常需要处理大量的数据，而 Android 和 iOS 端数据库则通常处理相对较小的数据量。
- 数据库连接方式：服务端数据库通常需要提供网络连接接口，而 Android 和 iOS 端数据库则通常通过本地文件操作来访问和操作数据库。
- 数据库管理：服务端数据库通常需要进行备份、恢复、优化等管理操作，而 Android 和 iOS 端数据库则通常由应用程序进行管理和维护。
- 数据库性能：服务端数据库通常需要具备高性能和高并发处理能力，而 Android 和 iOS 端数据库则通常不需要考虑这些问题。

3．移动端数据库的前景

当前移动端数据库的发展相对服务端还是比较缓慢的，主要原因包括如下几个方面。

- 移动设备资源受限：移动设备相比于 PC 和服务器等设备，其资源受限，包括存储容量、处理能力、电池寿命等方面，这就限制了移动端数据库的发展，使得数据库需要更加注重资源占用和性能优化。
- 数据安全问题：由于移动设备的易丢失和易被盗等特点，移动端数据库需要更加注重数据的安全性和隐私保护，这就限制了数据库的功能和灵活性，使得数据库无法像 PC 和服务器端那样灵活和自由。
- 数据同步问题：由于移动设备的网络环境不稳定、网络带宽有限等特点，移动端数据库需要更加注重数据同步和数据冲突解决。这就限制了数据库的复杂度和可扩展性，使得数据库需要更加注重数据同步和数据一致性。
- 应用场景不同：移动端数据库的应用场景同 PC 和服务器端不同，主要用于离线数据存储和轻量级数据处理等方面。这就限制了数据库的功能和复杂度，使得数据库需要更加注重轻量化和快速开发。

但是随着用户在移动端的使用体验逐步优化，移动端的数据库发展主要表现在以下几个方面。

- 大数据支持：随着移动设备的普及和用户需求的增长，移动端的数据库需要支持处理海量数据，因此，移动端数据库将越来越注重大数据的支持和处理能力。
- 云端存储：随着互联网技术的不断发展，云端存储已经成为越来越多移动应用的选择，移动端的数据库也将越来越注重云端存储，以提高数据的可靠性、可用性和安全性。
- 数据安全：随着移动应用的普及，用户的个人隐私数据和机密数据越来越受到关注，因此，移动端的数据库需要更加注重数据安全，包括数据加密、访问控制、备份和恢复等方面。
- 高性能和高并发：移动应用需要具备高性能和高并发处理能力，以提供更好的用户体验，因此，移动端的数据库需要更加注重数据读写性能和高并发访问能力。
- **智能化和自动化**：随着人工智能技术的不断发展，移动端的数据库也将越来越注重智能化和自动化，包括自动化数据分析、自动化备份和恢复、自动化数据清洗等方面。
- **跨平台支持**：由于移动设备的多样性和操作系统的不同，移动端的数据库需要更加注重跨平台支持，以提高数据的可移植性和可扩展性。

总的来说，移动端的数据库在未来的发展中将更加注重大数据支持、云端存储、数据安全、高性能和高并发、智能化和自动化以及跨平台支持等方面，以满足不断增长的用户需求和应用场景的需求。

不同的场景使用不同的数据库是大家普遍的认识。当然也有企业场景过多使用了不同产品的数据库，导致技术栈复杂。如果在每个场景并不是深度应用的前提下，多模和超融合数据库利用自身的引擎、插件或者功能支持，在原有数据库的场景上也支持了 NoSQL 等数据库的场景，有助于缓解这些问题。

1.4 数据库应用的发展趋势

数据库经历了 50 多年的发展，如今的发展可谓日新月异。由于国外数据库起步早，总体来说国外较为领先，引领着行业发展，国内业界则紧跟国外数据库技术发展趋势。

1.4.1 国外数据库应用发展趋势

通过世界公认的数据库排名网站 DB-Engines 可以了解国外数据库应用发展趋势，国际数据库流行排行榜如图 1-20 所示。截至 2023 年 4 月，在该数据库排名网站上收录的数据库产品多达 414 个。该网站的网址是 https://db-engines.com/en/ranking。

Rank Apr 2023	Rank Mar 2023	Rank Apr 2022	DBMS	Database Model	Score Apr 2023	Score Mar 2023	Score Apr 2022
1.	1.	1.	Oracle	Relational, Multi-model	1228.28	-33.01	-26.54
2.	2.	2.	MySQL	Relational, Multi-model	1157.78	-25.00	-46.38
3.	3.	3.	Microsoft SQL Server	Relational, Multi-model	918.52	-3.49	-19.94
4.	4.	4.	PostgreSQL	Relational, Multi-model	608.41	-5.41	-6.05
5.	5.	5.	MongoDB	Document, Multi-model	441.90	-16.89	-41.48
6.	6.	6.	Redis	Key-value, Multi-model	173.55	+1.10	-4.05
7.	7.	↑8.	IBM Db2	Relational, Multi-model	145.49	+2.57	-14.97
8.	8.	↓7.	Elasticsearch	Search engine, Multi-model	141.08	+2.01	-19.76
9.	9.	↑10.	SQLite	Relational	134.54	+0.72	+1.75
10.	10.	↓9.	Microsoft Access	Relational	131.37	-0.69	-11.41
11.	↑12.	11.	Cassandra	Wide column	111.81	-1.98	-10.19
12.	↓11.	↑14.	Snowflake	Relational	111.92	-3.27	+21.68
13.	13.	↓12.	MariaDB	Relational, Multi-model	95.93	-0.90	-14.38
14.	14.	↓13.	Splunk	Search engine	85.44	-2.54	-9.81
15.	↑16.	15.	Microsoft Azure SQL Database	Relational, Multi-model	79.06	+1.62	-6.72
16.	↓15.	16.	Amazon DynamoDB	Multi-model	77.45	-3.32	-5.46
17.	17.	17.	Hive	Relational	71.65	+0.74	-9.77
18.	18.	18.	Teradata	Relational, Multi-model	61.59	-2.14	-5.98
19.	19.		Databricks	Multi-model	60.97	+0.11	
20.	↑21.	↑24.	Google BigQuery	Relational	53.32	-0.12	+5.34

图 1-20 国际数据库流行排行榜

该排行榜可以被称为数据库领域的"兵器谱"。网站以多个数据库维度对各种数据库的流行度进行了排名,这个流行度来源于以下几个维度。

- 数据库系统在网络上被提及的次数。主要是指数据库名在含 Google、Bing 和 Yandex 搜索引擎上搜索到的数目。
- 对该数据库的感兴趣程度,主要指在谷歌趋势中被查询的频次。
- 专业技术讨论中提到该数据库的次数。数据来源于在知名开发者社区 Stack Overflow 和 DBA Stack Exchange 中被提问以及使用的次数。
- 各类招聘描述中对该数据库的提及次数。数据主要来源于主流招聘网站 Indeed 和 Simply Hired。
- 专业网站中使用的频率。数据主要来源于全球性职场社交平台 LinkedIn 及 Upwork。
- 社交平台的相关度。主要统计该数据库在 Twitter 中提到的次数。

排行榜仅反映该数据库在互联网上的流行度,与产品竞争力、市场份额、技术指标没有直接关系。

甲骨文公司的 Oracle 数据库,其技术一直作为数据库技术标准的风向标,它的演进方向也就是数据库的发展趋势。2022 年 10 月 18 号的 Oracle Cloud World 大会上,Oracle 宣布了令人期待的 Oracle Database 23c Beta 版。Oracle 宣称这是世界领先的融合数据库的最新版本,支持所有数据类型、任意工作负载和多种开发风格。这里说的融合数据库,就是多模的表述。在国内通常数据库厂商更加愿意使用超融合这个词来表达。

Oracle 提出了最新的数据库愿景:让开发和运行任何规模的任何应用和分析都非常容易,如图 1-21 所示,Oracle 23c 拥有许多新特性,支持多种数据类型,这些特性与数据类型使得程序开发变得简易,对于开发者、数据分析师而言都进一步简化了工作模式。HTAP 也是多模的一种,HTAP 对于分析师来说尤为重要。及时的查询与分析往往价值最高,T+1 的离线分析已经越来越不能满足分析师的日常工作了。有关 HTAP 的详细描述在第 6 章中介绍,这里不详细展开。

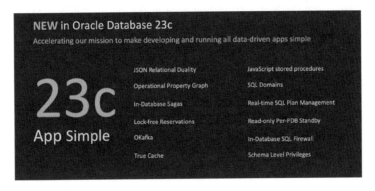

图 1-21 开发易用是未来的方向

让开发易用这个愿景也体现在数据库的多模特性上。如图 1-22 所示,越来越多的异构数据库的特性被一个数据库所集中,而这些数据库场景在一个数据库上的集中体现就是数据库的多模。比如，Oracle 23c 官方的宣传中最有亮点的能力就是划时代地提出了 JSON 二元性，后续又进一步改进为 JSON 关系统一性，JSON 和关系两种类型的数据实现了统一存储、统一访问。第一次在数据库领域将关系型和对象型两种架构的优势合二为一，既避免了开发人员从关系型数据库模型的角度去理解复杂星型模型，也避免了开发人员从文档对象型数据库的模型角度去掌握固有的冗余存储、数据更新的高代价和困难的并发控制能力，这个从 Oracle 数据库代码的底层改造酝酿数年的能力终于正式面世。Oracle Database 23c 提供了 300 余个新特性和增强功能，可为使用 JSON 场景、图数据库场景或微服务编写的应用程序实现突破性的易用功能，同时还增强了 SQL 使其更易于使用，并将 JavaScript 添加为存储过程语言。

图 1-22　数据库多模场景

开发人员是围绕着数据库进行开发的，而数据库简化应用程序开发无疑对使用者是重大的利好。所以 HTAP、多模、AI 和简化开发极有可能就是国外数据库的发展趋势。

目前，云数据库是国外竞争非常激烈的领域，AWS、微软、谷歌、甲骨文都是以云数据库为依托开展业务。数据库技术也是朝着云原生、分布式、多模、AI 自治、HTAP 等方向演进，有些数据库产品已经在部分或全面实现这些功能与特性，正在进行更加深入和精细化的提升。笔者认为，数据库技术的发展路线，除了多模以外都没有太大争议，在多模的技术路线上仍然有一些厂商坚持不同数据库保留独自的应用场景。

1.4.2　国内数据库应用发展趋势

国内数据库经过近些年的积累，目前发展趋势和国外并没有区别，只是在技术积累、用户认同和生态上还存在较大的差距。

这几年国产数据库也呈现出百花齐放、百家争鸣的状态。国内也有对标 DB-Engines 的数据库排行榜——墨天轮。墨天轮数据库排行榜如图 1-23 所示，截至 2023 年 4 月收录了 263 个国产数据库。

墨天轮数据库排行榜已经成为国内权威的排行榜。它的得分规则大致与 DB-Engines 相同，只是数据来源基本都是国内渠道，评分包括以下几方面。

第1章 漫谈数据库

最新解读报告：2023年4月中国数据库排行榜：达梦厚积薄发夺探花，亚信、星环勇毅笃行有突破　　2023年4月共263个数据库参与，点击查看排名规则更新

排行	上月	半年前	名称	模型	属性	三方评测	生态	专利	论文	得分
🏆	1	↑ 2	OceanBase +	关系型				151	18	688.36
🏆	2	↓ 1	TiDB +	关系型				26	44	609.16
🏆	↑ 4	3	达梦 +	关系型				381	0	553.46
4	↓ 3	4	openGauss +	关系型				562	65	521.20
5	5	↑↑ 7	人大金仓 +	关系型				232	0	410.37
6	↓ 6	↓ 5	PolarDB +	关系型				512	26	409.90
7	↑ 7	↓ 6	GaussDB +	关系型				562	65	407.62
8	8	8	TDSQL +	关系型				39	10	278.88
9	9	9	GBase +	关系型				152	0	252.22
10	10	10	AnalyticDB +	关系型				480	28	208.10
11	↑ 12	↑↑↑ 15	AntDB +	关系型				40	0	110.19
12	↓ 11	↓ 11	TDengine +	时序				4	0	92.06
13	13	13	GoldenDB +	关系型				114	-	75.21
14	14	↓↓ 12	SequoiaDB +	关系型				6	0	60.49
15	↑ 16	↑↑↑ 24	神舟通用 +	关系型				38	0	59.36
16	↑ 17	↑ 17	MogDB +	关系型				27	0	55.00
17	↑↑↑ 21	↑↑↑ 20	DolphinDB +	时序				-	-	54.68
18	↓↓ 15	↓↓ 14	TcaplusDB +	键值				22	3	53.80
19	19	↑↑↑ 22	SUNDB +	关系型				4	-	47.41
20	↓↓ 18	↑ 21	StarRocks +	关系型				2	0	46.35

图 1-23　墨天轮数据库排行榜

- 引入百度、必应、谷歌以及微信公众号文章（搜狗）当月搜索条目数，每个搜索引擎权重不同，如百度、公众号文章较高，最后按整体占比计算得分。
- 引入微信指数、百度指数、360趋势数据，通过搜索数据库关键字得到当月指数或趋势，计算得到每个数据库的平均指数，最后按整体占比计算得分。
- 信通院（工业和信息化部直属的科研单位）的第三方测评。
- 生态：包含社区平台、高校合作、培训认证、开放文档、代码开源、介质下载共六个维度。
- 论文：指厂商在国际数据库顶会、国际论坛、国际期刊发表的论文。
- 专利：厂商在国家或国际平台上可以公开查询的专利数，以体现数据库产品的创新能力。
- 招聘岗位数：根据数据库关键字搜索出每个数据库的岗位数量。
- 书籍：引入当当网相关书籍数据，根据数据库关键字搜索出每个数据库的书籍数量。
- Gartner市场份额排行及魔力象限。
- IDC市场份额排行。

与 DB-Engines 一样，墨天轮数据库排行榜仅反映某数据库在互联网上的流行度，与产品竞争力、市场份额、技术指标没有直接关系。

与国际数据库排行榜不同的是，国内排名前 20 的数据库几乎都是关系数据库。而国际数据库排名前 20 的数据库中，关系数据库、非关系数据库的键值数据库、列式数据库、文档型数据库都涉及了。在国际数据库排行榜（DB-Engines）中，图数据库暂列 21 位、时序数据库暂列 29 位。说明这两个细分场景的业务占比重很小。

从分类来说这些都属于数据库范畴，而搜索引擎也归入数据库范畴。希望通过这些介绍能让刚开始接触数据库的读者对数据库有一个更加全面的认识。

在 2021 年信通院发布的《数据库发展研究报告》中指出了数据库未来的七大趋势，这些趋势部分是国外数据库厂商已经实现的。所以这些趋势也是国内数据库未来的发展方向。

1．趋势一：多模数据库实现一库多用

在前文中已经提到了多模数据库，这里不再赘述。在数据库发展之初，只有关系数据库一种类型，所有数据都放在数据库中，包括结构化的交易记录、半结构化消息信息，以及非结构化的音视频。互联网时代数据复杂多样，应用程序对不同数据提出了不同存储要求，数据的多样性成为数据库平台面临的一大挑战。各种 NoSQL 数据库分别承担着不同的应用场景，关系数据库虽然占据着主要角色，但是半结构化的数据都向 NoSQL 迁移、非结构化的数据向对象存储迁移。一时间，关系数据库只保留了结构化数据。

不过这种状态没有维持多久，随着技术栈越来越复杂，大家发现开发、运营、运维、架构的困难也在增加，随之而来企业所付出的人力成本也不断攀升。尤其是在国内，相比较国外的技术栈应用存在更多浪费。不少 NoSQL 的应用是为了用而用，其实并不一定要去使用这些 NoSQL。甚至不了解为什么其他公司是这样使用的以及这些 NoSQL 的使用场景。

为了解决这种痛点，多模数据库的理念让企业不再需要为了不同场景、不同数据类型而应用不同的数据库产品。支持灵活的数据存储类型，将各种类型的数据进行集中存储、查询和处理，可以同时满足应用程序对于结构化、半结构化和非结构化数据的统一管理需求。现实让数据库再次回到之前 ALL In One 的路线上来。

多模对于开发是友好的，开发不必花费更多的异构数据库学习成本，多模的兼容性使得开发人员不需要考虑异构数据库的数据同步、传输和一致性。

多模对于运维是友好的，运维不必维护较多的异构数据库技术栈，降低了学习成本和维护成本。

多模对于架构来说是喜忧参半的。对于喜欢简单架构的架构师来说是福音，不必要管理复杂架构。对于喜欢复杂架构的架构师来说就存在一些困扰，架构精简发挥不出架构师的价值，数据库简化到单机状态后，甚至可能不需要架构了。

最终多模带来了总成本的下降。

2．趋势二：统一框架支撑分析与事务混合处理

这一趋势就是 OLTP 和 OLAP 的融合，国内各大厂商的数据库产品几乎都有 HTAP 的宣传，可以说这是一个标配。这个两年前的趋势现在已经是标准了，没有 HTAP 标签的数据库在国内很少见。

企业通常维护不同数据库以支持不同的任务，管理和维护成本高。这里的维护成本高不仅仅是在 OLTP 之外建立一个 OLAP 系统，关键是要将 OLTP 的数据实时准确地送到 OLAP，这个难度和成本是不可小觑的。以现在 ETL+Hadoop 的模式来说，ETL 需要人力去完成，数据加工需要人力，最后数据报表也需要人力。这背后是三个团队。因此，能够统一支持 OLTP 和 OLAP 的数据库成为众多企业的需求。而且 HTAP 对现有的 Hadoop 离线分析的架构会有较大的冲击，尤其是对 Hadoop 可有可无的中小企业本来说，HTAP 数据库对于 TB 级别的数据量有较好的表现，这也意味着中小企业可以不使用 Hadoop 技术就拥有了大数据系统。更加重要的是它省去了三个团队中的两个，并且做到了现有 ETL 无法做到的实时性和准确性。

ETL 本身就是应用在非实时的场景，而现在业务的诉求越来越向实时性靠拢。不能用上一个时代的技术解决这一个时代的问题。

3. 趋势三：运用 AI 实现管理自治

2019 年 6 月，Oracle 推出云上自治数据库 Autonomous Database；2020 年 4 月，阿里云发布"自动驾驶"级数据库平台 DAS；2021 年 3 月，华为发布了融入 AI 框架的 openGauss 2.0 版本。其均采用上述思想降低数据库集群的运维管理成本，保障数据库持续稳定、高效运行。

两年前发布的这个 AI 数据库的趋势目前依然是趋势，现在国内数据库的宣传也基本都把 AI 的特性作为产品的标配进行宣传。有了 AI 后是不是还需要 DBA 运维？这个我们在 1.5 节详细说明。

4. 趋势四：充分利用新兴硬件

阿里云的 PolarDB 数据库利用新硬件提升数据库的处理能力。PolarDB 核心技术实现如图 1-24 所示，阿里云采用了 NVMe 的硬件，使得磁盘访问达到了内存的访问性能。最近 10 年硬件技术突飞猛进，如今 SSD 硬盘已经成为数据库的标准配置。

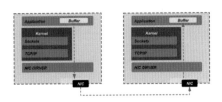

图 1-24 PolarDB 核心技术

许多硬件更新也是日新月异，英特尔傲腾（目前英特尔公司该产品已经停产）是英特尔公司发明的一种超高速内存新技术，兼容 NVMe，傲腾是该项技术的具体应用。如图 1-25

所示的 PMEM 技术，就是傲腾的非易失性内存，非易失性内存有着内存一样的效率，但是断电不易丢失。此前曾经用于 Oracle 的 Exadata 上。不过该硬件已经停产，所以从 X10M 开始，由持久内存 PMEM 改为 XRMEM（Exadata RDMA Memory），起着同样的数据读加速作用，XRMEM 就是数据库 SGA 的扩展。

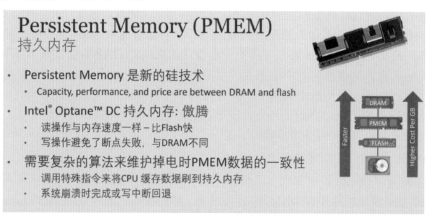

图 1-25　PMEM 非易失性内存

而远程内存直接访问（Remote Memory Direct Access，RDMA）技术的特点是无须操作系统或 CPU 参与，低延时，即可跨主机节点访问对方主机的内存数据。RDMA 技术原理如图 1-26 所示。RDMA 技术也同样应用在 Oracle 的 Exadata 上，目前在我国数据库产品中，已经开始使用非易失性内存和 RDMA 技术，PolarDB 仅仅是代表。

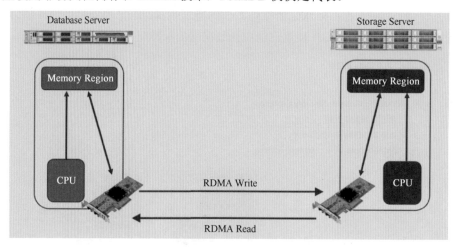

图 1-26　RDMA 的技术原理

5．趋势五：与云基础设施深度结合

2023 年 3 月 30 日，根据 TPC（Transaction Processing Performance Council）官网披露，腾讯云数据库 TDSQL 打榜 TPC-C，刷新了世界纪录，达 8.14 亿 tmpC（每分钟事务处理数）。

TPC-C 成绩榜单如图 1-27 所示。

Hardware Vendor	System	v tpmC	Price/tpmC	Watts/KtpmC	System Availability	Database	Operating System	TP Monitor	Date Submitted
Tencent Cloud	Tencent Database Dedicated Cluster (1650 Nodes)	814,854,791	1.27 CNY	NR	06/18/23	Tencent TDSQL v10.3 Enterprise Pro Edition with Partitioning and Physical Replication	Tencent tlinux 2.2	Nginx 1.16.1	03/24/23
ANT FINANCIAL	Alibaba Cloud Elastic Compute Service Cluster	707,351,007	3.98 CNY	NR	06/08/20	OceanBase v2.2 Enterprise Edition with Partitioning, Horizontal Scalability and Advanced Compression	Alibaba Aliyun Linux 2	Nginx 1.15.8	05/18/20
ANT FINANCIAL	Alibaba Cloud Elastic Compute Service Cluster	60,880,800	6.25 CNY	NR	10/02/19	OceanBase v2.2 Enterprise Edition with Partitioning, Horizontal Scalability and Advanced Compression	Alibaba Aliyun Linux 2	Nginx 1.15.8	10/01/19
ORACLE	SPARC SuperCluster with T3-4 Servers	30,249,688	1.01 USD	NR	06/01/11	Oracle Database 11g R2 Enterprise Edition w/RAC w/Partitioning	Oracle Solaris 10 09/10	Oracle Tuxedo CFSR	12/02/10
IBM	IBM Power 780 Server Model 9179-MHB	10,366,254	1.38 USD	NR	10/13/10	IBM DB2 9.7	IBM AIX Version 6.1	Microsoft COM+	08/17/10
ORACLE	SPARC T5-8 Server	8,552,523	.55 USD	NR	09/25/13	Oracle 11g Release 2 Enterprise Edition with Oracle Partitioning	Oracle Solaris 11.1	Oracle Tuxedo CFSR	03/26/13
ORACLE	Sun SPARC Enterprise T5440 Server Cluster	7,646,486	2.36 USD	NR	03/19/10	Oracle Database 11g Enterprise Edition w/RAC w/Partitioning	Sun Solaris 10 10/09	Oracle Tuxedo CFSR	11/03/09

图 1-27 TPC-C 最新榜单

TPC-C 被誉为数据库领域的"奥林匹克"。它是一个非常严苛的基准测试模型，考验完整的关系数据库系统全链路能力。TPC 官网网址为 https://www.tpc.org/tpcc/results/tpcc_results5.asp。

从 TDSQL 官方公布的数据可以看出，如图 1-28 所示，TDSQL 打榜所使用的硬件配置表。一共用了 1650 个数据节点和 3 个管理节点。每个数据节点都是 2 路处理器 48 核 96 线程，每台内存都达到了 768GB，而数据盘是 12×3.5TB 的 NVMe SSD。只有云才做得到调用如此多的存储和计算资源。

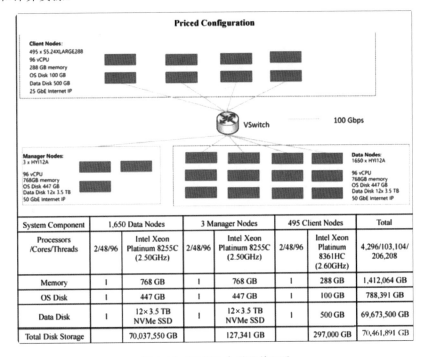

图 1-28 TDSQL 打榜硬件配置

6．趋势六：隐私计算技术助力安全能力提升

近年来，以同态加密等密码学技术为代表的软件解决方案和以可信执行环境（TEE）为代表的硬件方案为数据库安全设计提供了许多新思路。密码学方案的核心思路是整个运算过程都是在密文状态，通过基于数学理论的算法来直接对密文数据进行检索与计算。硬件方案的核心思路是将存放于普通环境（REE）的加密数据传递给 TEE 侧，并在 TEE 侧完成数据解密和计算任务。

7．趋势七：区块链数据库辅助数据存证溯源

区块链技术属于数据库范畴。区块链具有去中心化、信息不可篡改等特征，区块链数据库能够长期留存有效记录，数据库的所有历史操作均不可更改并能追溯，适用于金融机构、公安等行业的应用场景。

例如，Oracle 等一些数据库已经实现了区块链的特性，区块链作为数据库技术的一个分支，也开始融入主流数据库当中来了。

1.5　数据库与新兴数字技术

数据库是数据的载体。和数据打交道的技术，或多或少都用到了数据库技术。

1.5.1　数据库与大数据技术

谷歌在 2003—2006 年间发表了三篇论文，*MapReduce: Simplified Data Processing on Large Clusters*、*Bigtable: A Distributed Storage System for Structured Data* 和 *The Google File System*，介绍了 Google 如何对大规模数据进行存储和分析。这三篇论文开启了工业界的大数据时代。但是谷歌是一个商业公司，并不会把自己的产品免费提供出来。一个名叫 Doug Cuttin 的人按照论文内容自己用 Java 实现了一遍，使得 Hadoop 横空出世，而 Doug Cuttin 也被称为 Hadoop 之父。

Hadoop 的指导思想就是用廉价的服务器进行大量的分布式存储、分布式计算和并行计算。由于是分布式存储，所以可以进行水平扩展。一段时间内 Hadoop 成为大数据的代名词。使用数据分析的场景都向 Hadoop 靠拢，将 OLTP 的数据源源不断地送到 Hadoop 的"全家桶"中来。其组件包括：HDFS（分布式文件系统）、MapReduce（分布式计算框架）、HBase（分布式列式数据库）、ZooKeeper（分布式协作服务）、Hive（数据仓库）、Pig（ad-hoc 脚本）、Sqoop（数据 ETL/同步工具）、Flume（日志收集工具）、Mahout（数据挖掘算法库）、Oozie（工作流调度器）、Yarn（分布式资源管理器）、Spark（内存 DAG 计算模型），还包括 Kafka（分布式消息队列）等。这些技术栈构成了 Hadoop 的生态体系，从这个体系的组成可以看出，就是把 OLTP 数据库的数据送到 Hadoop 的文件系统中，然后进行分布式存储和分布式计算。Hadoop 也就顺其自然地承担起了 OLAP 的职责。由此可见，大数据技术的典型代表 Hadoop 与数据库技术关系极为紧密。

Hadoop 就是大数据的代名词，是开源的大数据基础软件与平台的基础。虽然 Hadoop 功能强大，为全球公认，但是在具体业务场景中，还缺乏很多功能，而 Hadoop 商业化就是完善这些功能，使其更好地应用于企业的业务场景。Hadoop 商业化最典型的公司就是 Hadoop 的三驾马车——Hortonworks、Cloudera 和 MapR，这三家大数据商业公司的名称与 logo 如图 1-29 所示。

图 1-29　Hortonworks、Cloudera 和 MapR

2018 年 4 月，两家大数据先驱 Cloudera 和 Hortonworks 宣布了双方称之为相对平等的合并，称将把两家公司合并为一个年收入约为 7.2 亿美元的新实体。相比于抱团取暖的 Cloudera 和 Hortonworks，MapR 的处境更为艰难了，甚至走到了"闭店裁员"的窘境。

Hadoop 商业化是它成败的关键，其本身产品复杂、维护难度高也是一方面的问题，一般公司需要一支团队来维护它的运作，这个成本是很高的。而中小企业很难负担这部分成本。

随着从对大数据一无所知，只知道人云亦云，盲目追崇概念，到现在用户发现实际使用过程中遇到的痛点越来越多。很多企业发现自己数据的积累其实并没有大数据场景。而随着硬件的进步，认为当初单体数据库处理不了，需要大数据技术才能处理的场景，现在都可以轻松在单体数据库上完成。

不少使用 Hadoop 的企业在实际工作中发现还是依靠单体数据库可以更加高效地完成任务。在实时性方面：比如有的公司的 MySQL 数据库一年只有几十 GB 的数据量，在 MySQL 数据库上通过索引只用几秒，甚至不到 1 秒就可以轻松完成最近一周的实时数据分析。但是由于使用了 Hadoop 系统，必须把这些数据搬迁到分布式系统中。无论是使用 Hive 还是 Impala 组件来实现统计分析，仅仅数据搬迁和同步的时间就不止 1 秒。在准确性方面：有的公司可能对 Hadoop 系统运算的准确性存疑，经常会先在 OLTP 的生产库查询一下，再到 Hadoop 系统去查询一下，如果两者一致，使用 Hadoop 系统的数据；如果两者不一致，那么必然是 Hadoop 系统错了，因为可能是 ETL 漏取数据或多取了数据。

如果说 ETL+Hadoop 系统为了报表而存在，那么它的意义是估算大致数据以及 T+1 的离线场景。可以在那些不紧迫的报表，或者对数据准确性不那么高，仅仅要求数量级上不要有偏差的场景使用它。而对于实时性高和准确性高的场景就完全不合适。

相同类型产品才能替换，在前面提及的 HTAP 中就可以看出，HTAP 在很多场景有替代 Hadoop 的趋势，当然不是所有场景，在非结构化处理的场景不是 HTAP 的替换范畴。Hadoop 的。

许多人因为 Hadoop 不是关系数据库所以误解它不是数据库技术，其实这个认识存在错误。Hive、Impala 这些组件都是数据库技术，只是作为数据的载体从传统关系数据库切换到了这些组件上。而在 DB-Engines 的定义上，Hive、Impala 也属于关系数据库，只不过不是传统关系型。

随着将来可能出现的新的名词和应用场景，只要是存储数据进行读写数据的都属于数据库。相反，有些数据库公司出品的产品由于不存储数据，比如 Oracle GoldenGate 就是作为一个数据传输管道，那么就属于中间件范畴，它们属于数据库技术但是不是数据库。

另外 Hadoop 不能代表大数据。它是大数据技术实现的一个分支，且这个分支中有部分技术变成了通用的技术，成为大数据技术的标配。但是，大数据技术还有很多其他分支，它们最终会演化成新的大数据实现方式。

1.5.2 数据库的延伸：区块链、物联网

1. 区块链

区块链的特点是不可篡改，如果要修改区块链中的信息，必须征得半数以上节点的同意并修改所有节点中的信息，而这些节点通常掌握在不同的主体手中，因此篡改区块链中的信息是一件极其困难的事。

相比于传统的网络，区块链具有两大核心特点：一是数据难以篡改，二是去中心化。那么每个节点其实就可以看作有一个数据库。一个区块链，尤其是公链，节点数是非常多的，要完成半数以上节点的变更才算完成，那么这些节点的完成周期就变得相对很长。所以有人对区块链的定义是"史上最慢的交易数据库"。

区块链有公链和联盟链。不同的开发商实现的方式不一样，有的用 MySQL 作为底层，有的用 PostgreSQL 作为底层，也有的用 SQLite 作为底层，还有的用 RocksDB（LSM 类存储引擎、数据库之一。它的特点在于写入的时候是 append only 的形式，跟日志一样只在文件后面追加）作为底层。

可见最终底层技术均为数据库，所以说区块链也属于数据库技术范畴。

2. 物联网

数字经济的发展离不开数字技术，随着数字技术的迅猛发展，数据库处理能力面临新的挑战，尤其是在物联网领域。由于终端设备众多、数据量巨大等特点，管理海量数据通常需要采用分布式数据库。

物联网是指连接各种智能设备的网络，是互联网从人到物的延伸。物联网系统具有异构性，主要表现为连接设备的多样性、通信网络的多样性和数据类型的多样性。连接设备的多样性意味着物联网涵盖了各种传感器、监控摄像头、智能仪表等设备，这些设备在数据传输和处理能力上存在差异。通信网络的多样性体现在物联网支持多种通信协议，以满足不同范

围和实时性要求的设备互动。数据类型的多样性指的是物联网环境中存在各种结构化、半结构化和非结构化数据，如监控图像和视频等。

由于物联网系统的异构性，根据应用需求，物联网数据的处理实时性和存储周期差异很大。典型的场景可以分为以下几类：设备监控和运维场景，根据采样频率的要求，一般需要秒级甚至毫秒级处理能力；在线业务场景需要按固定周期处理和存储数据；事务性应用场景通常需要分钟级的处理能力来快速处理数据；经营财务分析场景需要保存 1~3 年的数据并按天或按月提供统计报表。此外，基于大数据的人工智能应用场景需要全周期的设备数据，通常与数据仓库结合使用。数据库需要具备灵活的数据处理和存储能力，以适应不同场景的需求。

时序数据库是物联网系统的关键组成部分，它扮演着连接和管理物联网数据的重要角色。随着物联网的快速发展，包括时序数据库在内的数据库技术不断演进以满足物联网场景中的需求。传统的集中式数据库在物联网数据并发量大、采样频率高的场景下容易遇到性能瓶颈，并面临单点故障的风险。而分布式数据库逐渐成为物联网场景的理想选择。通过将数据分散存储在不同的服务器上，分布式数据库的每个节点可以独立处理数据请求，同时允许多个节点协同工作，从而提供了更高的性能、可靠性和扩展性。例如，在智能家居领域，分布式数据库提供了高效、可靠和智能的数据管理和分析能力，为实现智能家居的目标奠定了基础。

物联网的数据特点是：大量写入，几乎不改，而且允许少量的丢失。那么这些数据只做两件事，一个是存，另一个是取。物联网要做到快速大量写入，以及快速进行聚合分析。

物联网时序数据库自身支持数据分析，不需要和 Hadoop 对接。不过在实际工作中随着业务的发展可能会和在线业务系统发生数据交互。具体做法在物联网最佳实践（9.5 节）中论述。

1.6 与数据库长相厮守的 DBA

1.6.1 DBA 的定义与内涵

在这里首先定义一下 DBA。可能有的读者会说 DBA 就是 Database Administrator 的缩写。这本身并没有什么错，但是表达得不够全面。在 2000 年左右可以这样认为，但是在 2010 年左右就不能这样简单地定义了。随着数据增长和技术的演进，DBA 们现在要面对的不仅仅是一个数据库的安装与维护。

如今的 DBA 需要掌握多种类型的数据库，如 RDBMS、NoSQL、NewSQL。DBA 不可避免地要和操作系统与中间件打交道，所以也要具备这些技能和知识。DBA 也要处理光纤卡、存储这类硬件设备，还要涉及网络、交换机等网络领域。所以说，DBA 不仅仅是一个 Administrator，更加是一个 Architect——架构师。这个架构师一般称之为数据架构师，DataBase Architect。

以上表述其实还不足以表达 DBA 的内涵，因为 DBA 围绕着数据库展开工作，和开发打交道是不可避免的。如何审核开发写的 SQL、如何指导开发写 SQL，以及设计核心逻辑甚至把控需求也包含在其内。这个层次称之为 Analyst——分析师。

随着技术的发展，有些数据库已经开始智能化免运维的探索。Oracle 的自治数据库开启了这个领域的先河，也使得数据库的湖面再次泛起波澜。如果数据库可以免维护了，那么 DBA 做什么？

在这里举个例子：如果说到对数据库做优化，很多人可能还停留在 SQL 优化的层次。认为有管理员去改改参数或者说改写一下 SQL。这个是最低层次的，其实数据库除了内存参数（由于不知道数据库服务器所在的服务器内存有多大，所以要求用户自己定义）以外，很多参数都是对通用场景适配的参数。官方也是在这个通用参数下进行测试的。如果您的场景是通用场景，那么很多参数是不用优化的（专属场景可能有细微的调整）。如果某个参数有最佳值，那么该数据库发行商都会把这个参数做成默认值，甚至做成一个功能。而现在数据库的发展趋势不会在这个方面投入太多精力。

数据库优化的第二个层次是针对数据库对象（表设计、视图设计、列的数据类型、存储和函数等）的优化，如增加索引、微调表结构等。很多企业中 DBA 没有达到这个层次，这也是数据库自治中不能完全覆盖的。以 Oracle 为例，它已经实现了自动增加索引，但是目前甚至将来都不可能做到修改表结构。不幸的是现阶段许多企业无法理解为什么需要 DBA 去管开发，即使是 DBA 也觉得自己没有能力或没有资格管理开发。值得庆幸的是，越来越多的企业认识到了 DBA 应该去管开发，甚至管理业务需求，做到源头管理，有些企业已经这样做了。这样做的前提是 DBA 可以读开发代码，可以发现和指出问题的根结。

数据库优化的第三个层次是针对业务的优化，需要更改业务逻辑。也就是说指导业务，如果说第二个层次都做不到，那么第三个层次更加无法到达。好消息是依然有些企业意识到了这一点，企业的决策者发现了业务需求的不合理，而这些都在数据库层面表现了出来。那么让 DBA 去管理业务和需求再合适不过了。

1.6.2　DBA 的工作职责

从上面的定义就知道 DBA 都有哪些工作职责，这些职责可以在各大高端招聘或者猎头网站找到相关信息，部分职责项总结如下。

- 熟悉和了解不同类型数据库的体系结构和原理。
- 结合业务场景选择合适的数据库（如 RDBMS 和 NoSQL）。
- 根据数据库的应用需求，选择优化的硬件和网络环境。
- 根据数据库最佳实践进行高可用搭建并设置最优的备份策略。
- 结合业务场景审核需求。
- 结合业务场景进行数据库对象设计。
- 指导核心逻辑实现。
- 审核 SQL 质量，管控上线。
- 对数据库进行可观测性的监控。
- 持续优化 SQL。

以上是 DBA 的主要工作职责，实际工作中的零星事务也还有包括但不限于：数据库升级、数据库同步、数据库切换演练等。

1.6.3 DBA 需要具备的能力

每个岗位都有与之对应的能力要求，DBA 岗位能力要求是对应着其职责而定义的。随着 DBA 层次的不同，对应的能力要求也不同，一个资深的 DBA 需要具备以下 15 项能力。

- 至少掌握 1～2 种主流关系数据库的运维能力，包括但不限于体系结构了解、安装部署、高可用搭建、备份容灾、迁移升级、SQL 优化和故障诊断。
- 至少掌握 1～2 种 NoSQL 数据库的运维能力，包括但不限于安装部署、高可用搭建。
- 由于数据库是 IO 密集型产品，所以要求掌握对磁盘的总线、协议、接口以及阵列等硬件知识的能力。
- 由于数据库需要通过网络和客户端以及数据库各个节点之间通信，所以要求掌握网络通信、网络划分、路由配置等网络知识的能力。
- 由于数据库的数据存在数据表中，读写数据都是对数据库的表和索引等对象进行操作，所以要求掌握数据库对象（如表、索引、视图）的设计能力。
- SQL 优化可能涉及 SQL 改写，所以要求熟练掌握 SQL 等价改写的能力。
- 数据库是安装运行在操作系统之上的，所以要求掌握对 Linux/UNIX 操作系统的运维能力。
- 不可避免地需要在操作系统上进行定制化脚本，所以要求掌握一定的 Shell 编程能力。
- 面对定位的 SQL，需要根据 SQL 定位到开发源码指导开发改进。所以需要掌握一定的阅读 Java 或 Python 代码的能力。
- 几乎所有的数据库官方文档都是英文的，所以要求掌握较好的英文阅读能力。
- 许多开源数据库都是用 C 语言编写的，所以阅读开源数据库的代码，要求掌握一定的 C 语言的能力。
- 数据库的优化器是基于统计学建立起来的，对 SQL 的优化就是对统计学的实践，所以要求掌握一定的数学和算法知识，以及数据分析能力。
- 业务需求、数据库设计和代码实现是一个整体，指导设计整体达到最优，所以要求掌握一定的逻辑优化能力。
- 数据库安全至关重要，所以要掌握数据安全相关的能力。
- 由于要处理业务和开发之间的问题，所以要具备良好的沟通能力。这是所有要掌握技能中唯一一个非技术能力的软技能。

1.6.4 DBA 的价值

DBA 的主要价值就是维护数据库，间接维护整个系统的稳定性和业务连续性。如果把 IT 系统比作一个人体，硬件就是骨骼，网络就是神经，数据就是血液，那么把数据库比作心脏和大脑再适合不过了。而 DBA 就是心脑血管的医生，负责数据库的健康。在一个交易系统中，OLTP 数据库是最容易产生瓶颈的环节。因为其他环节是无状态的，可以横向扩展，而 OLTP 数据库虽然也可以扩展，但是受制于一致性的约束，不能无限制水平扩展，所以容易产生瓶颈。一般来说，DBA 是保健医生。如果日常保养得好，那么数据库健康对外服务，即使压力来了也能抗住请求压力。那么对 DBA 的要求就是保证系统能抗住请求压力。如果数据库能够抗住业务压力，那么就是合格的。如果数据库扛不住，其他环节再稳定，那么系统最终还是奔溃在短板上。

DBA 在数据库中扮演着力挽狂澜的角色。可以把 DBA 比作急救医生。在数据库出现问题的时候，顶住压力让数据库免于故障的影响，那么这个时候就是全体技术人员指望 DBA "妙手回春"的时候。

DBA 是核心敏感岗位。任何一家重视数据的公司对于数据库管理员的授权是慎之又慎的。一般公司对于敏感岗位都会签署保密协议，涉及数据库的 DBA 更是"重点盯防对象"。

1.6.5　DBA 是可以做一辈子的职业

基础软件包括操作系统、数据库、中间件等，是 IT 的基础产业之一。数据库作为基础软件，是一种极其复杂的技术集合体，处在信息系统的核心位置。数据库经过多年的发展，现在技术发展得很快。但是数据库的原理并没有改变，仅仅是随着硬件的提升而提升了性能与特性。

这点不仅仅是关系数据库，在 NoSQL 这类非关系数据库上也是如此。数据库本身是在数学和物理学基础上建立的，SQL（NoSQL 数据库也有很多支持了 SQL）是用户执行的高级语言。SQL 的作用是通过用户语言与底层物理数据库文件打交道。如果数学和物理学没有颠覆性改变的话，数据库原理也不会有较大的改变，那么这个领域的技能就不会被淘汰。

数据库安装与运行是在操作系统之上的，而现在的操作系统都是基于冯·诺依曼的计算机体系构建的。如果这个计算机体系不被打破，那么这个领域的技能也不会被淘汰。

同时，结合 DBA 的职责要进入业务了解场景、了解需求、审核需求、指导开发等，所以需要 IT 全生命周期的参与。即使这个系统不再有开发迭代了，只要系统不下线就还需要继续运维。

DBA 这个职业既有不容易被颠覆的知识体系，还有比业务和开发更长的生命周期，只要有 IT 系统，就有 DBA。

2023 年 ChatGPT 横空出世，对不少行业产生了冲击。当下也不乏有很多声音就是 DBA 是不是要被取代？这个话题的确有广泛的争议。

在数据库领域，Oracle 无疑处于霸主地位，至今为止无人撼动。业内共识是在集中式领域 Oracle 一骑绝尘。当然 Oracle 也有分布式的架构，只是在集中式领域已经是巅峰的存在了，说在很多设计和理念上领先其他数据库十年甚至二十年也不为过。早在 2018 年 Oracle 的 18c 版本中发布了自治数据库，其实从那时候起数据库就开始了自我革命，而不是 ChatGPT 出现后才开始的。随后很多数据库都这样设计，宣传自己的数据库是 AI 数据库、自治数据库。AI 自治已经到来了，那么 DBA 是不是失业了？

当下是云时代，云时代 DBA 还有存在的意义吗？

首先这里要先定义一下云。这里说的云是指公有云，比如 AWS 云、Azure 云、Oracle 云、谷歌云、阿里云、腾讯云、华为云等。而私有云不在这个范畴。为什么区别对待？这是有道理的。

私有云更加偏向资源池，没有公有云的特征属性。公有云的属性就是看云能解决什么，如能快速获得、弹性存储、弹性计算以及多云互通、跨境容灾等。但是私有的资源池很难做到。

云数据库的确解决了"部署、安装、备份、恢复"这些低价值工作。这些是 DBA 的基础工作。替代这些工作是云的价值。但是数据库不仅仅是 IaaS 的基础环境和 PaaS 的安装和监控。

多年的数据库运维经验得出，数据库的稳定性主要来自使用。占比几乎达到了 90% 以上，而磁盘空间、归档目录、参数设置、高可用问题等只占 10%。云解决了这 10% 的问题。而剩下的需要面对的是：

- 不合理的需求。
- 不合理的设计。
- 不合理的实现。

这些问题是需要去解决的。如今，云上的数据库会监控发现问题，能给的建议基本也是建立索引等，Oracle 多年前实现了自治数据库，也就是自动化建立索引。目前没有看到哪个数据库改表结构、改 SQL 逻辑、修改表的关联，甚至去反弹业务逻辑和需求的。

优化不仅仅是建立索引（当然这是必需的），很多工作要涉及实现、设计甚至是需求才能优化。比如发现一个 SQL，询问开发为什么要这样写？几乎 99% 的开发会说，需求如此，然后就没有然后了。而笔者的工作经历得出的结论是必须去触碰一下业务才能解决。有的时候，业务人员甚至业务的领导，得到的答复是业务需求不是这样的。这个时候通过改设计、改实现数据库性能会获得巨大的提升，不全面了解业务很难达到这样的效果。由于拿到了第一手的需求，甚至可以引导业务向着对数据库更加友好的方式提需求。然后就指导开发怎么去实现，怎么去写。

没有一个云数据库可以做到改表的设计以及 SQL 的逻辑，而且笔者认为可能未来 20~30 年都不会有。

公有云的本质是运维外包，且是一种具有规模效应的运维外包。公有云替代了 Database Administrator 的一部分工作，但是无法解决审核需求、参与设计和指导管理企业开发团队的相关工作。所以使用公有云的企业依旧需要 DBA 来处理需求、设计和实现的问题。

而私有云连数据库运维的底层工作都没有解决，所以私有云更加需要 DBA。当然如果 DBA 只会做基础工作也是不行的，终将被淘汰。DBA 必须具备改写 SQL、设计数据库对象、控制需求的能力。

第 2 章

如何提升数据库性能

本章内容针对开发人员关心的重要问题"如何提升数据库性能",讲解了索引的原理和针对不同场景、不同数据库的各类索引;如何根据业务需求构建相应的数据库的索引。当然性能提升不仅仅是靠索引,还涉及数据库对象的设计和 SQL 的写法,甚至还要上升到对需求的把控上。

本章内容
通过索引提升性能
通过 SQL 优化提升性能
避免数据库对象设计失误
从识别需求的合理性提升性能
减少 IO 操作提升数据库性能

2.1 通过索引提升性能

提升数据库性能可能是广大 DBA 最为关注的领域之一。DBA 经常会遇到随着数据量增多数据库性能下降或者大表查询缓慢的问题。事实上,数据库的查询效率和数据量的多少或表的大小基本没有关系。在任何一个关系数据库中,千万级别的数据量都不会有性能问题。凡是遇到问题的,基本上都没有使用到索引。

而正确使用索引可能让 SQL 性能提升 10~10000 倍。

2.1.1 索引的概念及原理

1. 索引的概念

索引这个词的英文是 Index,翻译成中文叫作目录。一本书可以根据目录中的页码快速找到所需的内容,这种通过目录去查找对应的内容的做法要比从头到尾逐页查找的效率高很多。目录是用来提高检索速度的,同样,索引的作用主要就是提升读取数据库数据性能。

举例说明,一本英文字典收录了几万个单词,它的目录是按照 26 个英文字母顺序排列的。如果要寻找 moon 这个单词,可以先找到以 m 开头的单词的目录(这一步已经降低了 96% 的目录查找量),再找第二位 o。很快能在仅有几页的目录中定位到这个单词具体的页码,然后翻到这一页就可以了。即使是一本 800 页的书,其目录也只有几页。通过目录去查找所需的内容比起把书全部翻一遍的效率至少提升 100 倍。

2. 索引的原理

数据库的索引就是数据表的目录,通过索引检索数据和通过目录查找页码是一样的道理。

数据库的索引与书籍的目录类似。就像书籍目录上记录了章节对应的页码一样,数据库的索引记录了指向具体数据的位置。有些数据块存储着实际的数据,而只有少数几个数据块存放了索引。

书籍的目录信息记录在页码所在的书页上,而数据库索引记录了指向物理地址的指针。一页目录上通常会包含多个章节的内容,而一个索引块上存放着映射数百行数据的指针,几个索引块就存放了几百条数据的指针。

3. 索引的容量

索引的容量是指索引(指向物理地址的指针)占据存储空间的大小,通常以 MB 或者 GB 为单位。这些指针在不同数据库不完全一致,一般是 6~8Byte。这里以 MySQL 为范例说明,假设一个数据块大小为 16KB(16×1024Byte=16384 Byte),那么这个 16KB 的数据块能存放多少指针呢?其实只要知道一个指针多长,那么用 16KB 去除以指针的长度就可以得出。MySQL 的索引其实还包括了主键,所以实际上的数据块能存放的指针数量计算应该是 16KB/(指针长度+主键长度)。这里假设主键长度是 8。16384/(6+8)≈1170,得出一个 16KB 的数据块上可以存储大约 1100 个指针,为了便于计算和理解,这里先按照 1000 计算,即一个索引块可以存储1000条指针，就像一本书的一页目录上对应了跨度为1000页的页码。这里是一个类比，因为在现实生活中超过1000页的书很少，一页目录的纸上也放不下。

索引的指针最终都指向具体的数据块，一个数据块能存储多少数据？这个其实和数据的大小有关系，就像一页A4纸上能写多少字和字体大小有关一样。这里假设一条记录是100Byte，那么一个数据块上可以存放16KB/0.1KB=160条数据。

这样算下来一个索引块可以对应1000个数据块，一个数据块能存放160条记录，那么一个索引块就能代表160×1000=160000条数据。

4. 索引的层级

索引的层级是指通过索引检索数据需要经过的路径节点。当数据量不断增加，索引块也不断增加达到一定程度的时候，就需要对索引块再次建立索引。所以索引可能因为数据量的不同而层级不同。基于以上的数据（一个索引块能代表160000条数据）来计算，索引每次增加一层，数据量就是之前的1000倍。一个高度为3的索引可以存放：1000(索引个数)×1000(索引个数)×160(每页行数)=160000000（1亿6千万，依次类推）。索引的层级结构如图2-1所示，可以看出从左到右就3层，1层与2层的分界线是几万的数据量对应的索引层级，2层与3层的分界线按照以上的数据推算，可能是几千万数据量对应的索引层级。

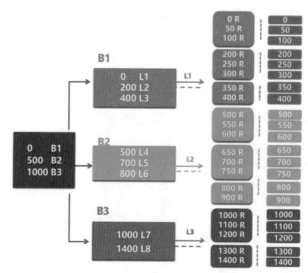

图2-1 索引的层级结构

想必大家都有去图书馆找书的经历。中国国家图书馆是中国最大的图书馆，藏书有几亿册。如果想找一本书，可以先根据书的类型确定书籍所在的楼与楼层，接着找到指定区域的书架，最后根据规则找到书在书架的第几层，在这一层中进行查找。这里的楼层、区域、书架、层级就是一级一级的索引。有了这样的检索路径，就不会随着书的数量增加而导致找书的效率变低。

5. 索引的效率

索引的效率是指从索引入口到最终找到数据的物理位置这个过程的寻址路径长度。在

图 2-1 中，可以把从索引入口到数据物理地址的路径看作一个横着放置的三角形，等底等高的三角形面积是一样的。底就是数据表中的数据行数，而高就是访问数据所走的路径。身在同一层的数量级下，从左到右的寻址路径是相等的。意味着极有可能即使差了一个数量级的数据体量下，通过索引检索同一级别的数据的效率是一样的，或者差距可以忽略不计。

接下来用一个量化的案例来说明索引层级的效率关系。

首先构造三张结构完全相同但是数据量不同的表，如图 2-2 所示，三个表的表结构相同，但是数据量不同。在三个表的相同列上分别建立索引，构建只有表名字和数据量不一样的初始化环境。可以看到三张表分别是：表名为 ONE 的表，数据行数为 10000 行；表名为 TEN 的表，数据行数为 100000 行；表名为 THOUSAND 的表，数据行数为 10000000 行。

```
SQL> select i.INDEX_NAME,i.INDEX_TYPE,i.TABLE_NAME,c.column_name,i.NUM_ROWS,i.VISIBILITY from
user_indexes i, user_ind_columns c where i.index_name=c.index_name and i.table_name = 'ONE';

INDEX_NAME              INDEX_TYPE    TABLE_NAME       COLUMN_NAME           NUM_ROWS VISIBILIT
--------------------    ----------    ------------     -------------        --------- ---------
IDX_ONE_ID              NORMAL        ONE              ID                       10000 VISIBLE

SQL> select i.INDEX_NAME,i.INDEX_TYPE,i.TABLE_NAME,c.column_name,i.NUM_ROWS,i.VISIBILITY from
user_indexes i, user_ind_columns c where i.index_name=c.index_name and i.table_name = 'TEN';

INDEX_NAME              INDEX_TYPE    TABLE_NAME       COLUMN_NAME           NUM_ROWS VISIBILIT
--------------------    ----------    ------------     -------------        --------- ---------
IDX_TEN_ID              NORMAL        TEN              ID                      100000 VISIBLE

SQL> select i.INDEX_NAME,i.INDEX_TYPE,i.TABLE_NAME,c.column_name,i.NUM_ROWS,i.VISIBILITY from
user_indexes i, user_ind_columns c where i.index_name=c.index_name and i.table_name = 'THOUSAND';

INDEX_NAME              INDEX_TYPE    TABLE_NAME       COLUMN_NAME           NUM_ROWS VISIBILIT
--------------------    ----------    ------------     -------------        --------- ---------
IDX_THOUSAND_ID         NORMAL        THOUSAND         ID                    10000000 VISIBLE
```

图 2-2 三个结构相同但数据量不同的表

如图 2-3 展示了不同数据量的索引层级差异。在图 2-3 中可以看到表名为 ONE 的表，索引层级 BLEVEL 为 1，数据行数为 10000 行；表名为 TEN 的表，索引层级 BLEVEL 为 1，数据行数为 100000 行；表名为 THOUSAND 的表，索引层级 BLEVEL 为 2，数据行数为 10000000 行。

```
SQL> select INDEX_NAME,INDEX_TYPE,TABLE_NAME,BLEVEL,NUM_ROWS,VISIBILITY from user_indexes where
table_name = 'ONE';

INDEX_NAME              INDEX_TYPE    TABLE_NAME           BLEVEL  NUM_ROWS VISIBILIT
--------------------    ----------    ------------        -------  -------- ---------
IDX_ONE_ID              NORMAL        ONE                       1     10000 VISIBLE

SQL> select INDEX_NAME,INDEX_TYPE,TABLE_NAME,BLEVEL,NUM_ROWS,VISIBILITY from user_indexes where
table_name = 'TEN';

INDEX_NAME              INDEX_TYPE    TABLE_NAME           BLEVEL  NUM_ROWS VISIBILIT
--------------------    ----------    ------------        -------  -------- ---------
IDX_TEN_ID              NORMAL        TEN                       1    100000 VISIBLE
```

图 2-3 不同数据量表的索引层级

```
SQL> select INDEX_NAME,INDEX_TYPE,TABLE_NAME,BLEVEL,NUM_ROWS,VISIBILITY from user_indexes where
table_name = 'THOUSAND';

INDEX_NAME              INDEX_TYPE   TABLE_NAME            BLEVEL    NUM_ROWS VISIBILIT
----------------------  -----------  --------------------  --------  -------- ---------
IDX_THOUSAND_ID         NORMAL       THOUSAND                   2    10000000 VISIBLE
```

图 2-3　不同数据量表的索引层级（续）

通过查询索引层级得知：ONE 表和 TEN 表，即 1 万行数据的表和 10 万行数据的表的索引层级都是 1，而 THOUSAND 表，即 1000 万行的数据表的索引层级仅仅比 1 万行的表多了一层。

分别在三个数据量级不同的表中，通过相同的列进行检索。使用到索引的时候，在返回相同的结果集的前提条件下比较检索数据的效率。

1 万行数据索引检索效率如图 2-4 所示：在 1 万行的 ONE 表中返回一行，用时约 10ms，发生了 4 次读取 IO 的操作。图中"已用时间"后打印出了"执行计划"，执行计划显示使用到了 IDX_ONE_ID 索引，估算返回 1 行。

```
SQL> select * from one where id=5555;

        ID NAME             AGE     HEIGHT      WEIGHT TIME
---------- ---------- ---------- ---------- ---------- -------------------------------
      5555 XRIzEFGkaz         18        180         180 10-8月 -23 05.20.10.086000 下午

已用时间：  00: 00: 00.01

执行计划
----------------------------------------------------------
Plan hash value: 1481126043

--------------------------------------------------------------------------------------------------
| Id  | Operation                            | Name        | Rows  | Bytes | Cost (%CPU)| Time     |
--------------------------------------------------------------------------------------------------
|   0 | SELECT STATEMENT                     |             |     1 |    82 |     2   (0)| 00:00:01 |
|   1 |  TABLE ACCESS BY INDEX ROWID BATCHED | ONE         |     1 |    82 |     2   (0)| 00:00:01 |
|*  2 |   INDEX RANGE SCAN                   | IDX_ONE_ID  |     1 |       |     1   (0)| 00:00:01 |
--------------------------------------------------------------------------------------------------

Predicate Information (identified by operation id):
---------------------------------------------------

   2 - access("ID"=5555)

Note
-----
   - dynamic statistics used: dynamic sampling (level=2)

统计信息
----------------------------------------------------------
          0  recursive calls
          0  db block gets
          4  consistent gets
          0  physical reads
          0  redo size
```

图 2-4　1 万行数据使用索引的查询效率

10 万行数据索引检索效率如图 2-5 所示：在 10 万行的 TEN 表中返回一行，用时约 10ms，发生了 4 次读取 IO 的操作。图中"已用时间"后打印出了"执行计划"，执行计划显示使用到了 IDX_TEN_ID 索引，估算返回 1 行。

```
SQL> select * from ten where id=55555;

        ID NAME                AGE     HEIGHT      WEIGHT TIME
---------- --------------- -------- ---------- ---------- -----------------------------
     55555 RpCmQprtBH            18        180        180 10-8月 -23 05.20.23.385000 下午

已用时间： 00: 00: 00.01

执行计划
----------------------------------------------------------
Plan hash value: 1802329469

--------------------------------------------------------------------------------------------
| Id  | Operation                           | Name       | Rows  | Bytes | Cost (%CPU)| Time     |
--------------------------------------------------------------------------------------------
|   0 | SELECT STATEMENT                    |            |     1 |    38 |     2   (0)| 00:00:01 |
|   1 |  TABLE ACCESS BY INDEX ROWID BATCHED| TEN        |     1 |    38 |     2   (0)| 00:00:01 |
|*  2 |   INDEX RANGE SCAN                  | IDX_TEN_ID |     1 |       |     1   (0)| 00:00:01 |
--------------------------------------------------------------------------------------------

Predicate Information (identified by operation id):
---------------------------------------------------

   2 - access("ID"=55555)

统计信息
----------------------------------------------------------
          0  recursive calls
          0  db block gets
          4  consistent gets
          0  physical reads
          0  redo size
```

图 2-5 10 万行数据使用索引的查询效率

1000 万行数据索引检索效率如图 2-6 所示：在 1000 万行的 THOUSAND 表中返回一行，用时约 10ms，发生了 4 次读取 IO 的操作。图中"已用时间"后打印出了"执行计划"，执行计划显示使用到了 IDX_THOUSAND_ID 索引，估算返回 1 行。

```
SQL> select * from thousand where id=5555555;

        ID NAME                AGE     HEIGHT      WEIGHT TIME
---------- --------------- -------- ---------- ---------- -----------------------------
   5555555 96UvSptPGW            18        180        180 10-8月 -23 04.58.16.551000 下午

已用时间： 00: 00: 00.01

执行计划
----------------------------------------------------------
Plan hash value: 2462535947
```

图 2-6 1000 万行数据使用索引的查询效率

```
| Id | Operation                          | Name            | Rows | Bytes | Cost (%CPU) | Time     |
|----|------------------------------------|-----------------|------|-------|-------------|----------|
|  0 | SELECT STATEMENT                   |                 |   1  |  39   |   4   (0)   | 00:00:01 |
|  1 |  TABLE ACCESS BY INDEX ROWID BATCHED| THOUSAND       |   1  |  39   |   4   (0)   | 00:00:01 |
| *2 |   INDEX RANGE SCAN                 | IDX_THOUSAND_ID |   1  |       |   3   (0)   | 00:00:01 |
```

Predicate Information (identified by operation id):
----

2 - access("ID"=5555555)

统计信息
----
          0 recursive calls
          0 db block gets
          5 consistent gets
          0 physical reads
          0 redo size

图 2-6　1000 万行数据使用索引的查询效率（续）

由于 1 万行和 10 万行数据的索引层级都是 1，1000 万行数据的索引层级是 2。所以通过以上三个执行计划，大家能看到，相同层数的 B 树索引命中查询的读取次数是相同的，都是 4。而 B 树层数上升一层，读取次数也仅仅增加一次，数据量明显增加的情况下，只要索引层数不大量上升，查询效率不会发生线性的变化。因为一次 IO 的时间都是毫秒级别，基本上可以忽略不计。

所以结论是很显然的。数据量多少和查询快慢没有必然关系，如果使用索引查询相同范围的数据并返回相同的结果，那么检索路径相差不大，最终检索效率也相差不大，甚至可能没有差别。但是如果不使用索引查询，那么就是数据量越多查询越慢。大家潜意识中的慢其实来源于 SQL 没有正确使用索引，而这一现象是普遍存在的。

在 OLTP 场景中使用好索引就可以解决绝大多数的问题，而这些问题的解决带来的收益不仅仅是执行速度变快，而更多的是让系统更加稳定。

2.1.2　索引的种类

索引有很多种，而日常只用到 1~2 种就已经可以满足业务的需要了。索引的类型和不同的数据库产品有关，有些产品支持的索引种类多，有些产品支持的索引种类少。不同的索引针对不同的场景，所以很多情况下，结合业务场景选型，索引就是其中一个关键的因素。下面介绍 9 种常见的索引类型。

1．B-树索引

B-树，即 B 树，还可以写成 B-Tree。B 树索引是关系数据库中使用最多的索引，没有之一。这里的 B 是 Balance 的意思，意为平衡。这个平衡非常重要，也就是说，如果数据不平衡（均匀），那么索引就不能达到最佳效果。但是在实际环境中数据倾斜是不可避免的，这种问题可以交给优化器来处理。B 树是一种自平衡的搜索树，它能够高效地支持按照关键字进行的查找、插入和删除操作。B+树是 B 树的一个升级版，相对于 B 树来说，B+树更充分地利用了节点的空间，让查询速度更加稳定。B*树是 B+树的变体，分配新节点的概率比 B+ 树要低，空间使用率更高。在 MySQL 的官方文档中对于 B-树索引的描述是：术语 B-Tree 的用途是作为索引设计的一般类别的参考。MySQL 存储引擎使用的 B 树结构可能被视为变体，因为在经典的 B 树设计中没有出现复杂的结构。按照这样来说，MySQL 的索引属于 B+Tree，但是在官方文档中并没有直接给出。

就现有资料显示，很多关系型数据库中使用的是 B+Tree，在 Oracle、MySQL 和 PostgreSQL 等数据库中实际查询索引的类型都显示为 BTree，并不会显示 B-Tree、B+Tree 或 B*Tree。这三种仅仅是在索引结构上略有差别，对于大部分使用者来说只要知道 B 树就可以了。

在"索引的层级"和"索引的效率"两个段落中，可以看出不同数据库处理大的数据量都是需要索引的，建立了正确的索引，数据量再多也不用担心检索数据变慢。与其说数据库能处理海量数据，不如说这些都是索引的功劳。

为方便读者理解，这里展开讲一下 B-Tree 和 B+Tree 的细微差别。

（1）B-Tree 索引结构

B-Tree 索引将数据按照一定的顺序存储在硬盘上，并通过多级节点构建一棵树形结构。每个节点中包含了一组关键字和指向子节点的指针，用来实现数据的快速查找。B-Tree 索引适用于随机查询和范围查询，并且对于涉及大量数据的数据库非常高效。真实的数据中要达到几十万、上百万的数量级才会出现索引的分级。这里用 0001~0007 几个自然数来假设达到索引分级的临界点后索引是怎么分级和存储的。

每个数字是一个数据集，代表一批数据，这些数据是组成足以改变索引结构的数据集合。如图 2-7 所示，B-Tree 初始化数据集 0001，数据集 0002 不断写入。随着数据增多，数据从只有 0001，变迁到现在有数据集 0001 和数据集 0002。此时，数据所对应的索引在一个层级。可以对照上一节中 1 万行的数据慢慢增加到 10 万行，数据量增加为原来的 10 倍，但是索引都是 1 层。

图 2-7　B-Tree 索引分裂步骤 1

当有数据集 0003 写入时，数据膨胀了，一层索引放不下了，层级也需要变高。达到一定程度会分裂。如图 2-8 所示，B-Tree 索引层级变成 2 层，0002 成为根节点，0002 也是临界值。因为索引是有序排列的，所以小于 0002 的数据的指针只可能在左边的索引块上，需要到左边索引块通过索引查找数据。同理，大于 0002 的数据的指针只可能在右边的索引块上，需要到右边索引块通过索引查找数据。

当再有 0004 集合写入时，先维持 0003 和 0004 两个集合平行，如图 2-9 所示，B-Tree 的 0003 和 0004 两个集合并列。

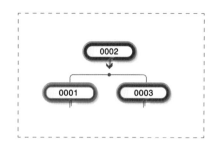

图 2-8 B-Tree 索引分裂步骤 2

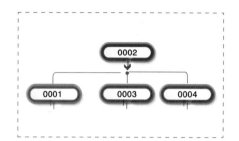

图 2-9 B-Tree 索引分裂步骤 3

当 0005 集合写入时，就再次重复 0003 集合写入时的状况。如图 2-10 所示，B-Tree 的第 2 层增加一个根节点。

然后 0006 集合写入，如图 2-11 所示，B-Tree 的 0005 和 0006 两个集合并列。

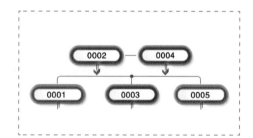

图 2-10 B-Tree 索引分裂步骤 4

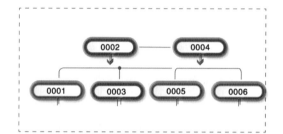

图 2-11 B-Tree 索引分裂步骤 5

最终 0007 集合写入，如图 2-12 所示，索引最终分裂为三层。

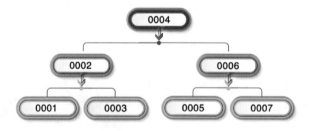

图 2-12 索引最终分裂成为三层

（2）B+Tree 索引结构

B+Tree 与 B-Tree 类似，但有一些显著区别。B+Tree 的特点是：所有数据均存储在叶子节点（叶子节点之间通过链表连接），而非叶子节点只存储关键字；叶子节点之间通过指针相互连接，形成一个有序链表，可以方便地进行范围查询。只不过从 0003 集合加入时，0001 节点可以在最下层和 0002 节点联通。如图 2-13 所示，B+Tree 变为 2 层（这里特别说明一下，根据官方文档这个连接双向的，即图中最下层的叶子节点是双向的）。

在 0004 集合加入后，0003 集合也同样上升到第二层。如图 2-14 所示，B+Tree 的 0004 节点加入，依然保持着底层节点的直接联通。

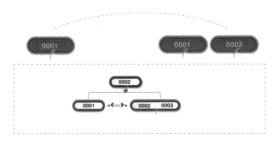

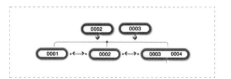

图 2-13　B+Tree 索引分裂步骤 1 和 2　　　　图 2-14　B+Tree 索引分裂步骤 3

0005 集合加入后，如图 2-15 所示，B+Tree 变为了 3 层，依然保持着对称且底层节点集合直接连接。

最后 0006 和 0007 集合加入，形成了如图 2-16 所示，B+Tree 有三层且第三层有两个节点的局面。

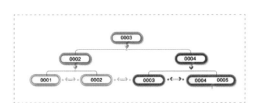

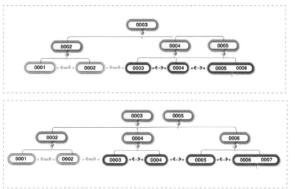

图 2-15　B+Tree 索引分裂步骤 4　　　　图 2-16　B+Tree 三层且第三层有两个节点

2．唯一索引

（1）唯一索引的定义

唯一索引（Unique Index）是一种数据库索引，用于确保数据表中的某个字段具有唯一性，即该字段的每个值在整个表中都是唯一的。可以是一列，也可以是几列组合而成。所以又叫唯一约束，用于防止数据重复，提供了唯一性约束。建立表并且设置唯一索引约束如图 2-17 所示。

```
SQL> create table uk_xxg (id int, a int,b varchar2(10));

表已创建。


SQL> create unique index uk1 on uk_xxg (id);

索引已创建。
```

图 2-17　建立表并且设置唯一索引约束

向表中写入两条数据，如图 2-18 所示，写入唯一测试数据。其中 b 列是字符串，在写入的时候应该加单引号，如果没有加单引号，数据库自动转换。

```
SQL> insert into uk_xxg values (1,1,'1');

已创建 1 行。

SQL> insert into uk_xxg values (2,2,2);

已创建 1 行。

SQL> commit;

提交完成。
```

图 2-18　写入唯一测试数据

查询表中数据，再次写入相同 ID 的数据，检查唯一约束的效果，如图 2-19 所示。

```
SQL> select * from uk_xxg;

        ID          A B
---------- ---------- ----------
         1          1 1
         2          2 2

SQL> insert into uk_xxg values (1,2,3);
insert into uk_xxg values (1,2,3)
*
第 1 行出现错误:
ORA-00001: 违反唯一约束条件 (XXG.UK1)
```

图 2-19　检查唯一约束的效果

从图 2-19 中的报错编码来看，这个是 Oracle 的编号唯一的错误码，可见防止数据重复是 Oracle 当时定义中最看中的。

（2）唯一索引的作用

使用唯一索引可以提高数据库的查询性能，因为唯一索引可快速确定某个字段的值是否存在，并消除了查询结果中重复的记录。

（3）唯一索引的使用场景

索引的唯一性体现在业务的唯一编码上，如 ID 号、手机号、邮箱等。通过唯一索引查询数据的效率和主键几乎一样。

（4）唯一索引的特点

唯一索引的特点是唯一索引列可以允许有 NULL 值。检查写入空值对唯一约束的影响，如图 2-20 所示。

从图 2-20 中可以看到，允许多个空值写入，因为在介绍 B-Tree 时讲过，Oracle 的 B-Tree 不存储空值，所以唯一约束不约束空。而主键是不能有空值的。

MySQL、PostgreSQL 与 Oracle 的索引对空值的处理略有不同，在唯一约束上完全一样，就不再重复了，读者可以自行验证。

```
SQL> insert into uk_xxg values ('',3,3);

已创建 1 行。
SQL> insert into uk_xxg values ('',3,4);

已创建 1 行。

SQL> commit;

提交完成。

SQL> select * from uk_xxg;

        ID          A B
---------- ---------- ----------
         1          1 1
         2          2 2
                    3 3
                    3 4
```

图 2-20 检查写入空值对唯一约束的影响

本书介绍的内容并不是专门针对哪一种数据库，所以会以多种数据库实践来验证。当然在某些场景下，会出现一些差异，这些是要注意的。

再回到 B-Tree 索引，用发散思维去考虑一下，如果 B-Tree 的索引中每个索引值都是唯一的，那么是不是可以认为唯一索引就是 B-Tree 树的特殊形式？每条记录都是唯一的 B-Tree，只是没有约束而已。

3. 主键索引

主键是一种数据库索引，用于确保数据表中的某个字段具有唯一性，即该字段的每个值在整个表中都是唯一的。可以是一列，也可以由几列组合而成。通过主键可以提高数据库的查询性能，因为主键可快速确定某个字段的值是否存在，并消除了查询结果中重复的记录。

主键索引通常用在需要唯一编码的场景，如 ID 号、手机号、邮箱等，或者是自增的 ID。

主键索引的特点不允许有 NULL 值。如图 2-21 所示，该表的 ID 是唯一的。去掉唯一约束（索引），改为主键（约束）索引。

```
Oracle Database 19c Enterprise Edition Release 19.0.0.0.0 - Production
Version 19.3.0.0.0

SQL> select * from uk_xxg;

        ID          A B
---------- ---------- ----------
         1          1 1
         2          2 2

SQL> desc uk_xxg;
 名称                                       是否为空? 类型
 ----------------------------------------- -------- ----------------------------
 ID                                                 NUMBER(38)
 A                                                  NUMBER(38)
 B                                                  VARCHAR2(10)
```

图 2-21 ID 唯一的样本数据

写入 ID 为空的数据，写入成功，如图 2-22 所示。

```
SQL> insert into uk_xxg (a,b) values (3,3);

已创建 1 行。

SQL> commit;

提交完成。

SQL> select * from uk_xxg;

        ID          A          B
---------- ---------- ----------
         1          1          1
         2          2          2
                    3          3
```

图 2-22 唯一数据的表写入空值

更改约束类型，将唯一索引变为主键，如图 2-23 所示。这样做的结果是数据库直接报错，更改失败，主键列不得有空值。

```
SQL> alter table uk_xxg add primary key (id);
alter table uk_xxg add primary key (id)
                                    *
第 1 行出现错误：
ORA-01449: 列包含 NULL 值；无法将其变更为 NOT NULL
```

图 2-23 将唯一索引变为主键

主键发生重复数据的例外。在工作中经常遇到这样的求助：要给一个表加主键，但是因为有重复数据加不上或者说有些数据前期有问题存在空值，导致主键无法建立。这是由于前期开发考虑不周，数据库设计的时候没有主键。

由于主键的唯一性，所以对表添加主键的过程是一个表级排序去重的过程。这个过程会有一定程度的锁定，不同数据库的做法不一样。无论是什么数据库执行这个过程的时间一定是随着数据量的增加而线性增加。对于正在运行的表来说这个代价相当大。对于这种场景来说，有停业务的解决办法和不停业务的解决办法。

多年前做公安系统时候，当时业务系统一个表都有至少几十亿条数据。这种场景如果因为疏忽没有加主键，造成了数据重复就很棘手。传统做法是去重以后再加主键，但是每秒都有成百上千的数据在写入，这个时候主键想加也加不上去，而且在去重过程中，还有新的重复源源不断产生。当时数据库使用的是 Oracle，Oracle 提供了一种解决方案，就是对之前的数据暂时不校验，只对新数据进行约束。通过这样的设置就可以先"止损"，使得新的数据后续不再重复，确保了业务的连续性，而之前的重复数据可以择机清理或者暂不处理。

整个特性通过下边的例子加以说明。建立存在重复数据的表，如图 2-24 所示。

检查重复数据，模拟给重复数据增加主键，结果必然失败，如图 2-25 所示。

模拟真实环境，对数据库设定延迟验证，如图 2-26 所示，只保证之后的数据符合主键定义，对之前的数据不进行验证。

```
SQL> create table  cf_xxg (id int, a int,b int);
```
表已创建。
```
SQL> insert into cf_xxg values (1,1,1);
```
已创建 1 行。
```
SQL> insert into cf_xxg values (1,2,2);
```
已创建 1 行。
```
SQL> insert into cf_xxg values (2,21,22);
```
已创建 1 行。
```
SQL> commit;
```
提交完成。

图 2-24　建立存在重复数据的表

```
SQL> select * from cf_xxg;

        ID          A          B
---------- ---------- ----------
         1          1          1
         1          2          2
         2         21         22

SQL> alter table cf_xxg add primary key (id);
alter table cf_xxg add primary key (id)
*
第 1 行出现错误:
ORA-02437: 无法验证 (XXG.SYS_C008997) - 违反主键
```

图 2-25　重复数据增加主键失败

```
SQL> alter table cf_xxg add primary key (id) deferrable enable novalidate;
```
表已更改。
```
SQL> select * from cf_xxg;

        ID          A          B
---------- ---------- ----------
         1          1          1
         1          2          2
         2         21         22

SQL> insert into cf_xxg values (3,33,33);
```
已创建 1 行。
```
SQL> commit;
```
提交完成。

图 2-26　数据库设定延迟验证

再次写入重复数据，如图 2-27 所示，新的重复数据不能写入，延迟验证生效，主键开始发挥作用，对之后的数据进行校验。

```
SQL> insert into cf_xxg values (2,44,44);
insert into cf_xxg values (2,44,44)
            *
第 1 行出现错误:
ORA-00001: 违反唯一约束条件 (XXG.SYS_C008998)


SQL> select * from cf_xxg;

        ID          A          B
---------- ---------- ----------
         1          1          1
         1          2          2
         2         21         22
         3         33         33
```

图 2-27　防止新数据重复写入

这个解决方案很好，但后续仍需要对之前的不合格数据进行处理，否则可能会对将来的运维产生一定的影响。举例来说，如果将来要对数据库进行导出，再导入的时候，就会有数据重复的报错。这样会使后续的人不明白为什么有主键的表会有重复数据，是出现了漏洞吗？其实不是，而是使用了数据库的特性。数据库的很多特性就是为了帮助和简化开发人员或者运维人员的工作量的。

4．复合索引

复合索引也称为组合索引，是一个以上字段组成的索引。单列字段构成的索引（约束）称为单列索引。

许多开发人员对于单列索引掌握得还不够好，对于复合索引来说掌握得就更差。有时候听到有些开发人员反馈，在开发人员视角中可能知道应该怎么建立复合索引，但是不确保这个复合索引能被使用到。如果怀着碰运气的想法建立了复合索引而且还用到了（不保证是不是最优，只是用到了），就觉得这个开发人员的水平很高，但是对于 DBA 来说这个太容易了。

一般来说，复合索引是用在多个条件过滤下的，如果 WHERE 后面单一谓词过滤后，结果集返回还是过大，则需要再加一个过滤条件才能发挥更好的效果。通常来说这样的情况下需要复合索引。

在使用复合索引时，WHERE 后面带有至少两个谓词过滤条件，且两个谓词对应的字段的区分度都很高。这里通过一个车牌筛查的案例加以说明。一个城市中的机动车达上百万辆，每年行车记录的数据量达上亿。在车辆管理场景中最常见的使用精准车牌号码加行车时间联合查询的 SQL 执行计划，如图 2-28 所示。

从图 2-28 可以看出，数据库结合"沪 A01234"这个车牌和 2022 年 12 月 16 日选择了 CAR1 索引。

当条件发生一点改变，比如没有看清楚全部车牌号码（最后两位没有看清），条件变成

"沪 A012%"这个车牌，时间还是 2022 年 12 月 16 日，那么车辆管理场景模糊车牌加行车时间的 SQL 的执行计划如图 2-29 所示。

```
SQL> explain plan for select * from car where carno like '沪A01234' and
t>=to_date('2022/12/16','yyyy/mm/dd') and t<=to_date('2022/12/16','yyyy/mm/dd');

Explained

SQL> select * from table(dbms_xplan.display);

PLAN_TABLE_OUTPUT
--------------------------------------------------------------------------------
Plan hash value: 42862083
--------------------------------------------------------------------------------
| Id  | Operation                            | Name | Rows | Bytes | Cost (%CPU)|
--------------------------------------------------------------------------------
|   0 | SELECT STATEMENT                     |      |    1 |   22  |   2   (0)  |
|   1 |  TABLE ACCESS BY INDEX ROWID BATCHED | CAR  |    1 |   22  |   2   (0)  |
|*  2 |   INDEX RANGE SCAN                   | CAR1 |    1 |       |   1   (0)  |
--------------------------------------------------------------------------------

Predicate Information (identified by operation id):
--------------------------------------------------------

   2 - access("CARNO"='沪A01234' AND "T"=TO_DATE(' 2022-12-16 00:00:00',
              'syyyy-mm-dd hh24:mi:ss'))

15 rows selected
```

图 2-28　精准车牌号码加行车时间联合查询的 SQL 执行计划

```
SQL> explain plan for select * from car where carno like '沪A012%' and
t>=to_date('2022/12/16','yyyy/mm/dd') and t<=to_date('2022/12/16','yyyy/mm/dd');

Explained

SQL> select * from table(dbms_xplan.display);

PLAN_TABLE_OUTPUT
--------------------------------------------------------------------------------
Plan hash value: 595990800
--------------------------------------------------------------------------------
| Id  | Operation                            | Name | Rows | Bytes | Cost (%CPU)|
--------------------------------------------------------------------------------
|   0 | SELECT STATEMENT                     |      |    1 |   22  |   3   (0)  |
|   1 |  TABLE ACCESS BY INDEX ROWID BATCHED | CAR  |    1 |   22  |   3   (0)  |
|*  2 |   INDEX RANGE SCAN                   | CAR2 |    1 |       |   2   (0)  |
--------------------------------------------------------------------------------

Predicate Information (identified by operation id):
--------------------------------------------------------

   2 - access("T"=TO_DATE(' 2022-12-16 00:00:00', 'syyyy-mm-dd hh24:mi:ss') AND
              "CARNO" LIKE '沪A012%')
       filter("CARNO" LIKE '沪A012%')

16 rows selected
```

图 2-29　模糊车牌号码加行车时间的 SQL 的执行计划

从图 2-29 中可以看出，数据库结合"沪 A012%"这个车牌和 2022 年 12 月 16 日选择了 CAR2 索引。

通过查看如图 2-30 所示的 CAR 表的表结构和索引定义得知，CAR1 的索引是 CARNO（车牌）+T（时间），而 CAR2 的索引是 T（时间）+CARNO（车牌）。

```
SQL> desc car;
Name   Type          Nullable Default Comments
-----  ------------- -------- ------- --------
ID     INTEGER       Y
CARNO  VARCHAR2(10)  Y
T      DATE          Y


SQL> select dbms_metadata.get_ddl('INDEX',upper('car1')) from dual ;

DBMS_METADATA.GET_DDL('INDEX',UPPER('CAR1'))
--------------------------------------------------------------------------------

  CREATE INDEX "XXG"."CAR1" ON "XXG"."CAR" ("CARNO", "T")
  PCTFREE 10 INITR


SQL> select dbms_metadata.get_ddl('INDEX',upper('car2')) from dual ;

DBMS_METADATA.GET_DDL('INDEX',UPPER('CAR2'))
--------------------------------------------------------------------------------

  CREATE INDEX "XXG"."CAR2" ON "XXG"."CAR" ("T", "CARNO")
  PCTFREE 10 INITR
```

图 2-30 CAR 表的表结构和索引定义

从这里可以看出根据结果集的不同，选择了不同顺序的复合索引。当车牌精确时，CAR1 的索引第一列返回数据相对较少，高效；当车牌模糊时，CAR2 的索引第一列返回数据相对较少，高效。

5．函数索引

函数索引是指在索引上添加函数，这种是一种特殊的处理方式。由于 SQL 中对谓词使用了函数转换，导致无法使用到索引，所以需要在索引上添加与之对应的函数，使得 SQL 可以使用索引。

函数索引通常用于数据类型和 SQL 不匹配的场景，此时 SQL 程序可能已经无人维护或者数据类型无法变更。在这种错误前提条件下，只有用函数索引才能最大限度上提高检索效率。

通过下面案例展示函数索引的应用场景。数据样本如图 2-31 所示，重点关注时间（TIME）字段。

```
SQL> select * from xxghs;

      ID A         TIME
--------- ---------- --------------
      1 1         16-7月 -23
      2 2         15-7月 -23
```

图 2-31 函数索引数据样本

```
 3  3        14-7月 -23
 4  4        13-7月 -23
 5  5        12-7月 -23
 6  6        11-7月 -23
 7  7        10-7月 -23
 8  8        09-7月 -23
 9  9        08-7月 -23
10 10        07-7月 -23
```

已选择 10 行。

图 2-31　函数索引数据样本（续）

案例中的表名为 xxghs，其表结构如图 2-32 所示，TIME 列为时间类型。

```
SQL> desc xxghs;
 名称                              是否为空？ 类型
 -------------------------------- -------- --------------
 ID                                         NUMBER(38)
 A                                          VARCHAR2(10)
 TIME                             NOT NULL  DATE
```

图 2-32　待实验表的表结构

由于开发人员不熟悉数据库，SQL 会出现如图 2-33 所示的对索引列进行函数运算，导致索引失效的情况。这是因为在 Time 列上加上了 to_char 函数，使之发生了函数转换，索引失效。执行计划显示执行了 TABLE ACCESS FULL 的全表扫描。

```
SQL> explain plan for  select * from xxghs t where to_char(time,'yyyy-mm-dd')='2023-07-15';

已解释。

SQL> SELECT * FROM TABLE(DBMS_XPLAN.DISPLAY);

PLAN_TABLE_OUTPUT
--------------------------------------------------------------------------------
Plan hash value: 434114425

--------------------------------------------------------------------------
| Id | Operation         | Name  | Rows | Bytes | Cost (%CPU)| Time     |
--------------------------------------------------------------------------
|  0 | SELECT STATEMENT  |       |   1  |   29  |    3   (0) | 00:00:01 |
|* 1 |  TABLE ACCESS FULL| XXGHS |   1  |   29  |    3   (0) | 00:00:01 |
--------------------------------------------------------------------------

Predicate Information (identified by operation id):
---------------------------------------------------

PLAN_TABLE_OUTPUT
--------------------------------------------------------------------------------

   1 - filter(TO_CHAR(INTERNAL_FUNCTION("TIME"),'yyyy-mm-dd')='2023-07-1
       5')
```

图 2-33　函数运算导致索引失效

由于没有源代码或者其他特殊原因，无法修改应用程序中的 SQL。只能建立函数索引，如图 2-34 所示，此时在建立索引的命令中也做了对应的 to_char 函数。

```
SQL> create index ct2 on xxghs (to_char(TIME, 'yyyy-mm-dd'));


SQL> explain plan for  select * from xxghs t where to_char(time,'yyyy-mm-dd')='2023-07-15';
已解释。
SQL> SELECT * FROM TABLE(DBMS_XPLAN.DISPLAY);
PLAN_TABLE_OUTPUT
--------------------------------------------------------------------------------
Plan hash value: 2478637636

--------------------------------------------------------------------------------
--------------
| Id  | Operation                           | Name  | Rows  | Bytes | Cost (%CPU
)| Time     |

--------------------------------------------------------------------------------
--------------

PLAN_TABLE_OUTPUT
--------------------------------------------------------------------------------
|   0 | SELECT STATEMENT                    |       |     1 |    36 |     2  (0
)| 00:00:01 |

|   1 |  TABLE ACCESS BY INDEX ROWID BATCHED| XXGHS |     1 |    36 |     2  (0
)| 00:00:01 |

|*  2 |   INDEX RANGE SCAN                  | CT2   |     1 |       |     1  (0
)| 00:00:01 |

--------------------------------------------------------------------------------
--------------

PLAN_TABLE_OUTPUT
--------------------------------------------------------------------------------

Predicate Information (identified by operation id):
---------------------------------------------------

   2 - access(TO_CHAR(INTERNAL_FUNCTION("TIME"),'yyyy-mm-dd')='2023-07-15')
```

图 2-34　建立函数索引

再次执行 SQL，看到执行计划已经使用到了新建立的 CT2 的带有 TO_CHAR 的函数索引，执行了 INDEX RANGE SCAN 的范围扫描。

函数索引就是在违反原则的情况下的补救措施，将本来应该的 SQL 适配索引原则，改为了索引适配 SQL。

6．位图索引

位图索引（Bitmap Index）是一类特殊的数据库索引技术，该类索引使用 bit 数组进行存储与计算操作。位图索引采用一种基于位图存储结构的索引方式，它使用二进制的位来表示索引。对于一个数据表，位图索引将所有可能出现的值都列出，每列对应一个二进制数。若某行数据的某个字段值与对应的列匹配，则在该列对应的二进制数中标记这一行的位置，否则不进行标记。位图索引解决了在数组处理和位运算处理上的问题。

位图索引的特点是字段的唯一值是少量且可以枚举的，而且是几乎不变更的，比如性别、血型等。

位图索引可以用于对区分度不高的字段的统计分析需求，可以弥补 B-Tree 索引在这方面的不足。B-Tree 索引对列的区分度有要求，区分度越高越好，如果区分度低，那么通过索引返回数据较多，可能索引就失效了。这种场景下就不适合 B-Tree 索引。

以下是使用位图索引的例子。一般来说，SQL 语句中如果没有 WHERE 条件，那么执行计划就会是全表扫描，无过滤条件的执行计划如图 2-35 所示。

```
SQL> explain plan for select * from wt FETCH FIRST 10 ROWS ONLY;

已解释。

SQL> select * from table(dbms_xplan.display);

PLAN_TABLE_OUTPUT
--------------------------------------------------------------------------
Plan hash value: 3755492315

--------------------------------------------------------------------------
| Id  | Operation             | Name | Rows | Bytes | Cost (%CPU)| Time     |
--------------------------------------------------------------------------
|   0 | SELECT STATEMENT      |      |   10 |   400 |     2   (0)| 00:00:01 |
|*  1 |  VIEW                 |      |   10 |   400 |     2   (0)| 00:00:01 |
|*  2 |   WINDOW NOSORT STOPKEY|     |   10 |   190 |     2   (0)| 00:00:01 |
|   3 |    TABLE ACCESS FULL  | WT   |   10 |   190 |     2   (0)| 00:00:01 |
--------------------------------------------------------------------------

PLAN_TABLE_OUTPUT
--------------------------------------------------------------------------
Predicate Information (identified by operation id):
--------------------------------------------------------------------------

   1 - filter("from$_subquery$_002"."rowlimit_$$_rownumber"<=10)
   2 - filter(ROW_NUMBER() OVER ( ORDER BY  NULL )<=10)

已选择 16 行。
```

图 2-35　无过滤条件的执行计划

如果 WHERE 后谓词的条件区分度很低，那么也不会用到索引，返回数据约占比 50%，导致全表扫描，如图 2-36 所示。

```
SQL> explain plan for select count(*) from wt where org='A';

Explained

SQL> select * from table(dbms_xplan.display);
```

图 2-36　返回数据约占比 50%导致全表扫描

```
PLAN_TABLE_OUTPUT
--------------------------------------------------------------------------------
Plan hash value: 2751217529
--------------------------------------------------------------------------------
| Id  | Operation          | Name | Rows  | Bytes | Cost (%CPU)| Time     |
--------------------------------------------------------------------------------
|   0 | SELECT STATEMENT   |      |     1 |     2 | 44589   (1)| 00:00:02 |
|   1 |  SORT AGGREGATE    |      |     1 |     2 |            |          |
|*  2 |   TABLE ACCESS FULL| WT   |   11M |   22M | 44589   (1)| 00:00:02 |
--------------------------------------------------------------------------------

Predicate Information (identified by operation id):
---------------------------------------------------
   2 - filter("ORG"='A')

14 rows selected
```

图 2-36　返回数据约占比 50% 导致全表扫描（续）

为区分度低的列，建立位图（BitMap）索引。如图 2-37 所示，开启并行建立 BitMap 索引，其中 parallel 为并行执行，加速索引建立。

```
SQL> create bitmap index bt on wt (org)  parallel 12;
```

索引已创建。

图 2-37　并行建立 BitMap 索引

位图索引在区分度极低的场景下，使用到了索引进行聚合。位图索引的执行计划如图 2-38 所示。

```
SQL> explain plan for select count(*) from wt where org='A';

Explained

SQL> select * from table(dbms_xplan.display);

PLAN_TABLE_OUTPUT
--------------------------------------------------------------------------------
Plan hash value: 2544984479
--------------------------------------------------------------------------------
| Id  | Operation                     | Name | Rows  | Bytes | Cost (%CPU)| Time   |
--------------------------------------------------------------------------------
|   0 | SELECT STATEMENT              |      |     1 |     2 |   247   (0)| 00:00: |
|   1 |  SORT AGGREGATE               |      |     1 |     2 |            |        |
|   2 |   BITMAP CONVERSION COUNT     |      |   11M |   22M |   247   (0)| 00:00: |
|*  3 |    BITMAP INDEX SINGLE VALUE  | BT   |       |       |            |        |
--------------------------------------------------------------------------------

Predicate Information (identified by operation id):
---------------------------------------------------
   3 - access("ORG"='A')

15 rows selected
```

图 2-38　位图索引的执行计划

执行该 SQL，即使不用缓存也不用列式计算，在最普通的数据库环境中，千万级别的数据量也只要 20ms 即可完成计算。在千万级别数据量下，位图索引真实执行效果如图 2-39 所示。

```
SQL> select count(*) from wt where org='A';

  COUNT(*)
----------
  10000000
```

已用时间: 00: 00: 00.02

图 2-39 位图索引真实执行效果

通过隐藏索引的技术，将索引设置为优化器不可见。该场景在 Oracle 11g 和 MySQL 8 等数据库上都已实现，其目的是考虑到如果删除索引发生性能问题，再次建立索引成本较高，所以先把索引设置为优化器不可见，检查没有这个索引对数据库影响大不大，确定没有影响后再删除。设置索引对优化器不可见如图 2-40 所示。

```
SQL> alter index bt invisible;

Index altered
```

图 2-40 设置索引对优化器不可见

隐藏索引后，对 ORG 字段进行分组统计。可见如图 2-41 所示，执行计划变成了无索引下全表扫描聚合。

```
SQL> explain plan for select count(*),org from wt group by org;

Explained


Executed in 0.07 seconds


SQL> select * from table(dbms_xplan.display);

PLAN_TABLE_OUTPUT
--------------------------------------------------------------------------------
Plan hash value: 733428446

--------------------------------------------------------------------------------
| Id  | Operation          | Name | Rows  | Bytes | Cost (%CPU)| Time     |
--------------------------------------------------------------------------------
|   0 | SELECT STATEMENT   |      |     4 |     8 | 46334   (5)| 00:00:02 |
|   1 |  HASH GROUP BY     |      |     4 |     8 | 46334   (5)| 00:00:02 |
|   2 |   TABLE ACCESS FULL| WT   |   47M|   89M| 44510   (1)| 00:00:02 |
--------------------------------------------------------------------------------

9 rows selected
```

图 2-41 优化器不可见再次导致全表扫描聚合

把隐藏的索引释放开，使得索引对优化器可见，如图 2-42 所示。

再次使用 WHERE 条件下的全表扫描聚合，如图 2-43 所示，在位图索引下全表扫描聚合。可以发现执行计划最终用了位图索引扫描。

```
SQL> alter index bt visible;

Index altered

Executed in 0.023 seconds
```

图 2-42 恢复索引对优化器可见

```
SQL> explain plan for select count(*),org from wt group by org;

Explained

Executed in 0.022 seconds


SQL> select * from table(dbms_xplan.display);

PLAN_TABLE_OUTPUT
--------------------------------------------------------------------------------
Plan hash value: 2336920617
--------------------------------------------------------------------------------
| Id  | Operation                   | Name | Rows  | Bytes | Cost (%CPU)| Time     |
--------------------------------------------------------------------------------
|   0 | SELECT STATEMENT            |      |     4 |     8 |   980   (0)| 00:00:01 |
|   1 |  SORT GROUP BY NOSORT       |      |     4 |     8 |   980   (0)| 00:00:01 |
|   2 |   BITMAP CONVERSION COUNT   |      |   47M |   89M |   980   (0)| 00:00:01 |
|   3 |    BITMAP INDEX FULL SCAN   | BT   |       |       |            |          |
--------------------------------------------------------------------------------

10 rows selected
```

图 2-43 位图索引下全表扫描聚合

对有 4700 万行数据的表进行全表聚合统计，真实耗时为 115ms。如图 2-44 所示，位图索引真实执行的执行效率约为 115ms。

```
SQL> select count(*),org from wt group by org;

  COUNT(*) ORG
---------- ----------
  10000000 A
  10000000 B
  15000000 C
  12000000 D

Executed in 0.115 seconds
```

图 2-44 位图索引真实执行效率

7. 部分索引

　　MySQL 和 PostgreSQL 都有部分索引的实现，但二者不同。MySQL 的部分索引只将字段的部分内容作为索引，而不是整个字段作为索引的索引。比如一个字段有 30 个字符，只将其左侧连续的部分字符作为索引。MySQL 的部分索引使得索引占用空间更加小，检索速度更快。部分索引主要用在从字段的前半部分就可以准确定位而无须定位全部内容的场景。而 PostgreSQL 的部分索引是在索引后边加 where 条件，只对符合条件的记录建立索引，针对部分行。以下是应用 MySQL 的部分索引特性的案例。

如图 2-45 所示，初始化 MySQL 部分索引。

```
mysql> create table xuexiaogang (id int ,name1 varchar(10),name2 varchar(10));
Query OK, 0 rows affected (0.02 sec)

mysql> insert into xuexiaogang values  (1,'123456789','123456789');
Query OK, 1 row affected (0.01 sec)

mysql> insert into xuexiaogang values  (2,'223456789','223456789');
Query OK, 1 row affected (0.00 sec)

mysql> insert into xuexiaogang values  (3,'323456789','323456789');
Query OK, 1 row affected (0.00 sec)

mysql> insert into xuexiaogang values  (4,'423456789','423456789');
Query OK, 1 row affected (0.00 sec)

mysql> insert into xuexiaogang values  (5,'523456789','523456789');
Query OK, 1 row affected (0.00 sec)

mysql> insert into xuexiaogang values  (6,'623456789','623456789');
Query OK, 1 row affected (0.00 sec)

mysql> insert into xuexiaogang values  (7,'723456789','723456789');
Query OK, 1 row affected (0.00 sec)

mysql> insert into xuexiaogang values  (8,'823456789','823456789');
Query OK, 1 row affected (0.00 sec)

mysql> insert into xuexiaogang values  (9,'923456789','923456789');
Query OK, 1 row affected (0.00 sec)
```

图 2-45　初始化 MySQL 部分索引

如图 2-46 所示，从 MySQL 部分索引的数据样本中可以看出 name1 列和 name2 列的特点是只有前几列不一样，后面都一样，即只要前两位就可以定位数据（实际场景类似于手机归属地，前 7 位就可以定位手机号段来自地市级的不同运营商）。

```
mysql> select * from xuexiaogang;
+------+-----------+-----------+
| id   | name1     | name2     |
+------+-----------+-----------+
|    1 | 123456789 | 123456789 |
|    2 | 223456789 | 223456789 |
|    3 | 323456789 | 323456789 |
|    4 | 423456789 | 423456789 |
|    5 | 523456789 | 523456789 |
|    6 | 623456789 | 623456789 |
|    7 | 723456789 | 723456789 |
|    8 | 823456789 | 823456789 |
|    9 | 923456789 | 923456789 |
+------+-----------+-----------+
9 rows in set (0.00 sec)
```

图 2-46　MySQL 部分索引的数据样本

分别为 name1 和 name2 建立索引，建立普通索引和部分索引的命令如图 2-47 所示。其中 name2(3) 为部分索引，意思是取这个字段的前 3 位做索引，而不是用整个字段。

```
mysql> create index t1 on xuexiaogang (name1);
Query OK, 0 rows affected (0.02 sec)
Records: 0  Duplicates: 0  Warnings: 0

mysql> create index t2 on xuexiaogang (name2(3));
Query OK, 0 rows affected (0.03 sec)
Records: 0  Duplicates: 0  Warnings: 0
```

图 2-47　建立普通索引和部分索引的命令

在普通索引执行计划和部分索引执行计划的对比（见图 2-48）中可以看出 key_len 有所不同，但是执行效果几乎一样，都用到了索引，返回 1 行数据。

```
mysql> explain select * from xuexiaogang where name1='223456789';
+----+-------------+-------------+------------+------+---------------+------+---------+-------+------+----------+-------+
| id | select_type | table       | partitions | type | possible_keys | key  | key_len | ref   | rows | filtered | Extra |
+----+-------------+-------------+------------+------+---------------+------+---------+-------+------+----------+-------+
|  1 | SIMPLE      | xuexiaogang | NULL       | ref  | t1            | t1   | 43      | const |    1 |   100.00 | NULL  |
+----+-------------+-------------+------------+------+---------------+------+---------+-------+------+----------+-------+
1 row in set, 1 warning (0.02 sec)

mysql> explain select * from xuexiaogang where name2='223456789';
+----+-------------+-------------+------------+------+---------------+------+---------+-------+------+----------+-------------+
| id | select_type | table       | partitions | type | possible_keys | key  | key_len | ref   | rows | filtered | Extra       |
+----+-------------+-------------+------------+------+---------------+------+---------+-------+------+----------+-------------+
|  1 | SIMPLE      | xuexiaogang | NULL       | ref  | t2            | t2   | 15      | const |    1 |   100.00 | Using where |
+----+-------------+-------------+------------+------+---------------+------+---------+-------+------+----------+-------------+
1 row in set, 1 warning (0.00 sec)
```

图 2-48　普通索引执行计划和部分索引执行计划对比

key_len 的数值与字符集关系很大，在 MySQL 5.7 以前使用的是 utf8 字符集，一个字符占 3 个字节，MySQL 8 使用的是 utf8mb4 字符集，一个字符占 4 个字节。此外，MySQL 规定如果是可变长度的字段，那么需要再加两个字节长度来存放可变长度的字段。如果字段可以为空，再增加一个字节长度来存放可为空的信息。

所以 name1 的 key_len 是 10×4+2+1=43，name2 设置了部分索引，只用了 3 个字符，所以是 3×4+2+1=15。

部分索引的特点是在索引对表进行收敛的情况下，继续收敛。假设一个表有 50 个字段，大小约为 100GB。如果理想化一点，每个字段一样大小。那么在其中一列上建立索引，不用回表，索引覆盖。基于假设，这样其实只查了 1/50 的总表，也就是 100GB 的 1/50，即 2GB。所以说即使全索引扫描也是比全表快 50 倍。因为只查这个字段等于这个字段类比成为一个大表的小表。假设一个表有 50 个字段，每个字段大小一样，查一个字段相当于一个表的 1/50。那么这样的部分索引等于又在 1/50 的基础上继续缩小，又会得到更大的提升。

8．全文索引

全文索引是一种用于查询文本数据的索引技术。它可以通过对文本内容进行分析，并建立倒排索引来实现对文本数据的高效查询。解决在关系数据库中前后都加%全模糊场景下不能使用 B-Tree 索引的问题。当 WHERE 后面谓词前后使用%时是无法利用 B-Tree 的。关系数据库在简单场景下的全文索引也可以实现类似 Elasticsearch 的全文索引功能。在没有全文索引或对实时性、一致性要求很高的场景下，利用关系数据库的全文索引实现模糊匹配也不失为一种解决方案。

关系数据库的全文索引和 Elasticsearch 还有一定的差距。在语义解析、同义词等方面处理得不够好。

曾经遇到过这样的求助，需求是要查公司名。但是公司名字是带括号的，括号可能是全角的也可能是半角的。其实这种问题经常遇到，许多用户似乎习惯在输入时采用任意的方式，只是在查询时进行区分。其实这是不对的。就像垃圾分类是在小区垃圾桶进行？还是在垃圾处理厂最后一道手续的时候进行？显然前期分类比较合适。

这方面在以前的工作中有过最佳实践，在做公安系统的项目时，录入路口、路段、坐标等都是有严格规定的。所以在页面上用户可以随意输入，但是输入框内会自动全部转换成统一格式，以便统一存储。这样从设计到实现，不仅规范而且高效。

在 MySQL 数据库中建立全文索引字段的数据表，如图 2-49 所示。重点是 fulltext 这个关键字，with parser ngram 也是语法关键字。

```
mysql> create table a1 (id int auto_increment primary key,user varchar(20), fulltext(user) with parser ngram);
Query OK, 0 rows affected (0.20 sec)
```

图 2-49　包含全文索引的 MySQL 数据表

模拟数据如图 2-50 所示（数据是随便编的，不一定有这些公司，主要是为了区分全角和半角的括号），从图中可以看到 MySQL 全文样本数据中的括号有些是全角，有些是半角。

```
mysql> select * from a1;
+----+----------------------------------+
| id | user                             |
+----+----------------------------------+
| 1  | 网易（杭州）有限公司             |
| 2  | 网易(上海)科技有限公司           |
| 3  | 阿里巴巴（中国）网络技术有限公司 |
| 4  | 百度(广州)科技有限公司           |
| 5  | 京东（北京）有限公司             |
| 6  | 新浪(北京)网络技术有限公司       |
| 7  | 小米（中国）有限公司             |
| 8  | 网易（广州）有限公司             |
| 9  | 阿里巴巴（杭州）网络技术有限公司 |
| 10 | 百度(上海)科技有限公司           |
| 11 | 京东（广州）有限公司             |
| 12 | 小米（深圳）有限公司             |
+----+----------------------------------+
12 rows in set (0.00 sec)
```

图 2-50　MySQL 全文样本数据

如果命令是如图 2-51 所示的全模糊查询公司名，即采用前后%的通配符来查询公司名，那么查询出来的数据会比实际上多。因为需求是查询一个属地的公司，而此时查出来了三个属地，与需求不符。

如果命令是如图 2-52 所示的全模糊查询属地名，即采用传统前后%的通配符查询属地名，那么查出来的数据包含了其他的公司名，与需求不符。

```
mysql> select * from a1 where user like '%网易%';
+----+--------------------------------+
| id | user                           |
+----+--------------------------------+
|  1 | 网易（杭州）有限公司           |
|  2 | 网易(上海)科技有限公司         |
|  8 | 网易（广州）有限公司           |
+----+--------------------------------+
3 rows in set (0.00 sec)
```

图 2-51　全模糊查询公司名

```
mysql> select * from a1 where user like '%广州%';
+----+--------------------------------+
| id | user                           |
+----+--------------------------------+
|  4 | 百度(广州)科技有限公司         |
|  8 | 网易（广州）有限公司           |
| 11 | 京东（广州）有限公司           |
+----+--------------------------------+
3 rows in set (0.00 sec)
```

图 2-52　全模糊查询属地名

于是采用 MySQL 特有的全文索引，如图 2-53 所示的 MySQL 全文索引分割，既实现了需求的功能，也排除了全角和半角的问题。

```
mysql> select * from a1 where match(user) against ('+网易 +广州' IN BOOLEAN MODE);
+----+--------------------------------+
| id | user                           |
+----+--------------------------------+
|  8 | 网易（广州）有限公司           |
+----+--------------------------------+
1 row in set (0.00 sec)
```

图 2-53　MySQL 全文索引分割

9. 反转索引

反转索引是将正常的键值首尾调换后再进行存储的索引。反转索引最主要的用途是防止热块争用，让顺序写入的序列号产生的键值能在存储时打散。也可间接应对前%这种半模糊不能使用 B-Tree 索引的问题。反转索引通常用于模糊查询。

一般来说有时候需要模糊查询，查找模糊的格式是后面为%，关系数据库的模糊查找（通配符放在最后）如图 2-54 所示。

```
SQL> explain plan for select * from thousand where name like 'abc%';

Explained


SQL>  SELECT * FROM TABLE(DBMS_XPLAN.DISPLAY);

PLAN_TABLE_OUTPUT
```

图 2-54　关系数据库的模糊查找

```
-------------------------------------------------------------------
Plan hash value: 3711133080
-------------------------------------------------------------------
| Id | Operation                            | Name                | Rows | Byt
|  0 | SELECT STATEMENT                     |                     |   4  |  1
|  1 |  TABLE ACCESS BY INDEX ROWID BATCHED | THOUSAND            |   4  |  1
|* 2 |   INDEX RANGE SCAN                   | SYS_AI_b190b0jhqcfyv|   4  |

Predicate Information (identified by operation id):
---------------------------------------------------
   2 - access("NAME" LIKE 'abc%')
       filter("NAME" LIKE 'abc%')

15 rows selected
```

图 2-54 关系数据库的模糊查找（续）

前%模糊是关系数据库的大忌。如图 2-55 所示的关系数据库的前%模糊，无论该列有没有索引都是全表扫描。

```
SQL> explain plan for select * from thousand where name like '%abc';

Explained

SQL> SELECT * FROM TABLE(DBMS_XPLAN.DISPLAY);

PLAN_TABLE_OUTPUT
-------------------------------------------------------------------
Plan hash value: 4023600978
-------------------------------------------------------------------
| Id | Operation         | Name     | Rows | Bytes | Cost (%CPU)| Time     |
|  0 | SELECT STATEMENT  |          | 250K | 8544K |  7671   (1)| 00:00:01 |
|* 1 |  TABLE ACCESS FULL| THOUSAND | 250K | 8544K |  7671   (1)| 00:00:01 |

Predicate Information (identified by operation id):
---------------------------------------------------
   1 - filter("NAME" LIKE '%abc' AND "NAME" IS NOT NULL)

13 rows selected
```

图 2-55 关系数据库的前%模糊

因为字段第一位是%，意味着第一位可以是任意数值，如字母、数字、下画线、特殊符号等。那么意味着全表的数据都有可能是查询的结果，所以这是一个全表操作。

直到现在仍然有很多开发人员不知道为什么字段第一位%会是全表查询。也有很多人问过笔者，对于 name like '%abc%'怎么办？是要写成 name like '%abc' or name like 'abc%'这样吗？

其实关系数据库中可以使用 name like 'abc%'这种写法，因为这种场景 B-Tree 索引可以轻松解决。

但是对于前%模糊是不好处理的，需要一些技术手段。如图 2-56 所示，建立反转索引测试表，并且模拟写入 100 万数据后收集统计信息。此处用 Oracle 做范例，PostgreSQL 的步骤和 Oracle 一模一样，不再占用篇幅说明。而 MySQL 目前没有此场景的应对方案。

```
SQL> create table xuexiaogang3 (id int ,a int ,b varchar2(10));

Table created

SQL> create index x3 on xuexiaogang3 (id);

Index created

SQL> exec xxg3;存储过程写入100万数据

PL/SQL procedure successfully completed

SQL> exec dbms_stats.gather_table_stats(ownname=>'XXG',tabname=>'XUEXIAOGANG3');

PL/SQL procedure successfully completed
```

图 2-56　反转索引测试表的创建和初始化

如果查询 B 列，由于 B 列没有索引，所以如图 2-57 所示的无索引列的执行计划，必然是全表扫描。

```
SQL> explain plan for select * from xuexiaogang3 where b='VSnjadwGhW';

Explained


SQL> SELECT * FROM TABLE(DBMS_XPLAN.DISPLAY);

PLAN_TABLE_OUTPUT
--------------------------------------------------------------------------------
Plan hash value: 4013022863
--------------------------------------------------------------------------------
| Id  | Operation         | Name         | Rows  | Bytes | Cost (%CPU)| Time     |
--------------------------------------------------------------------------------
|   0 | SELECT STATEMENT  |              |     1 |    21 |   998   (1)| 00:00:01 |
|*  1 |  TABLE ACCESS FULL| XUEXIAOGANG3 |     1 |    21 |   998   (1)| 00:00:01 |
--------------------------------------------------------------------------------

Predicate Information (identified by operation id):
---------------------------------------------------
   1 - filter("B"='VSnjadwGhW')

13 rows selected
```

图 2-57　无索引列的执行计划

即使有索引场景下，按照 B-Tree 索引的常识，前%不能使用索引，如图 2-58 所示。

```
SQL> explain plan for select * from xuexiaogang3 where b like '%GhW';

Explained


SQL> SELECT * FROM TABLE(DBMS_XPLAN.DISPLAY);
```

图 2-58　前%不能使用索引

```
PLAN_TABLE_OUTPUT
-------------------------------------------------------------------------
Plan hash value: 4013022863
-------------------------------------------------------------------------
| Id  | Operation         | Name        | Rows  | Bytes | Cost (%CPU)| Time     |
-------------------------------------------------------------------------
|   0 | SELECT STATEMENT  |             | 50000 | 1025K |  1000   (2)| 00:00:01 |
|*  1 | TABLE ACCESS FULL | XUEXIAOGANG3| 50000 | 1025K |  1000   (2)| 00:00:01 |
-------------------------------------------------------------------------

Predicate Information (identified by operation id):
-------------------------------------------------------------------------
   1 - filter("B" LIKE '%GhW' AND "B" IS NOT NULL)

13 rows selected
```

图 2-58 前%不能使用索引（续）

接下来建立反转索引，执行 create index xd3 on xuexiaogang3 (reverse(b))。如图 2-59 所示的建立反转索引，可以看到优化器也能识别前%。注意：这里是在索引列前面加了 REVERSE 这个关键字。Reverse 是颠倒、反转的意思，其实从广义上来说反转索引是一种特殊的函数索引，这个函数就是 reverse。

```
create index xd3 on xuexiaogang3 (reverse(b));

索引已创建。

SQL> explain plan for select * from xuexiaogang3 where reverse(b) like  reverse ('%GhW');

Explained


SQL> SELECT * FROM TABLE(DBMS_XPLAN.DISPLAY);

PLAN_TABLE_OUTPUT
-------------------------------------------------------------------------
Plan hash value: 504417197
-------------------------------------------------------------------------
| Id  | Operation                           | Name        | Rows | Bytes | Cos
-------------------------------------------------------------------------
|   0 | SELECT STATEMENT                    |             |    2 |   64  |
|   1 |  TABLE ACCESS BY INDEX ROWID BATCHED| XUEXIAOGANG3|    2 |   64  |
|*  2 |   INDEX RANGE SCAN                  | XD3         |    2 |       |
-------------------------------------------------------------------------

Predicate Information (identified by operation id):
-------------------------------------------------------------------------
   2 - access(REVERSE("B") LIKE 'WhG%')
       filter(REVERSE("B") LIKE 'WhG%')

15 rows selected
```

图 2-59 建立反转索引的执行计划

接下来执行 select * from xuexiaogang3 where reverse(b) like reverse ('%GhW')。最终执行

结果如图 2-60 所示，达到了预期。

```
SQL> select * from xuexiaogang3 where reverse(b) like  reverse ('%GhW');

        ID                                              A B
---------- --------------------------------------------------
    245298                                         245298 aPqfJVMGhW
    196097                                         196097 dbocqVQGhW
    834328                                         834328 wGOITbZGhW
    385866                                         385866 zoWSLoaGhW
    230016                                         230016 xaUNPojGhW
     92407                                          92407 MDtyKVnGhW
    267975                                         267975 lemlFjnGhW
         2                                              2 VSnjadwGhW
    849019                                         849019 IQaFErzGhW

9 rows selected
```

图 2-60　使用反转索引返回结果

以上介绍的都是常见索引，而并非全部的索引。知道了主要索引的使用场景后，就可以结合不同场景使用不同的索引。通常来说一个表上的索引能够覆盖 99%的读请求场景就是设计比较到位的，如果能覆盖 100%是最好的。

举个例子：一个电商网站的订单表是一个主要业务流程表。可以想象一下，无论是用户查询还是投诉跟踪，订单号是被频繁用到的查询。而订单号从业务角度是唯一的，所以这个作为唯一键或者主键就是一个索引。而买家登录查看自己的订单（由于可能有几年的历史了），查看自己最近 1 个月的订单，那么这种场景就是"买家代码+时间"的复合索引（只有买家代码也是可以的，但是不如"买家代码+时间"高效，尤其是数据量大的时候）。当然这里还可能涉及卖家，那么还需要"卖家代码+时间"的复合索引。

对于平台的分析师，想分析最近一周某些买家或卖家的购买情况，就需要使用"时间+买家代码"的复合索引或者"时间+卖家代码"的复合索引。当然也可以只有时间的单列索引不用复合索引，这个要结合实际情况分析。

这样一来就有 4~5 个索引了。虽然不一定能保证满足每个需求，但是绝大多数场景会被满足，极少数不是主要业务场景的特殊情况要具体问题具体分析。

总的指导思想就是根据需求设定表和索引，不能先有表再去适配业务。

如果出现所有索引都不能覆盖主要业务场景的情况，那么说明有两个可能：表设计错了或者需求不合理。

2.1.3　规避索引使用的误区

索引通过算法解决的是效率问题，而不能解决逻辑问题。以下总结了使用索引要规避的 9 个误区。

误区 1：系统上线无任何索引，让用户可以任意维度（字段）进行单一检索。直到在使用系统的过程中用户抱怨某些操作或者功能缓慢甚至不可用，再去定位这些操作或者功能背后的 SQL。发现如果是索引缺失，再去逐一变更增加索引。这种一般发生在研发与业务不对等的组织架构中，开发人员处于弱势，业务提什么做什么。发现索引缺失便添加索引，如果加了索引也没有解决就让 Hadoop 去处理。而实际上很多问题在 Hadoop 上面也解决不了。这就需要对于需求的合理性进行评估，好好设计表和编写 SQL，这样无须使用 Hadoop，并且极有可能其处理速度远比 Hadoop 处理的速度快。

误区 2：复合索引从左到右层级递减。这种理解一般源于人的角度，省、市、区（县）、乡镇、街道、人员依次逐级查找。所以索引就是 4~5 级的设定，而实际上对计算机这种二进制的系统而言，通过算法直接找身份证这种唯一或者区分度很高的字段效率是最高的。在一个集团内，无须一级子公司、分子公司、部门、小组这样查找员工。直接使用工号查找效率最高。

误区 3：基于经常改变的列建立索引。产生这种误区的原因可能是业务人员提出只看已成功或者未成功的数据。在这种需求下，开发人员为了满足状态类的查询需求，就只对该状态建立索引。其实这样做不一定能起到使用索引的效果。即使特殊场景下可能会用到索引，但是频繁更新索引的值会导致索引的聚簇因子发生改变，性能会越来越差，最终可能还是无法使用到索引。而最佳实践是状态类的字段不建立索引，建立索引的列都是不会进行修改的列，如订单号、姓名、身份证号等。

误区 4：区分度低的字段建立 B-Tree 索引。这种理解类似于误区 3，其范围比误区 3 大。误区 3 一般是指状态只有 2~3 个区分度，属于区分度极低。而区分度低是指该字段有一定的区分度，虽然不止 3 个，但是也不超过 10 个，依然是可以枚举的。比如说，一个集团下属 5 个公司，按照公司建立索引。虽然公司不是经常改变的列，但是每个公司的数据约占全集团总数据五分之一。数据库依然可能是全表扫描。

误区 5：索引一定比全表快。这种理解过度夸大了索引的作用。索引的作用是在数据库的上百亿条数据中快速找一条或一类数据时，做到精准定位，效率提升，要是在 100 亿数据中找 80 亿条数据这样的海量数据查询，索引就不适用了，因为当返回结果集大于全表的一定比例后，就会触发全表扫描。此时，全表扫描的速度比全表+全索引要快一些。

误区 6：组合索引的顺序无关紧要。有这种误解是混淆了优化器对于 WHERE 条件后面谓词的前后顺序无关的概念。WHERE 条件谓词出现的先后顺序是无关的，但是需要用到组合索引的第一列才有效果。所以组合索引的第一列很关键，如果 WHERE 条件后面谓词全都没有对应到组合索引的第一个，那么几乎就是用不到索引了。Oracle 独有的跳跃扫描的特殊场景除外。

误区 7：组合索引的列越多越好。组合索引一般是 2~3 列就可以起到定位效果，如果 3 列还没有起到过滤定位的效果，那么后续再多的列也没有什么效果了。

误区 8：索引越多越好。这种错误的思维是严防死守的被动防御，会出现列列是索引，甚至各种索引的组合排列，最后导致索引比表大，但是依然性能低下。索引多未必能提高查询速度，但是一定会降低写入速度。只有正确的索引才能起到提升速度的效果。

误区 9：索引列上输入一个值用到了索引，另外一个值也一定能用到索引。真实情况是未必。当数据发生倾斜的时候，倾斜的数据是无法用到索引的。因为只有返回结果集小于 5%（不同数据库的这个阈值不一样）才会较好地使用索引。而超过一定的比例则会发生全表扫描。

比如在人口库查找某个少数民族的人数可以用到索引，但是查找汉族人数是无法用到索引的，因为返回数据量占数据总量的95%。

2.1.4　通过索引实现海量数据中的高效查询

在实际工作中，有的项目中单表的数据就达到了上百亿行的数据量级，很多业务场景中需要 DBA 实现在这种海量数据中的高效查询。在百亿级别的海量数据中实现高效查询一定不能使用全表查询，否则不仅仅是查询快慢的问题，甚至会出现宕机的问题。实际上，用好索引，实现百亿级别的海量数据高效查询是轻而易举的。以下通过笔者亲身经历的一个案例加以说明。

有一次某市公安部门提出要抓套牌车辆。所谓套牌车就是车牌号码一样，但是可能车的外观不一样。当然也有一些套牌车辆和被套牌车辆可能车型、外观高度相仿。

套牌和假牌最大的区别是：假牌照在车管所数据库中查不到，而套牌车的车牌在车管所数据库中是真实存在的，但该车牌被人复制套用，存在一牌多车的情况。公安机关想通过车辆的行车数据定位套牌车的活动区域，以有针对性地布防。

这里要给读者做一个数据规模的说明。该市大约拥有 300 万辆机动车。而大部分机动车每天经过 30 个（上班 15 个、下班 15 个）道路监控设备（路口和路段），可以被记录到行车数据。每天产生的行车数据可能将近 1 亿条，半年的车辆行驶数据会达到几十亿乃至上百亿条。如果要通过查询行车数据查处套牌车辆，也就是说要在有几十亿乃至上百亿条数据的一个表中进行筛选。

那怎么去通过数据辨别套牌车呢？既然套牌车的车辆外观（如颜色、车型）不一样。那么是不是可以把所有的车辆数据的车牌、颜色、车型等通过大数据技术（如 Hadoop）进行图像比对？这样做不是不可以，但是通过大数据技术进行图像筛查，数据实时同步是一个大问题。Hadoop 最适合的场景是离线计算，如果希望每天做一次全量同步再进行计算，这个过程可能耗时几天都未必有结果，效率很低。更何况抓套牌车辆的需求是希望秒级出结果。

把车管库所有的机动车车牌的所有行车数据调出来做对比是否可行呢？这样的做法效率极其低下。因为对于一个拥有几百万辆机动车的城市来说，套牌车可能只占其机动车保有总量的十万分之一，即大约十几辆或者几十辆。这样去对比的话，即使可以得出结果，那么 99.999% 的运算都是无效运算。提出的需求是要找出异常的车辆数据，而这样做的逻辑是排除所有正常车辆的数据，而正常的车辆数据几乎等于全量数据了。

所以关键点是需要排查出异常车辆的特征值。这个时候就需要合理的业务逻辑了，公安部门再次细化需求：套牌车一般都是原车主接到罚单或者其他违法记录后，申诉发现自己的车牌可能被套牌了。所以这是一个精准化的定位，只要将这个申诉的车牌某个时间段的行车记录全部查出来，对比时间即可。例如该车在 2010 年 1 月 10 日上午 9 点在城东，那么 8 点 50 到 9 点或者说 9 点到 9 点 10 分，就不可能在城北、城南或者城西，因为早高峰期间道路非常拥堵，同一辆车无法在短时间内到达相距很远的两地。所以只要查询车牌号码在公安车辆监控数据库一段时间的行车数据就能得到套牌车的行驶轨迹。

具体运算方法示例如下：

```
Hphmnew   varchar2（10）；号牌号码
Jgsjnew    date；经过时间
Dddmnew   varchar2（10）；地点代码
```

下面是套牌车辆甄别的 SQL 语句：

```
Select hphm,jgsj,dddm into hphmnew,jgsjnew,dddmnew from tgcl where hphm='Z' and jgsj>sysdate-10/1440 and jgsj<sysdate and dddm='X' or dddm='Y';
```

Tgcl 为机动车车辆数据表；X 和 Y 两个参数分别表示两个不同的地点，实际运算时是外部输入的参数，如 X=人民广场，Y=莘庄；sysdate 表示一个时间点，sysdate-1/24 表示该时间点前一个小时的时间点；Z 表示一个车牌号码，实际运算时是外部输入的参数，如沪 A12345。

其中号牌号码、经过时间、地点代码都是有索引的，所以 SQL 可以直接利用已经建立的索引，通过索引进行数据查找，而不用全表遍历查找。即使这个表数据达到了 100 亿条的级别，这类查找都可以在 1s 内完成计算，既可以对历史记录进行分析，也可以对 10min 以内的数据进行实时的比对。最终以上场景的 SQL 都可以在 1s 内完成检索。而如果使用 ETL+Hadoop 的方式来实现，首先无法做到实时，其次 Hadoop 将计算任务发送到每个计算节点的时间已经不止 1s，所以在数据库设计合理以及 SQL 实现合理的情况下，其效率远超 Hadoop。

其实对于绝大多数企业和应用场景来说，用户希望能精确定位和返回需要的数据。而这种查询场景恰恰是索引技术所擅长的。利用索引进行数据检索，就避免了全表扫描，效率极高。其关键点在于结合业务场景精心设计索引。在此前提下，绝大多数业务通过单机数据库都可以满足使用要求。甚至不需要采用分布式数据库技术，更加不需要使用 Hadoop 等技术。

2.1.5 自动化索引：数据库自治的趋势

索引是 OLTP 数据库的关键所在，通常来说需要开发人员或者 DBA 结合业务场景精心设计所需要的索引，并且在上线之初进行手工建立，如果需要调整也由 DBA 手工进行调整。目前数据库在向着自治化、智能化的方向演进，对于 B-Tree 索引，Oracle 已经实现根据 SQL 运行的条件自动建立索引，无须人工介入。这也是数据库 AI 的体现，实现了自动化运维。下面用案例说明一下索引自治的特性，即无须人工建立索引，由数据库根据业务上发生的 SQL 自动建立必要的索引，不过该特性官方建议在一体机下使用。

如图 2-61 所示，首先，创建一个普通表 i 作为自动化实验表。

```
SQL> create table i (id int,name varchar2(10),t date);
```

表已创建。

图 2-61　自动化实验表创建

用 Java 代码给实验表写入 100 万行数据，如图 2-62 所示。这 100 万行数据属于凭空杜撰的实验数据，只是为了达到一定的数量级。当然读者也可以用存储过程实现模拟，存储过程实现起来更加简单。

```java
package db;

import java.sql.*;
import java.util.Date;
import java.util.Random;

public class oracle_ten_thousand {
    public static void main(String[] args) {
        Connection conn = null;
        PreparedStatement pstmt_insert = null;
        String str="abcdefghijklmnopqrstuvwxyzABCDEFGHIJKLMNOPQRSTUVWXYZ0123456789";
        Random random1=new Random();
        try{
            Class.forName("oracle.jdbc.driver.OracleDriver");
            conn = DriverManager.getConnection( url: "jdbc:oracle:thin:@10.60.143.101:1521/tu", user: "xxg", password: "xxg");
            conn.setAutoCommit(false);//关闭自动提交
            pstmt_insert = conn.prepareStatement( sql: "INSERT INTO i (id, name, t) VALUES (?,?,?)");
            Date startTime = new Date();
            for (int i=0; i < 1000000; i++){
                StringBuffer sb = new StringBuffer();
                for (int j=0; j<6; j++){
                    int number=random1.nextInt(str.length());
                    char charAt = str.charAt(number);
                    sb.append(charAt);
                }
                pstmt_insert.setInt( parameterIndex: 1,i);
                pstmt_insert.setString( parameterIndex: 2,sb.toString());
                pstmt_insert.setDate( parameterIndex: 3,new java.sql.Date(System.currentTimeMillis()));
                pstmt_insert.addBatch();
            }
            pstmt_insert.executeBatch();
            conn.commit();//手动提交
            Date endTime = new Date();
            System.out.println("百万条写入总耗时: " + (endTime.getTime()-startTime.getTime()) + "ms");
        }catch (Exception e){
            e.printStackTrace();
        }finally {
            try {
                if (pstmt_insert != null) pstmt_insert.close();
                if (conn != null) conn.close();
            }catch (Exception e){
                e.printStackTrace();
            }
        }
    }
}
```

```
Run: oracle_ten_thousand
/Library/Java/JavaVirtualMachines/zulu-8.jdk/Contents/Home/bin/java ...
百万条写入总耗时: 23216ms

Process finished with exit code 0
```

图 2-62 Java 程序模拟写入数据

实验表和数据都准备完毕，下面检查数据库的参数。在 cdb 中执行 "alter system set "_exadata_feature_on"=true scope=spfile;"，如图 2-63 所示，使用命令行打开一体机隐含参数（自动化索引参数）。出现 "scope=spfile" 这也意味着这个参数要重启数据库才能生效。

```
SQL> alter system set "_exadata_feature_on"=true scope=spfile;

系统已更改。
```

图 2-63　打开一体机隐含参数

重启数据库以后，把所有 PDB（程序数据库文件）打开，如图 2-64 所示需要重启数据库操作。现在该重启后的数据库实例就具备了自动化索引的功能了。自动化索引主要是确保 IO 先可以扛过全表扫描的冲击，这也是为什么在一体机上可以这样做。一体机强大的 IO 确保了 SQL 可以在无索引的情况下不至于产生故障。

```
SQL> shutdown immediate;
数据库已经关闭。
已经卸载数据库。
ORACLE 例程已经关闭。
SQL> startup;
ORACLE 例程已经启动。

Total System Global Area 1.0117E+10 bytes
Fixed Size                  9161688 bytes
Variable Size            1023410176 bytes
Database Buffers          486539264 bytes
Redo Buffers                7614464 bytes
In-Memory Area           8589934592 bytes
数据库装载完毕。
数据库已经打开。

SQL> alter pluggable database all open;

插接式数据库已变更。
```

图 2-64　重启数据库操作

把数据库自动索引的功能打开涉及如下参数的三种策略。

1）启用自动索引。其命令如图 2-65 自动索引配置所示，使其创建并且生效。

```
SQL> EXEC DBMS_AUTO_INDEX.CONFIGURE('AUTO_INDEX_MODE','IMPLEMENT');

PL/SQL 过程已成功完成。
```

图 2-65　自动索引配置

2）启用自动索引，但只作为不可见自动索引，无法被 SQL 所使用。其命令如图 2-66 设置自动索引不可见所示，使其创建但是不生效。

```
SQL> EXEC DBMS_AUTO_INDEX.CONFIGURE('AUTO_INDEX_MODE', 'REPORT ONLY');
```

图 2-66　设置自动索引不可见

3）关闭自动索引。其命令如图 2-67 设置关闭自动索引所示。

```
SQL> EXEC DBMS_AUTO_INDEX.CONFIGURE('AUTO_INDEX_MODE', 'OFF');
```

图 2-67　设置关闭自动索引

有些公司有严格的把控，觉得即使数据库有这个自动索引的功能，但是 DDL（数据定义语言）还是最好自己设置，所以不轻易使用这个功能。这里演示采用第一种启用自动索引的策略，检查最终参数配置如图 2-68 所示。

```
SQL> show pdbs;

    CON_ID CON_NAME                       OPEN MODE  RESTRICTED
---------- ------------------------------ ---------- ----------
         5 TU                             READ WRITE NO
SQL> SELECT con_id, parameter_name, parameter_value FROM cdb_auto_index_config;

    CON_ID PARAMETER_NAME                 PARAMETER_VALUE
---------- ------------------------------ --------------------
         5 AUTO_INDEX_COMPRESSION         OFF
         5 AUTO_INDEX_DEFAULT_TABLESPACE
         5 AUTO_INDEX_MODE                IMPLEMENT
         5 AUTO_INDEX_REPORT_RETENTION    31
         5 AUTO_INDEX_RETENTION_FOR_AUTO  373
         5 AUTO_INDEX_RETENTION_FOR_MANUAL
         5 AUTO_INDEX_SCHEMA
         5 AUTO_INDEX_SPACE_BUDGET        50
```

图 2-68　检查最终参数配置

执行如下格式化命令：

```
COLUMN parameter_name FORMAT A40
COLUMN parameter_value FORMAT A20
SELECT con_id, parameter_name, parameter_value FROM cdb_auto_index_config;
```

为了使自动索引生效，这里需要收集表的统计信息。如果表的统计信息为空，那么自动化索引无法建立，因为空表是不需要索引的。检查表的统计信息如图 2-69 所示，执行"select table_name, num_rows from dba_tab_statistics where table_name='I';"检查对应的表。

```
SQL> select table_name, num_rows from dba_tab_statistics where table_name='I';

TABLE_NAME             NUM_ROWS
-------------------- ----------
I
```

图 2-69　检查表的统计信息

如果统计信息为空的，需要使用"exec dbms_stats.gather_table_stats('XXG', 'I', estimate_percent =>100, cascade=>true, degree => 16);"语句收集统计信息。其中第一个参数 XXG 是 schema 的名字，第二个参数是表名字，第三个参数是收集百分比，当表很大时采

样 10%就可以，第四个参数是收集表的统计信息时把索引的统计信息一并收集，第五个参数是并行度，CPU 越多，设置的并行度就可以越高，执行的速度也就越快。上述语句执行结果如图 2-70 所示。

```
SQL> begin
  2  dbms_stats.gather_table_stats
  3  ('XXG','A',estimate_percent =>100,cascade=>true,degree => 16);
  4  end;
  5  /
PL/SQL 过程已成功完成。
```

图 2-70　收集统计信息命令及结果

最后检查表统计信息收集完毕后的情况，如图 2-71 所示。

```
SQL> select table_name,num_rows from
  2  dba_tab_statistics where table_name='I';

TABLE_NAME   NUM_ROWS
----------   ----------
I            1000000
```

图 2-71　检查表统计信息命令及结果

接下来对实验表 i 进行表的相关对象检查，先确认表 i 上无索引，如图 2-72 所示。

```
SQL> select i.INDEX_NAME,i.INDEX_TYPE,i.TABLE_NAME,c.column_name,i.NUM_ROWS,i.VISIBILITY from user_indexes i, user_ind_columns c where i.index_name=c.index_name and i.table_name = 'I';

未选定行
```

图 2-72　检查实验表当前索引情况

此时对表 i 的 name 列进行查询必然是全表扫描，无索引全表扫描如图 2-73 所示。

```
SQL> explain plan for select * from I where name='IFUWlQ';

已解释。

SQL> SELECT * FROM TABLE(DBMS_XPLAN.DISPLAY);

PLAN_TABLE_OUTPUT
--------------------------------------------------------------------------
Plan hash value: 2668636144

--------------------------------------------------------------------------
| Id  | Operation         | Name | Rows | Bytes | Cost (%CPU)| Time     |
--------------------------------------------------------------------------
|   0 | SELECT STATEMENT  |      |    1 |    20 |  1927   (1)| 00:00:01 |
|*  1 |  TABLE ACCESS FULL| I    |    1 |    20 |  1927   (1)| 00:00:01 |
--------------------------------------------------------------------------
```

图 2-73　无索引全表扫描的执行计划

模拟现实环境对该表的一列进行查询，再次使用 Java 代码进行循环 1 万次的查询，如图 2-74 所示。

```java
package db;

import java.sql.*;
import java.util.Date;

public class oracle_query_test {
    public static void main(String[] args) {
        Connection conn = null;
        try {
            Date date_start = new Date();
            Class.forName("oracle.jdbc.driver.OracleDriver");
            conn = DriverManager.getConnection( url: "jdbc:oracle:thin:@10.60.143.101:1521/tu", user: "xxg", password: "xxg");
            Statement sql = conn.createStatement();
            for (int i = 1; i <= 10000; i++) //执行的次数
                sql.executeQuery( sql: "SELECT * from i WHERE name='IFUWLQ'");

            Date date_end = new Date();
            System.out.print("查询完成: " + (date_end.getTime()-date_start.getTime()));
        }
        catch (Exception e) {
            e.printStackTrace();
        }finally {
            //关闭连接，释放资源
            try {
                if (conn != null) conn.close();
            }
            catch (SQLException e) {
                e.printStackTrace();
            }
        }
    }
}
```

```
/Library/Java/JavaVirtualMachines/zulu-8.jdk/Contents/Home/bin/java ...
查询完成: 190813
Process finished with exit code 0
```

图 2-74　使用 Java 代码循环查询

结束后检查表 i 的索引情况。如图 2-75 自动化索引检查所示，索引是 SYS 用户建立的，带了 AI 开头说明是 AI 推荐的索引。根据输入的查询条件 where 的谓词 name 建立了 name 列的索引。

```
SQL> select i.INDEX_NAME,i.INDEX_TYPE,i.TABLE_NAME,c.column_name,i.NUM_ROWS,i.VISIBILITY from all_indexes i,
user_ind_columns c where i.index_name=c.index_name and i.table_name = 'I';

INDEX_NAME                  INDEX_TYPE      TABLE_NAME      COLUMN_NAME      NUM_ROWS VISIBILIT
--------------------------  --------------  --------------  --------------  --------- ---------
SYS_AI_gf8kk842rtwca        NORMAL          I               NAME              1000000 VISIBLE
```

图 2-75　自动化索引检查

再次检查执行计划，自动化索引的执行计划如图 2-76 所示。

```
SQL> explain plan for select * from I where name='IFUWlQ';
已解释。

SQL> SELECT * FROM TABLE(DBMS_XPLAN.DISPLAY);
PLAN_TABLE_OUTPUT
-----------------------------------------------------------------------------------------
Plan hash value: 2093416052

-----------------------------------------------------------------------------------------
| Id  | Operation                            | Name                  | Rows | Bytes | Cost (%CPU)| Time     |
-----------------------------------------------------------------------------------------
|   0 | SELECT STATEMENT                     |                       |    1 |   20  |    5   (0)| 00:00:01 |
|   1 |  TABLE ACCESS BY INDEX ROWID BATCHED | I                     |    1 |   20  |    5   (0)| 00:00:01 |
|*  2 |   INDEX RANGE SCAN                   | SYS_AI_gf8kk842rtwca  |    1 |       |    3   (0)| 00:00:01 |
-----------------------------------------------------------------------------------------
```

图 2-76　自动化索引的执行计划

如果自动建立的索引因某些原因需要删除，按照常规方法对索引进行维护，首先将索引设置为不可见。如图 2-77 所示，自动化索引设置不可见失败。

```
SQL> alter index SYS_AI_gf8kk842rtwca invisible;
alter index SYS_AI_gf8kk842rtwca invisible
*
第 1 行出现错误:
ORA-01418: 指定的索引不存在
```

图 2-77　设置不可见失败

提示索引不存在，首先应想到的是用户权限的问题。查看一下索引的 owner，由 owner 账户做指定修改，检查权限指定 schema 执行如图 2-78 所示，依然无法操作。

```
SQL> select owner,index_name,auto,tablespace_name from dba_indexes natural where auto='YES';

OWNER    INDEX_NAME              AUT TABLESPACE_NAME
-------- ----------------------- --- ---------------
XXG      SYS_AI_8gvkk6g731ud5    YES USERS
XXG      SYS_AI_dzjsnubmd6g0k    YES USERS
XXG      SYS_AI_0ps20xm50czy1    YES USERS
XXG      SYS_AI_bmk67470fqcwb    YES USERS
XXG      SYS_AI_b190b0jhqcfyv    YES USERS
XXG      SYS_AI_c5ms4nfbs5f0q    YES USERS
XXG      SYS_AI_fhcgc0pqah0zn    YES USERS
XXG      SYS_AI_6r9qc0cfc1gf9    YES USERS
XXG      SYS_AI_8t6y460bdgg3q    YES USERS
XXG      SYS_AI_gf8kk842rtwca    YES USERS

已选择 10 行。

SQL> show user;
USER 为 "XXG"

SQL> alter index XXG."SYS_AI_gf8kk842rtwca" invisible;
alter index XXG."SYS_AI_gf8kk842rtwca" invisible
*
第 1 行出现错误:
ORA-65532: 无法变更或删除自动创建的索引
```

图 2-78　检查权限指定 schema 执行

如果进行删除操作，也是无法进行的。指定 schema 进行删除失败，如图 2-79 所示。

```
SQL> drop index XXG."SYS_AI_gf8kk842rtwca";
drop index XXG."SYS_AI_gf8kk842rtwca"
                *
第 1 行出现错误：
ORA-65532: 无法变更或删除自动创建的索引
```

图 2-79　指定 schema 进行删除失败

根据错误代码，查到官方对于自动创建的索引也是不建议修改的：

ORA-65532: cannot alter or drop automatically created indexes.
Cause: An attempt was made to alter or drop an automatically created index.
Action: Do not perform the action on an automatically created index.

那么只有找其他方法进行删除了。可以使用迁移索引如图 2-80 所示。新建一个表空间，将索引指向这个新的表空间，然后删除表空间。最后可以看到这条索引已经被成功删除。

```
SQL> create tablespace autoindex datafile '/u01/app/oracle/oradata/ORACLE19C/tu/ai.dbf' size 2G;
表空间已创建。

SQL> select name from v$tablespace;

NAME
----------------------------------
SYSTEM
SYSAUX
UNDOTBS1
TEMP
USERS
AUTOINDEX

已选择 6 行。

SQL> alter index "SYS_AI_gf8kk842rtwca" rebuild tablespace autoindex online;

索引已更改。

SQL> drop tablespace autoindex including contents;
表空间已删除。

SQL> select i.INDEX_NAME,i.INDEX_TYPE,i.TABLE_NAME,c.column_name,i.NUM_ROWS,i.VISIBILITY from user_indexes i, user_ind_columns c where
i.index_name=c.index_name and i.table_name = 'I';

INDEX_NAME              INDEX_TYPE  TABLE_NAME  COLUMN_NAME      NUM_ROWS VISIBILIT
----------------------  ----------  ----------  -----------      -------- ---------
SYS_AI_8t6y460bdgg3q    NORMAL      I           ID                2000000 VISIBLE
```

图 2-80　迁移索引

可能许多文章都介绍过如何让索引自动建立，但没考虑过自动索引的维护。如果采用自动索引对删除索引的维护难度加大了，也许是实验数据库版本的问题或者是没有使用一体机的原因。

数据库索引的自动化，使得 DBA 们可以更加注重业务逻辑的把控，就像套牌车的案例。如何配合业务或引导业务使用索引，这才是 DBA 后续的工作方向，否则动辄使用全量计算会使得索引的作用失效。

2.2　通过 SQL 优化提升性能

SQL 优化是 DBA 必须熟练掌握的一项技能。SQL 优化能够解决大量实际工作中的问题。

本节内容会通过多个案例说明如何通过 SQL 优化提升数据库性能。

2.2.1　SQL 优化实现高速执行任务

通过 2.1 节对索引的介绍，相信读者都知道在需求合理的情况下（所谓需求合理就是尽可能利用到索引，而不是全量查询），通过索引都是可以快速完成检索和统计的，即使没有列式数据库的加持。

结合索引的使用，再通过优化 SQL 查询语句的查询逻辑，在实践中是可以实现高速执行查询任务的。以下就通过两个例子"快速（实时）生成报表"和"高速执行任务"加以说明。

1. SQL 优化快速（实时）生成报表

多年前，有一次笔者单位的某个开发人员小 A 找到我，说他们的报表生成得比较慢。小 A 的部门每天要执行 1 次报表生成，但生成一次报表要 15 小时。不仅用户接受不了，小 A 自己都接受不了。小 A 问能不能优化一下？能否一个小时就完成报表生成？在小 A 看来，15 个小时的报表如果能 1 个小时完成就很好了，这相当于效率提升了 15 倍。

接下来笔者详细询问了小 A 读取数据的逻辑，发现了问题的关键点：取值范围过大。

其中一个查询条件是 time<sysdate-1/24，即统计一个小时以前的数据。这就是问题所在，在逻辑上犯了一个错误，边界范围与实际范围不等价。小 A 期望的是上次统计后到一个小时以前的时间段。但是实际执行的查询逻辑是只有范围的上限（即一个小时以前），没有范围的下限（即上次统计时间），显然范围超出了期望的预期。按照实际的取数逻辑：哪怕是 1949 年也是在一个小时以前，这样的话取值范围过大。既然每天统计，那么理论上统计的应该只是当天的。所以时间上应该是在一天的区间内。笔者把这个查询条件修改为 time>trunc（sysdate）and time<sysdate-1/24。优化了导致查询缓慢的最根本的逻辑问题。

此外，笔者发现小 A 要执行的 SQL 查询还包含几个毫无必要的关联和排序，这都会增加查询时间。于是笔者去掉了几个表的关联和几个不必要的排序。最后该 SQL 对应的报表生成时间从 15 个小时缩减至 6 秒。

当小 A 看到这个结果时候惊讶不已，他没想到生成报表的时间可以缩短到秒级，查询速度整整提升了 9000 倍！

其实生成报表是日常工作中都会用到的场景，有日报、周报和月报等各类报表。以下通过一个具体案例来说明对于一个千万条数据量的查询，通过对 SQL 查询语句的优化，报表也可以很快生成（秒级甚至毫秒级），达到实时查询的效果。本案例基于 PostgreSQL 数据库，是一个模拟订单系统，有 1000 万条订单数据。

（1）用存储过程模拟写入 1000 万条订单数据

1）首先建立一个随机函数（该函数为自定义函数，有两个参数，start_num 和 end_num，都为整型，用于生成这两个整数之间的一个随机整数，目的是让数据看起来更加像真实业务产生的数据，而不是一个静态值），具体代码如下。

```
CREATE OR REPLACE FUNCTION random_int(
start_num INT,
```

```
    end_num INT
)
RETURNS INT
AS $BODY$
BEGIN
  RETURN FLOOR(start_num + random() * (end_num - start_num + 1));
END;
$BODY$
  LANGUAGE plpgsql VOLATILE;
```

2）建立一个模拟的订单表，其表结构如图 2-81 所示，包含订单 ID、买家代码、金额、卖家代码、时间这些基本要素字段。

```
xxg=# \d dd
                       Table "public.dd"
   Column   |            Type             | Collation | Nullable | Default
------------+-----------------------------+-----------+----------+---------
 ddid       | character varying(20)       |           |          |
 buyercode  | character varying(18)       |           |          |
 money      | integer                     |           |          |
 sellercode | character varying(18)       |           |          |
 time       | timestamp without time zone |           |          |
```

图 2-81　模拟订单表表结构

3）建立一个存储过程（函数）bu()，其作用是循环 100 万次 INSERT，模拟 100 万条订单数据写入订单表。具体 SQL 代码如下。

```
create or replace function bu() returns
boolean AS
$BODY$
declare iinteger;
  begin
i:=1;
  FOR i IN 1..1000000 LOOP
    INSERT INTO dd (ddid,buyercode,money,sellercode,time) VALUES ('DD'||lpad (cast(i as varchar),18,'0'),'B'||lpad(cast (trunc(100+i/100) as varchar), 7,'0'),random_int(1,10000), 'S'||lpad(cast (trunc(100+i/100) as varchar), 7, '0'),'2020-12-02 08:00:00'::timestamp+ interval '1 day'+ ('-' || i || ' minutes')::interval);
    end loop;
    return true;
end;
$BODY$
LANGUAGE plpgsql;
```

4）多次修改执行时间（2020-12-02 08:00:00）等初始化条件后，执行 SQL 语句 "select *

from bu() as tab;"进行数据写入。

最终执行完存储过程后,查询订单表数据,模拟订单完成,如图 2-82 所示。订单表有 1600 多万条数据。

```
xxg=# select * from bu() as tab
 tab
-----
 t
(1 row)

Time: 2404.981 ms (00:02.405)

xxg=# select count(*) from dd;
  count
----------
 16690004
(1 row)

Time: 1583.136 ms (00:01.583)
xxg=#
```

图 2-82 模拟订单完成

(2)分别对需要统计的维度建立索引

一般统计都带有周期性,所以统计的维度必须含有时间维度,在本例中为 time 字段。由于订单是买卖双方达成的,所以分析数据时可能会分别涉及买家和卖家维度。所以分别以"时间+买家维度"和"时间+卖家维度"建立索引。统计 3 个月的订单总数如图 2-83 所示。

```
xxg=# create index timebuyercode on dd (time,buyercode);
CREATE INDEX
Time: 25120.749 ms (00:25.121)
xxg=# create index timesellercode on dd (time,sellercode);
CREATE INDEX
Time: 26971.159 ms (00:26.971)


xxg=# select count(*),to_char(time,'yyyy-mm') from dd where time>'2019-07-01'
xxg-# and time<'2019-10-01' group by to_char(time,'yyyy-mm');
 count | to_char
-------+---------
 92040 | 2019-07
 92115 | 2019-08
 89099 | 2019-09
(3 rows)
```

图 2-83 统计 3 个月的订单总数

该项统计 1 秒多即完成,每个月大约有 9 万个订单,平均每天 3000 个。这充分说明,在索引和需求明确的前提条件下，30 万数据量级的查询是可以秒级完成的。如果是日报之类的更少数据量的报表查询就更快了。例如，统计每天订单数大于 50 的买家（buyercode）和卖家（sellercode），卖家维度日报统计如图 2-84 所示，买家维度日报统计如图 2-85 所示。

```
xxg=# select count(*),sellercode from dd where time>'2019-07-17'
xxg-# and time<'2019-07-18' group by sellercode having count(*)>50;
 count | sellercode
-------+------------
   100 | S0004666
   100 | S0004670
   100 | S0007371
   100 | S0004672
   100 | S0007375
   100 | S0004675
   100 | S0004671
    80 | S0007376
   100 | S0004674
   100 | S0004665
   100 | S0007367
   100 | S0004669
   100 | S0007370
   100 | S0004668
   100 | S0004664
   100 | S0007365
    53 | S0004663
   100 | S0007372
   100 | S0007364
   100 | S0004676
   100 | S0007373
   100 | S0004667
   100 | S0007368
   100 | S0007374
    87 | S0004677
    59 | S0007362
   100 | S0007363
   100 | S0007369
   100 | S0004673
   100 | S0007366
(30 rows)

Time: 6.178 ms
```

图 2-84　卖家维度日报统计

```
xxg=# select count(*),buyercode from dd where time>'2019-07-17'
xxg=# and time<'2019-07-18' group by buyercode having count(*)>50;
 count | buyercode
-------+-----------
   100 | B0004671
   100 | B0004664
   100 | B0007375
   100 | B0007365
   100 | B0007367
   100 | B0007364
   100 | B0004674
   100 | B0004673
   100 | B0007368
   100 | B0007371
   100 | B0007366
    87 | B0004677
    80 | B0007376
   100 | B0007363
   100 | B0004666
   100 | B0004669
   100 | B0004670
    53 | B0004663
   100 | B0004668
   100 | B0007374
   100 | B0007372
   100 | B0007370
   100 | B0007369
   100 | B0004676
   100 | B0004672
   100 | B0004665
   100 | B0004667
   100 | B0004675
    59 | B0007362
   100 | B0007373
(30 rows)

Time: 6.089 ms
```

图 2-85　买家维度日报统计

这两个查询都在不到 7ms 的时间就完成了。所以报表查询时间不仅可以达到秒级，甚至可以达到毫秒级。这基本上可以算是实时出报表了。

快速生成报表的关键是梳理需求和对应索引。尤其是日报,已经定义了是当天的数据,那么就和历史数据无关,当天的数据量不会很大。只是平时见过很多需求要求统计当天的数据,但是 SQL 写的是全表查询,实际查询了所有历史数据,或者由于 SQL 查询没有用到索引,再或者是根本没建立索引,最后导致报表生成越来越慢。

在小数据量下,即使没有 HTAP 或者没有列存数据库的加持下,只要优化 SQL 查询逻辑,也能实现快速的报表生成。

2. SQL 优化实现高速执行任务

有一次开发人员小 B 找到笔者解决一个定时查询任务执行过慢的问题。这个定时任务是找出新增用户,新增用户的定义是当天第一次下单的用户。该定时任务每天执行一次。这里涉及两种可能性,第一种可能性是用户当天注册下单;第二种可能性,之前注册(比如上周)过,当天第一次下单。小 B 的做法是:枚举出已有的用户 ID(用 groupby 语句),然后排除已有用户(用 not in 语句)。如果有相同 ID 的用户再次下订单,根据逻辑就会被排除,不属于新用户。如果是新的 ID 写入(这样就不用考虑当天注册还是过去注册的因素),就会被作为新用户筛选出来。最后结果是,如果没有新用户 ID,明天继续执行这个任务;如果有新 ID 出现,把新 ID 继续加到 SQL 的 not in 里面。

小 B 的逻辑不能算错,但不是最优的。他的逻辑必然是一个全表遍历的过程,所以随着数据量的增多,查询会越来越慢,而且属于这个 not in 筛选的 ID 会很多,数量可能有几千,也可能有几万,笔者发现的时候已经达到 6000 多了。试想一下,如果微信也这样每天查看新增用户,那么 not in 后面要写 9 亿~10 亿个 ID。

解决小 B 这个问题的关键点就是:**用户最早的订单时间是当天,即只对当天订单涉及的用户进行检索,依据当天用户的最早交易时间与当天的时间相比较即可。**

以下继续通过上一节建立的模拟交易系统来说明。

- 沿用上一节订单表的 1000 万条订单数据。
- 沿用上一节订单表的索引。

由于 time 字段和 buyercode 有索引,所以查询当天数据时,即使数据有一万条也是可以用到索引的,用到索引就和数据量多少没有关系。千万订单当天执行计划和耗时如图 2-86 所示,用到了索引,执行时间约为 0.1ms。真实环境返回数据时间可能多一些,但是也是在 1s 内。

```
xxg=# explain select * from dd where time>CURRENT_DATE;
                                    QUERY PLAN
---------------------------------------------------------------------------
 Index Scan using timesellercode on dd  (cost=0.56..488.78 rows=124 width=49)
   Index Cond: ("time" > CURRENT_DATE)
(2 rows)


xxg=# select * from dd where time>CURRENT_DATE;
```

图 2-86 千万订单当天执行计划和耗时

```
 ddid       | buyercode | money | sellercode |         time
------------+-----------+-------+------------+----------------------------
 DD20230530 | B2000135  |  999  | S001       | 2023-05-30 17:55:42.29685
 DD20230531 | B0000135  |  333  | S001       | 2023-05-30 18:02:07.031728
(2 rows)

Time: 0.928 ms
```

图 2-86　千万订单当天执行计划和耗时（续）

下面的 SQL 语句是关键，如图 2-87 所示。

```
select buyercode, min(dd.time) as mtime from dd  where buyercode in
(select buyercode
            from (select buyercode, count(*) from dd
where time > CURRENT_DATE group by buyercode) A)
group by buyercode
having min(dd.time) > CURRENT_DATE;
```

图 2-87　关键的 SQL 语句

1）执行第 3 行的子查询，即可获得当天几千条订单的用户分组，可能会是几十个或几百个 buyercode。

2）执行第 2~3 行语句，目的仅仅是把 buyercode 取出来，而不需要 count(*)，如果是 MySQL 数据库，可以直接得出结果（因为 MySQL 的分组语法不太严谨，这样的 SQL 可以执行，也可以得到预期的数据，但是其他数据库聚合函数后面必须带分组函数）。所以 Oracle 和 PostgreSQL 需要这样执行。

3）执行第 1~4 行语句，获取当天 buyercode 的所有最小时间。因为 buyercode+time 有索引，所以直接获得。这是解决问题的必要数据。

4）执行第 5 行的过滤语句，也是解决问题的关键，若当天 buyercode 的**最小时间大于当天 0 点，那么就是新用户，否则就不是**。

最终新用户识别执行计划如图 2-88 所示，由于使用到了索引，那么最终即使全表有 5000 万行数据，执行效率也不会随着数据量的增加而降低。

```
xxg=# explain select buyercode, min(dd.time) as mtime from dd  where buyercode in
xxg-# (select buyercode
xxg(#            from (select buyercode, count(*) from dd
xxg(# where time > CURRENT_DATE group by buyercode) A)
xxg-# group by buyercode
xxg-# having min(dd.time) > CURRENT_DATE;
                              QUERY PLAN
-----------------------------------------------------------------------
 GroupAggregate  (cost=104.99..81001.66 rows=1759 width=17)
   Group Key: dd.buyercode
```

图 2-88　新用户识别执行计划

```
            Filter: (min(dd."time") > CURRENT_DATE)
      -> Nested Loop  (cost=104.99..78977.04 rows=389096 width=17)
            -> GroupAggregate  (cost=104.43..106.28 rows=123 width=17)
                  Group Key: dd_1.buyercode
                  -> Sort  (cost=104.43..104.74 rows=124 width=9)
                        Sort Key: dd_1.buyercode
                        -> Index Only Scan using timebuyercode on dd dd_1  (cost=0.56..100.12 rows=124 width=9)
                              Index Cond: ("time" > CURRENT_DATE)
            -> Index Only Scan using buyercodetime on dd  (cost=0.56..609.59 rows=3163 width=17)
                  Index Cond: (buyercode = (dd_1.buyercode)::text)
(12 rows)
```

图 2-88　新用户识别执行计划（续）

最终新用户识别执行效率如图 2-89 所示，耗时约 5ms，效果和预期一致。

```
xxg=# select buyercode, min(dd.time) as mtime from dd  where buyercode in
xxg-# (select buyercode
xxg(#              from (select buyercode, count(*) from dd
xxg(# where time > CURRENT_DATE group by buyercode) A)
xxg-# group by buyercode
xxg-# having min(dd.time) > CURRENT_DATE;
 buyercode |         mtime
-----------+---------------------------
 B2000135  | 2023-05-30 17:55:42.29685
(1 row)


Time: 5.142 ms
```

图 2-89　新用户识别执行效率

所以只要通过合理的设计和实现，报表查询可以很高效，是有可能不随着数据量的增加而变慢的。以上的 SQL 优化思想在主流数据库上的效果可以得到体现。

2.2.2　慎用分页有效提升性能

所谓分页，是指通过 SQL 查询语句来获取大量数据时，将查询结果按照一定规则分割为多个页面展示，以提高用户体验和查询效率。查询分页的方式可以通过 LIMIT 和 OFFSET 关键字实现。比如在日常工作中 DBA 通常会遇到一种需求，即查看输入结果集的总页数，这就需要通过分页加以实现。当返回的结果集过大的时候，采用分页确实是一个绝佳的思路，使用得当时可以很好地提升性能。

但是你听过分页造成的故障吗？没错，分页也会造成系统的故障，而且是重大故障。当然这是由特殊原因造成的。在实践中，需要慎用分页，否则不但起不到提升性能的作用，有时候还会起到反效果。

以查询大量数据利用分页实现部分数据快速返回为例，在数据库中实现该分页的 SQL 语句如下。

```
select * from t where rownum<10（Oracle 的写法）
select * from t fetch first 10 rows only（Oracle 的特殊写法）
select * from t limit 10（MySQL、TiDB 和 PostgreSQL 的写法）
```

即使是几十亿数据甚至 PB 级别的数据库，上述的分页查询语句也可以实现毫秒级别返回。其原因是并未发生真实结果集查询，而是快速定位结果集的第一条物理地址，然后依次返回 10 行后，立即停止，而不需要返回所有结果集。对数据库来说是极轻量级操作。

而在实际工作环境中，还有这样一种分页查询的方式，SQL 语句如下。

```
select count(1) from (SELECL *  FROM  T WHERE …)
```

这句 SQL 语句由两部分组成，括号内的子查询（即大写的 SQL 语句）是用户输入条件后，程序生成的 SQL 语句；而小写部分就是在大写的子查询语句外层上套了一层 COUNT 语句。由于子查询优先执行也必须完全执行出结果才能执行父查询，所以这是对结果集的全集进行了一次真实的求和运算。

实际上，这条语句不是开发人员写的，而是 Mybatis 框架自带的。Mybatis 是一款持久层框架，它支持定制化 SQL、存储过程以及高级映射。避免了使用大量的 JDBC 代码和手动设置参数以及获取结果集。Mybatis 分页的实现方式是 pagehelper 组件，使用该组件就是在原始的分页 SQL 语句上了套了一层 COUNT，其本质是一个分页组件，在结果集套上了 COUNT 就是其中的封装设计。正因为它封装了这部分获取结果集的操作，所以当结果集较大的时候，会产生一定的问题。

笔者第一次看到这个 SQL 语句的时候因为不了解开发框架，所以对此很不解，为什么要这么写？理想的分页 SQL 语句没有子查询，而是直接使用索引定位到逻辑地址或物理地址，然后获得偏移量，从而达到分页的效果。不过后来学习了一些开发知识后才了解到这是开发框架所生成的，而这部分不受开发人员控制，甚至许多开发人员对此也不了解。

最终笔者向开发人员进行了讲解。该分页组件的逻辑是对整个现有的 SQL 语句进行封装，在现有的 SQL 语句上进行一次嵌套，整个 SQL 语句作为一个子查询，那么势必需要将子查询执行完毕才能执行父查询。也就是说，执行该 SQL 语句对结果集进行了真实的查询，而且是把全部数据扫描了一遍。

其实常规理解的分页是在原始 SQL 尾部追加限制完成的，而不是通过嵌套形式完成的。无论数据多少，只要进行了分页，则读取数据量只有 MB 级别，甚至 KB 级别。读取的数据量非常少，甚至不用读取磁盘，效率非常高。

而该分页组件读取的数据量是 GB 级别甚至可能是 TB 级别，读取的数据几乎全是磁盘，效率很低，计算量很大。该分页组件这样设计，其原因或许是需求对总页数的精确计算。

如果仅仅这样，那还不足以让大家吃惊。下面要讲的才是让人不能接受的，那就是分页组件的 COUNT 不是执行一次就结束了，而是每次当用户单击下一页，它都会再次执行一次。

举个例子：对于数据库中 1000 万行的单表，符合条件的数据有 10 万行。不要求显示总条数：查 10 页，每页 10 条。实际累计扫描 100 条。

同等环境中，如果用分页组件实现要求显示总条数：查 10 页，每页 10 条。实际累计扫描 10×10 万+10×10=1000100 条。

在真实环境中，这个造成的影响通常还会被放大，因为真实环境中存在表关联再普遍不过了，所以这里还涉及多表关联的问题（详见第 8 章中的多表关联）。

数据量为 1000 万的 N 表关联，符合条件的数据有 10 万条。关联比 1∶1，不要求显示总条数：查 10 页，每页 10 条。实际累计扫描 100 条×N 表，可能是几百条数据。若 N=5，就是 500 条。

同等环境中，如果用分页组件实现要求显示总条数：查 10 页，每页 10 条。实际累计扫描 10(单击了 10 次)×10 万(每次真实查询了 10 万行)×N 表的个数(由于是 N 表关联所以 N 个表每个表扫描一次)+10×10(真实的每页 10 行，最终输出 10 次 10 行)=N×100 万+100(这 100 可以忽略不计)。若 N=5，就是 5000100 条。

然而现实中多表关联的 1∶1 比例太过于理想，真实环境是 1 对多。那么这个结果集被再次放大。

千万级别多表多倍关联分页与组件分页对比所示，1000 万的 N 表关联，符合条件的数据有 10 万。关联比 1∶M，不要求显示总条数：查 10 页，每页 10 条。实际累计扫描 100 条×N×M 表的个数，可能是几千条。若 N=5，M=10，就是 5000 条。

同等环境中，如果用分页组件实现要求显示总条数：查 10 页，每页 10 条。实际累计扫描 10×10 万×N 表的个数×M 的比例=M×N×100 万条+10×10。若 N=5，M=10，就是 50000100 条。这时候查询量已经差了 1 万倍了。

这里还有两个重量级的因素没有做加权：

- 如果业务功能对应的 SQL 语句没有用到索引，使得查询的是全表呢（以上还是基于建立索引、索引正确而且用到了索引的理想场景）？有时候索引正确，但是达到一定的返回比例，优化器认为还是使用全表更为高效。
- 结果集中的数据需要排序呢？百万级、千万级的排序会更为复杂。

所以在真实的环境中，有时候就是因为这种场景连续单击几次下一页，就造成了大数据量的 SQL 并发执行。因为单击一次下一页只需要 1s，而后台的 SQL 语句要几秒甚至几十秒才能执行完毕，所以就造成了 IO 竞争、CPU 竞争。

笔者亲身经历过一次这样的故障：查询合同信息，但是由于合同的金额没有在合同主表上，需要关联合同明细表，将每个合同的明细累加值计算出来进行显示。同时也因为合同主表上缺乏业务需要的字段，不得不去再度关联所需字段的表。造成了每行合同数据都是十个表关联才能得出完整信息。即使这十张表都是一对一的关系，也意味着读取十张表的各一行记录。而实际上合同主表和合同明细大约是 1∶1000 的关系。所以当最终查询 10 万行合同数据的时候，数据库运算大约是 1 亿 100 万次。而每次单击下一页，就执行 1 亿次的运算。最终导致了服务器 IO 使用率过高，产生严重的 IO 等待，CPU 使用率达到 100%，最终数据库宕机。

有读者不禁要问有什么好的解决办法吗？参考建议有以下几点：
- 如果沿用框架分页组件，那么一定要确保结果集很小，比如最终查出几百条数据。这样即使有 N 表关联和 1∶M 的对应关系，最终 SQL 语句发生的关联扫描数据行数也不会很多，可以控制在不超过一万行量级。
- 放弃使用框架分页组件，单独撰写 COUNT。
- 直接修改需求，时间范围不受限制，不显示总页数，可以不断单击下一页，实现真正意义上的分页。

此外，补充一个因为必须要显示总页数，而不得不对结果集进行 COUNT 的方案。即对结果集单独撰写 COUNT，该实现方案有一个前提：多表关联的对应关系为 1∶1 的场景。

例如，在两表关联为一对一关系下，对比下面两个 SQL。

SQL1:
select a.a1 b.b1, a.a2, a.a3, a.a3, a.a4 from a,b where a.id=b.id and... a.id=XXX
SQL2:
select a.a1 a.a2, a.a3, a.a3, a.a4, from a where a.id=XXX

这两个 SQL 查询的结果集单从行数上来说是一致的。这样单独去执行 SQL2 结果集的行数统计，可以减少一半的运算量，这对于行式存储的数据库来说收益是很大的。SQL1 仅仅是两个表的关联，真实的数据库中，5~10 个表的关联也是普遍存在的，如果可以（结合实际情况）尽可能减少关联，就有可能获得性能的提升。

SQL1 这种场景是在设计之初考虑不周，后续为了一个字段关联一个表，这样的设计埋下了隐患，后期只能尽可能缓解。

2.2.3 从认知上杜绝低效 SQL

说杜绝低效 SQL 有点绝对，更加准确的表述应该是及早干预，基本阻止。笔者曾经在公众号上写过一篇《标准与理念》的文章，主旨就是及早干预。

《魏文王问扁鹊》的故事大家都耳熟能详了，魏文王问扁鹊说："你们家兄弟三人，都精于医术，到底哪一位最好呢？"扁鹊答说："长兄最好，中兄次之，我最差。"魏文王再问："那么为什么你最出名呢？"扁鹊答说："我长兄治病，是治病于病情发作之前。由于一般人不知道他事先能铲除病因，所以他的名气无法传出去，只有我们家的人才知道。我中兄治病，是治病于病情初起之时。一般人以为他只能治轻微的小病，所以他的名气只及于本乡里。而我扁鹊治病，是治病于病情严重之时。一般人都看到我在经脉上穿针管来放血、在皮肤上敷药等大手术，所以以为我的医术高明，名气因此响遍全国。"

文中意思是：事后控制不如事中控制，事中控制不如事前控制，但是大多数的企业经营者均未能认识到这一点，等到错误的决策造成了重大的损失后才寻求弥补，有时是亡羊补牢，为时已晚。

笔者听过很多有关数据库发生故障的案例,其实都是发展要到"扁鹊"这种要"起死回生"的境地才去处理的,笔者自己也经历过很多。一个系统一开始就看到有出现问题的苗头,但这个时候提出来,开发人员往往不以为然,等到了宕机的时候,再希望有人能力挽狂澜。虽然问题有时候也解决了,但实际上这些问题原本大可不必产生。

在一个 IT 系统中常见的问题有哪些?

- GC(垃圾收集)时间过长,Java 虚拟机堆内存中的对象比较多或者比较大。
- OOM,从数据库取了太多的不应该取的数据。
- CPU 使用率过高,Java 的实现逻辑有问题或者执行了数据库全表扫描。
- 数据库 IO 读写量高,执行数据库全表扫描。
- 连接数过高,执行数据库全表扫描,或者事务产生了大量的行锁,甚至产生了表级锁定。
- 磁盘空间满,数据库所在服务器的磁盘空间耗尽、数据库的表空间耗尽或者是数据库归档日志的空间耗尽等。

以上这些问题不能说覆盖了全部问题,但是基本涉及了日常数据库发生的 90% 的问题。产生这些问题的原因绝大多数都是因为对于 SQL 如何操作数据库的认知不到位,操作数据不当导致的。对于这些问题如何预防?笔者给出的方案是建立数据库的开发规范,这个规范有别于开发人员的 Java 开发规范。数据库的 SQL 开发规范关注的是 SQL 的质量。

笔者将 SQL 开发总结为《数据库设计开发规范》,包括对表、索引、视图、函数、存储过程、触发器等数据库对象的设计规范和 SQL 的实现逻辑设计规范。其中部分对象,如需求的确定、索引以及更新频度和数据的生命周期等,应该在需求评审阶段就已经确定,业务场景与数据库表的关联设计见表 2-1。需要说明的是,这里的设计必须由专业的数据库从业者来进行。

表 2-1 业务场景与数据库表的关联设计

表名	中文名字	主键/索引	更新频度	记录数	保持时间	需求	SQL
Table1	经过车辆	车牌/时间	每秒写入 1000 条数据	预计约 100 亿条	3 个月	场景 1 场景 2	Select Insert
Table2							

填写这样的表格后,说明从需求、设计到实现这三个环节的认知已经被统一了。对应的 SQL 也就基本编写完成,达到了大部分的管理要求。

当今的信息化的平台对业务连续性都有较高的要求。因此,越来越多的企业管理者也开始重视 SQL 的质量,要求在上线前进行 SQL 审核。而且从认知上意识到,不是在上线发布时才开始审核,而是在设计之初就进行审核,这一点从表 2-1 可以看到。

如果在开发之初不能建立上述表格,那么开发对于数据如何写入和读取不明确,根本不知道会有什么需求,从而导致 SQL 质量上存在较大问题,那么整个项目或者系统运行过程中就会出现无穷无尽的问题。

基于以上的工作再进行上线前的审核以及上线后的监控,那么监控后的干预会成为极少数的偶发事件，而不是常态。当然有的企业 DBA 无权干预开发，这就只能事后"救火"了。

好在许多企业已经逐步意识到 SQL 质量的重要性，正在积极地改进。我身边就有企业领导提出了"写烂 SQL 要有负罪感"。操作数据库的人，要对数据库有敬畏之心，因为每一个命令都不能出现错误，有的时候一个失误都会导致重大问题，笔者始终觉得 SQL 写得不好是万恶之源。

正常情况下，MySQL 数据库或者 PostgreSQL 数据库，在一个主流的物理机内，保守估计都有每秒约 10 万次执行更新单行数据的能力，至于点查（根据唯一索引或者主键查询）至少也有每秒约 100 万次的查询单行数据的能力，当然这样的速度前提是 SQL 写得好。

那 SQL 写得不好的呢？写得不好的，执行时间会显著增加，甚至达到小时级别。所以一定要从思想上认识到杜绝低效 SQL 的重要性。

2.3　避免数据库对象设计失误

数据库对象主要是指表结构、表之间关联关系、视图、索引等。这些对象设计得好，数据库就会展现出优异的性能。而若这些设计出现偏差和失误，则会给数据库带来严重的性能问题，也会对日后的开发维护带来巨大的工作量。

2.3.1　避免不必要的多表关联导致的低效查询

在工作中经常遇到，由于前期设计不合理、考虑不足，导致查询需要跨多表的情况。只为了其中的部分数据，或是为了过滤一个额外的字段，往往会出现额外关联表。这样的设计导致最终查询效率极差，以下通过一个例子来说明不必要的多表关联所造成的问题。

建立 A、B、C、D 四个表，都通过统一的 ID 作为关联字段，每个表有自身的标志字段，分别为 a、b、c、d。其中，A、B 表都是一对一的关系，A 表和 B 表模拟不同场景产生的数据或是不同业务生成的数据，它们是业务需要进行关联查询的两张主要的表。C 表与 A 表为 1：100 的父子关系，其中 C 表存储了数字数据，查询场景中往往用于统计查询。而 D 表是一个标记表，可以理解为状态或是标记，只有 0 或 1 两个值。那么能够想到的一个常用的场景便是关联 A、B 表进行查询，同时要汇总 C 表的数据，并且过滤 D 表中的某一个状态，常见多表关联查询如图 2-90 所示。

```
mysql> select aa.id id, aa.a a, bb.b b, sum(cc.c) sum, dd.d d from aa join bb
    -> on aa.id=bb.id join cc on aa.id=cc.id join dd on aa.id=dd.id
    ->where dd.d=1 and aa.id<10 group by aa.id,aa.a,bb.b, dd.d;
+------+------------+------------+-------+------+
| id   | a          | b          | sum   | d    |
+------+------------+------------+-------+------+
|    0 | 1u3meTc7yr | RN4WlbjlaZ |  4950 |    1 |
|    1 | idL17meIy3 | kKxsqHPin5 | 14950 |    1 |
|    3 | Iqd2cCj8nE | x4abrjiGcm | 34950 |    1 |
```

图 2-90　常见多表关联查询

```
|    5 | ETtCX6wLlL   | NmCFIzJoST   | 54950 |    1 |
|    6 | kUI5vLmf5Q   | lGW82WNfHA   | 64950 |    1 |
|    9 | pcBeJCmYux   | OV0A1niPi6   | 94950 |    1 |
+------+--------------+--------------+-------+------+
6 rows in set (0.02 sec)
```

图 2-90 常见多表关联查询（续）

该 SQL 是一个跨了 4 个表的查询，查询了 D 表的 d 列等于 1 的结果集。多表关联的慢日志如图 2-91 所示，查询这些数据，扫描到了这 4 张表的全表数据，累计 10300 行，返回 44 行。

```
# Time: 2023-06-16T02:35:18.240442Z
# User@Host: root[root] @ localhost []  Id:   111
# Query_time: 0.027506  Lock_time: 0.000013 Rows_sent: 6  Rows_examined: 10300
SET timestamp=1686882918;
select aa.id id, aa.a a, bb.b b, sum(cc.c) sum, dd.d d from aa join bb
on aa.id=bb.id join cc on aa.id=cc.id join dd on aa.id=dd.id
where dd.d=1 and aa.id<10 group by aa.id,aa.a,bb.b, dd.d;
```

图 2-91 多表关联的慢日志

对于这类关联多表的查询，往往可以通过更合理的表设计来规避低效查询。上面的模拟场景中，若 C 表的数据会高频度使用统计查询的方式，通常最佳实践是在主表上加入数据统计的字段，即通过落地统计数据的方式减少数据库的计算压力。而对于 D 表这类状态或是标记表，则可以直接并入主表，作为一列存在。落地计算结果和冗余状态字段如图 2-92 所示。

```
mysql> alter table aa add column sumc int;
Query OK, 0 rows affected (0.03 sec)
Records: 0  Duplicates: 0  Warnings: 0

mysql> update aa set sumc=(select sum(c) from cc where cc.id=aa.id);
Query OK, 100 rows affected (1.08 sec)
Rows matched: 100  Changed: 100  Warnings: 0

mysql> alter table aa add column d int;
Query OK, 0 rows affected (0.09 sec)
Records: 0  Duplicates: 0  Warnings: 0

mysql> update aa set d=(select dd.d from dd where aa.id=dd.id);
Query OK, 100 rows affected (0.04 sec)
Rows matched: 100  Changed: 100  Warnings: 0
```

图 2-92 落地计算结果和冗余状态字段

经过设计优化过后，重新设计表查询如图 2-93 所示，SQL 达成同样查询效果。

```
mysql> select aa.id id, aa.a a, bb.b, aa.sumc sum, aa.d from aa join bb
    ->on aa.id=bb.id where aa.d=1 and aa.id<10;
+------+------------+------------+-------+------+
| id   | a          | b          | sum   | d    |
+------+------------+------------+-------+------+
|    0 | 1u3meTc7yr | RN4WlbjlaZ |  4950 |    1 |
|    1 | idL17meIy3 | kKxsqHPin5 | 14950 |    1 |
|    3 | Iqd2cCj8nE | x4abrjiGcm | 34950 |    1 |
|    5 | ETtCX6wLlL | NmCFIzJoST | 54950 |    1 |
|    6 | kUI5vLmf5Q | lGW82WNfHA | 64950 |    1 |
|    9 | pcBeJCmYux | OV0A1niPi6 | 94950 |    1 |
+------+------------+------------+-------+------+
6 rows in set (0.00 sec)
```

图 2-93 重新设计表查询

没有建立任何索引的情况下，去除不必要关联慢日志如图 2-94 所示，日志显示了 SQL 扫描了 200 条，返回了 44 行。在结果一致的情况下，查询效率提升了 50 倍。

```
# Time: 2023-06-16T02:36:16.950452Z
# User@Host: root[root] @ localhost []  Id:    111
# Query_time: 0.001801  Lock_time: 0.000016 Rows_sent: 6  Rows_examined: 200
SET timestamp=1686882976;
select aa.id id, aa.a a, bb.b, aa.sumc sum, aa.d from aa join bb
on aa.id=bb.id where aa.d=1 and aa.id<10;
```

图 2-94 去除不必要关联慢日志

基于这样的设计，在每次 C 表有新增记录的时候，在一个事务中需要同步修改 A 表的 sumc 列。在实际工作中，这样做的效率提升远不止 50 倍，可能是上百上千倍的提升。

2.3.2 避免动态计算结果没有单独存储导致的低效查询

所谓数据落地是指：以周期形式或者触发形式对相关业务数据进行 SUM、COUNT 的汇总，并且存储在另外的表上。常见的周期形式有车流量统计：每个小时对获得的车辆数据进行统计，并且存储到一个小时表中。每天定时对小时表的统计数据再次累加获得相应的日流量数据，并且存储到一个日流量表中。依次类推，如果最后要统计年度流量，只需要将月统计表中的 12 个月的月统计结果相加即可。常见的触发形式有短信余额提醒：您收到转账 1 万元，当下账户余额 10 万元。每次交易流水表数据新增的情况下，同步变更余额表中的对应的余额值。

笔者遇到过很多类似如下的 SQL 写法：

```
SELECT SUM(XXX) FROM 资金流水表 WHERE USER=1234;
```

这句 SQL 的含义就是每次从资金流水中计算出用户的余额。资金流水表中有收入记为+，支出记为-。累计计算流水如图 2-95 所示，code1 的余额是 5-2+4=7，code2 的余额是 1+8-23=6。

```
mysql> select * from h;
+------+------+-------+
| id   | code | money |
+------+------+-------+
|    1 |    1 |     5 |
|    2 |    1 |    -2 |
|    3 |    2 |     1 |
|    4 |    2 |     8 |
|    5 |    1 |     4 |
|    6 |    2 |    -3 |
+------+------+-------+
6 rows in set (0.00 sec)

mysql> select code,sum(money) from h group by code;
+------+------------+
| code | sum(money) |
+------+------------+
|    1 |          7 |
|    2 |          6 |
+------+------------+
2 rows in set (0.00 sec)
```

图 2-95　累计计算流水

这样做除了开发简单这一点优势外，并没有其他任何好处，而且有几大致命问题。

- 问题 1：随着时间的推移，数据越来越多。每个用户的流水是累积的，每次返回余额等于将这个用户的历史操作做了一次快进回放。虽然是快进但是总体来讲是越来越慢的。
- 问题 2：如果处理大客户的数据可能一次余额展示就非常慢。
- 问题 3：如果交易的流水表进行数据归档，那么余额就不准了。

大约 10 年前第一次见到这样写的 SQL 语句时，笔者告知开发人员这样写是错误的，应该建立一张用户账户余额表，用来记录每次交易完毕后的余额数据，否则随着时间的推移和用户交易次数的累加，后续用户查询自己的余额会越来越慢。但是该开发人员不以为然。过了两年后，开发人员找到笔者说，现在计算余额的速度确实越来越慢了，应该怎么处理？

这实际上就是一个没有将必要的数据计算单独存储导致的低效查询的典型案例。因为每次遍历交易流水花费的时间是线性增加的，所以处理速度越来越慢。

笔者给开发人员的建议是，将余额在另外一个表中进行存储。具体的方案通过下面一个例子来加以说明。案例使用的数据库是 MySQL，初始化数据为 9999999 行数据，业务流水表和用户账户表中都涉及了用户代码、交易金额这些必要的字段。

首先通过流水表中某个 code 的全部收支的流水金额获得用户余额，计算流水表获得用户余额如图 2-96 所示。扫描了 100217 行数据历时 767ms，用户余额是 502539557。

```
mysql> select count(*) from t;
+----------+
| count(*) |
+----------+
|  9999999 |
+----------+
1 row in set (0.68 sec)

mysql> select sum(money) from t where code='T00028';
+------------+
| sum(money) |
+------------+
|  502539557 |
+------------+
1 row in set (0.77 sec)

# User@Host: root[root] @ localhost []  Id:   103
# Query_time: 0.767251  Lock_time: 0.000301 Rows_sent: 1  Rows_examined: 100217
SET timestamp=1685684245;
select sum(money) from t where code='T00028';
```

图 2-96　计算流水表获得用户余额

此时将 t 表的数据按照用户维度，进行一次初始化分组计算，并且将初始化的分组聚合结果全部存储到 Tabhis 表中，余额存储单独查询如图 2-97 所示。对于 Tabhis 表中的 code 条件查询只查询该表的一行数据，其效果类似于 Redis 表的 KV 点查，一个 code 对应一个 Key，一个 sum_money 对应一个 Value。

```
mysql> select * from tabhis where code='T00028';
+--------+-----------+
| code   | sum_money |
+--------+-----------+
| T00028 | 502539557 |
+--------+-----------+
1 row in set (0.00 sec)

# User@Host: root[root] @ localhost []  Id:   103
# Query_time: 0.000883  Lock_time: 0.000347 Rows_sent: 1  Rows_examined: 1
SET timestamp=1685684233;
select * from tabhis where code='T00028';
```

图 2-97　余额存储单独查询

在数据汇聚的单独存储的设计思路中，模拟查询的效果只查询一行，耗时 0.8ms。从实际执行效率来说也和 Redis 的单条查询相差不大。比起没有将余额单独存储、简单计算过往流水的实现方式来说，访问汇聚存储表的查询性能提升了 900 多倍。后续如果有流水变动，只需要在一个事务中做流水和更新历史表两个操作，即可以确保数据的一致性。对流水表新增数据变更如图 2-98 所示，在一个事务中流水表 t 新增 1 条数据，同时更新历史落地表 Tabhis 中对应 code 的值。新增之前流水表 t 的初始数据为 9999999 行，新增之后为 10000000 行。

```
mysql> begin;
Query OK, 0 rows affected (0.00 sec)

mysql> insert into t (code,money,time) values ('T00028',3000,'2021-10-10');
Query OK, 1 row affected (0.01 sec)

mysql> update tabhis set sum_money=sum_money+3000 where code='T00028';
Query OK, 1 row affected (0.01 sec)
Rows matched: 1  Changed: 1  Warnings: 0

mysql> commit;
Query OK, 0 rows affected (0.01 sec)

mysql> select count(*) from t;
+----------+
| count(*) |
+----------+
| 10000000 |
+----------+
1 row in set (0.68 sec)
```

图 2-98　对流水表新增数据变更

如果用流水表 t 去计算 code 为 T00028 的 sum_money，则扫描行数从初始化的 100217 行变成现在的 100218 行。而如果访问历史落地表 Tabhis 去读取 code 为 T00028 的 sum_money（因为访问的是静态值，没有计算过程），永远访问 1 行。写入变更后两种方式对比如图 2-99 所示。

```
mysql> select * from tabhis where code='T00028';
+--------+-----------+
| code   | sum_money |
+--------+-----------+
| T00028 | 502542557 |
+--------+-----------+
1 row in set (0.00 sec)

# Query_time: 0.715350  Lock_time: 0.000341 Rows_sent: 1  Rows_examined: 100218
SET timestamp=1685684336;
select sum(money) from t where code='T00028';

mysql> select sum(money) from t where code='T00028';
+------------+
| sum(money) |
+------------+
|  502542557 |
+------------+
1 row in set (0.72 sec)

# Query_time: 0.000754  Lock_time: 0.000318 Rows_sent: 1  Rows_examined: 1
SET timestamp=1685684348;
select * from tabhis where code='T00028';
```

图 2-99　写入变更后两种方式对比

以上的解决方案完美解决了之前的三大问题。

问题 1 的数据累积解决了。每次只访问静态数据，无须累积得出。

问题 2 的大客户数据解决了。不同用户的交易次数不均衡不影响效率，都只有一行静态数据。用户余额的查询场景，可满足每秒几百次并发访问请求。效果约等于存储在 Redis 中，但是比 Redis 的一致性要强。

问题 3 的数据归档迁移问题解决了。即使流水表中的数据被归档到其他数据库后，用户余额不会因为缺失部分流水数据而统计不准确。

问题 1 和 2 的争论焦点是动态计算还是将计算的结果单独存储。开发人员可能和 DBA 各执一词。开发人员认为动态计算的开发工作量小，DBA 认为存储计算结果后静态读取的方法数据库性能最佳。两种观点各有利弊。

不过当一个表的行数达到一定程度时，管理者会从管理角度出发做数据归档，归档的数据被迁移到离线的历史库中。这个时候开发人员就必须回到动态计算的数据结果要不要单独存储的问题上来。

其实该问题不仅仅是余额场景会遇到，库存、资金、流量统计等场景都会遇到，应用上述思路解决该问题很轻松。

2.3.3 避免没有明确需求查询条件导致的低效查询

没有明确需求是指用户对于大表、宽表的数据，缺乏明确获取数据的模型或目的，以任意一个字段而且仅有一个字段作为过滤条件的、无特定目的的查询。

在工作中遇到的很多消耗大量资源的操作都来源于类似的综合查询，某 CRM 系统的工单查询界面如图 2-100 所示。该工单综合查询做得比较规范，受理开始时间和结束时间是必选项。

图 2-100　某 CRM 系统的工单查询界面

但是许多企业的综合查询中没有必选条件，而是以任意维度进行查询。例如在上图中取消了时间限制，这时再按照"用户品牌"或"是否已发短信"字段作为条件来查询，那么返回的数据量可能占总数的五分之一或三分之一，在这种场景下，查询将不会使用到索引，而是全表查询。因为"用户品牌"字段可能没有建立索引，而"是否已发短信"这种字段，通过前面 2.1 节可以知道即使建立索引也不会用到索引。

目前这种不限制必选条件的情况非常普遍，其原因是业务没想好需要查询什么，所以没有不清楚从什么维度去检索，也就无法明确需求。一旦没有必选项，那么只要选择的某个字段没有索引，那么就会全表扫描，进而导致问题的产生。

软件开发过程中，明确需求是非常重要的。所以一定要事先约定好检索维度（列），根据这些维度（列）建立索引。在检索维度（列）确定以后，其他条件可以随便选择。因为先决条件已经限定了，那么范围不会扩大。比如上图选择了 1 天，那么后面随便选择条件，甚至不选，其结果集也就是一天的数据而已。在一天的几万或者几十万行数据表中过滤数据，基本没有什么压力。

2.3.4　避免过度分表

过度分表是指在一个表中可以完成对象设计的情况下将表拆开，或者直接设计多个表同时写入，读取时候组合展示。

在工作中多次遇到开发人员自己设计的表但是性能不佳的情况。

例如，有一次开发人员小 C 找到笔者，给笔者看一段 SQL。其中有一处是两表关联，一个表是业务表，另一个表从名字上看是错误日志表。然后两个表进行了 not exists 的关联。案例模拟表如图 2-101 所示，ten 表为业务表，而 err 表为错误日志表。

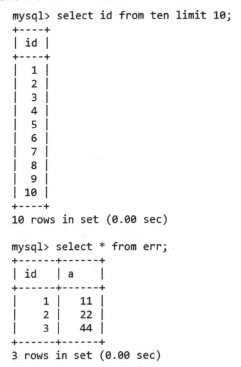

图 2-101　案例模拟表

开发人员小 C 的设想是当操作业务表 ten 的时候，如果业务逻辑上对这条数据进行处理失败，那么就标记这条记录异常，将这条记录存储到 err 日志表中。上图中可以看到，在前 10 条中有 3 条数据已经同步在 err 表中。小 C 要查询业务上没有异常的数据，SQL 就会写成图 2-102 所示的样子，得到了预期的数据。

```
mysql> select id  from ten where not  exists  (select id from err t where t.id=ten.id ) and id<10;
+----+
| id |
+----+
|  4 |
|  5 |
|  6 |
|  7 |
|  8 |
|  9 |
+----+
6 rows in set (0.01 sec)
```

图 2-102　排除异常数据

如果继续执行发现下一条也会发生异常，那么会继续写入 err 表。业务逻辑错误继续写入日志表如图 2-103 所示，再次排除异常数据的 SQL 就可以继续过滤出正常的业务数据。

```
mysql> insert into err values (4,44);
Query OK, 1 row affected (0.01 sec)

mysql> select id  from ten where not  exists  (select id from err t where t.id=ten.id ) and id<10;
+----+
| id |
+----+
|  5 |
|  6 |
|  7 |
|  8 |
|  9 |
+----+
5 rows in set (0.00 sec)
```

图 2-103　业务逻辑错误继续写入日志表

其实这样的设计逻辑有些复杂，模型不够简洁。实际上，对业务表的数据进行操作后，如果想标记其在业务上正确与否，只需要在业务表上增加一个状态位。在业务表上增加字段记录状态如图 2-104 所示，设定一个默认值，如果异常标记为 1。那么在实验中模拟上述 1～4 条为异常数据。

```
mysql> alter table ten add err varchar(5) default '0';
Query OK, 0 rows affected (0.18 sec)
Records: 0  Duplicates: 0  Warnings: 0

mysql> update ten set err ='1' where id in (1,2,3,4);
Query OK, 0 rows affected (0.00 sec)
Rows matched: 3  Changed: 0  Warnings: 0
```

图 2-104　在业务表上增加字段记录状态

筛选业务逻辑正常的 SQL，改为单表实现过滤正常业务数据，如图 2-105 所示。

两种方法的效果一样，但是对比一下后台的执行情况就发现了区别。原 SQL 的执行计划和扫描行数如图 2-106 所示，执行计划略显复杂，扫描 17 行返回 5 行。

```
mysql> select * from ten where err='0' and id<10;
+----+--------+------+--------+--------+-------------------------+-----+
| id | name   | age  | weight | height | time                    | err |
+----+--------+------+--------+--------+-------------------------+-----+
|  5 | 3d7WOO |   18 |    150 |    180 | 2023-04-19 14:07:47.862141 | 0 |
|  6 | wYc4eI |   18 |    150 |    180 | 2023-04-19 14:07:47.862141 | 0 |
|  7 | 0I0aXt |   18 |    150 |    180 | 2023-04-19 14:07:47.862141 | 0 |
|  8 | ARG4dA |   18 |    150 |    180 | 2023-04-19 14:07:47.862141 | 0 |
|  9 | XOctqk |   18 |    150 |    180 | 2023-04-19 14:07:47.862141 | 0 |
+----+--------+------+--------+--------+-------------------------+-----+
5 rows in set (0.01 sec)

mysql>
```

图 2-105　单表实现过滤正常业务数据

```
mysql> explain select id from ten where not exists (select id from err t where t.id=ten.id) and id<10;
+----+--------------+------------+------------+-------+---------------+------------------+---------+---------+------+----------+-------------------------+
| id | select_type  | table      | partitions | type  | possible_keys | key              | key_len | ref     | rows | filtered | Extra                   |
+----+--------------+------------+------------+-------+---------------+------------------+---------+---------+------+----------+-------------------------+
|  1 | SIMPLE       | ten        | NULL       | range | PRIMARY       | PRIMARY          | 4       | NULL    |    5 |   100.00 | Using where; Using index|
|  1 | SIMPLE       | <subquery2>| NULL       | eq_ref| <auto_distinct_key> | <auto_distinct_key> | 5 | x.ten.id |   1 | 100.00 | Using where; Not exists |
|  2 | MATERIALIZED | t          | NULL       | index | IDX_T_ID      | IDX_T_ID         | 5       | NULL    |    5 |   100.00 | Using index             |
+----+--------------+------------+------------+-------+---------------+------------------+---------+---------+------+----------+-------------------------+
3 rows in set, 2 warnings (0.00 sec)

# Time: 2023-06-17T04:47:50.439608Z
# User@Host: root[root] @ localhost []  Id: 118
# Query_time: 0.001341  Lock_time: 0.000013 Rows_sent: 5  Rows_examined: 17
SET timestamp=1686977270;
select id from ten where not exists (select id from err t where t.id=ten.id) and id<10;
```

图 2-106　原 SQL 的执行计划和扫描行数

而新的 SQL 的情况，其执行计划和扫描行数如图 2-107 所示，执行计划极简，扫描 9 行返回 5 行。

```
mysql> explain select id from ten where err='0' and id<10;
+----+-------------+-------+------------+-------+---------------+---------+---------+------+------+----------+-------------+
| id | select_type | table | partitions | type  | possible_keys | key     | key_len | ref  | rows | filtered | Extra       |
+----+-------------+-------+------------+-------+---------------+---------+---------+------+------+----------+-------------+
|  1 | SIMPLE      | ten   | NULL       | range | PRIMARY       | PRIMARY | 4       | NULL |    9 |    10.00 | Using where |
+----+-------------+-------+------------+-------+---------------+---------+---------+------+------+----------+-------------+
1 row in set, 1 warning (0.00 sec)

# Time: 2023-06-17T05:58:10.728540Z
# User@Host: root[root] @ localhost []  Id: 118
# Query_time: 0.001344  Lock_time: 0.000013 Rows_sent: 5  Rows_examined: 9
SET timestamp=1686981490;
select id from ten where err='0' and id<10;
```

图 2-107　新 SQL 的执行计划和扫描行数

两者返回行数一样，是因为其等价。而扫描行数相差了将近一倍（真实环境中的一倍可能是几百行，也可能是几百万行，差距会非常巨大）。

2.3.2 节讲了在某些场景下单表维度处理数据性能不佳，需要增加一个维度来解决问题。2.3.4 节讲了在某些场景下多表维度处理数据性能不佳，需要减少一个维度解决问题。其实这个并不矛盾，兵无常态、水无常势，一切都是结合实际情况而定，没有一个固定的公式。原则上，站在数据库的角度采用数据库消耗最小的设计方案就是最好的设计。

2.4 从识别需求的合理性提升性能

人们常说，技术是用来支持业务需求的。不过有时候需求未必全部是合理的，可能存在多个需求的互斥、在当下技术下需求无法实现以及需求不合理的诸多种情况。所以一个需求应该通过评估其合理性，得到确认后再进入设计环节，对合理的需求进行数据库的设计，才能从根本上确保并提升性能。

2.4.1 拒绝无效需求

无效需求通常指不合理需求、无意义需求，以及无必要的需求等实现代价极大甚至无法实现且收益极低的需求。这里实现代价极大就意味着数据库处于高负荷甚至是超负荷的工作状态中，极易导致数据库发生故障。

在 SQL 编写过程中，常见的范围运算有 ">" "<" "in" "or" 等。通常 column between and 的写法和（column>and column<）的写法非常普遍。而 column=1 or column=2…这样的语句可能也会被改写成为 columnin (1,2,3,…)。

上述语句条件要么是相同列或不同列的逻辑且的关系，要么是一个列的逻辑或的关系。只要这一列有索引，用到索引检索数据，都会得到高效的执行，也就是不会产生全表查询。

但是有一种情况要引起大家注意，即两个不同列的逻辑或的关系，如 select *from ten where a=1 or b=1；这样的 SQL 语句。

面对这样的需求是要慎之又慎的。因为这样的需求通常可能是需求方的逻辑没有考虑清楚。因为逻辑与的运算不同，逻辑或的运算代表前后两个表达式任意一个为真即可，通常说的 SQL 注入就是类似这样的写法：

```
select * from user where name='张三' or 1=1
```

这里的 1=1 恒定为真，查询的结果不是 name='张三'的数据，而是整个 user 表的数据。以下面这个例子来说明再清楚不过了。

在 14 亿的人口库中查询姓名是薛晓刚且性别是男的人，这种逻辑且只要姓名列有索引就行，返回数据仅仅几行。而如果查询姓名是薛晓刚或性别是男的人，这种逻辑或不仅姓名列要有索引，性别列也要有索引。即使性别列有索引也未必有效果，因为最终符合条件的大约有 7 亿行（因为男女比例大致为 1：1）。这是需求的初衷吗？几乎不可能有这种需求，大概率是需求的逻辑没想清楚。所以当发现仅有一个逻辑或运算时，一定要确定需求的逻辑是否合理，判断需求是不是有效。

用全国人口的海量数据进行此类查询的需求举例属于极端情况，可能一般业务没有这么大的数据量。但请看下面的例子，这是一个 and 运算和 or 运算执行计划比较的例子，通过这个例子可以说明两个不同列的逻辑或的关系会导致全表扫描，以至于查询效率极低或者毫无价值。

首先来看单一列检索的执行计划,单表单一条件检索如图 2-108 所示,非常明确地使用到了 id 列的索引。

```
SQL> explain plan for select * from ten where id<100;

Explained


SQL> SELECT * FROM TABLE(DBMS_XPLAN.DISPLAY);

PLAN_TABLE_OUTPUT
---------------------------------------------------------------------------
Plan hash value: 1802329469
---------------------------------------------------------------------------
| Id  | Operation                            | Name       | Rows | Bytes | Cost |
---------------------------------------------------------------------------
|   0 | SELECT STATEMENT                     |            |  120 |  4320 |   33 |
|   1 |  TABLE ACCESS BY INDEX ROWID BATCHED | TEN        |  120 |  4320 |   33 |
|*  2 |   INDEX RANGE SCAN                   | IDX_TEN_ID |  120 |       |    2 |
---------------------------------------------------------------------------

Predicate Information (identified by operation id):
---------------------------------------------------
   2 - access("ID"<100)

14 rows selected
```

图 2-108 单表单一条件检索

上述 SQL 语句中,使用逻辑与 and 的运算关系进一步缩小返回结果集。单表逻辑与检索如图 2-109 所示,数据库的优化器依然选择了索引扫描。

```
SQL> explain plan for select * from ten where id<100 and name='abc';

Explained


SQL> SELECT * FROM TABLE(DBMS_XPLAN.DISPLAY);

PLAN_TABLE_OUTPUT
---------------------------------------------------------------------------
Plan hash value: 2092749114
---------------------------------------------------------------------------
| Id  | Operation                            | Name                  | Rows | Byt
---------------------------------------------------------------------------
|   0 | SELECT STATEMENT                     |                       |    1 |
|*  1 |  TABLE ACCESS BY INDEX ROWID BATCHED | TEN                   |    1 |
|*  2 |   INDEX RANGE SCAN                   | SYS_AI_fhcgc0pqah0zn  |    1 |
---------------------------------------------------------------------------

Predicate Information (identified by operation id):
---------------------------------------------------
   1 - filter("ID"<100)
   2 - access("NAME"='abc')

15 rows selected
```

图 2-109 单表逻辑与检索

此时 name 列如果没有索引，那么优化器会果断地选择有索引的 id 列。如果 name 列有索引就会像上图所示，选择两个索引中更优的索引。如果 id 列和 name 列为联合索引，那么优化器会选择联合索引。

如果 id 列与 name 列的条件为逻辑或的关系。那么情况就不一样了，请看如图 2-110 所示的逻辑或关系的执行计划，SQL 会先后用到两个索引。

```
SQL> explain plan for select * from ten where id<100 or name='abc';

Explained

SQL> SELECT * FROM TABLE(DBMS_XPLAN.DISPLAY);

PLAN_TABLE_OUTPUT
-------------------------------------------------------------------------
Plan hash value: 499813655
-------------------------------------------------------------------------
| Id  | Operation                              | Name                 | Rows | Byt
-------------------------------------------------------------------------
|   0 | SELECT STATEMENT                       |                      |  121 |  43
|   1 |  TABLE ACCESS BY INDEX ROWID BATCHED   | TEN                  |  121 |  43
|   2 |   BITMAP CONVERSION TO ROWIDS          |                      |
|   3 |    BITMAP OR                           |                      |
|   4 |     BITMAP CONVERSION FROM ROWIDS      |                      |
|   5 |      SORT ORDER BY                     |                      |
|*  6 |       INDEX RANGE SCAN                 | IDX_TEN_ID           |
|   7 |     BITMAP CONVERSION FROM ROWIDS      |                      |
|*  8 |      INDEX RANGE SCAN                  | SYS_AI_fhcgc0pqah0zn |
-------------------------------------------------------------------------

Predicate Information (identified by operation id):
---------------------------------------------------
   6 - access("ID"<100)

PLAN_TABLE_OUTPUT
-------------------------------------------------------------------------
       filter("ID"<100)
   8 - access("NAME"='abc')

22 rows selected
```

图 2-110 逻辑或关系的执行计划

那么如果有一个列没有索引会如何？从 ten 的表结构来看，age 列没有建立索引。

将上述 SQL 写成逻辑或的 SQL，保证有一列没有索引。逻辑或下有一列无索引如图 2-111 所示，可以看到其结果是全表扫描。也就是说，在逻辑或的情况下，需要每个参与运算的列都有索引。A=1 或者 B=1 或者 C=1 要求 A、B、C 三个列都要有索引，一旦一列没有索引，就会发生全表扫描。而 D=1 或者 D=2 或者 D=3 可以改写成 Din(1,2,3)，是因为恰好这些都是同一列。

```
SQL> explain plan for select * from ten where id<100 or age=12;

Explained
```

图 2-111 逻辑或下有一列无索引

```
SQL> SELECT * FROM TABLE(DBMS_XPLAN.DISPLAY);

PLAN_TABLE_OUTPUT
--------------------------------------------------------------------------
Plan hash value: 2565709720
--------------------------------------------------------------------------
| Id  | Operation          | Name | Rows  | Bytes | Cost (%CPU)| Time     |
--------------------------------------------------------------------------
|   0 | SELECT STATEMENT   |      |   121 |  4356 |   205   (1)| 00:00:01 |
|*  1 |  TABLE ACCESS FULL | TEN  |   121 |  4356 |   205   (1)| 00:00:01 |
--------------------------------------------------------------------------

Predicate Information (identified by operation id):
---------------------------------------------------
   1 - filter("ID"<100 OR "AGE"=12)

13 rows selected
```

图 2-111　逻辑或下有一列无索引（续）

在日常工作中，笔者不止一次在巡检数据库时发现 SQL 中包含类似 where Acode='X' or Bcode='Y' 的语句，这种 SQL 语句经常会因为某一列缺失索引导致全表扫描从而降低数据库性能引发故障。从图 2-110 中也可以得出，技术上也可以为每一列都建立索引来确保 SQL 避免全表扫描。但是这样的做法仅仅是技术上的补救，并没有深入业务得彻底解决问题。对于这种特殊的逻辑，就需要结合实际数据去分析，是需求有问题还是设计有问题。通常可以对 Acode 和 Bcode 中值得怀疑的一列进行统计。

笔者曾经有一次就对该表的 Bcode 字段进行了统计，得到了不可思议的结果。发现 Bcode 字段仅有几条数据，其余数据都是空值。对于一个拥有上千万条数据的表，一个字段仅有几条数据说明这个 SQL 以及其背后对应的需求几乎没有存在的意义。对于这个结果自然要和开发人员进行确认，一般来说开发人员也不清楚其原因，最后去咨询业务方。最终的答复是该功能三年前已经不用了，所以近三年的数据中这个字段是空值。这就是典型的无效需求。

以上案例最后都可以通过去掉逻辑或的运算关系使得 SQL 性能得到提升，但是背后的问题值得深思。那就是对需求的把控，有时候无意义的需求或者伪需求的实现，可能在不经意间就导致了系统的性能衰减，进而可能引发故障。

2.4.2　正确理解开发需求

这里的开发需求，其实是业务对开发人员提的需求。首先需求不一定都是正确的，需要开发人员自己识别。这里说的正确是指需求的合理性。而合理的需求还需要得到正确的理解。

不知道大家有没有听过这个段子"字体大一点，再大一点，LOGO 也要大"，概括一句就是"字要大"。很多设计师就去把字体放大了。可是达到要求了吗？没有。字要大是为强化宣传内容的辨识度，字要大是表象，其实质是要醒目。调整字体大小没有作用，凸显对比度才是解决问题的关键。

在数据库的开发中也经常会对于需求的理解不准确的问题。

例如，下面这两句 SQL。

SQL1：SELECT * FROM T WHERE 谓词=XXXX order by 时间 DESC;

SQL2：SELECT * FROM T WHERE 谓词=XXXX and 时间 between 半年前 and 现在 order by 时间 DESC;

SQL1 的含义是查询一个条件为 XXXX 的所有数据，并且按照记录的生成时间递减的顺序进行排列。排序后展示最新的数据。

SQL2 的含义是查询一个条件为 XXXX 且记录生成在半年以内的数据，并且按照记录的生成时间递减的顺序进行排列。排序后展示最新的数据。

既然需求总是要看最新的数据。那么是不是可以增加逻辑与的条件（只查询最近的数据），缩写结果集的范围，而不需要查询全部的数据（一年前或者几年前）？

这类 SQL 在日常中会遇到很多，对于 SQL1 这样的情况需要核实最真实的需求，就像"字要大"是不是最真实的想法。

当然 SQL1 可能执行得很慢，非常有可能涉及大量数据的排序。开发人员小 D 就曾找到笔者问发生这种情况怎么办？笔者就问小 D，既然需求涉及最新的数据，比如一天有 1000 个订单，那么加一个时间范围"一周"怎么样？这个时候小 D 说，业务人员没有提及。笔者问小 D："那你问过吗？"小 D 说他没问。要理解真实的诉求，必须去询问清楚，不能含混模糊。

过了一段时间 SQL1 的问题解决了，小 D 又找到笔者说遇到了 SQL2 的场景。SQL2 的业务不接受 1 周的时间约束，他们业务周期长。小 D 说他已经先设置时间范围为一年。结果是需求实现时查询一年数据比较慢，业务人员抱怨慢。随后小 D 和业务人员不断进行协商，小 D 提出查询三个月如何？业务人员说不行，时间要保证半年，这样不断尝试。其实有的时候 SQL2（不是 100%都可以）的场景有更好的解决方式。既然就时间范围不能达成统一意见，就不要限制时间范围了，不限制时间查询如图 2-112 所示。

```
xxg=# select * from dd where buyercode='B0000123' limit 5;
        ddid          | buyercode | money | sellercode |        time
----------------------+-----------+-------+------------+---------------------
 DD0000000000000002333 | B0000123 |   942 | S0000123   | 2021-09-25 08:00:00
 DD0000000000000002303 | B0000123 |  4279 | S0000123   | 2021-09-23 08:00:00
 DD0000000000000002348 | B0000123 |  5474 | S0000123   | 2021-09-19 08:00:00
 DD0000000000000002323 | B0000123 |  6969 | S0000123   | 2021-09-19 08:00:00
 DD0000000000000002390 | B0000123 |  8490 | S0000123   | 2021-09-18 08:00:00
(5 rows)

Time: 1.229 ms
xxg=# select * from dd where buyercode='B0000123' limit 5 offset 5;
        ddid          | buyercode | money | sellercode |        time
----------------------+-----------+-------+------------+---------------------
 DD0000000000000002314 | B0000123 |  7480 | S0000123   | 2021-09-17 08:00:00
 DD0000000000000002318 | B0000123 |  4609 | S0000123   | 2021-09-17 08:00:00
 DD0000000000000002351 | B0000123 |  8477 | S0000123   | 2021-09-14 08:00:00
 DD0000000000000002347 | B0000123 |  1672 | S0000123   | 2021-09-12 08:00:00
 DD0000000000000002394 | B0000123 |  3253 | S0000123   | 2021-09-11 08:00:00
(5 rows)

Time: 1.045 ms
```

图 2-112 不限制时间查询

这里采用的是 PostgreSQL 数据库，此方法在主流数据库中效果是类似的。从图中可以看到，查询这个 buyercode 的时候没有限制时间，采用分页一页一页往下翻。甚至都没有指明排序，就已经按照业务诉求倒序排好了。这是因为这里建立了一个倒序的索引，倒序索引无须指定排序，如图 2-113 所示。

```
xxg=# \d dd
                    Table "public.dd"
   Column   |            Type             | Collation | Nullable | Default
------------+-----------------------------+-----------+----------+---------
 ddid       | character varying(20)       |           |          |
 buyercode  | character varying(18)       |           |          |
 money      | integer                     |           |          |
 sellercode | character varying(18)       |           |          |
 time       | timestamp without time zone |           |          |
Indexes:
    "bcodetimedesc" btree (buyercode, "time" DESC)
```

图 2-113 倒序索引无须指定排序

可以对比带上排序的效果，如图 2-114 所示。两者结果无差异，数据一致。

```
xxg=# select * from dd where buyercode='B0000123' order by time desc limit 10;
         ddid          | buyercode | money  | sellercode |        time
-----------------------+-----------+--------+------------+---------------------
 DD0000000000000002333 | B0000123  |    942 | S0000123   | 2021-09-25 08:00:00
 DD0000000000000002303 | B0000123  |   4279 | S0000123   | 2021-09-23 08:00:00
 DD0000000000000002348 | B0000123  |   5474 | S0000123   | 2021-09-19 08:00:00
 DD0000000000000002323 | B0000123  |   6969 | S0000123   | 2021-09-19 08:00:00
 DD0000000000000002390 | B0000123  |   8490 | S0000123   | 2021-09-18 08:00:00
 DD0000000000000002314 | B0000123  |   7480 | S0000123   | 2021-09-17 08:00:00
 DD0000000000000002318 | B0000123  |   4609 | S0000123   | 2021-09-17 08:00:00
 DD0000000000000002351 | B0000123  |   8477 | S0000123   | 2021-09-14 08:00:00
 DD0000000000000002347 | B0000123  |   1672 | S0000123   | 2021-09-12 08:00:00
 DD0000000000000002394 | B0000123  |   3253 | S0000123   | 2021-09-11 08:00:00
(10 rows)

Time: 1.709 ms
```

图 2-114 显示指定排序

其实微博和朋友圈也是类似这样的设计，没有限制。只要用户有耐心，那可以无限制地查下去。SQL2 适合指定简单的 1~2 个谓词的条件过滤（不能有很多或复杂的条件），而这个谓词又可以和时间作为联合索引，小 D 面临的这个场景就适合。

2.4.3 拒绝不合理的需求

在 2.2.2 节的故障案例中，导致这个问题的源头其实是不合理的需求。即使当初查询几百万数据并没有直接导致故障，但是存在一个问题：1000 万条数据，每页一般即使有 100 行也要 10 万页，一般人可能去查看这 10 万页内容吗？答案当然是否定的，看 10 页左右就关闭查询页面了，可以说并没有起到查询全部数据的作用。后台几千万次甚至可能上亿次的运算白做了。这种可以用一个形象的例子来比喻：到饭店把所有饭菜都点了，上菜后只吃了一个菜，夹了一口吃完了，那么所有的饭菜都浪费了。

这儿再举一个很典型的不合理需求的例子。

有一次开发人员小 D 找到笔者，要对一个业务场景的 SQL 进行优化。目前该 SQL 执行起来非常慢。该 SQL 如下：

SELECT * FROM T WHERE TEL LIKE '%123456%' OR IDCARD LIKE '%123456%' OR CODE LIKE '%123456%' OR ADDRESS LIKE '%123456%'…

这个 SQL 语句产生的原因是：需求给了一个字符串，但是不说这个字符串代表什么含义，这类似谷歌和百度的搜索，属于不特定的字段查询，这种 SQL 几乎每家公司都会出现。首先不可能每一列都建立一个索引，而且即使建立了索引，对于 where 条件后谓词模糊查询时，给谓词前后都增加%通配符也是无法用到索引的。似乎每家公司的业务方都认为安装一个 MySQL 就可以实现百度的功能。这是不切实际的。百度是搜索引擎，而 OLTP 数据库虽然也支持全文索引，但是那是在一个列上的支持，而不是所有列。这种明显属于不合理的需求，显然是要拒绝的。

问题分析到这里已经不是优化 SQL 可以解决的了，必须告知业务方应该给出计算总页数需求的合理范围，而对于多列谓词前后加%通配符进行模糊查询的场景需要至少明确必选的一列作为索引，不能以任意维度作为单一筛选条件。

2.4.4 引导业务改善需求

在设计 F-16 的时候，军方对 F-16 的设计需求是速度在 2～2.5 马赫的低成本轻型战斗机。要知道飞行速度从一倍音速提升到 2 倍音速（一马赫就是一倍音速）空气阻力会是原来的 4 倍。考虑因此需要的动力以及机身上重量的苛刻条件，当时的技术很难达到。

作为武器制造商的设计团队追问美国国防部，军方为什么有这样的需求？提出这个需求是真实意图吗？如果是为什么定为 2 倍音速？

经过需求方（军方）和开发方（制造商）的讨论，得出答案是"为了快速撤离战场"。军方最真实的想法就是这个战斗机执行完任务以后就安全返回，不能被地方的地对空导弹等打下来。

了解到了这个真实的需求以后，设计团队通过提升推力重量比改善了战斗机的加速性能和机动性能，用灵巧性取代了最初对速度的要求。

这个军方就类似于平时的业务方，开发方就类似于日常的开发人员。而在工作中有多少开发人员会去问业务方为什么要这样提？在遇到需求不合理的情况下有没有站在数据库稳定性的角度上引导业务方调整需求、优化需求甚至拒绝需求？

引导业务方改善需求，其实质是引导需求方明确需求和细化需求的过程。例如在 SQL 中不乏出现 SELECT * FROM TABLE WHERE STATUS=0 ODRER BY TIME DESC；这样的情况，其目的是查询某个名为 STATUS 的字段为 0 的所有数据，然后排序只看生成时间最新的数据。通过 2.1 节得知，一般来说 STATUS 字段的值可能会经常改变，不太适合作为索引条件，而且这种区分度较低的字段可能建立索引也未必能用得到。那么这个 SQL 很可能就要全表扫描,并且排序。其实从这个 SQL 最后的排序中可以看到 ODRER BY TIME DESC,从而推断业务方一定提出了要看最新的数据,否则开发人员不太会加一个这样的排序。

根据时间排序的关键字得出需求方的真实意图,这也是引导需求方的突破口。既然业务方要看最新的数据,那么设计人员、开发人员可以参考 2.4.2 节的建议,引导需求方加一个时间范围,比如 3 天内、一周内、半个月内或者 1 个月内。这样就极有可能用到时间索引。最终实现需求但又不会多做无用功。

有时候需求方没有主动说明的,需要设计人员和开发人员主动去问。如果问到的需求并没有好的解决手段,那么就要看看最原始的需求是不是可以通过变通的方法实现或变相实现,再或者是达到与之相差不大但是也能接受的效果。F-16 的案例中用灵巧性完成了最原始的需求,这对大家都是一个启迪。

2.5 减少 IO 操作提升数据库性能

数据库在几十年的发展历程中最大的瓶颈是存储设备的 IO 吞吐能力。这是由于传统存储磁道寻址慢(详细介绍请看 3.2.4 节关于 IO 吞吐的介绍)。几乎所有的数据库优化都是为了减少 IO 吞吐而做的,而数据库的体系结构也是基于减少 IO 交互而设计的。只要解决或缓解 IO 吞吐问题,即可有效提升数据库的性能。

一个系统中数据是流动的,这个流动的过程就是数据库的读、写,经常需要将大量的数据写入数据库,用于后续的分析和查询。然而,不同的数据写入方式可能会对数据库的性能和效率产生不同的影响。几乎所有的开发人员都知道批量的写入数据比单条写入会有效率上的提升,但是对于其提升程度以及不同变量所产生的影响没有具体的量化概念。为了探究这一问题,以下几个小节围绕减少 IO 交互提升数据库性能的关键点,通过大量数据批量写入数据库与单条数据写入的对比,说明两种写入方式在不同场景下的优劣,会用到多种数据库、多种驱动或存储过程,以评估它们在性能、稳定性、易用性等方面的差异。从而为应用开发人员和数据库开发人员提供参考。

2.5.1 正确认识数据库的性能

本例采用数据库自动的功能实现数据的批量写入,其实质就是将多次 IO 操作合并成一次 IO 操作,大大减少了 IO 写入的次数,有效提升了数据库的性能,使得单体数据库即可满足一般中小企业的业务要求。

(1) MySQL 存储过程

MySQL 存储过程为 2.3.2 节模拟的 1000 万条数据写入数据库的存储过程源代码。以下语句定义了一个整型的变量 v,为变量 v 赋初始值 1。然后开启循环,循环的内容是一个 insert 语句。每循环一次 v 自增加 1,当 v 达到 1000 万时,循环终止。

```
DELIMITER $$
CREATE   PROCEDURE t1()
BEGIN
```

```
    DECLARE v int ;
    SET v = 1;
    LOOP_LABEL:LOOP
        insert into t (code,money,time) values (concat('T',lpad(floor (RAND()*100),
5,'0')),RAND()*10000,DATE_SUB('2020-01-01',INTERVAL RAND()*1000000 minute));
        SET v = v+1;
        IF v >= 10000000 THEN
            LEAVE LOOP_LABEL;
        END IF;
    END LOOP;
END $$
DELIMITER ;
```

MySQL 存储过程一千万条数据批量写入如图 2-115 所示，耗时约 1096s。每秒写入 9000 多行，该 MySQL 所在的虚拟机存储为普通机械硬盘。

```
mysql> DELIMITER ;
mysql>
mysql> begin;
Query OK, 0 rows affected (0.00 sec)

mysql> call t1;
Query OK, 1 row affected (18 min 16.38 sec)

mysql> commit;
Query OK, 0 rows affected (20.59 sec)

mysql> select count(*) from t;
+----------+
| count(*) |
+----------+
|  9999999 |
+----------+
1 row in set (0.84 sec)
```

图 2-115　MySQL 存储过程一千万条数据批量写入

（2）PostgreSQL 存储过程

PostgreSQL 存储过程为 2.2.2 节模拟的 1000 万数据写入数据库的存储过程源代码。

```
create or replace function bu() returns
boolean AS
$BODY$
declare iinteger;
begin
i:=1;
FOR i IN 1..1000000 LOOP
INSERT INTO dd (ddid,buyercode,money,sellercode,time) VALUES ('DD'|| lpad(cast(i as
```

```
varchar),18,'0'),'B'||lpad(cast (trunc(100+i/100) as varchar), 7,'0'),random_int(1,10000),
'S'||lpad(cast (trunc(100+i/100) as varchar), 7,'0'),'2020-12-02 08:00:00'::timestamp+
interval '1 day'+ ('-' || i || 'minutes')::interval);
    end loop;
    return true;
    end;
    $BODY$
    LANGUAGE plpgsql;
```

PostgreSQL 存储过程一千万批量写入如图 2-116 所示，耗时约 370s。每秒写入 27000 多行，该 PostgreSQL 所在的虚拟机存储为普通机械硬盘。

```
x=# begin;
BEGIN
Time: 0.346 ms
x=*# select * from bu() as tab;
 tab
-----
 t
(1 row)

Time: 369702.808 ms (06:09.703)
x=*# commit;
COMMIT
Time: 1.305 ms
x=# select count(*) from dd;
  count
----------
 10000000
(1 row)

Time: 4731.658 ms (00:04.732)
```

图 2-116 PostgreSQL 存储过程一千万批量写入

通过以上数据可以得出，一般的关系数据库虽然受制于 WAL 的机制，虽然每秒写入几百万条数据比较困难，但是每秒写入 1 万～3 万条数据的速度也是比较轻松的。而这种性能基本满足大部分企业的需求，数据写入不会成为瓶颈。大的互联网公司由于有上亿的用户，会遇到数据写入瓶颈，需要 RabbitMQ 或 Kafka 之类的消息队列进行削峰填谷的缓冲，而中小企业每天的数据写入量较小，假设每天写入量为 10 万，这意味着平均每秒写入 1 条数据。在这种场景下是不需要消息队列来缓冲的，否则进队列再出队列会降低系统整体的写入能力。而且多了几个传输环节，发生故障的概率也就相应增加了。

2.5.2　减少 IO 交互——批量写入数据（MySQL）

减少 IO 交互、有效提升数据库性能的方法之一就是批量写入数据。以下几节内容就将围绕这一点进行介绍。本节首先介绍如何用 Java 语言批量将数据写入 MySQL 数据库。

1．准备工作：安装开发工具 IntelliJ IDEA

启动开发工具，新建一个工程，IDEA 新建工程第一步如图 2-117 所示。

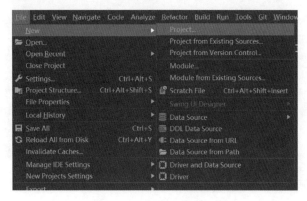

图 2-117　IDEA 新建工程第一步

单击 Project 后，弹出对话框如图 2-118 所示的新建工程第二步。选择左侧高亮的 Java，单击"Next"按钮。

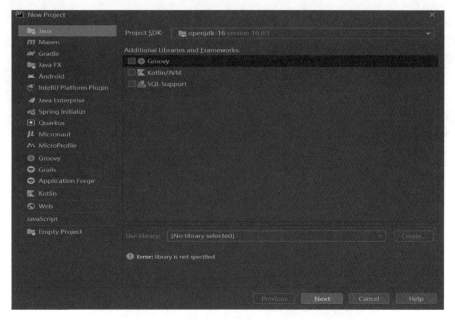

图 2-118　新建工程第二步

之后来到新建工程第三步，如图 2-119 所示，单击"Next"按钮。

在为 New Project 命名时，在选的目录下起一个名字。本示例中 D:\code 是一个路径，mysql-t 是一个名字，IDEA 工具会自动在 D:\code 下创建一个空的文件夹（不要用已有的文件夹，更加不能是有文件的文件夹，让工具自己创建）。新建工程第四步如图 2-120 所示，单击"Finish"按钮。

图 2-119 新建工程第三步

图 2-120 新建工程第四步

2. 编写代码：声明和配置

然后在弹出的对话框中选择 New Window，如图 2-121 所示。

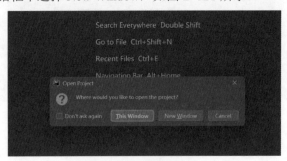

图 2-121 单击新窗口

在如图 2-122 所示的新建 JAVA CLASS 的界面中，右键单击 src 文件夹，在弹出的一级菜单中选择 New，继而在弹出的二级菜单中选择 Java Class。

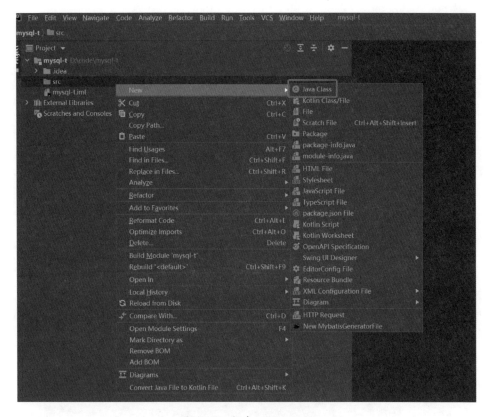

图 2-122 新建 JAVA CLASS

在随后弹出的类名定义的对话框中,输入笔者提供的类名"mysql_insert_single_test"就可以。如果想自定义的话,那么对应代码中 MAIN 后面的名字也要做对应修改。如图 2-123 所示,在红框地方输入 mysql_insert_single_test,然后单击 Class。

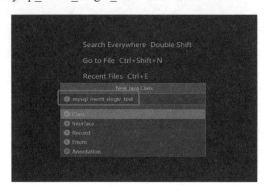

图 2-123 New Java Class

由于是用 JDBC 连接数据库,所以还需要把 JDBC 的 JAR 包添加进来。添加 JDBC 驱动如图 2-124 所示。

图 2-124 添加 JDBC 驱动

单击 Project Structure 后,在弹出的对话框中选择 Modules,如图 2-125 所示添加 JAR 包,单击红色框中的"+"号,然后选择 1 JARS or Directories。

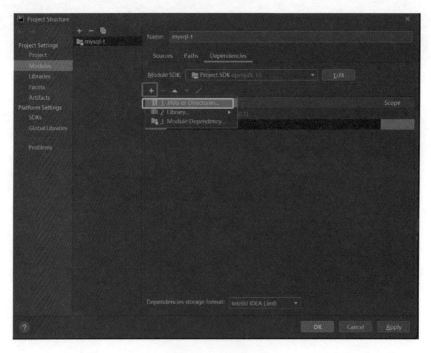

图 2-125 添加 JAR 包

选择事先准备好的 JDBC 驱动,添加 JDBC 驱动,如图 2-126 所示。

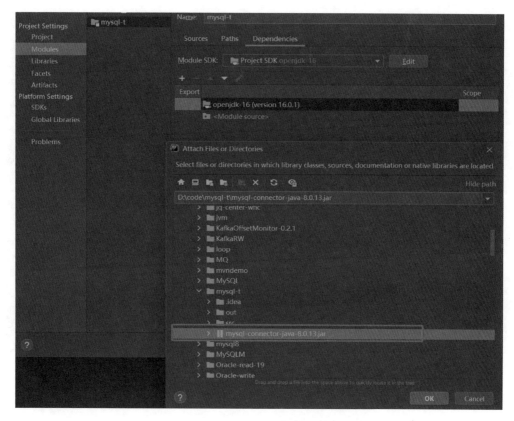

图 2-126　添加 JDBC 驱动

3．数据库准备：建立表

下面要做的工作是对 MySQL 数据库建立测试表，MySQL 测试表结构如图 2-127 所示。

```
mysql> desc ten;
+--------+-------------+------+-----+-------------------+-------------------+
| Field  | Type        | Null | Key | Default           | Extra             |
+--------+-------------+------+-----+-------------------+-------------------+
| id     | int         | NO   | PRI | NULL              | auto_increment    |
| name   | varchar(30) | YES  |     | NULL              |                   |
| age    | int         | YES  |     | NULL              |                   |
| weight | int         | YES  |     | NULL              |                   |
| height | int         | YES  |     | NULL              |                   |
| time   | timestamp(6)| NO   |     | CURRENT_TIMESTAMP(6) | DEFAULT_GENERATED |
+--------+-------------+------+-----+-------------------+-------------------+
6 rows in set (0.01 sec)
```

图 2-127　MySQL 测试表结构

4．代码调试

将本节的代码复制进去即可，然后右键菜单，单击如图 2-128 所示的编译和执行代码的红框中的按钮，就可以对 MySQL 数据库进行写入的压力测试了。

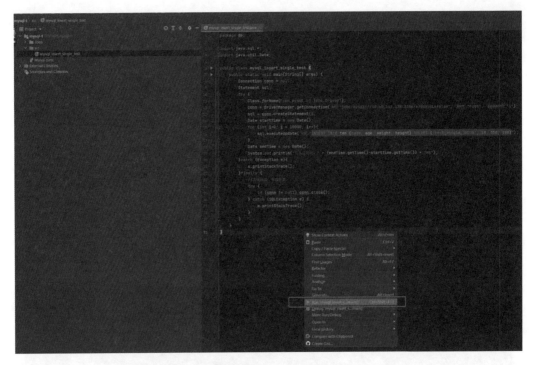

图 2-128 编译和执行代码

结果是：使用开发工具单条提交，执行 1 万次，耗时 211s，平均每秒约执行 500 条。MySQL 单次提交的效率如图 2-129 所示。

图 2-129 MySQL 单次提交效率

MySQL 数据库 1 万条数据单条提交 JAVA 代码如下所示。

```java
package db;
import java.sql.*;
import java.util.Date;
public class mysql_insert_single_test {
    public static void main(String[] args) {
        Connection conn = null;
        Statement sql;
        try {
            Class.forName("com.mysql.cj.jdbc.Driver");
            conn = DriverManager.getConnection("jdbc:mysql://10.60.143.138:3306/x?useSSL=false", "root", "1");
            sql = conn.createStatement();
            Date startTime = new Date();
            for (int i=0; i< 10000; i++){
                sql.executeUpdate("INSERT INTO ten (name, age, weight, height) VALUES ('test_single_10000', 18, 150, 180)");
            }
            Date endTime = new Date();
            System.out.println("写入总耗时: " + (endTime.getTime()-startTime.getTime()) + "ms");
        }catch (Exception e){
            e.printStackTrace();
        }finally {
            //关闭连接，释放资源
            try {
                if (conn != null) conn.close();
            } catch (SQLException e) {
                e.printStackTrace();
            }
        }
    }
}
```

上述代码贴近真实业务场景，在机械硬盘环境下，每秒 500 条的写入虽然不算快，但是也不算慢，远高于一般中小企业的 OLTP 数据库写入要求。

数据库除了单条写入就是批量写入了，在同等环境下，MySQL 的批量写入运行结果如图 2-130 所示。

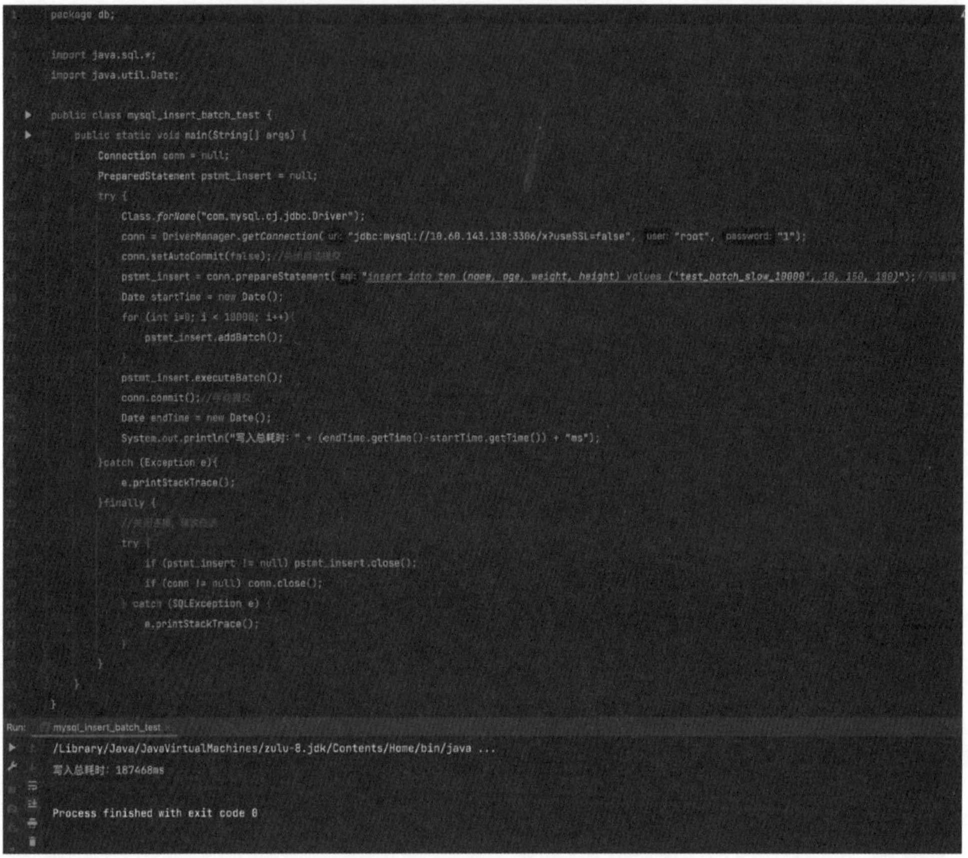

图 2-130 MySQL 批量写入 1

结果是从办公网客户端批量写入 1 万条数据，用时 187s，比单条写入的效率稍快。下面的代码中加粗部分是批量与单条的区别，请注意观察。

```
package db;
import java.sql.*;
import java.util.Date;
public class mysql_insert_batch_test {
    public static void main(String[] args) {
        Connection conn = null;
        PreparedStatementpstmt_insert = null;//这里不一样
        try {
            Class.forName("com.mysql.cj.jdbc.Driver");
            conn = DriverManager.getConnection("jdbc:mysql://10.60.143.138:3306/x?useSSL=false", "root", "1");
```

```
        conn.setAutoCommit(false);//关闭自动提交,这里不一样
        pstmt_insert = conn.prepareStatement("insert into ten (name, age, weight, height) values
        ('test_batch_slow_10000', 18, 150, 180)");//预编译,这里不一样
        Date startTime = new Date();
        for (int i=0; i< 10000; i++){
            pstmt_insert.addBatch();//这里不一样
        }
        pstmt_insert.executeBatch();//这里不一样
        conn.commit();//手动提交
        Date endTime = new Date();
        System.out.println("写入总耗时: " + (endTime.getTime()-startTime.getTime()) + "ms");
    }catch (Exception e){
        e.printStackTrace();
    }finally {
        //关闭连接，释放资源
        try {
            if (pstmt_insert != null) pstmt_insert.close();//这里不一样
            if (conn != null) conn.close();
        } catch (SQLException e) {
            e.printStackTrace();
        }
    }
}
```

批量提交是为了减少内存和磁盘的交互，速度应该是成倍提升，很明显这个结果没有达到预期。这里有一个开发技巧，即增加一个参数 **rewriteBatchedStatements=true**，可以实现批量提交。这个参数的作用是"拼接"地批量提交，即 MySQL 特有的"insert into values (),(),(),()…"。如果不用这个参数，则是通用的 SQL 语法，insert into values ();insert into values ();…，如图 2-131 所示。

图 2-130 和图 2-131 的两种做法同为批量写入，但是写入速度相差很大，图 2-131 的批量操作时间约为 1s，效率比起单条提交提升了将近 200 倍。在遇到开发写入慢的情况可以排查一下是否存在这种情况。

如果到后台日志中查看，会发现第一次 MySQL 单条提交的日志如图 2-132 所示。

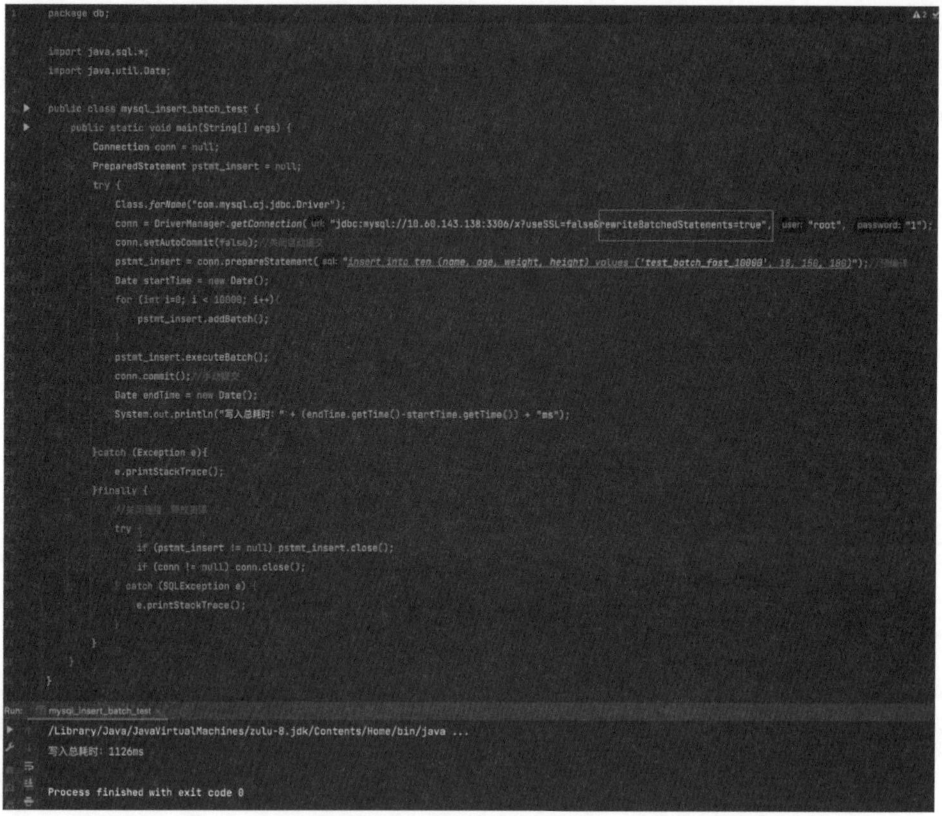

图 2-131　MySQL 批量写入 2

```
# User@Host: root[root] @ [10.60.218.2]  Id:  118
# Query_time: 0.000106  Lock_time: 0.000000 Rows_sent: 0  Rows_examined: 0
SET timestamp=1681289844;
SET autocommit=1;
# Time: 2023-04-12T08:57:24.754190Z
# User@Host: root[root] @ [10.60.218.2]  Id:  118
# Query_time: 0.003507  Lock_time: 0.000009 Rows_sent: 0  Rows_examined: 0
SET timestamp=1681289844;
INSERT INTO ten (name, age, weight, height) VALUES ('test_single_10000', 18, 150, 180);
# Time: 2023-04-12T08:57:24.762670Z
# User@Host: root[root] @ [10.60.218.2]  Id:  118
# Query_time: 0.002214  Lock_time: 0.000004 Rows_sent: 0  Rows_examined: 0
SET timestamp=1681289844;
INSERT INTO ten (name, age, weight, height) VALUES ('test_single_10000', 18, 150, 180);
# Time: 2023-04-12T08:57:24.776772Z
```

图 2-132　MySQL 单条提交日志

```
# User@Host: root[root] @ [10.60.218.2] Id: 118
# Query_time: 0.002210  Lock_time: 0.000004 Rows_sent: 0 Rows_examined: 0
SET timestamp=1681289844;
INSERT INTO ten (name, age, weight, height) VALUES ('test_single_10000', 18, 150, 180);
```

图 2-132 MySQL 单条提交日志（续）

可见，在开启自动提交后，单条提交在 MySQL 内是一条一条执行的，在实验环境下，每次执行加提交的平均耗时约为 0.002s，这意味着在排除其他变量的影响下，在这台实验服务器上由 Java 驱动的单条写入的速度就是每条 2ms。这里请注意是这台实验机器。因为写入的瓶颈并不完全在数据库，而是和操作系统以及存储 IO 有莫大的关系。即使单条数据写入，如果 IO 能力很强，那么这个写入速度也会得到提升。

接下来看第一次 MySQL 批量提交日志，如图 2-133 所示。

```
SET autocommit=0;
insert into ten (name, age, weight, height) values ('test_batch_10000', 18, 150, 180);
# Time: 2023-04-12T09:03:20.222903Z
# User@Host: root[root] @ [10.60.218.2] Id: 119
# Query_time: 0.000335  Lock_time: 0.000003 Rows_sent: 0 Rows_examined: 0
SET timestamp=1681290200;
insert into ten (name, age, weight, height) values ('test_batch_10000', 18, 150, 180);
# Time: 2023-04-12T09:03:20.243006Z
# User@Host: root[root] @ [10.60.218.2] Id: 119
# Query_time: 0.000375  Lock_time: 0.000005 Rows_sent: 0 Rows_examined: 0
SET timestamp=1681290200;
insert into ten (name, age, weight, height) values ('test_batch_10000', 18, 150, 180);
# Time: 2023-04-12T09:05:55.861748Z
# User@Host: root[root] @ [10.60.218.2] Id: 119
# Query_time: 0.000300  Lock_time: 0.000004 Rows_sent: 0 Rows_examined: 0
SET timestamp=1681290355;
insert into ten (name, age, weight, height) values ('test_batch_10000', 18, 150, 180);
# Time: 2023-04-12T09:05:55.879799Z
# User@Host: root[root] @ [10.60.218.2] Id: 119
# Query_time: 0.000320  Lock_time: 0.000003 Rows_sent: 0 Rows_examined: 0
SET timestamp=1681290355;
insert into ten (name, age, weight, height) values ('test_batch_10000', 18, 150, 180);
```

图 2-133 MySQL 第一次批量提交日志

由日志可见，关闭自动提交后，由 Java 做批量执行。在 MySQL 中确实能够达到批量提交的效果，总执行时长也有一定程度的降低，每条执行的耗时平均能够做到 0.001s 以下，理论上每秒能够做到 1000 条以上的数据写入，即使加上 Java 预编译以及批量提交的耗时，总耗时依然能够做到比单条提交快。而采用了 rewriteBatchedStatements=true 写法后，在日志中看到的 SQL 是拼接的。MySQL 第二次批量提交日志如图 2-134 所示。

```
# Time: 2023-04-11T07:21:13.282854Z
# User@Host: root[root] @ [10.60.218.2] Id: 112
# Query_time: 0.000116 Lock_time: 0.000000 Rows_sent: 0 Rows_examined: 0
SET timestamp=1681197673;
SET autocommit=0;
# Time: 2023-04-11T07:21:13.453690Z
# User@Host: root[root] @ [10.60.218.2] Id: 112
# Query_time: 0.107718 Lock_time: 0.000032 Rows_sent: 0 Rows_examined: 0
SET timestamp=1681197673;
insert into ten (name, age, weight, height) values ('test', 18, 150, 180),('test', 18, 150, 180),('test', 18, 150, 180),...
# Query_time: 0.002519 Lock_time: 0.000000 Rows_sent: 0 Rows_examined: 0
SET timestamp=1681197673;
commit;
```

图 2-134　MySQL 第二次批量提交日志

在日志中可以看到，MySQL 中一千条分开执行的数据导入，实际是以一条 "insert into x values (x1),(x2),(x3),…" 形式的语句执行的，其执行时间仅约为 100ms，也就是说，一次批量写入批量提交的动作在 MySQL 中被优化成了一次单条批量写入单条提交的动作。

通过研究和跟踪驱动源码后发现，之所以产生了上述优化，是因为当遇到 rewriteBatchedStatements=true 的选项设置时，会执行一个特别的处理方法 executeBatchSerially()，而这个方法会将程序的批量写入、更新或是删除优化为一条带有大量参数的写入（数据量大的情况下会拆分为多个批次），这是驱动的作用。

2.5.3　减少 IO 交互——批量写入数据（Oracle）

在数据库领域，Oracle 无疑是最好的产品。在同样的环境中对比一下单次和批量写入的性能。建立一张与前一个 MySQL 案例相同结构的实验表，看看不同数据库对于相同来源的命令是否有不同的处理方式。Oracle 测试数据表结构如图 2-135 所示。

```
SQL> desc ten;
 名称                                      是否为空? 类型
 ----------------------------------------- -------- ----------------------------
 ID                                                 NUMBER(20)
 NAME                                               VARCHAR2(30)
 AGE                                                NUMBER(3)
 HEIGHT                                             NUMBER(4)
 WEIGHT                                             NUMBER(4)
 TIME                                               TIMESTAMP(6)
```

图 2-135　Oracle 测试数据表结构

注意：Oracle 中 JDBC 的 JAR 和 MySQL 不一样，需要替换这个文件。其他步骤和 MySQL 的开发准备工作一样，不再赘述。Oracle 循环写入，每循环一次提交一次，循环 10000 次的执行结果如图 2-136 所示。

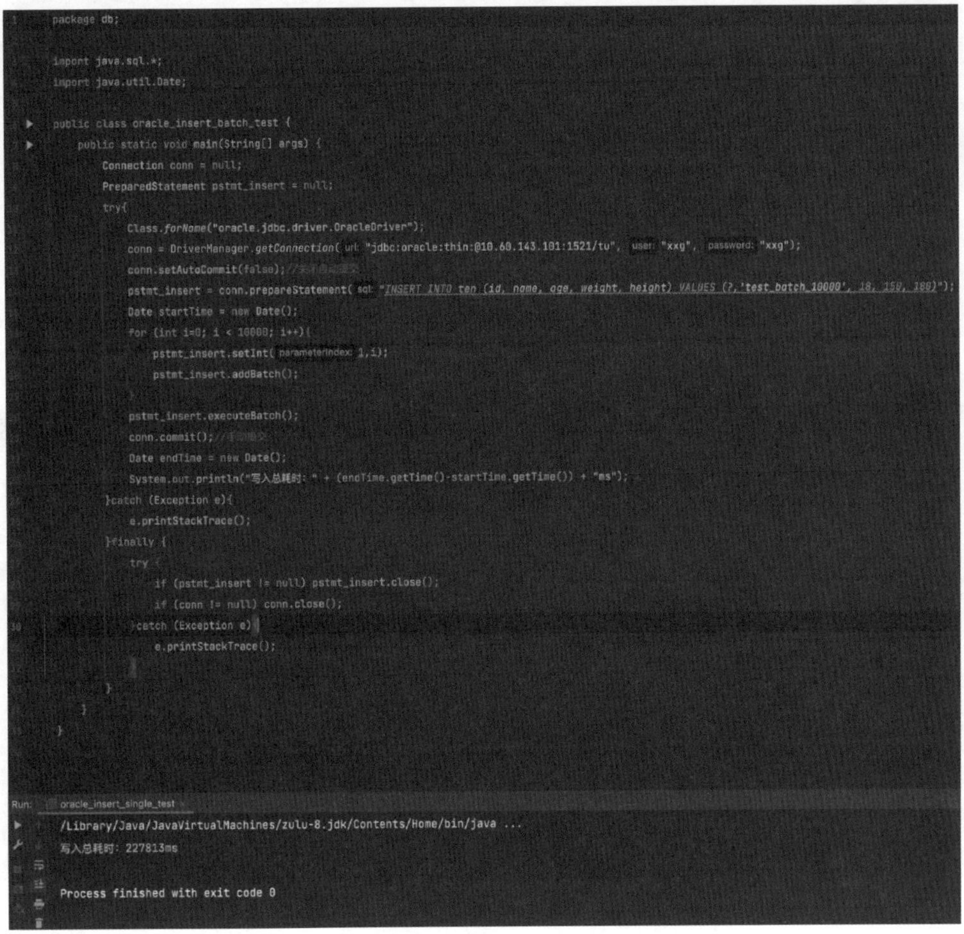

图 2-136　Oracle 单条提交循环 10000 次写入

从图 2-136 可以看出，10000 次循环单条提交，累计用时约 227s，比 MySQL 同样场景执行速度快一些。

下面看看批量提交表现如何？Oracle 批量执行 1 万条数据写入如图 2-137 所示。

执行的结果中，10000 次批量执行写入 220ms 完成。也就是说，每秒批量写入 10000 次即使是机械硬盘的硬件环境，这个性能也是很出色的，执行任务毫无压力。

2.5.4　减少 IO 交互——批量写入数据（PostgreSQL）

近几年，PostgreSQL 上升势头明显，国内许多国产数据库要么是基于 MySQL 的，要么是基于 PostgreSQL 的。本节介绍 Java 通过 JDBC 驱动连接 PostgreSQL 并进行读写。需要准备 PostgreSQL 的 JDBC 驱动，Java 程序的建立方式见前文，通过同样的方式导入 JDBC 的 JAR 包。如图 2-138 所示，ten 表为 PostgreSQL 的测试数据表的表结构。

```java
package db;

import java.sql.*;
import java.util.Date;

public class oracle_insert_batch_test {
    public static void main(String[] args) {
        Connection conn = null;
        PreparedStatement pstmt_insert = null;
        try{
            Class.forName("oracle.jdbc.driver.OracleDriver");
            conn = DriverManager.getConnection( url:"jdbc:oracle:thin:@10.60.143.101:1521/tu", user: "xxg", password: "xxg");
            conn.setAutoCommit(false);//关闭自动提交
            pstmt_insert = conn.prepareStatement( sql:"INSERT INTO ten (id, name, age, weight, height) VALUES (?,'test_batch_10000', 18, 150, 180)");
            Date startTime = new Date();
            for (int i=0; i < 10000; i++){
                pstmt_insert.setInt( parameterIndex: 1,i);
                pstmt_insert.addBatch();
            }
            pstmt_insert.executeBatch();
            conn.commit();//手动提交
            Date endTime = new Date();
            System.out.println("写入总耗时: " + (endTime.getTime()-startTime.getTime()) + "ms");
        }catch (Exception e){
            e.printStackTrace();
        }finally {
            try {
                if (pstmt_insert != null) pstmt_insert.close();
                if (conn != null) conn.close();
            }catch (Exception e){
                e.printStackTrace();
            }
        }
    }
}
```

```
Run:  oracle_insert_batch_test
  /Library/Java/JavaVirtualMachines/zulu-8.jdk/Contents/Home/bin/java ...
  写入总耗时: 220ms

  Process finished with exit code 0
```

图 2-137　Oracle 批量 1 万条数据写入

```
xxg=# \d ten
                         Table "public.ten"
  Column |         Type          | Collation | Nullable |       Default
---------+-----------------------+-----------+----------+-------------------
 id      | integer               |           | not null |
 name    | character varying(20) |           |          |
 age     | integer               |           |          |
 height  | integer               |           |          |
 weight  | integer               |           |          |
 time    | timestamp without time zone |     |          | CURRENT_TIMESTAMP
Indexes:
    "ten_pkey" PRIMARY KEY, btree (id)
```

图 2-138　PostgreSQL 测试数据表的表结构

编写 Java 代码如下，完成连接以及基础查询。PostgreSQL 单条提交循环 10000 次的效果如图 2-139 所示。从实际数据来看，累计耗时约 97s。

```java
package db;

import java.sql.*;
import java.util.Date;
import java.util.Random;

public class postgresql_insert_test {
    public static void main(String[] args) {
        Connection conn = null;
        PreparedStatement pstmt_insert = null;
        String str="abcdefghijklmnopqrstuvwxyzABCDEFGHIJKLMNOPQRSTUVWXYZ0123456789";
        Random random1=new Random();
        try{
            Date date_start = new Date();
            Class.forName("org.postgresql.Driver");
            conn = DriverManager.getConnection( url: "jdbc:postgresql://10.60.143.32:1922/xxg", user: "postgres", password: "postgres");
            pstmt_insert = conn.prepareStatement( sql: "INSERT INTO ten (id, name, age, height, weight) VALUES (?,?,18, 150, 180)");
            for (int i=0; i<10000; i++){
                StringBuffer sb = new StringBuffer();
                for (int j=0; j<6; j++){ //六位随机大小写+数字数学符号
                    int number=random1.nextInt(str.length());
                    char charAt = str.charAt(number);
                    sb.append(charAt);
                }
                pstmt_insert.setInt( parameterIndex: 1, x: i+10000);
                pstmt_insert.setString( parameterIndex: 2, sb.toString());
                pstmt_insert.executeUpdate();
            }
            Date date_end = new Date();
            System.out.print("写入完成: " + (date_end.getTime()-date_start.getTime()) + "ms");

        }catch (Exception e){
            e.printStackTrace();
        }finally {
            //关闭连接，释放资源
            try {
                if(pstmt_insert != null) pstmt_insert.close();
                if (conn != null) conn.close();
            } catch (SQLException e) {
                e.printStackTrace();
            }
        }
    }
}
```

```
Run:   postgresql_insert_test ×
   /Library/Java/JavaVirtualMachines/zulu-8.jdk/Contents/Home/bin/java ...
   写入完成: 97565ms
   Process finished with exit code 0
```

图 2-139　PostgreSQL 单条提交循环 10000 次

下面看看批量写入表现如何？批量写入的关键点是关闭自动提交的设置。Java 批量写入 PostgreSQL 数据库如图 2-140 所示，在图中第 17 行为 setAutoCommit(false)。

图 2-140　Java 批量写入 PostgreSQL 数据库

程序结束，累计耗时 1.3s。通过命令行验证数据，检查 PostgreSQL 数据库批量写入，如图 2-141 所示。

```
xxg=# select count(*) from ten;
 count
--------
 100000
(1 row)
```

图 2-141　检查 PostgreSQL 数据库批量写入

至此，完成 Java 通过 JDBC 驱动对于 PostgreSQL 的读写操作。

通过上述几个例子可以看出，无论使用何种数据库，批量执行写入都比单次写入快。尽管单次执行慢，但是也能每秒写入 500 条数据。

2.5.5　精简架构

在一线互联网公司的消费互联网场景下，会有并发（每秒几十万条甚至上百万条数据）点查读取数据库的压力，会用到 Redis 等缓存技术，Redis 也的确适用于这些场景。但是对于一般中小企业来说，读取压力并不是很大。通过 2.1 节可以知道，使用了索引以后数据的检索效率是不会随着数据量增加而变慢的。

以 MySQL 和 Redis 为例来进行点查 QPS（每秒查询率）的比较。假设有一个登录场景，对于中小企业来说，有几万人同时登录比较有压力。这里同时是指的同一秒，而不是一个时间段。有许多人担心系统登录存在瓶颈。下面用实例来说明这个担心是不必要的，实验分别采用 MySQL、Oracle 等数据库说明。

在 MySQL 数据库建立用户表和权限表，如图 2-142 所示。

```
mysql> create table user
    -> (id int auto_increment,userid varchar(10),password varchar(10),primary key(id));   用户表
Query OK, 0 rows affected (0.06 sec)

mysql> create table qx
    -> (id int ,userid varchar(10),qx varchar(10));    权限表
Query OK, 0 rows affected (0.06 sec)
```

图 2-142　MySQL 数据库建立用户表和权限表

模拟生成 50 万条用户数据。在前面三个小节演示了 Java 代码，本小节采用的方式是存储过程，存储过程代码如下。

```
DELIMITER //
CREATE PROCEDURE us()
BEGIN
    DECLARE v INT;
    SET v = 0;
loop_label: LOOP
```

```
            INSERT INTO user (userid, password) VALUES (CONCAT('t', v), v);
            SET v = v + 1;
            IF v >= 500000 THEN
                LEAVE loop_label;
            END IF;
        END LOOP;
END //
```

用户登录过程是获取用户名密码以及权限的一个过程，用户名密码以及权限设计得是否高效非常重要，本例以最合理、最简化的设计为前提，不考虑不合理设计。其中包含 CONCAT 的两句代码分别是模拟获取用户名密码过程和获取权限过程，代码如下。

```
DELIMITER //
BEGIN
    DECLARE v int;
    DECLARE a varchar(10);
    DECLARE b varchar(10);
    DECLARE c varchar(10);
    DECLARE d varchar(10);
    SET v = 0;

loop_label: LOOP
        SELECT userid, password INTO a, b FROM user WHERE userid = CONCAT('t', v);
        SELECT userid, qx INTO c, d FROM qx WHERE userid = CONCAT('t', v);
        SET v = v + 1;

        IF v >= 500000 THEN
            LEAVE loop_label;
        END IF;
    END LOOP;
END //
```

执行的结果如图 2-143 所示，50 万用户在单 CPU 下串行读取两个表（点查，每次返回一行）用时约 50s。即每秒 1 万次读取操作，每次读 user 和 qs 表各一条数据，QPS 估算达到了 2 万。

```
top - 16:09:57 up 15 days,  3:45,  2 users,  load average: 0.29, 0.18, 0.26
Tasks: 165 total,   1 running, 163 sleeping,   1 stopped,   0 zombie
%Cpu0  :100.0 us,  0.0 sy,  0.0 ni,  0.0 id,  0.0 wa,  0.0 hi,  0.0 si,  0.0 st
%Cpu1  :  0.0 us,  0.0 sy,  0.0 ni,100.0 id,  0.0 wa,  0.0 hi,  0.0 si,  0.0 st
%Cpu2  :  0.3 us,  0.3 sy,  0.0 ni, 99.3 id,  0.0 wa,  0.0 hi,  0.0 si,  0.0 st
%Cpu3  :  0.0 us,  0.0 sy,  0.0 ni,100.0 id,  0.0 wa,  0.0 hi,  0.0 si,  0.0 st
%Cpu4  :  0.0 us,  0.0 sy,  0.0 ni,100.0 id,  0.0 wa,  0.0 hi,  0.0 si,  0.0 st
```

图 2-143　50 万用户串行访问数据

```
%Cpu5  :  0.0 us,  0.0 sy,  0.0 ni,100.0 id,  0.0 wa,  0.0 hi,  0.0 si,  0.0 st
%Cpu6  :  0.0 us,  0.0 sy,  0.0 ni,100.0 id,  0.0 wa,  0.0 hi,  0.0 si,  0.0 st
%Cpu7  :  0.0 us,  0.0 sy,  0.0 ni,100.0 id,  0.0 wa,  0.0 hi,  0.0 si,  0.0 st
MiB Mem :  15820.9 total,   8829.4 free,    778.5 used,   6212.9 buff/cache
MiB Swap:      0.0 total,      0.0 free,      0.0 used.  14634.6 avail Mem

    PID USER      PR  NI    VIRT    RES    SHR S  %CPU  %MEM     TIME+ COMMAND
 191561 mysql     20   0 3046180 552988  41332 S 100.0   3.4  30:28.09 mysqld

mysql> call usqx;
Query OK, 1 row affected (49.50sec)
```

图 2-143　50 万用户串行访问数据（续）

以上还仅仅是 MySQL 数据库在单 CPU 下，如果 N 个 CPU 还可以发挥 N 倍的效果。而 Redis 由于读写是单线程的，无法使用到十几个 CPU，所以一般场景下，MySQL 数据库足以应付此场景。

第 3 章

如何运维好数据库

数据库的运维是广大 DBA 的重要工作,而处理数据库故障是运维工作中最重要的一项,故障陪伴着 DBA 成长。本章将讲述笔者多年来遇到的各类典型的数据库故障,从根源上分析数据库故障的产生原因,从而运维好数据库。

本章内容
运维好数据库的关键:处理故障
分析处理数据库故障的关键点
数据库故障处理的典型案例分析

3.1 运维好数据库的关键：处理故障

数据库运行过程中或多或少会遇到故障，如何快速定位和处理这些故障显得尤为关键。因为数据库的故障影响面会很大，可能会让整个 IT 系统进入停滞状态。资深 DBA 们并不担心故障，每一次处理故障都是经验值的增长。如果这个故障是遇到过的，那么在处理故障的熟练程度上就更进一步；如果这个故障是没遇到过的，那就是一次挑战、一次成长。

3.1.1 常见的数据库故障类型

数据库是一个复杂的系统，出现的故障千奇百怪，即使工作十几年的人依然有可能遇到新故障。这些新故障可能是新产品、新特性产生的，也有可能是自己的经验不足没有遇到的。

本节要讨论的是几乎每个数据库系统都会遇到的常规故障，覆盖了日常绝大多数的故障类型，这些故障的出现从未停止过。

1. CPU 使用率过高

CPU 使用率有平均值和峰值两个维度。CPU 使用率过高通常是指执行 SQL 查询引起 CPU 峰值使用率达到 50%以上，这个峰值可能是短暂的几秒钟，也可能是持续的一段时间。如果 CPU 使用率瞬间峰值达到 80%以上，可能短暂几秒对系统不是致命的，但是如果峰值一直持续就会影响系统的稳定性。对数据库而言，不是 CPU 使用率达到 100%才存在问题，其实当 CPU 使用率达到 50%的时候就会出现等待的情况。所以不能等到传统意义上的使用率达到 80%才预警，而是要降低这个阈值。

如果 CPU 使用率的平均值达到 80%以上，那么数据库就处于一种亚健康状态。处于这种状态下的系统稍微增加一点业务量，CPU 使用率就会达到峰值，就有可能出现数据库不可用的风险。

CPU 使用率高通常是数据库磁盘 IO 使用率高引起的，虽然并不绝对，但是有一个较强的关联度。因为关系数据库从一开始就是使用把数据放在内存中的方式，使得操作数据变得更快。此外，只有磁盘承担持久化的责任，而这就使得读写磁盘的数据库活动会话数量比较多，这些都会体现在磁盘 IO 使用率上。所以数据库磁盘 IO 使用率是一个很关键的指标。

一个会话在没有执行完成时是活动的，完成后该会话就进入了休眠状态，所以数据库的会话数量多要具体看是不是活动的会话数量多。

当活动会话持续增多，多到 CPU 处理不过来，只能排队等待的时候，数据库就会出现响应缓慢甚至无法响应的情况。许多数据库连接数被配置了 10 万，其实一般的中小企业的数据库，平时的活动连接数都是个位数。如果遇到了活动连接数达到几十个，那么此时数据库很可能已经有了很严重的问题。如果活动连接数达到了上百个，基本数据库已经无法正常对外服务了。

2. 慢 SQL 导致系统几乎不可用

引起 CPU 使用率高的原因主要就是慢 SQL 导致的，所以 SQL 慢虽然不一定会导致数据库崩溃，但是很容易造成系统响应变慢，请求超时，超过周期性时长的上限。每秒要求 10 次的并发，那么 SQL 执行的时间最好控制在 0.1s 以内。如果一个数据库的定时任务是 5 分钟执行一次，那么这个任务对应的 SQL 执行时间就不能超过 5 分钟，否则就会间接造成系统的不可用。

3. 磁盘满导致数据库不可用

这种故障虽然相比较前几种故障来说出现概率小很多，但是有时也会遇到。磁盘满属于低级故障，但是比起其他故障来说却更加严重。一旦发生这类情况，无论哪种数据库都会产生不可写的报错。通常来说，这种报错重启并不能解决，必须人工介入。

（1）归档空间满

正式环境的数据库一般都做了容灾高可用，而这就必须将归档日志打开，以实现从库或备库通过读取归档日志重现日志变化。MySQL 数据库和 PostgreSQL 数据库直接将归档日志存放在指定的操作系统目录中，随着日志越积越多而没有及时清理最终会导致操作系统空间耗尽，归档无法写入，在线日志不能生成归档，最终所有数据库写操作停止。Oracle 数据库设定了一个归档目录并且指定了该目录的大小，相当于在操作系统的空间限制之上又多了一个限制。同理，当这个空间耗尽，归档也会无法写入，导致在线日志不能生成归档，导致数据库写操作停止。如今一般有经验的 DBA 都会设置定时清理过期的归档，极少发生类似故障。

（2）表空间满

在 MySQL 数据库和 PostgreSQL 数据库上几乎不存在表空间满的问题。主要是在 Oracle 数据库上设定表空间的时候，采用了默认的数据文件不自动扩展。这里着重说明，不自动扩展是对的。如果采用自动扩展，那么数据文件会随着写入数据自动增长到 32GB 的上限后而无法扩展，最终出现表空间满的问题。如果一个表空间有多个数据文件，所有数据文件齐头并进地扩展，极有可能把操作系统磁盘耗尽后，再次扩展最终出现表空间满的问题。而这个时候 DBA 干预已经来不及了。如今数据库都会有各式各样的监控，其中一个监控项就是表空间使用率。通常定义表空间使用率不应该超过 85%。

（3）磁盘满

除了归档日志和数据文件本身会造成磁盘满，其他数据库相关操作也可能导致磁盘满。例如，Oracle 数据库中的事件文件、trace 文件、alert 文件、审计文件等各式各样的文件常年积攒就会占据数据库的磁盘空间，形成隐患。另外，在 MySQL 数据库上由于不合理的需求或设计导致大事务，使得临时表空间特别大，除了临时表空间，还有临时目录，这些都会占据有限的磁盘空间。当磁盘空间耗尽时，会返回相关的失败信息。所以数据库所在的服务器上也要监控到每个层级的操作系统目录。

其实数据库故障远不止这些，以上这些是常见故障。常见故障都不是难题，而罕见故障几乎都是难题。在日常工作中，很难遇到难以解决的故障，主要是这些常见故障时常出现。

3.1.2 数据库故障的危害

数据库作为整个 IT 系统的"心脏"，一旦出现问题，工作不正常或者停止工作，整个系统会全部陷入瘫痪，所以数据库故障对一个 IT 系统来说是致命的打击。

为了将这种危害降到最低，会有很多针对数据库的保障措施来避免故障，如双活高可用、灾备高可用以及同城双机房、两地三中心等。相关监管机构也对于业务连续性的 RPO（Recovery Point Objective，数据恢复点目标，主要指的是业务系统所能容忍的数据丢失量）和 RTO（Recovery Time Object 从灾难发生到业务系统恢复服务功能所需要的最短时间周期）有着严格的要求。

现如今，互联网和物联网普及，对大型服务提供商的业务连续性要求非常高。比如第三方支付系统、大型制造业企业的流水线和出入库系统、运营商和医院等公共事业的 IT 系统，如不可用，会造成很严重的社会影响。这里说的系统不可用不一定是数据库造成的，但是很大概率会因为数据库故障导致不可用,因为其他环节都属于硬件和网络层面等基础设施环境，而这些基础设施相对成熟稳定，既没有频繁的变更，也几乎没有定制化的使用要求。

3.2 分析处理数据库故障的关键点

引起数据库故障的因素有操作系统层面、存储层面，还有断电断网的基础环境层面（以下称外部因素），以及应用程序操作数据库和人为操作数据库这两个层面（以下称内部因素）。这些故障中外部因素发生的概率较小，可能几年都未发生过一起。许多 DBA 从入职到离职都可能没有遇到过外部因素导致的故障，但是内部因素导致的故障可能每天都产生。内部因素中占据主导的是应用程序导致的故障，可以说数据库故障的主要元凶就是应用程序的 SQL 写得不够好。

3.2.1 分析数据库故障的主要原因——SQL

一个正常工作的数据库不会无缘无故发生问题，只有当数据库运行一段代码或一系列代码造成数据库过载的时候才会发生故障。这里的程序代码就是 SQL，这里的过载可能是 IO 过载、CPU 过载、内存溢出、会话数满、无法和数据库建立连接、程序响应缓慢甚至无法响应，极端情况还导致整个操作系统无法响应，无法登录到数据库所在的操作系统进行干预维护。这样的案例比比皆是，虽然也有人为误操作的情况，但是出现的概率和频次远远小于应用程序导致问题的概率和频次。所以数据库发生故障，首先要分析 SQL 语句，这是主因。

SQL 是由开发人员编写的，但是责任不完全是开发人员的。

1）SQL 的成因：SQL 是为了实现特定的需求而编写的，那么需求的合理性是第一位的。一般来说，在合理的需求下即使有问题的 SQL 也是可以挽救的。但是如果需求不合理，那么就为 SQL 问题埋下了隐患。

2）SQL 的设计：这里的设计主要是数据库对象的设计。即使是合理的需求，如果在数据库设计层面没有把控的环节或者保证，那么很多优化就会大打折扣。

3）SQL 的实现：这部分是开发人员所涉及的工作。但是这已经是流程的末端，这个时候改善的手段依然有，但是属于挽救措施。

在笔者多年的工作中，数据库故障主要来源于三个方面：不合理的需求、不合理的设计和不合理的实现。而这些如果从管理和流程上明确规定由经验丰富的 DBA 介入，那么对数据库故障的源头是有很大的控制作用的。

3.2.2 分析 SQL 的主要问题——处理速度慢

在分析 SQL 语句时，SQL 运行缓慢绝对是最主要的问题，没有之一。笔者亲身经历过的案例 99.99%都属于这类问题，而且广泛存在于不同行业中。不是每个企业每天都会遇到断电或服务器硬件损坏，但是几乎每个企业每天都会遇到被 SQL 运行缓慢困扰的情况。这个问题似乎就是一个根深蒂固的顽疾。之所以出现这种状况，主要是这个问题非常容易在开始阶段被忽视，随着系统运行的数据量和使用频度的增加逐渐出现。所以一般都是后期甚至是系统交付以后才发现。这些情况就像慢性病一样潜伏在数据库中，一旦遇到访问量增加，虽然没有让数据库直接宕机，但是也会让请求不断延迟，导致超时或中断。虽然数据库还能勉强提供服务，但是有可能间接造成系统的不可用。或者慢的程度不断累积从量变到质变，最终导致数据库直接宕机。

3.2.3 分析数据库参数对数据库的影响

数据库的参数有很多，有几个对于性能来说比较重要。一个是全局内存参数，这里指的是 Oracle 数据的 SGA、PGA 等参数，MySQL 数据库的 innodb_buffer_pool_size 等参数，和 PostgreSQL 数据库的 shared_buffers 等参数，这些参数是需要使用者自己设置的。作为通用数据库，许多参数官方是基于通用场景设置的，不去调整也可以。而以上提到的内存参数一定要修改，因为数据库不知道操作系统能分配多少资源给到自己。比如，Oracle 可以知道所在的服务器有多少 CPU 和内存，这一点从 AWR 上就可以看出来，甚至给出 SGA 和 PGA 的设置建议。不过以上内存参数也需要 DBA 进行合理的设置，设置到最佳状态，不仅可以让数据尽可能加载到内存处理，也可以更好地利用数据库的一些特性。如 Oracle 的 In-Memory 和 MySQL 的 memory 引擎等。

对于非通用场景的数据库参数，可以进行针对性的调整，以起到更好的效果。但是这种调整应该是极少数参数，而不是大部分参数。一般来说，除了内存参数和 IO 参数，其他参数的修改不会起到质的飞跃。如果有这种经过微调就能发挥关键作用的参数，官方一定会设置成默认的，如果没有设置成默认的，那么一定会做成一种功能，比如 Oracle 从 11g 开始的直接路径读这个功能。当一次 SQL 大于一定的 IO 阈值，那么数据库不再将这些数据放到内存中，就让 SQL 在磁盘上读取这些数据。避免大量 IO 将数据库的热数据从内存中挤出，而后续高频次的读又要重新从磁盘上加载数据。而这对于经常全表扫描的 SQL 或开发能力不佳的企业来说就是性能低下的。虽然这种现象在国内较为普遍，但是也并不意味着关闭这个功能就一定合适。个人认为是越早发现这些问题并改造，比起先容忍这些问题以至于到了后期根深蒂固无法改正要好。

当然，与其纠结很多参数的设置方面，不如把精力更多地放在如何设计和优化 SQL 上。这种收益比起修改参数的性价比更加高。这里并不是不建议大家去调整参数，而是鼓励大家去研究每个参数的作用。只是除了个别几个参数以外，大部分参数不能够起到力挽狂澜、扭转乾坤的作用。曾经 Oracle 的 RWP 团队在分享时候说过，官方给的性能指标都是基于默认参数提供的。

3.2.4 分析硬件短板带来的问题——IO 吞吐能力与 CPU 计算能力

在数据库技术中，硬件短板通常指的是 IO 吞吐能力。数据库的 IO 是指数据库系统通过输入/输出（I/O）操作与存储介质（如硬盘）进行数据读取和写入的过程。其实数据库演进过程在很大程度就是围绕着存储设备的 IO 能力而做出的数据库体系设计。由于数据库的读写压力主要的瓶颈在于硬件的 IO 能力，所以数据库很多优化手段也是针对减少 IO 操作方向而做的。数据库努力克服的短板也是针对 IO 的吞吐量和存储容量，这里说的数据库泛指所有数据库。当 IO 的吞吐能力成为瓶颈以后，CPU 再多也很难发挥出效果，CPU 会陷入空闲等待之中，从而浪费了大量的计算资源。

在数据库操作中，增、删、改、查是最基本也是最常见的指令。而这些指令的实质就是调动磁盘的 IO 和内存的 IO。常见的 IO 操作包括从磁盘读取数据到内存以供查询或更新，以及将数据写回磁盘以持久化存储。而数据库的 IO 能力主要取决于磁盘的性能。同样，SQL 在不同的硬件上表现形式不一样，在高性能 IO 的设备上可以顺利执行完成；在低 IO 性能的设备上就执行缓慢甚至无响应，导致程序奔溃甚至数据库无响应，最终发展成中断服务的宕机故障。在不少企业中，由于历史原因很多应用程序性能低下导致问题频出，但是没有办法进行重写，于是将数据库迁移到高性能 IO 设备的服务器上。原有的故障问题均得以解决。其主要原因就是用新硬件的性能弥补老硬件的短板。下面介绍一下其中的原理和知识体系。

在 2000 年左右安装和配置服务器的时候，经常看到磁盘有 7200 转，10000 转和 15000 转这些标识。这个转是说磁盘一分钟转多少次。硬盘进化历史如图 3-1 所示，图 3-1 中最左侧的硬盘需要寻址，这种称为机械硬盘（HDD）。

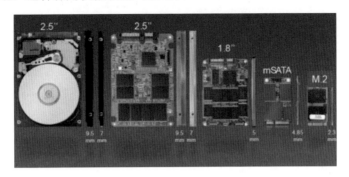

图 3-1　硬盘进化历史

但是机械硬盘读写速度慢，其慢的主要原因就是磁头不停转圈地寻址。为了解决这个寻址问题，存储厂商研发处理固态硬盘，俗称 SSD。机械硬盘与固态硬盘如图 3-2 所示，图 3-2 中左边是机械硬盘要转圈寻址，右侧是固态硬盘不用磁头寻址，所以会快很多。

图 3-2 机械硬盘与固态硬盘

如果对硬件不了解,可能以为所有 SSD 都一样,其实不然。

以某云服务提供商举例,其对云数据库仅提供本地 SSD 和弹性 SSD,没有 HDD。主要原因为数据库是 IO 密集型产品,这点上无论 OLTP 还是 OLAP 都是一样的,对于 OLTP 来说更加突出。对于 IO 密集型的场景来说,类似 NAS 这样的是不能够满足要求的。对于数据库而言,满足高 IOPS 是用户的第一道瓶颈,只有满足了高 IOPS 之后,用户才能把瓶颈点从存储移向内存或处理器等。如果用户不能满足高 IOPS,那么使用内核再多的处理器也是浪费,因此,数据库只能用固态硬盘这一种解决方案。为了更好地使用 CPU,必须提高 IO 能力。

而 SSD 分为 SATA SSD 和 NVMe SSD。NVMe SSD 要比 SATA 的 SSD 快得多。未来,SSD 一定是会从 SATA 走向 NVMe,这是两种相同的介质,但是出口协议不同,从而导致性能相差 4~5 倍。可以大胆推测将来的数据库使用 SSD 是一种标配,而这种标配是不应该成本过高的。

HDD 响应时间为 5ms 左右,基本上在数据库领域也是一次 IO 响应的时间。而通过索引找到一次数据,当内存没有数据时,在上百亿数据量的情况下,索引也就只有 4 层,只需 4 次 IO 响应,只要 20ms 响应时间。再通俗一点说一个 SQL 用到了索引,精确地返回了少量的数据,机械硬盘场景中执行时间在 20ms 之内是非常正常的。而 PCIe 的闪存卡的响应时间是小于 0.1ms 的。

读者可能要问,PCIe 是指什么?下面就详细讲解 PCIe,它是接口,还可能是总线。

集成电路上,一个硬件和另外一个硬件需要一根线(也可能是许多线)来连接,很多线组合在一起就叫总线。

一个硬件和另外一个硬件通信需要通过一定频率的信号,这种叫作协议。

硬件和硬件的集成叫作接口。

这三者缺一不可,即使协议很快且接口效率很高,但是总线如果跟不上成为瓶颈的话,速度也上不去。

总线处理能力很好,但是协议不佳或接口效率低,因为出现了不匹配的瓶颈,那么总体表现得也会很差。但是集成电路一般不用考虑这些,因为都是生产商设置好的,基本是匹配的。

这里列举一下这三者的对应关系,协议总线接口对应表如图3-3所示。

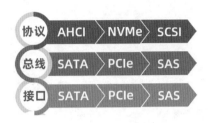

图3-3 协议总线接口对应表

协议有AHCI协议、NVMe协议和SCSI协议三种协议。

总线有SATA总线、PCIe总线和SAS总线三种总线。

接口有SATA接口、PCIe接口和SAS接口。

由于以上这些的存在,就造就了很多种组合,如AHCI协议SATA总线SATA接口的;也有AHCI协议SATA总线PCIe接口的;也会有NVMe协议PCIe总线PCIe接口等。

下面列举几个组合的理论上限,之所以是理论上限,是因为最终还是看设计和具体使用的材料。协议总线接口组合理论上限如图3-4所示。

接口	总线	协议	传输速率
SAT	SATA3.0	AHCI	600MB/s
M.2B&M-Key (PCIe接口的一种)	SATA3.0	AHCI	12GB/s
M.2M-Key	PCIe3.0×4	AHCI	4GB/s
M.2M-Key	PCIe4.0×4	NVMe	8GB/s

图3-4 协议总线接口组合理论上限

以上都是SSD,可以看到SSD之间传输速率相差十几倍。笔者从前对非SSD的机械硬盘做过测试,顺序读写效率大约是每秒5MB。就上面的数据而言,即使是最差的SSD其传输速率也是机械硬盘的100倍,更不要说好的SSD的速率了。所以可能在实际工作中发现同样是SSD会相差很大。

技术也在不断进步,可能受限于作者知识领域的限制,没有能够深入说明。也可能在日新月异的技术更迭中还有新的颠覆性理论和技术出现,会打破以上的说明。

以上的知识是为了下面做铺垫,为什么数据库虚拟机部署、物理机部署和一体机部署差别这么大?虚拟机多了一层虚拟化,已经在磁盘的这些协议、总线和接口的方面上进行了一定程度的打折。而物理机部署的时候这些优势就体现出来了。如今一体机已经不是新鲜事物了,Oracle、阿里、腾讯等都推出了自己的一体机。一体机从上到下都做了最佳的适配,而且一体机内部的计算节点和存储节点的网络也做了最佳的适配。阿里PolarDB一体机如图3-5所示。

图 3-5 阿里 PolarDB 一体机

Oracle 的 Exadata 一体机如图 3-6 所示，本书写的是 X9M 的四分之一配置。可能马上就过期了，会有新一代产品出来。

型号	Exadata X9M-2 HC 1/4配
Exadata 数据库服务器	2台
数据库服务器总CPU Cores	128个
数据库服务器总内存容量	2TB （最大扩展到4TB）
内部高速互联交换机（36端口100GB/s RoCE网络交换机）	2台
管理交换机（Management Switch）	1台
Exadata 分布式智能存储服务器（HC高容量配置）	3台
存储服务器CPU总核数	96个用于SQL卸载处理
持久化内存（TB）	4.5TB
NVMe PCIe闪存卡(TB)	76.8 TB
SAS-3磁盘容量(TB)	648 TB
两副本冗余后可用容量(TB)	245.4 TB
三副本冗余后可用容量(TB)	192.4 TB
最高SQL闪存带宽	135GB/s
最高SQL PMEM 读取IOPS	560万
最高SQL 闪存 写入 IOPS	184.2万

- 一体机具备极高的存储IO带宽和极高IOPS吞吐能力，可以有效解决在传统X86/VM+SAN架构下因IO瓶颈造成的数据库性能问题。
- 一体机提供了192TB可用存储容量（三副本冗余后），不仅可以满足各业务系统当前的存储空间需求（约10TB以上），也可以满足未来五年各业务系统数据增长所需扩展的存储空间。

图 3-6 Oracle 的 Exadata 一体机

数据库问题中绝大多数由 SQL 造成，而 SQL 的问题绝大多数又是 IO 瓶颈导致。一般来说，优化 SQL 的主要思路都是解决如何减少 IO（无论是逻辑 IO 还是物理 IO）去考虑的。所以要么大力设计和优化 SQL，要么提升硬件短板来解决 IO 瓶颈。

3.3 数据库故障处理的典型案例分析

笔者有多年的数据库维护经历，遇到过成百上千次的故障事件，其中不乏很多典型案例。

通过对这些典型的数据库故障进行深度分析后，发现各种因素中占比最大的还是开发人员对数据库知识了解不够、SQL 掌握不好以及对数据库的不了解。

3.3.1 一次 Elasticsearch 误删除的故障分析

一次有个从事运维的朋友小 E 在网上求助，他因为误操作将 Elasticsearch 集群的数据全部清除了。在了解了前因后果以后，笔者发现这个删除操作是无法恢复的，只能告诉他从数据源的地方重新拉取一下，尽可能把损失降到最低，这个案例非常有警示作用。下面模拟复现一下这个误操作。Elasticsearch 数据库中原有数据情况如图 3-7 所示，该数据库中只有一个索引（在 Elasticsearch 中的索引就是通常意义上大家理解的表）。

```
[root@dev-dbteam-143107 ~]# elasticsearch-sql-cli
                    asticElasticE
              ElasticE sticEla
        sticEl ticEl         Elast
     lasti Elasti           tic
   cEl       ast            icE
  icE         as             cEl
  icE         as             cEl
  icEla      las              El
  sticElasticElast          icElas
 las       last            ticElast
El        asti            asti   stic
El        asticEla      Elas      icE
El      Elas cElasticE ticEl       cE
Ela     ticEl      ticElasti       cE
 las    astic          last       icE
  sticElas            asti       stic
   icEl              sticElasticElast
   icE                sticE  ticEla
   icE                 sti    cEla
   icEl                sti     Ela
    cEl                sti     cEl
     Ela             astic   ticE
      asti         ElasticElasti
        ticElasti lasticElas
          ElasticElast

                 SQL
                 6.8.6

sql> show tables;
    name       |    type
---------------+---------------
xuexiaogang    |BASE TABLE
```

图 3-7　Elasticsearch 数据库中原有数据情况

模拟写入多个索引，构造案例环境如图 3-8 所示，已经写入了 3 个索引，分别是 library、rary 以及 trar。

```
[root@dev-dbteam-143107 ~]# curl -XPUT http://10.60.143.107:9200/library/books/1 -H 'content-Type:application/json'  -d '{"title": "计算机", "name":"计算机基础", "publish_date": 1592463877763, "price": 32.5}'
```

图 3-8　构造案例环境

```
        {"_index":"library","_type":"books","_id":"1","_version":1,"result":"created",
"_shards":{"total":2,"successful":1,"failed":0},"_seq_no":0,"_primary_term":1}[root@
dev-dbteam-143107 ~]#


    [root@dev-dbteam-143107 ~]#
    [root@dev-dbteam-143107 ~]# curl -XPUT http://10.60.143.107:9200/rary/books/2 -H
'content-Type:application/json' -d '{"title": "计算机", "name":"计算机基础", "publish_date":
1592463877763, "price": 32.5}'
        {"_index":"rary","_type":"books","_id":"2","_version":1,"result":"created","_shards":
{"total":2,"successful":1,"failed":0},"_seq_no":0,"_primary_term":1}[root@dev-dbteam-
143107 ~]#


    [root@dev-dbteam-143107 ~]#
    [root@dev-dbteam-143107 ~]# curl -XPUT http://10.60.143.107:9200/trar/books/3 -H
'content-Type:application/json' -d '{"title": "计算机", "name":"计算机基础", "publish_date":
1592463877763, "price": 32.5}'
        {"_index":"trar","_type":"books","_id":"3","_version":1,"result":"created","_shards":
{"total":2,"successful":1,"failed":0},"_seq_no":0,"_primary_term":1}[root@dev-dbteam-
143107 ~]#
```

图 3-8 构造案例环境（续）

此时通过 elasticsearch-sql-cli 使用 SQL 进行查询，如图 3-9 所示，看到有多张数据库表，其中包括 library、rary 以及 trar 是以表的形式存在的。也可以使用 select 语句分别去查询这几张表的数据。

```
sql> show tables;
      name        |     type
------------------+---------------
library           |BASE TABLE
rary              |BASE TABLE
trar              |BASE TABLE
xuexiaogang       |BASE TABLE

sql> select * from rary;
      name        |    price      | publish_date    |    title
------------------+---------------+-----------------+---------------
计算机基础         |32.5           |1592463877763    |计算机

sql> select * from trar;
      name        |    price      | publish_date    |    title
------------------+---------------+-----------------+---------------
计算机基础         |32.5           |1592463877763    |计算机
```

图 3-9 使用 SQL 语法查询 Elasticsearch

下面模拟小 E 最初的想法，他想删除包含 rar 的索引（表）的数据，也许是业务要求清理一下过期数据。于是小 E 期望执行如图 3-10 所示的 Elasticsearch 期望的模糊删除，如果执行这个命令，那么所有包含 rar 这三个字符的索引（表）的数据全部删除，但是 xuexiaogang 的索引数据应该保留。

```
[root@dev-dbteam-143107 ~]# curl -XDELETE http://10.60.143.107:9200/*rar*
{"acknowledged":true}[root@dev-dbteam-143107 ~]#

sql> show tables;
     name         |     type
------------------+----------------
xuexiaogang       |BASE TABLE
```

图 3-10　Elasticsearch 期望的模糊删除

但是很不幸，在实际操作中小 E 执行的命令如图 3-11 所示，他在第一个 * 后面不小心多了一个空格。在笔者这个模拟机的版本上有一个报错，在小 E 的生产环境版本中是没有这个报错的，这个可能是不同版本的原因。

```
[root@dev-dbteam-143107 ~]# curl -XDELETE http://10.60.143.107:9200/* rar*
{"acknowledged":true}curl: (6) Could not resolve host: rar*; Unknown error
[root@dev-dbteam-143107 ~]#

sql> show tables;
     name         |     type
------------------+----------------
```

图 3-11　实际执行的 Elasticsearch 命令

但是无论是不是有报错，最终的结果都是一样的，那就是连 xuexiaogang 这个索引也删除了。即这个空格导致了整个 Elasticsearch 的数据全部清空了。

其实在运维过程中可能由于一个不经意或不小心就可能造成故障。本案例就是一个典型的因为疏忽误删除全库数据的教训，在执行维护命令时一定要仔细再仔细。

3.3.2　一次配置文件丢失导致的 MySQL 数据库故障分析

某天一个开发工程师在微信上向笔者求助：他的应用程序无法连接数据库了。这个数据库并不是归笔者维护的，只是友情帮忙检查。第一个想到的是问他故障现象，描述为 Tomcat 显示报错，无法连接数据库。遇到这种问题，给大家一个排除故障的思路，第一反应是去验证反馈信息的准确性。比如让其他人去连一下数据库看看能不能连上；此外排查一下网络连接问题，比如用 telnet 工具测试数据库的端口正常与否。

排查结果是这位开发工程师使用 telnet 测试端口不通，而其他人可以连接数据库。这个问题的诡异点就在此，应用的 IP 地址和数据库的 IP 地址是同网段，所以不存在防火墙等问题。数据库不是以观察者角度不同而不同的。既然部分人为操作数据库可以连接，部分人为操作不可以，这就需要验证数据库本身是否存在问题了。

笔者登录到数据库服务器，检查数据库运行状态，如图 3-12 所示。

图 3-12　数据库运行状态检查

从当时的截图看，数据库没有正常运行。也许启动过了但是启动失败，也可能是运行遇到问题了，所以两方面的人反馈的不一样。不过凭借经验，还是倾向于数据库现在不能正常使用且不是可以连接的状态。既然如此，那么先重启一下数据库，尝试解决问题。执行了重启数据库命令后，那位开发工程师依旧反映应用无法连接数据库，而运维工程师则可以连接成功。

这就让人更迷惑了。难道说是数据库系统运行的服务器的问题？于是笔者建议开发工程师重启服务器。服务器操作系统重启之后，数据库仍然没有启动成功，而且通过进程检查命令 ps .ps -eflgrep mysql 发现没有进程，连上图的数据库运行状态失败的信息都没有了。这是正式环境，瞬间压力来了。

此时一定要冷静，越是这种时候越考验 DBA 的知识掌握程度、技术能力、心理承受能力、应变能力和从细微地方发现问题的能力。

数据库启动失败第一个想法先看看配置文件。数据库应该是和操作系统的后台一起启动的。找不到数据目录或权限不对等，都是最常见的数据库启动失败的原因。数据库启动失败一定是什么配置失败了。

采用 find / -name my.cnf 命令查找配置文件路径，因为有时候可能有定制化的安装要求，这个配置文件未必在默认路径。但是检查的结果让人很震惊，因为全磁盘都找不到配置文件。不管是 MySQL 的哪个版本，或者不管是什么路径，这个参数名字是一定的。而如今全磁盘都没找到，难怪数据库无法启动！这就是问题，数据库的参数文件没有了，这是为什么？

这种情况笔者也是第一次遇到。其实所谓的经验就是不断地遇到新的问题，然后尝试用已有的知识体系去解决它或用已知的联系未知的推断去尝试。尤其是后者对能力的提升有很大的帮助。

冷静几秒后，决定去案发地——默认路径再找找，看看能发现什么线索。事实也证明这个排查思路是对的，因为找到了一个 my.cnf.rpmsave 文件。看到这个文件眼前一亮。不知道读者朋友看到这个文件名有什么联想吗？从文件名看 rpm+save 作为后缀，还可以推断这个是 rpm 安装产生的配置文件。

在已经安装好的 MySQL 数据库上再次运行安装，安装程序在分配这个数据库自带的配置文件（my.cnf）的时候发现默认的路径下已经存在了这个文件，那么就会把已有的 my.cnf 备份一下，产生 my.cnf.rpmsave。

一般这样的情况下，会有一个 my.cnf 文件，但是奇怪的是现在没有。既然知道这些情况，那么可以先把这个 my.cnf.rpmsave 改成默认名字，保证能重启恢复数据库。

果不其然，将 my.cnf.rpmsave 改成 my.cnf，数据库启动成功。运维人员表示可以连接使用数据库，开发人员这个时候也排查到了原因，就是网络造成的问题。他在排查时用另外一个应用服务器 telnet 数据库端口可以成功连接，但是他使用原来的应用服务器 telnet 数据库端口就连接不上。

故障解决了，问题产生的原因也知道了。可以判断是有人在操作时留下的一个隐患，只是这个机器一直没重启，一旦重启就暴露了问题。而在运行过程中有人在服务器上又安装了一次数据库，安装不成功，导致参数文件被转移到旁边了。

以上问题可以在实验环境模拟复原。如图 3-13 所示，检查 MySQL 数据库服务在运行，以及获取 MySQL 的运行时间（后续可以检查数据库重启后时间重置）。

```
[root@base ~]# service mysqld status
mysqld (pid  4408) is running...

[root@base ~]# mysql -uroot -p1 -e "show status like '%uptime%'"
mysql: [Warning] Using a password on the command line interface can be insecure.
+---------------------------+--------+
| Variable_name             | Value  |
+---------------------------+--------+
| Uptime                    | 6502   |
| Uptime_since_flush_status | 6502   |
+---------------------------+--------+
```

图 3-13　检查 MySQL 数据库运行

将默认配置文件模拟重复安装，将名字改为 my.cnf.rpmsave，并且使得默认配置文件不存在。无参数情况下 MySQL 数据库重启如图 3-14 所示，启动成功。

```
[root@base ~]# mv /etc/my.cnf /etc/my.cnf.rpmsave
[root@base ~]# service mysqld stop
Stopping mysqld:                                           [  OK  ]
[root@base ~]# service mysqld start
Starting mysqld:                                           [  OK  ]
[root@base ~]# mysql -uroot -p1 -e "show status like '%uptime%'"
mysql: [Warning] Using a password on the command line interface can be insecure.
+---------------------------+--------+
| Variable_name             | Value  |
+---------------------------+--------+
| Uptime                    | 7      |
| Uptime_since_flush_status | 7      |
+---------------------------+--------+
[root@base ~]#
```

图 3-14　无参数情况下 MySQL 数据库重启

把数据库停止以后重启操作系统，如图 3-15 所示。

```
[root@base ~]# service mysqld stop
Stopping mysqld:                                           [  OK  ]
[root@base ~]# reboot

Broadcast message from root@base
        (/dev/pts/2) at 15:31 ...

The system is going down for reboot NOW!
[root@base ~]# Connection closing...Socket close.

Connection closed by foreign host.

Disconnected from remote host(10.60.143.134-mysql8.0.28-meb) at 15:31:41.

Type `help' to learn how to use Xshell prompt.
```

图 3-15 重启操作系统

重启操作系统后，检查数据库日志，如图 3-16 所示，发现数据库没有启动成功的标志。

```
[root@base ~]# tail -20f /var/lib/mysql/base.err
2023-05-09T07:30:17.470365Z 0 [Warning] [MY-010909] [Server] /usr/sbin/mysqld: Forcing close of thread 14  user: 'root'.
2023-05-09T07:30:20.857504Z 0 [System] [MY-010910] [Server] /usr/sbin/mysqld: Shutdown complete (mysqld 8.0.26)  MySQL Community Server - GPL.
2023-05-09T07:30:55.141798Z 0 [System] [MY-010116] [Server] /usr/sbin/mysqld (mysqld 8.0.26) starting as process 5024
2023-05-09T07:30:55.168867Z 1 [System] [MY-013576] [InnoDB] InnoDB initialization has started.
2023-05-09T07:30:55.855379Z 1 [System] [MY-013577] [InnoDB] InnoDB initialization has ended.
2023-05-09T07:30:56.252014Z 0 [Warning] [MY-013129] [Server] A message intended for a client cannot be sent there as no client-session is attached. Therefore, we're sending the information to the error-log instead: MY-001681 - 'InnoDB Memcached Plugin' is deprecated and will be removed in a future release.
InnoDB MEMCACHED: Memcached uses atomic increment
2023-05-09T07:30:56.438238Z 0 [Warning] [MY-013746] [Server] A deprecated TLS version TLSv1 is enabled for channel mysql_main
2023-05-09T07:30:56.438919Z 0 [Warning] [MY-013746] [Server] A deprecated TLS version TLSv1.1 is enabled for channel mysql_main
2023-05-09T07:30:56.469095Z 0 [Warning] [MY-010068] [Server] CA certificate ca.pem is self signed.
2023-05-09T07:30:56.470171Z 0 [System] [MY-013602] [Server] Channel mysql_main configured to support TLS. Encrypted connections are now supported for this channel.
2023-05-09T07:30:56.569006Z 0 [System] [MY-011323] [Server] X Plugin ready for connections. Bind-address: '::' port: 33060, socket: /var/run/mysql/mysqlx.sock
2023-05-09T07:30:56.569213Z 0 [System] [MY-010931] [Server] /usr/sbin/mysqld: ready for connections. Version: '8.0.26'  socket: '/var/lib/mysql/mysql.sock'  port: 3306  MySQL Community Server - GPL.
2023-05-09T07:31:32.924492Z 0 [System] [MY-013172] [Server] Received SHUTDOWN from user <via user signal>. Shutting down mysqld (Version: 8.0.26).
2023-05-09T07:31:35.834598Z 0 [System] [MY-010910] [Server] /usr/sbin/mysqld: Shutdown complete (mysqld 8.0.26)  MySQL Community Server - GPL.
2023-05-09T07:32:09.046662Z 0 [System] [MY-010116] [Server] /usr/sbin/mysqld (mysqld 8.0.26) starting as process 2114
2023-05-09T07:32:09.261952Z 1 [System] [MY-013576] [InnoDB] InnoDB initialization has started.
2023-05-09T07:32:10.394605Z 1 [System] [MY-013577] [InnoDB] InnoDB initialization has ended.
2023-05-09T07:32:11.036113Z 0 [Warning] [MY-013129] [Server] A message intended for a client cannot be sent there as no client-session is attached. Therefore, we're sending the information to the error-log instead: MY-001681 - 'InnoDB Memcached Plugin' is deprecated and will be removed in a future release.
InnoDB MEMCACHED: Memcached uses atomic increment
^C
```

图 3-16 检查数据库日志

这种状态实质上是数据库"假死"，状态上是在启动，但是无法正常对外服务。检查 MySQL 服务和进程如图 3-17 所示。MySQL 服务是停止的，但是 MySQL 进程和守护进程是启动的。这里略微和案例有点不同，案例中是无任何进程。这个微小的差别不影响实验过程和论证结果。

```
[root@base ~]# service mysqld status
mysqld is stopped
[root@base ~]# ps -ef|grep mysql
root       1956  1025  0 15:32 ?        00:00:00 /bin/sh /etc/rc5.d/S64mysqld start
root       1982  1956  0 15:32 ?        00:00:00 /bin/sh /usr/bin/mysqld_safe --datadir=/var/lib/mysql --
socket=/var/lib/mysql/mysql.sock --pid-file=/var/run/mysqld/mysqld.pid --basedir=/usr --user=mysql
mysql      2114  1982  1 15:32 ?        00:00:01 /usr/sbin/mysqld --basedir=/usr --datadir=/var/lib/mysql --plugin-
dir=/usr/lib64/mysql/plugin --user=mysql --log-error=base.err --pid-file=/var/run/mysqld/mysqld.pid --
socket=/var/lib/mysql/mysql.sock
root       2762  2255  0 15:33 pts/1    00:00:00 grep mysql
[root@base ~]# kill -9 1982
[root@base ~]# kill -9 2114
[root@base ~]# ps -ef|grep mysql
root       3189  2255  0 15:34 pts/1    00:00:00 grep mysql
```

图 3-17 检查 MySQL 服务和进程

使用 kill-9 的命令杀掉两个僵死进程，先处理守护进程，再处理数据库进程。然后还原配置文件如图 3-18 所示，在还原参数后，数据库重启成功。数据全部可以访问，至此此案例分析完毕。

```
[root@base ~]# mv /etc/my.cnf.rpmsave /etc/my.cnf
[root@base ~]# service mysqld restart
Stopping mysqld:                                           [  OK  ]
Starting mysqld:                                           [  OK  ]
[root@base ~]# service mysqld status
mysqld (pid  3539) is running...
[root@base ~]# mysql -uroot -p1 -e "show status like '%uptime%'"
mysql: [Warning] Using a password on the command line interface can be insecure.
+-------------------------------+-------+
| Variable_name                 | Value |
+-------------------------------+-------+
| Uptime                        | 20    |
| Uptime_since_flush_status     | 20    |
+-------------------------------+-------+
[root@base ~]# mysql  -e "show databases"
+--------------------+
| Database           |
+--------------------+
| information_schema |
| innodb_memcache    |
| m                  |
| mysql              |
| performance_schema |
| sys                |
| test               |
| x                  |
| x1                 |
| xuexiaogang        |
| xuexiaogang1       |
| xuexiaogang2       |
| xxg                |
+--------------------+
```

图 3-18 还原配置文件

本案例的特点是运维管理混乱，轻视或者无视运维工作的重要性，造成数据库常年无人维护。而且有充足的证据（RPMSAVE）表明有人在服务器上随意安装数据库，造成了配置文件的丢失，造成的故障环境比较罕见，需要 DBA 了解数据库的体系结构才能处理好故障。

3.3.3 一次 In-Memory 丢失引起的故障分析

在 1.3.3 节中介绍列式数据库时提到了 Oracle 的 In-Memory，它在有亿条数据的大型数据库中可以实现秒级分析统计型查询。其实 In-Memory 除了在列式数据库的分析场景中表现优秀，执行在内存中的全表查询时，也因为数据驻留在内存中，能够非常快地完成全表扫描的动作。

在笔者的日常工作中，很多场景一天运行上亿次的 SQL 查询，每次查询都会扫描全表的几万条数据，在使用了 In-Memory 特性后单次执行都是在微秒级别，并且对数据库几乎没什么冲击。但有一次在 In-Memory 的使用过程中也发生了意料之外的故障。

有一天，笔者被告知测试数据库访问不到了，DBA 检查发现数据库被意外关闭了。于是对测试数据库手工启动了一下。第二天，包括开发和测试等使用者反馈测试数据库卡顿现象严重。DBA 检查了 CPU、IO 以及 OGG 等相关环节后告知笔者，数据库所在服务器 CPU 占用不高、但是 IO 占用量大，连接会话数多，应用程序连接数据库出现连接不成功的现象。笔者在经过一系列查询之后，从 AWR 中看到 SQL 执行得并不快，比起之前这些全表关联查询而且频次很高的 SQL 都是微秒级别的查询速度来说，现在这些 SQL 都是秒级。找到相关 SQL 检查其执行计划，发现是全表扫描 TABLEACCESSFULL，而事先预判的表已经加载到 In-Memory 对应的执行计划应该是 TABLEACCESSINMEMORYFULL。于是想到了可能是测试数据库重启后未能自动加载相关数据表。解决方式是立即将这些表全部手工加载。加载以后的效果立竿见影，开发和测试等相关人员反馈所有问题全部没有了。

通过手工加载数据表到内存解决了这个问题，所以得出问题的原因就是数据表从内存中被清理了。而这个清理就是因为数据库实例的重启，而且重启后没有自动加载。

故障排除后，笔者进一步思考，In-Memory 在数据库重启后能不能自动加载？若是可以，一旦数据库实例重启，就不需要人工介入了。答案是可以的。在 21c 中，Oracle 支持了自主的 In-Memory 管理，通过一个简单的初始化参数 inmemory_automatic_level 设置，DBA 将不再需要人工指定将哪些数据表放置在内存中，数据库可以自动判断需要将哪些对象加入或驱逐出 In-Memory 的列式存储中，检查 inmemory_automatic_level 参数如图 3-19 所示。

```
SQL> show parameter inmemory_automatic_level;

NAME                                 TYPE        VALUE
------------------------------------ ----------- ------------------------------
inmemory_automatic_level             string      HIGH
```

图 3-19　inmemory_automatic_level 参数检查

这里扩展一下，In-Memory 的特性中不仅仅是自动加载，并行执行关系也很大。In-Memory 聚合运算比较如图 3-20 所示。

```
SQL> select count(id) from m;

  COUNT(ID)
----------
 100000004

已用时间:  00: 00: 00.08
SQL> select sum(id) from m;

    SUM(ID)
----------
5.0000E+15

已用时间:  00: 00: 06.13
SQL> select sum(a) from m;

     SUM(A)
----------
5.0000E+15

已用时间:  00: 00: 05.40
SQL> select avg(a) from m;

     AVG(A)
----------
50000002.5

已用时间:  00: 00: 04.14
```

图 3-20 In-Memory 聚合运算比较

读者可以发现除了 count 用时 80ms 以外,其余对不同列的 sum、avg 这样的聚合运算耗时都是 4~6s。最开始笔者使用 In-Memory 时不能理解,都是计算为什么差别如此之大。为此还问过官方的技术人员,官方给出了如图 3-21 所示的等待分析。

```
SQL> alter session set timed_statistics = true;
会话已更改。

SQL>alter session set statistics_level=all;

会话已更改。

SQL> alter session set max_dump_file_size = unlimited;

会话已更改。

SQL> alter session set events '10046 trace name context forever, level 12';
会话已更改。

SQL> set arraysize 5000;

SQL> select sum(a) from m ;


已用时间:  00: 00: 00.01
alter session set events '10046 trace name context off';
会话已更改。

select value from v$diag_info where name='Default Trace File';
```

图 3-21 等待分析

```
    SUM(A)
----------
5.0000E+15

已用时间:  00: 00: 04.84
SQL>
会话已更改。

已用时间:  00: 00: 00.00
SQL>
VALUE
--------------------------------------------------------------------------------
/u01/app/oracle/diag/rdbms/oracle19c/oracle19c/trace/oracle19c_ora_8603.trc

已用时间:  00: 00: 00.00
SQL>
```

图 3-21　等待分析（续）

最终答复是从原始日志可以看到虽然 tkprof 显示的是 fetch 的时间，但其实是 fetch 之前发生的。这部分没有记录的时间说明是 CPU 在做相关操作。而 execute 时间很少，基本为 0，说明通过 IM 取数据是非常快的。所以，官方认为使用 In-Memory 和不使用相差的时间不太大。因为主要的时间消耗不在 In-Memory 数据获取上，而是在 CPU 的消耗上。虽然官方没给出需要的答案，但是给了笔者巨大的启示。

再次对上亿加载到 In-Memory 的表进行 sum 聚合，观测 CPU 的运行情况，单一 CPU 的 In-Memory 计算如图 3-22 所示。从图 3-22 中看到只有 CPU8 在近乎满负荷地工作。

```
top - 10:12:10 up 34 days, 44 min,  2 users,  load average: 0.19, 0.13, 0.10
Tasks: 528 total,   2 running, 526 sleeping,   0 stopped,   0 zombie
%Cpu0  :  0.0 us,  0.0 sy,  0.0 ni,100.0 id,  0.0 wa,  0.0 hi,  0.0 si,  0.0 st
%Cpu1  :  0.3 us,  0.0 sy,  0.0 ni, 99.7 id,  0.0 wa,  0.0 hi,  0.0 si,  0.0 st
%Cpu2  :  1.0 us,  0.0 sy,  0.0 ni, 99.0 id,  0.0 wa,  0.0 hi,  0.0 si,  0.0 st
%Cpu3  :  0.0 us,  0.0 sy,  0.0 ni,100.0 id,  0.0 wa,  0.0 hi,  0.0 si,  0.0 st
%Cpu4  :  0.3 us,  0.0 sy,  0.0 ni, 99.7 id,  0.0 wa,  0.0 hi,  0.0 si,  0.0 st
%Cpu5  :  1.7 us,  0.0 sy,  0.0 ni, 98.3 id,  0.0 wa,  0.0 hi,  0.0 si,  0.0 st
%Cpu6  :  1.0 us,  0.0 sy,  0.0 ni, 99.0 id,  0.0 wa,  0.0 hi,  0.0 si,  0.0 st
%Cpu7  :  0.0 us,  0.0 sy,  0.0 ni,100.0 id,  0.0 wa,  0.0 hi,  0.0 si,  0.0 st
%Cpu8  : 98.3 us,  0.3 sy,  0.0 ni,  1.4 id,  0.0 wa,  0.0 hi,  0.0 si,  0.0 st
%Cpu9  :  0.3 us,  0.3 sy,  0.0 ni, 99.3 id,  0.0 wa,  0.0 hi,  0.0 si,  0.0 st
%Cpu10 :  0.7 us,  0.3 sy,  0.0 ni, 99.0 id,  0.0 wa,  0.0 hi,  0.0 si,  0.0 st
%Cpu11 :  0.0 us,  0.0 sy,  0.0 ni,100.0 id,  0.0 wa,  0.0 hi,  0.0 si,  0.0 st
%Cpu12 :  0.3 us,  0.0 sy,  0.0 ni, 99.7 id,  0.0 wa,  0.0 hi,  0.0 si,  0.0 st
%Cpu13 :  0.7 us,  0.0 sy,  0.0 ni, 99.3 id,  0.0 wa,  0.0 hi,  0.0 si,  0.0 st
%Cpu14 :  0.0 us,  0.0 sy,  0.0 ni,100.0 id,  0.0 wa,  0.0 hi,  0.0 si,  0.0 st
%Cpu15 :  0.0 us,  0.0 sy,  0.0 ni,100.0 id,  0.0 wa,  0.0 hi,  0.0 si,  0.0 st
KiB Mem : 32927412 total,  1370092 free,  4051224 used, 27506096 buff/cache
KiB Swap:  4194300 total,  4166644 free,    27656 used. 21896764 avail Mem
```

图 3-22　单一 CPU 的 In-Memory 计算

分析可得，这个计算现在是串行的，没有发挥出多 CPU 的并行能力。将表改为并行，Oracle 表并行设置如图 3-23 所示。并且再次执行 sum 聚合，发现执行时间从 6s 多降低到 750ms，执行速度提升了 10 倍左右。

```
SQL> alter table m parallel;

表已更改。

已用时间:  00: 00: 00.03
SQL> select sum(a) from m ;

    SUM(A)
----------
5.0000E+15

已用时间:  00: 00: 00.75
SQL>
```

图 3-23 Oracle 表并行设置

在并行设置下，CPU 的工作如图 3-24 所示，充分发挥出数据库的潜力和优势。

通过这个案例得出结论：

1）In-Memory 在 count 场景下一亿条数据可以秒出，甚至不到一秒（因为列式存储，而且还是内存计算）。而实际上它对于多表关联（目前是 2 表关联）查询的提升也是 50 到 100 倍的。

```
Tasks: 560 total,  18 running, 542 sleeping,   0 stopped,   0 zombie
%Cpu0  : 60.5 us,  1.3 sy,  0.0 ni, 38.2 id,  0.0 wa,  0.0 hi,  0.0 si,  0.0 st
%Cpu1  : 60.5 us,  1.7 sy,  0.0 ni, 37.9 id,  0.0 wa,  0.0 hi,  0.0 si,  0.0 st
%Cpu2  : 60.4 us,  1.7 sy,  0.0 ni, 38.0 id,  0.0 wa,  0.0 hi,  0.0 si,  0.0 st
%Cpu3  : 59.9 us,  2.3 sy,  0.0 ni, 37.8 id,  0.0 wa,  0.0 hi,  0.0 si,  0.0 st
%Cpu4  : 60.2 us,  2.0 sy,  0.0 ni, 37.8 id,  0.0 wa,  0.0 hi,  0.0 si,  0.0 st
%Cpu5  : 60.9 us,  1.3 sy,  0.0 ni, 37.7 id,  0.0 wa,  0.0 hi,  0.0 si,  0.0 st
%Cpu6  : 59.7 us,  2.3 sy,  0.0 ni, 38.0 id,  0.0 wa,  0.0 hi,  0.0 si,  0.0 st
%Cpu7  : 60.1 us,  2.0 sy,  0.0 ni, 38.0 id,  0.0 wa,  0.0 hi,  0.0 si,  0.0 st
%Cpu8  : 59.5 us,  3.7 sy,  0.0 ni, 36.9 id,  0.0 wa,  0.0 hi,  0.0 si,  0.0 st
%Cpu9  : 60.1 us,  2.0 sy,  0.0 ni, 38.0 id,  0.0 wa,  0.0 hi,  0.0 si,  0.0 st
%Cpu10 : 60.1 us,  2.0 sy,  0.0 ni, 38.0 id,  0.0 wa,  0.0 hi,  0.0 si,  0.0 st
%Cpu11 : 60.1 us,  2.3 sy,  0.0 ni, 37.5 id,  0.0 wa,  0.0 hi,  0.0 si,  0.0 st
%Cpu12 : 60.4 us,  1.7 sy,  0.0 ni, 38.0 id,  0.0 wa,  0.0 hi,  0.0 si,  0.0 st
%Cpu13 : 59.6 us,  2.0 sy,  0.0 ni, 38.4 id,  0.0 wa,  0.0 hi,  0.0 si,  0.0 st
%Cpu14 : 59.2 us,  2.6 sy,  0.0 ni, 38.2 id,  0.0 wa,  0.0 hi,  0.0 si,  0.0 st
%Cpu15 : 60.3 us,  2.0 sy,  0.0 ni, 37.7 id,  0.0 wa,  0.0 hi,  0.0 si,  0.0 st
```

图 3-24 CPU 并行工作

2）HATP 的确很重要，避免了读写分离和复杂的架构。如图 3-25 所示的混合事务与分析，对于一亿级别的数据量，不到一秒即可完成查询。

3）In-Memory 在 sum 和 avg 下需要并行执行才能更好地发挥作用。

```
SQL> select count(*) from m;

  COUNT(*)
----------
 100000004

已用时间:  00: 00: 00.21

SQL> insert into m values (100000005,100000005,100000005,sysdate);
```

图 3-25 混合事务与分析

已创建 1 行。

已用时间: 00: 00: 00.09
SQL> commit;

提交完成。

已用时间: 00: 00: 00.00
SQL> select count(*) from m;

　COUNT(*)

 100000005
已用时间: 00: 00: 00.08

图 3-25　混合事务与分析（续）

3.3.4　一次疑似分区查询异常的故障分析

某个工作日，笔者发现 zabbix 监控显示某台数据库负载情况异常，如图 3-26 所示。

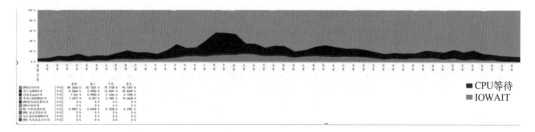

图 3-26　zabbix 监控 MySQL 数据库负载

从图中可以看到 CPU 等待和 IOWAIT 曲线不同程度偏高。虽然监控指标没有到达满负荷，但是作为一个专业人员，笔者很敏锐地发现这个指标不正常。IOWAIT CPU 在等待指令要求的磁盘 I/O 完成的这个时间段内，所有 CPU 处于空闲状态的时间百分比。意思就是：CPU 的进程在等待硬件资源（如磁盘读写）的响应，而此时因为大量磁盘读取没有完成，所以没有给 CPU 完成的信号，CPU 进程就继续在等磁盘读取结束的指令，这个过程被称为"不可中断睡眠"（Uninterruptible Sleep）的状态。在这种状态下，进程无法被任何信号中断，即使管理员尝试终止它也不行，其实是一种特殊形式的 CPU 空闲。在此 CPU 的等待队列上有线程在等待 IO 完成。

对于这个问题，笔者常用的分析思路是：对于 MySQL 数据库来说应该把慢日志打开，并且设置 100ms 的记录阈值（这个结合公司或企业的实际情况而定，个人建议设置为 100ms）。这样出现问题后可以用 pt-query-digest 这个工具对日志进行统计分析。具体命令是 pt-query-digest 待分析的日志文件名>/绝对路径/统计分析报名文件名。MySQL 统计慢日志如图 3-27 所示。通过统计分析报告发现，Query 2 这句 SQL 执行异常。

从早上 8 点 10 分至 8 点 41 分，总共执行 6 次，平均每次耗时 657s。进一步排查发现该 SQL 是对一张千万级数据量的表进行查询操作，该查询命中索引，又带有 limit 参数，从执行计划上看不出端倪，但是在生产中出现如此执行效率不正常。

```
# Query 2: 0.00 QPS, 2.10x concurrency, ID 0x4B5D42599F69D0D4444E3F28C58D4118 at byte 169490775
# This item is included in the report because it matches —limit.
# Scores: V/M = 42.39
# Time range: 2023-05-12T08:10:25 to 2023-05-12T08:41:40
# Attribute    pct    total    min    max    avg    95%    stddev   median
# ============ ===   ======  ======  ======  ======  ======  ======  ======
# Count         0       6
# Exec time    24    3942s    390s   865s    657s    833s    167s    759s
# Lock time     0      3ms   326us  856us   537us   839us   199us   633us
# Rows sent     0      10       0      2    1.67    1.96    0.73    1.96
# Rows examine  3   71.63M  11.92M 11.95M  11.94M  11.87M      0   11.87M
# Query size    0   32.70k   5.45k  5.45k   5.45k   5.45k      0    5.45k
# String:
# Hosts
# Users
```

图 3-27　MySQL 统计慢日志

SQL 语句简化并去除敏感变量后如图 3-28 所示。

```
select
    *
from temp
WHERE STATUS != '00'
    and   BATCH_ID=1005617185
    and   FLAG='32'
ORDER BY SECOND_ID DESC limit 0,10
```

图 3-28　脱敏慢 SQL

其中，BATCH_ID 列与 SECOND_ID 列都有常规索引，解释该 SQL 后可以看到执行计划中使用了 SECOND_ID 列的索引，但是实际查询却发现查询了近乎全表的数据，查询效率也非常低下，慢 SQL 的日志信息如图 3-29 所示。

```
# Time: 2023-05-12T08:10:25.314030+08:00
# User@Host:
# Query_time: 389.656238  Lock_time: 0.000856 Rows_sent: 2  Rows_examined: 12502207
```

图 3-29　慢 SQL 的日志信息

在前一晚发布变更时，为方便后续归档，对表进行过维护。在数据归档的同时，建立了以时间为键的分区表。此时其分区的动作便成了首要怀疑目标，查询引发大量扫描是不是因为分区与这种场景的查询冲突导致？

建立实验表进行模拟。如图 3-30 所示，建立 MySQL 分区表排查。

```
mysql> show create table xxg_p\G
*************************** 1. row ***************************
       Table: xxg_p
Create Table: CREATE TABLE `xxg_p` (
  `id` int NOT NULL AUTO_INCREMENT,
```

图 3-30　建立 MySQL 分区表排查

```
  `code` varchar(10) DEFAULT NULL,
  `n` int DEFAULT NULL,
  `time` datetime NOT NULL,
  PRIMARY KEY (`id`,`time`),
  KEY `code` (`code`),
  KEY `n` (`n`)
) ENGINE=InnoDB AUTO_INCREMENT=334281 DEFAULT CHARSET=utf8mb4 COLLATE=utf8mb4_0900_ai_ci
/*!50100 PARTITION BY RANGE (to_days(`time`))
(PARTITION p0 VALUES LESS THAN (738886) ENGINE = InnoDB,
 PARTITION p1 VALUES LESS THAN (738917) ENGINE = InnoDB,
 PARTITION p2 VALUES LESS THAN (738945) ENGINE = InnoDB,
 PARTITION p3 VALUES LESS THAN (738976) ENGINE = InnoDB,
 PARTITION p4 VALUES LESS THAN (739006) ENGINE = InnoDB,
 PARTITION p5 VALUES LESS THAN (739037) ENGINE = InnoDB,
 PARTITION p6 VALUES LESS THAN (739067) ENGINE = InnoDB,
 PARTITION p7 VALUES LESS THAN (739098) ENGINE = InnoDB,
 PARTITION p8 VALUES LESS THAN (739129) ENGINE = InnoDB) */
1 row in set (0.00 sec)
```

图 3-30 建立 MySQL 分区表排查(续)

表中模拟写入约 30 万行数据,其中 code 模拟生产中的 BATCH_ID 字段,有大量重复数据,n 模拟生产中的 SECOND_ID 字段,基本由唯一数据构成,同时为非主键索引。

对表进行相似的查询实验,分区表使用索引并且指明其他列倒序返回,如图 3-31 所示。

```
mysql> select * from xxg_p where code='A' order by n desc limit 10;
+-------+------+-------+---------------------+
| id    | code | n     | time                |
+-------+------+-------+---------------------+
| 20000 | A    | 19999 | 2023-01-14 21:19:00 |
| 19999 | A    | 19998 | 2023-01-14 21:18:00 |
| 19998 | A    | 19997 | 2023-01-14 21:17:00 |
| 19997 | A    | 19996 | 2023-01-14 21:16:00 |
| 19996 | A    | 19995 | 2023-01-14 21:15:00 |
| 19995 | A    | 19994 | 2023-01-14 21:14:00 |
| 19994 | A    | 19993 | 2023-01-14 21:13:00 |
| 19993 | A    | 19992 | 2023-01-14 21:12:00 |
| 19992 | A    | 19991 | 2023-01-14 21:11:00 |
| 19991 | A    | 19990 | 2023-01-14 21:10:00 |
+-------+------+-------+---------------------+
10 rows in set (1.41 sec)# Query_time: 0.002046  Lock_time: 0.000013 Rows_sent: 10  Rows_examined: 10
```

图 3-31 分区表使用索引并且指明其他列倒序返回

该 SQL 的执行计划是,使用索引并且指明其他列倒序返回的执行计划,如图 3-32 所示。

```
mysql> explain select * from xxg_p where code='A' order by n desc limit 10;
+----+-------------+-------+-------------------------+-------+---------------+------+---------+------+------+----------+------------------------------------+
| id | select_type | table | partitions              | type  | possible_keys | key  | key_len | ref  | rows | filtered | Extra                              |
+----+-------------+-------+-------------------------+-------+---------------+------+---------+------+------+----------+------------------------------------+
|  1 | SIMPLE      | xxg_p | p0,p1,p2,p3,p4,p5,p6,p7,p8 | index | code          | n    | 5       | NULL | 327  | 3.05     | Using where; Backward index scan   |
+----+-------------+-------+-------------------------+-------+---------------+------+---------+------+------+----------+------------------------------------+
1 row in set, 1 warning (0.00 sec)
```

图 3-32 使用索引并且指明其他列倒序返回的执行计划

可以看出完全是一个正常的查询,使用到了 code 列的索引。但是这个查询实际耗时约 1s,对于这个查询来说,耗时如此之多肯定是不正常的。经过检查后台的 MySQL 的慢日志(见图 3-33 所示),该 SQL 虽然只返回 10 行,但是扫描了 28 万余行。约等于全表扫描了。这样的扫描量,再加上看似没问题的执行计划,已经达到了模拟复现生产异常查询的目的了。

```
# Time: 2023-05-14T09:09:20.341905Z
# User@Host: root[root] @ localhost []  Id:   60
# Query_time: 1.407572  Lock_time: 0.000014 Rows_sent: 10  Rows_examined: 285300
SET timestamp=1684055358;
select * from xxg_p where code='A' order by n desc limit 10;
```

图 3-33　MySQL 慢日志截图

接下来做简单的控制变量的实验，基本就能定位问题了，首先建立一张结构一模一样的表，但是排除分区，看看是不是由分区造成的问题，表结构如图 3-34 所示，对照分区表建立的非分区表（NP 的含义是 nopartition）。

```
mysql> show create table xxg_np\G
*************************** 1. row ***************************
       Table: xxg_np
Create Table: CREATE TABLE `xxg_np` (
  `id` int NOT NULL AUTO_INCREMENT,
  `code` varchar(10) DEFAULT NULL,
  `n` int DEFAULT NULL,
  `time` datetime NOT NULL,
  PRIMARY KEY (`id`,`time`),
  KEY `ncode` (`code`),
  KEY `nn` (`n`)
) ENGINE=InnoDB AUTO_INCREMENT=501001 DEFAULT CHARSET=utf8mb4 COLLATE=utf8mb4_0900_ai_ci
1 row in set (0.00 sec)
```

图 3-34　对照分区表建立的非分区表

再次执行同样的查询语句，非分区表使用索引并且指明其他列倒序返回，如图 3-35 所示。

```
mysql> select * from xxg_np where code='A' order by n desc  limit 10;
+-------+------+-------+---------------------+
| id    | code | n     | time                |
+-------+------+-------+---------------------+
| 20000 | A    | 19999 | 2023-01-14 21:19:00 |
| 19999 | A    | 19998 | 2023-01-14 21:18:00 |
| 19998 | A    | 19997 | 2023-01-14 21:17:00 |
| 19997 | A    | 19996 | 2023-01-14 21:16:00 |
| 19996 | A    | 19995 | 2023-01-14 21:15:00 |
| 19995 | A    | 19994 | 2023-01-14 21:14:00 |
| 19994 | A    | 19993 | 2023-01-14 21:13:00 |
| 19993 | A    | 19992 | 2023-01-14 21:12:00 |
| 19992 | A    | 19991 | 2023-01-14 21:11:00 |
| 19991 | A    | 19990 | 2023-01-14 21:10:00 |
+-------+------+-------+---------------------+
10 rows in set (1.81 sec)
```

图 3-35　非分区表使用索引并且指明其他列倒序返回

可以看到 SQL 耗时将近 2s。从实验已经能够基本确定，这个异常查询的出现和分区表变更没有关系。可以排除是运维的问题。再看看后台日志验证，非分区表慢日志查询如图 3-36 所示。

```
# Time: 2023-05-14T12:07:54.495826Z
# User@Host: root[root] @ localhost []  Id:   62
# Query_time: 1.812342  Lock_time: 0.000011 Rows_sent: 10  Rows_examined: 460020
SET timestamp=1684066072;
select * from xxg_np where code='A' order by n desc  limit 10;
```

图 3-36　非分区表慢日志查询

从图 3-36 可以看出即使在不分区的表上，同样的查询有的效率也很低。继续通过对语句的排查检查到底是语句中的哪部分对查询效率造成了影响。

第一步去掉排序，观察执行查询效率。不带排序执行查询的效率如图 3-37 所示。

```
mysql> select * from xxg_p where code='A' limit 10;
+----+------+---+---------------------+
| id | code | n | time                |
+----+------+---+---------------------+
|  1 | A    | 0 | 2023-01-01 00:00:00 |
|  2 | A    | 1 | 2023-01-01 00:01:00 |
|  3 | A    | 2 | 2023-01-01 00:02:00 |
|  4 | A    | 3 | 2023-01-01 00:03:00 |
|  5 | A    | 4 | 2023-01-01 00:04:00 |
|  6 | A    | 5 | 2023-01-01 00:05:00 |
|  7 | A    | 6 | 2023-01-01 00:06:00 |
|  8 | A    | 7 | 2023-01-01 00:07:00 |
|  9 | A    | 8 | 2023-01-01 00:08:00 |
| 10 | A    | 9 | 2023-01-01 00:09:00 |
+----+------+---+---------------------+
10 rows in set (0.00 sec)

# Time: 2023-05-16T01:37:39.191634Z
# User@Host: root[root] @ localhost []  Id:    78
# Query_time: 0.005317  Lock_time: 0.000024 Rows_sent: 10  Rows_examined: 10
SET timestamp=1684201059;
select * from xxg_p where code='A' limit 10;
```

图 3-37 不带排序执行查询的效率

去掉排序后该 SQL 用时 5ms，扫描过程是扫描 10 行，返回了 10 行。采用排除法分析是这个排序引发了问题。通过控制变量法，想到了加上排序，但是换个排序字段。带主键排序执行效率如图 3-38 所示。

```
mysql> select * from xxg_p where code='A' order by id desc limit 10;
+--------+------+-----+---------------------+
| id     | code | n   | time                |
+--------+------+-----+---------------------+
| 334280 | A    | 999 | 2023-05-15 00:00:00 |
| 334279 | A    | 998 | 2023-05-15 00:00:00 |
| 334278 | A    | 997 | 2023-05-15 00:00:00 |
| 334277 | A    | 996 | 2023-05-15 00:00:00 |
| 334276 | A    | 995 | 2023-05-15 00:00:00 |
| 334275 | A    | 994 | 2023-05-15 00:00:00 |
| 334274 | A    | 993 | 2023-05-15 00:00:00 |
| 334273 | A    | 992 | 2023-05-15 00:00:00 |
| 334272 | A    | 991 | 2023-05-15 00:00:00 |
| 334271 | A    | 990 | 2023-05-15 00:00:00 |
+--------+------+-----+---------------------+
10 rows in set (0.00 sec)

# Time: 2023-05-16T01:40:08.077441Z
# User@Host: root[root] @ localhost []  Id:    78
# Query_time: 0.001889  Lock_time: 0.000012 Rows_sent: 10  Rows_examined: 10
SET timestamp=1684201208;
select * from xxg_p where code='A' order by id desc limit 10;
```

图 3-38 带主键排序执行效率

从图 3-38 中也看到执行结果，耗时不到 2ms。那这两种排序的区别在哪里？分别看看执行计划。不同排序 SQL 的执行计划对比如图 3-39 所示。

```
mysql> explain select * from xxg_p where code='A' order by id desc limit 10;
+----+-------------+-------+-----------------------+------+---------------+------+---------+-------+--------+----------+---------------------+
| id | select_type | table | partitions            | type | possible_keys | key  | key_len | ref   | rows   | filtered | Extra               |
+----+-------------+-------+-----------------------+------+---------------+------+---------+-------+--------+----------+---------------------+
|  1 | SIMPLE      | xxg_p | p0,p1,p2,p3,p4,p5,p6,p7,p8 | ref  | code          | code | 43      | const | 167187 |   100.00 | Backward index scan |
+----+-------------+-------+-----------------------+------+---------------+------+---------+-------+--------+----------+---------------------+
1 row in set, 1 warning (0.00 sec)

mysql> explain select * from xxg_p where code='A' order by n desc limit 10;
+----+-------------+-------+-----------------------+-------+---------------+------+---------+------+------+----------+------------------------------------+
| id | select_type | table | partitions            | type  | possible_keys | key  | key_len | ref  | rows | filtered | Extra                              |
+----+-------------+-------+-----------------------+-------+---------------+------+---------+------+------+----------+------------------------------------+
|  1 | SIMPLE      | xxg_p | p0,p1,p2,p3,p4,p5,p6,p7,p8 | index | code          | n    | 5       | NULL |   20 |    50.00 | Using where; Backward index scan   |
+----+-------------+-------+-----------------------+-------+---------------+------+---------+------+------+----------+------------------------------------+
1 row in set, 1 warning (0.00 sec)
```

图 3-39　不同排序 SQL 的执行计划对比

从图中可以看到两种查询使用到了不同的索引，由此猜测是排序使 MySQL 选择使用不同的索引，从而导致了效率问题。

如果排序使用的列是无索引的列呢？按常识应该效率低于有索引的情况。排序无索引的列如图 3-40 所示，删除 n 列上的索引。

```
mysql> alter table xxg_p drop index n;
Query OK, 0 rows affected (0.09 sec)
Records: 0  Duplicates: 0  Warnings: 0

mysql> select * from xxg_p where code='A' order by n desc limit 10;
+--------+------+-------+---------------------+
| id     | code | n     | time                |
+--------+------+-------+---------------------+
| 670280 | A    | 19999 | 2023-05-29 00:00:00 |
| 670279 | A    | 19998 | 2023-05-29 00:00:00 |
| 670278 | A    | 19997 | 2023-05-29 00:00:00 |
| 670277 | A    | 19996 | 2023-05-29 00:00:00 |
| 670276 | A    | 19995 | 2023-05-29 00:00:00 |
| 670275 | A    | 19994 | 2023-05-29 00:00:00 |
| 670274 | A    | 19993 | 2023-05-29 00:00:00 |
| 670273 | A    | 19992 | 2023-05-29 00:00:00 |
| 670272 | A    | 19991 | 2023-05-29 00:00:00 |
| 670271 | A    | 19990 | 2023-05-29 00:00:00 |
+--------+------+-------+---------------------+
10 rows in set (0.13 sec)
```

图 3-40　排序无索引的列

从实际实验的结果来看，是有点出乎意料的。在去除索引后，反而得到了极高的效率提升。从该 SQL 执行的日志，指定无索引排序日志，如图 3-41 所示。

```
# Time: 2023-05-16T02:51:42.296493Z
# User@Host: root[root] @ localhost []  Id:    79
# Query_time: 0.126978  Lock_time: 0.000015 Rows_sent: 10  Rows_examined: 20010
SET timestamp=1684205502;
select * from xxg_p where code='A' order by n desc limit 10;
```

图 3-41　指定无索引排序日志

从图中看到 SQL 的扫描行数已经明显减少了,再看该 SQL 的执行计划,如图 3-42 所示。

```
mysql> explain select * from xxg_p where code='A' order by n desc limit 10;
+----+-------------+-------+-------------------------+------+---------------+------+---------+-------+-------+----------+----------------+
| id | select_type | table | partitions              | type | possible_keys | key  | key_len | ref   | rows  | filtered | Extra          |
+----+-------------+-------+-------------------------+------+---------------+------+---------+-------+-------+----------+----------------+
|  1 | SIMPLE      | xxg_p | p0,p1,p2,p3,p4,p5,p6,p7,p8 | ref | code         | code | 43      | const | 37274 |   100.00 | Using filesort |
+----+-------------+-------+-------------------------+------+---------------+------+---------+-------+-------+----------+----------------+
1 row in set, 1 warning (0.00 sec)
```

图 3-42 指定无索引排序执行计划

看到 SQL 用上了 code 索引,根据 code 索引得到 20000 行,又根据排序得到 10 行,总共扫描 20010 行,最终返回 10 行。该执行计划与使用 ID 排序是相同的执行计划。即使扫描了 code 字段为相同值的所有数据,但也远远优于全表扫描。

将索引再加回,进行正序排序查询实验,用来验证该问题是不是由于倒序引发的问题。重建索引后执行计划和日志如图 3-43 所示。

通过以上信息可以看到,在正序排序的情况下,即使使用了该排序字段的索引也能够进行高效率查询。

那查到这里,这个问题已经基本可以定位了。在查询中,如果对大量结果进行倒序排序,同时排序规则为非查询条件中的索引字段,就会出现查询扫描了接近全表数据,查询效率极差的情况。

这是因为倒序排序时 MySQL 选择了效率极低的排序字段索引所致,可以认为是 MySQL 优化器的选择不够最优的极端案例。

```
mysql> alter table xxg_p add index n (n);
Query OK, 0 rows affected (2.29 sec)
Records: 0  Duplicates: 0  Warnings: 0

mysql> explain select * from xxg_p where code='A' order by n asc limit 10;
+----+-------------+-------+-------------------------+-------+---------------+------+---------+------+------+----------+-------------+
| id | select_type | table | partitions              | type  | possible_keys | key  | key_len | ref  | rows | filtered | Extra       |
+----+-------------+-------+-------------------------+-------+---------------+------+---------+------+------+----------+-------------+
|  1 | SIMPLE      | xxg_p | p0,p1,p2,p3,p4,p5,p6,p7,p8 | index | code         | n    | 5       | NULL |   95 |    10.49 | Using where |
+----+-------------+-------+-------------------------+-------+---------------+------+---------+------+------+----------+-------------+
1 row in set, 1 warning (0.00 sec)

mysql> select * from xxg_p where code='A' order by n asc limit 10;
+--------+------+---+---------------------+
| id     | code | n | time                |
+--------+------+---+---------------------+
| 650281 | A    | 0 | 2023-05-15 00:00:00 |
| 650282 | A    | 1 | 2023-05-15 00:00:00 |
| 650283 | A    | 2 | 2023-05-15 00:00:00 |
| 650284 | A    | 3 | 2023-05-15 00:00:00 |
| 650285 | A    | 4 | 2023-05-15 00:00:00 |
| 650286 | A    | 5 | 2023-05-15 00:00:00 |
| 650287 | A    | 6 | 2023-05-15 00:00:00 |
| 650288 | A    | 7 | 2023-05-15 00:00:00 |
| 650289 | A    | 8 | 2023-05-15 00:00:00 |
| 650290 | A    | 9 | 2023-05-15 00:00:00 |
+--------+------+---+---------------------+
10 rows in set (0.01 sec)

# Time: 2023-05-16T05:12:33.506377Z
# User@Host: root[root] @ localhost []  Id:    87
# Query_time: 0.002298  Lock_time: 0.000012 Rows_sent: 10  Rows_examined: 68
SET timestamp=1684213953;
select * from xxg_p where code='A' order by n asc limit 10;
```

图 3-43 重建索引后执行计划和日志

知道了原因,解决方案也就应运而生。首先尝试对排序字段建立相应的倒序索引,如图 3-44 所示。

```
mysql> alter table xxg_p add index nd(n desc);
Query OK, 0 rows affected (2.99 sec)
Records: 0  Duplicates: 0  Warnings: 0

mysql> explain select * from xxg_p where code='A' order by n desc limit 10;
+----+-------------+-------+----------------------------+-------+---------------+------+---------+------+------+----------+-------------+
| id | select_type | table | partitions                 | type  | possible_keys | key  | key_len | ref  | rows | filtered | Extra       |
+----+-------------+-------+----------------------------+-------+---------------+------+---------+------+------+----------+-------------+
|  1 | SIMPLE      | xxg_p | p0,p1,p2,p3,p4,p5,p6,p7,p8 | index | code          | nd   | 5       | NULL |  909 |    10.44 | Using where |
+----+-------------+-------+----------------------------+-------+---------------+------+---------+------+------+----------+-------------+
1 row in set, 1 warning (0.00 sec)

mysql> select * from xxg_p where code='A' order by n desc limit 10;
+--------+------+-------+---------------------+
| id     | code | n     | time                |
+--------+------+-------+---------------------+
| 670280 | A    | 19999 | 2023-05-29 00:00:00 |
| 670279 | A    | 19998 | 2023-05-29 00:00:00 |
| 670278 | A    | 19997 | 2023-05-29 00:00:00 |
| 670277 | A    | 19996 | 2023-05-29 00:00:00 |
| 670276 | A    | 19995 | 2023-05-29 00:00:00 |
| 670275 | A    | 19994 | 2023-05-29 00:00:00 |
| 670274 | A    | 19993 | 2023-05-29 00:00:00 |
| 670273 | A    | 19992 | 2023-05-29 00:00:00 |
| 670272 | A    | 19991 | 2023-05-29 00:00:00 |
| 670271 | A    | 19990 | 2023-05-29 00:00:00 |
+--------+------+-------+---------------------+
10 rows in set (1.58 sec)

# Time: 2023-05-16T05:46:15.043523Z
# User@Host: root[root] @ localhost []  Id:    87
# Query_time: 1.577126  Lock_time: 0.000013 Rows_sent: 10  Rows_examined: 260038
SET timestamp=1684215973;
select * from xxg_p where code='A' order by n desc limit 10;
```

图 3-44　对指定排序字段建立倒序索引

从实验结果来看，即使使用到了倒序索引，仍然无法提升效率。

接下来采用建立组合（复合）索引的方式，同时包含条件列和排序列。谓词列和排序列都建立组合索引，执行效率如图 3-45 所示。

```
mysql> alter table xxg_p add index cn(code,n);
Query OK, 0 rows affected (4.15 sec)
Records: 0  Duplicates: 0  Warnings: 0

mysql> explain select * from xxg_p where code='A' order by n desc limit 10;
+----+-------------+-------+----------------------------+------+---------------+------+---------+-------+-------+----------+-----------------------------------+
| id | select_type | table | partitions                 | type | possible_keys | key  | key_len | ref   | rows  | filtered | Extra                             |
+----+-------------+-------+----------------------------+------+---------------+------+---------+-------+-------+----------+-----------------------------------+
|  1 | SIMPLE      | xxg_p | p0,p1,p2,p3,p4,p5,p6,p7,p8 | ref  | code,cn       | cn   | 43      | const | 39594 |   100.00 | Backward index scan; Using index  |
+----+-------------+-------+----------------------------+------+---------------+------+---------+-------+-------+----------+-----------------------------------+
1 row in set, 1 warning (0.00 sec)

mysql> select * from xxg_p where code='A' order by n desc limit 10;
+--------+------+-------+---------------------+
| id     | code | n     | time                |
+--------+------+-------+---------------------+
| 670280 | A    | 19999 | 2023-05-29 00:00:00 |
| 670279 | A    | 19998 | 2023-05-29 00:00:00 |
| 670278 | A    | 19997 | 2023-05-29 00:00:00 |
| 670277 | A    | 19996 | 2023-05-29 00:00:00 |
| 670276 | A    | 19995 | 2023-05-29 00:00:00 |
| 670275 | A    | 19994 | 2023-05-29 00:00:00 |
| 670274 | A    | 19993 | 2023-05-29 00:00:00 |
| 670273 | A    | 19992 | 2023-05-29 00:00:00 |
| 670272 | A    | 19991 | 2023-05-29 00:00:00 |
| 670271 | A    | 19990 | 2023-05-29 00:00:00 |
+--------+------+-------+---------------------+
10 rows in set (0.00 sec)

# Time: 2023-05-16T05:54:32.071635Z
# User@Host: root[root] @ localhost []  Id:    87
# Query_time: 0.001352  Lock_time: 0.000012 Rows_sent: 10  Rows_examined: 10
SET timestamp=1684216472;
select * from xxg_p where code='A' order by n desc limit 10;
```

图 3-45　谓词列和排序列都建立组合索引

这个组合索引的建立使得执行达到了预期效果，即使是倒序查询的情况下，依然能够做到扫描 10 条就输出 10 条的最优效率，用空间换来了时间。

至此，这一次生产中出现的故障就已经算是完美解决了。

对本次故障分析进行一下复盘。

在数据极度不均衡的场景下，通过索引查询，查找到数据倾斜较大的数据范围，同时又指定其他索引且不是主键的倒序排序，MySQL 的优化器不能很好地处理，会引发较严重的性能问题。

同样的场景，如果换作 Oracle——拥有可以说是业界最强优化器的数据库，是不是能够完美处理？

在 Oracle 中建立相同结构的测试表，插入相同数据，也建立相同索引。Oracle 数据倾斜指定非索引列倒序排序执行结果如图 3-46 所示。

```
SQL> select * from xxg_auto_part where code='A' order by n desc FETCH FIRST 10 ROWS ONLY;

    ID CODE              N TIME
---------- ---------- ---------- ---------------
     20000 A              20000 14-1月 -23
     19999 A              19999 14-1月 -23
     19998 A              19998 14-1月 -23
     19997 A              19997 14-1月 -23
     19996 A              19996 14-1月 -23
     19995 A              19995 14-1月 -23
     19994 A              19994 14-1月 -23
     19993 A              19993 14-1月 -23
     19992 A              19992 14-1月 -23
     19991 A              19991 14-1月 -23

已选择 10 行。

已用时间：  00: 00: 00.04
SQL>
```

图 3-46　Oracle 数据倾斜指定非索引列倒序排序执行结果

执行 SQL 查询耗时 40ms，没有遇到 MySQL 上的错误选择索引的问题。

通过查看如图 3-47 所示的数据倾斜指定非索引列倒序排序执行计划。看到了在 Oracle 数据库中，该 SQL 查询仅仅使用了 code 列的索引，估算数据 19075 行与实际的 20000 行基本符合。

```
SQL> explain plan for select * from xxg_auto_part where code='A' order by n desc FETCH FIRST 10 ROWS ONLY;

已解释。

已用时间：  00: 00: 00.05
SQL> SELECT * FROM TABLE(DBMS_XPLAN.DISPLAY);

PLAN_TABLE_OUTPUT
--------------------------------------------------------------------------------
Plan hash value: 1728078835

--------------------------------------------------------------------------------
| Id  | Operation                                  | Name          | Rows  | Bytes | Cost (%CPU)| Time     | Pstart| Pstop |
--------------------------------------------------------------------------------
|   0 | SELECT STATEMENT                           |               |    10 |   680 |   148   (2)| 00:00:01 |       |       |
|*  1 |  VIEW                                      |               |    10 |   680 |   148   (2)| 00:00:01 |       |       |
|*  2 |   WINDOW SORT PUSHED RANK                  |               | 19075 |  465K |   148   (2)| 00:00:01 |       |       |
|   3 |    PARTITION RANGE ALL                     |               | 19075 |  465K |   146   (0)| 00:00:01 |     1 |1048575|
|   4 |     TABLE ACCESS BY LOCAL INDEX ROWID BATCHED| XXG_AUTO_PART | 19075 |  465K |   146   (0)| 00:00:01 |     1 |1048575|
|*  5 |      INDEX RANGE SCAN                      | PCODE         | 19075 |       |    61   (0)| 00:00:01 |     1 |1048575|
--------------------------------------------------------------------------------
```

图 3-47　数据倾斜指定非索引列倒序排序执行计划

本案例最初着手点是怀疑数据库的分区导致性能下降继而引起故障，但是通过对比分区和非分区的结果发现故障并非由分区导致，险些误判。最终通过实验复现是由于特殊的数据倾斜产生的场景导致。在如此极端的数据倾斜场景下，Oracle 的优化器和 MySQL 的优化器做出了不一样的优化方式。这也再次给了大家一个启示,任何数据库的转换并不是说语法兼容就可以了,语法兼容只是表明上看得到的,还有很多是深层次看不到的。

3.3.5 一次数据库归档导致的故障分析

各种数据库如果需要搭建高可用,那么必然需要开启归档。无论是 Oracle 的 archivelog,还是 MySQL 的 binlog 或是 PostgreSQL 的归档日志,只要开启归档,必然涉及归档目录或归档空间满。一旦发生此问题,数据库从机制上来说就会陷入写入失败的情况。

数据库归档空间满,可以说是一个低级但是又经常遇到的问题。解决的方式只有一个:就是加强监控。从运维角度,有经验的 DBA 都会给归档日志设置保留时间。比如在 Oracle 数据库的主库上有 RMAN,可以删除已经传输完毕的脚本。在 PostgreSQL 数据库的主库上有专门的 pg_archivecleanup,MySQL 数据库可以通过设置如图 3-48 所示的 Binlog 过期参数,来确保日志被及时清理。

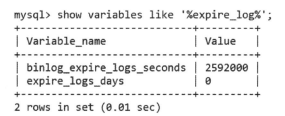

图 3-48　Binlog 过期参数

有了这些脚本和参数设定,是不是就可以高枕无忧了?答案是否定的。不止一次故障中得到的教训就是,一定要对开发人员的工作进行审核(虽然很多企业没有这个环境也没有这个观念)。

在笔者从业的各个阶段都遇到过明明已经预估了日志的增加速度,也有脚本定期删除过期的归档日志,但仍然发生日志占满磁盘空间的情况。Oracle 的脚本中若传输至从库的日志已经完成应用了,那么这些日志完成了它的使命,可以删除了。同样,PostgreSQL 和 MySQL 的归档日志在删除前也应该检查从库的应用情况。但是生产中还是会发生由于开发的功能上线前没有审核,本来应该执行更新一条数据,如 Updatetablesetcolumn=xxxwhereid=? 这样的 SQL 在编写的时候漏掉了 where 条件,导致本来更新一行,现在更新全表。而原本的逻辑是循环更新 1000 次,共更新 1000 条。那么现在全表是 100 万条数据,实际更新 10 亿条。虽然数据只是被重复不停地更新,数据量没有变化,但是这些更新全部都落地到归档日志中传输到从库,从库也必然需要重放 1000 次的 100 万条更新。这个期间归档日志的增加速度远超过预先设定的定期删除速度,使得数据库归档空间很快用完了。

关系数据库都是遵循日志先写的原则,一旦提交,一般来说就会要写入归档日志,有的数据库可以配置不是每次提交都写日志,但是不推荐。而且数据库的日志都是顺序性写入。空间不足(可能是分配的空间不足或者是磁盘空间耗尽)最终导致归档日志无法写入。所以最终所有数据库的写入操作都被卡在了提交阶段。以 Oracle 为例,出现 ORA-00257: archiver error。

这种场景处理起来较为容易，只要删除归档日志即可。

Oracle 的处理方式是：事先就应该完成编辑一个脚本 deleteachive.sh，然后设置成 Linux 操作系统的定时任务。以下脚本为删除一个小时以前的归档日志。

```
[root@adg1 oracle]# cat deleteachive.sh

su - oracle -c "rman target / <<EOF
crosscheck archivelogall;
delete nopromptarchivelog all completed before 'sysdate-1/24';
crosscheck backup;
report obsolete;
DELETE OBSOLETE;"
```

设置每 4 小时清理一次

```
[root@adg1 oracle]# crontab -l
0 */4 * * * /home/oracle/deleteachive.sh >> /home/oracle/recorder.log 2>&1
```

由于处理故障优先，无法等到定时任务执行清理。故需要手工清理 Oracle 归档日志，直接在命令行执行/home/oracle/deleteachive.sh 进行清理，如图 3-49 所示。

```
[root@adg1 oracle]# /home/oracle/deleteachive.sh

恢复管理器: Release 19.0.0.0.0 - Production on 星期四 6月 8 20:46:28 2023
Version 19.3.0.0.0

Copyright (c) 1982, 2019, Oracle and/or its affiliates.  All rights reserved.

已连接到目标数据库: O19C (DBID=2893056646)

RMAN>
使用目标数据库控制文件替代恢复目录
分配的通道: ORA_DISK_1
通道 ORA_DISK_1: SID=27 设备类型=DISK
对归档日志的验证成功
归档日志文件名 = /u01/app/oracle/recovery_area/O19C/1_2042_1106770120.dbf RECID=3816 STAMP=1138993974
已交叉检验的 1 对象


RMAN>
释放的通道: ORA_DISK_1
分配的通道: ORA_DISK_1
通道 ORA_DISK_1: SID=27 设备类型=DISK
db_unique_name 为 O19C1 的数据库的归档日志副本列表
=================================================================
关键字    线程序列    S 时间下限
-------  ----  -------  -  ------------------
3816     1     2042      A 2023:06:0813:34:02
         名称: /u01/app/oracle/recovery_area/O19C/1_2042_1106770120.dbf

已删除的归档日志
归档日志文件名 = /u01/app/oracle/recovery_area/O19C/1_2042_1106770120.dbf RECID=3816 STAMP=1138993974
1 对象已删除
```

图 3-49　手工清理 Oracle 归档日志

这样数据库就可以继续顺序地写入日志，优先缓解线上的压力。其实类似的操作在其他数据库也是一样的。MySQL 日志保留策略如图 3-50 所示，expire_logs_days 是保存多少天，达到时间，数据库即可自动清理。0 表示无限制存储。到了 MySQL 8 新增了 binlog_expire_logs_seconds 参数，定义到了秒。

```
mysql> show variables like '%expire_logs%';
+----------------------------+---------+
| Variable_name              | Value   |
+----------------------------+---------+
| binlog_expire_logs_seconds | 2592000 |
| expire_logs_days           | 0       |
+----------------------------+---------+
2 rows in set (0.01 sec)
```

图 3-50 MySQL 日志保留策略

MySQL 数据库遇到类似情况的处理方式是：直接清理 binlog。先登录数据库获得当前 binlog 的编号。MySQL 二进制日志编号获取如图 3-51 所示，可以看到当前编号是 20，在数据目录下的确有 18、19 和 20 号三个日志。

```
mysql> show master status\G
*************************** 1. row ***************************
             File: mysql-bin.000020
         Position: 94795263
     Binlog_Do_DB:
 Binlog_Ignore_DB:
Executed_Gtid_Set: 41d10bcf-8f25-11ed-b604-005056964c25:1,
5d90d84e-8f26-11ed-bcb2-005056964c25:1,
913c87ec-8f24-11ed-b349-005056964c25:1-6,
a1ee0f84-8f25-11ed-988d-005056964c25:1,
bca44c90-8f26-11ed-a1b0-005056964c25:1-100618
1 row in set (0.00 sec)


-rw-r----- 1 mysql mysql     12582912 Jun  5 14:54 ibdata1
-rw-r----- 1 mysql mysql     50331648 Jun  5 14:54 ib_logfile0
-rw-r----- 1 mysql mysql     50331648 Jun  2 17:36 ib_logfile1
-rw-r----- 1 mysql mysql     12582912 May 28 11:14 ibtmp1
drwxr-x--- 2 mysql mysql         4096 Apr 27 13:40 innodb_memcache
drwxr-x--- 2 mysql mysql         4096 May 28 11:13 #innodb_temp
drwxr-x--- 2 mysql mysql         4096 Jun  5 14:53 m
drwxr-x--- 2 mysql mysql         4096 Jan  8 15:33 mysql
-rw-r----- 1 mysql mysql          954 May 28 11:13 mysql-bin.000018
-rw-r----- 1 mysql mysql   2120012452 May 31 20:07 mysql-bin.000019
-rw-r----- 1 mysql mysql     94795263 Jun  5 14:54 mysql-bin.000020
-rw-r----- 1 mysql mysql           57 May 31 20:07 mysql-bin.index
-rw-r----- 1 mysql mysql          176 May 28 11:12 mysqld-auto.cnf
```

图 3-51 MySQL 二进制日志编号获取

执行清理日志命令，清理 MySQL 二进制日志命令如图 3-52 所示，编号 20 之前的日志命令全部被清空了。

```
mysql> PURGE MASTER LOGS TO 'mysql-bin.000020';
Query OK, 0 rows affected (0.15 sec)


-rw-r----- 1 mysql mysql  50331648 Jun  5 14:54 ib_logfile0
-rw-r----- 1 mysql mysql  50331648 Jun  2 17:36 ib_logfile1
-rw-r----- 1 mysql mysql  12582912 May 28 11:14 ibtmp1
drwxr-x--- 2 mysql mysql      4096 Apr 27 13:40 innodb_memcache
drwxr-x--- 2 mysql mysql      4096 May 28 11:13 #innodb_temp
drwxr-x--- 2 mysql mysql      4096 Jun  5 14:53 m
drwxr-x--- 2 mysql mysql      4096 Jan  8 15:33 mysql
-rw-r----- 1 mysql mysql  94795263 Jun  5 14:54 mysql-bin.000020
-rw-r----- 1 mysql mysql        19 Jun  8 20:43 mysql-bin.index
-rw-r----- 1 mysql mysql       176 May 28 11:12 mysqld-auto.cnf
-rw-r----- 1 mysql mysql  26214400 Jun  5 14:54 mysql.ibd
```

图 3-52　清理 MySQL 二进制日志命令

PostgreSQL 的处理方式是直接清理归档日志，命令如图 3-53 所示，切换到当前归档日志目录 /pgdata/pg14/pg_root/pg_wal。找到当前日志较新的一个日志名，执行 PG 自带归档日志清理程序 pg_archivecleanup 执行清理，注意 pg_archivecleanup 后面有一个 . 表示当前路径。清理后类似于 MySQL，该日志之前的归档日志都被清理了。

```
-rw------- 1 pg14 pg14 16M May 31 20:12 000000010000000200000001F
-rw------- 1 pg14 pg14 16M May 31 20:12 0000000100000002000000020
-rw------- 1 pg14 pg14 16M May 31 20:13 0000000100000002000000021
-rw------- 1 pg14 pg14 16M May 31 20:13 0000000100000002000000022
-rw------- 1 pg14 pg14 16M May 31 20:13 0000000100000002000000023
-rw------- 1 pg14 pg14 16M May 31 20:14 0000000100000002000000024
-rw------- 1 pg14 pg14 16M May 31 20:14 0000000100000002000000025
-rw------- 1 pg14 pg14 16M May 31 20:14 0000000100000002000000026
-rw------- 1 pg14 pg14 16M May 31 20:12 0000000100000002000000027
-rw------- 1 pg14 pg14 16M May 31 20:12 0000000100000002000000028
-rw------- 1 pg14 pg14 16M May 31 20:12 0000000100000002000000029
-rw------- 1 pg14 pg14 16M May 31 20:13 000000010000000200000002A
-rw------- 1 pg14 pg14 16M May 31 20:13 000000010000000200000002B
-rw------- 1 pg14 pg14 16M May 31 20:14 000000010000000200000002C
-rw------- 1 pg14 pg14 16M May 31 20:14 000000010000000200000002D
-rw------- 1 pg14 pg14 16M May 31 20:15 000000010000000200000002E
drwx------ 2 pg14 pg14   6 May 19  2022 archive_status
[pg14@MySQL8M pg_wal]$ pwd
/pgdata/pg14/pg_root/pg_wal

[pg14@MySQL8M pg_wal]$ pg_archivecleanup . 000000010000000200000002D
[pg14@MySQL8M pg_wal]$ ll
total 32M
-rw------- 1 pg14 pg14 16M May 31 20:14 000000010000000200000002D
-rw------- 1 pg14 pg14 16M May 31 20:15 000000010000000200000002E
drwx------ 2 pg14 pg14   6 May 19  2022 archive_status
```

图 3-53　PostgreSQL 清理归档日志

以上是不同数据库的归档清理方式，用于紧急处理故障。当然最好的方式是加强监控，避免故障来临时手忙脚乱。切记，以上都是通过数据库命令进行日志删除，不能在操作系统上直接用 rm 的命令删除，否则会带来各种各样的问题。

3.3.6 一次数据库 binlog 写入失败的故障分析

多年前有个朋友发给笔者一份故障报告，请笔者帮他看看问题分析是否到位。报告内容是这样的：该数据库服务器中 tmp 文件系统内历史遗留文件较多，文件系统总容量空间为 4GB，剩余空间少于 1GB。该数据库由于 tmp 文件空间不足无法生成新的 binlog，故而发生重启。短时间重启过程中，MHA 判断主库故障，控制进行主从切换。应对措施是，紧急对/tmp 进行扩容，将/tmp 下的文件清理。

通过以上描述，读者朋友应该能知道这是一个 MySQL 数据库的故障。最初笔者对故障报告的理解是 binlog 放在了 tmp 目录下。感觉十分奇怪，因为一般来说没有这种做法。而且后面的应对措施中解决方式让笔者觉得好像就是日志文件放在了 tmp 目录下。于是与对方又沟通了一次，了解到其表达的意思是因为 tmp 目录满了，无法在默认目录下生成 binlog，而不是把 binlog 放在了 tmp 下。而且运维人员也做了实验，tmp 满了依然可以在默认目录产生新的日志文件。所以以上报告可以说在表述上有严重问题，而且也没搞清楚问题产生的根本原因。

于是笔者远程检查了出问题前后数据库的健康状况，发现很不乐观。SQL 问题较大，就在出问题前还发生过近 2000 万数据的全表遍历，非常消耗资源。但是这也不是产生问题的原因。

笔者通过查阅 MySQL 的技术文档有所收获。MySQL 的体系设定了在大事务（默认数据量 32KB 变化）情况下，日志不是直接写入日志文件，而是先放在 tmp 下临时汇总一下。等提交后合并到默认的 binlog。所以在 error 中出现的无法生成 binlog 其实是无法生成临时 binlog，而不是无法生成正式 binlog 的意思。而且一旦完成合并，立即删除临时 binlog，所以不好重现。而 tmp 目录满了以后，临时 binlog 生成不了了，所以出现了故障。那么数据库进行实例的关闭，这是设计初衷，不是漏洞。而由于后台有 mysqlsafe 的守护进程，使得数据库退出后再次触发它启动，所以表现形式为数据库重启。

下面笔者做了个复现的实验让大家了解一下这个过程。首先，如图 3-54 所示，反复执行 SQL：insert into m select * from m；这样这个 m 表就以 2 倍的速度不断递增，数据量达到一定程度形成大事务。

```
mysql> insert into m select * from m;
Query OK, 1572864 rows affected (15.23 sec)
Records: 1572864  Duplicates: 0  Warnings: 0
```

图 3-54 反复执行 SQL 模拟大事务

开始执行插入时很快。但是当达到 70 多万条数据，再次翻倍写入的时候，随着数据写入的增多就开发变慢。执行一次要 15s。而这个时候去检查/tmp 目录就可以看到如图 3-55 所示 binlog 产生了临时文件，两个带着#的文件。随着 insert 的结束，事务结束，这两个文件很快消失了。

```
srwxrwxrwx  1 mysql mysql           0 May 16 15:01 mysql3307.sock
srwxrwxrwx  1 mysql mysql           0 May 16 15:01 mysql3308.sock
srwxrwxrwx  1 mysql mysql           0 May 16 15:01 mysql3309.sock
srwxrwxrwx  1 mysql mysql           0 May 16 15:01 mysql3310.sock
drwx------  2 mysql mysql        4096 Mar 28 14:17 orbit-gdm
drwx------. 2 mysql mysql        4096 Feb  5 18:10 pulse-B3IOzhjmB2dd
drwx------  2 mysql mysql        4096 Mar 28 14:17 pulse-z0VpfhfBFewd
-rw-rw----  1 mysql mysql    11010048 May 19 18:59 #sql_7e45_0.MYD
-rw-rw----  1 mysql mysql        1024 May 19 18:59 #sql_7e45_0.MYI
drwx------  2 root  root         4096 May 19 14:17 vmware-root
[root@base tmp]# ll
total 20
drwx------. 2 mysql mysql 4096 Dec  7  2015 keyring-BhfGbX
srwxrwxrwx  1 mysql mysql    0 May 16 15:01 mysql3307.sock
srwxrwxrwx  1 mysql mysql    0 May 16 15:01 mysql3308.sock
srwxrwxrwx  1 mysql mysql    0 May 16 15:01 mysql3309.sock
srwxrwxrwx  1 mysql mysql    0 May 16 15:01 mysql3310.sock
drwx------  2 mysql mysql 4096 Mar 28 14:17 orbit-gdm
drwx------. 2 mysql mysql 4096 Feb  5 18:10 pulse-B3IOzhjmB2dd
drwx------  2 mysql mysql 4096 Mar 28 14:17 pulse-z0VpfhfBFewd
drwx------  2 root  root  4096 May 19 14:17 vmware-root
```

图 3-55 binlog 产生临时文件

通过以上的实验（当时数据库版本是 MySQL 5.7）过程发现，事务会临时在 /tmp 下产生文件，如果 /tmp 目录满了，的确会影响到 binlog 的生成。之前模拟到 tmp 满了依然可以在默认目录产生新的日志文件，是因为写入少量数据是不会产生临时文件的。产生临时文件和一个参数有关，binlog_cache_size，该参数默认 32KB。

Innodb 对于未提交事务产生的二进制日志先缓存，事务提交后写入到二进制文件，binlog_cache_size 基于会话，所以不能得设置太大，但是设置得太小，如果事务的记录大于 binlog_cache_size 会把缓存中的日志写入到临时文件，会降低性能，所以要小心设置。binlog 的 cache 和使用情况如图 3-56 所示，没有达到 binlog_cache_size 的小事务就不会用到磁盘，仅仅 Binlog_cache_use 会增加，表示使用缓冲的次数。当达到一定程度就会有落盘的动作。可以通过这些状态值评估数据库当前运行的状况。

```
mysql> show status like '%binlog_cache%';
+-----------------------+-------+
| Variable_name         | Value |
+-----------------------+-------+
| Binlog_cache_disk_use | 14    |
| Binlog_cache_use      | 107   |
+-----------------------+-------+
2 rows in set (0.00 sec)

mysql> insert into h select * from h;
Query OK, 640 rows affected (0.02 sec)
Records: 640  Duplicates: 0  Warnings: 0

mysql> insert into h select * from h;
Query OK, 1280 rows affected (0.04 sec)
```

图 3-56 binlog 的 cache 和使用情况

```
Records: 1280  Duplicates: 0  Warnings: 0

mysql> show status like '%binlog_cache%';
+-----------------------+-------+
| Variable_name         | Value |
+-----------------------+-------+
| Binlog_cache_disk_use | 14    |
| Binlog_cache_use      | 109   |
+-----------------------+-------+
2 rows in set (0.01 sec)

mysql> insert into h select * from h;
Query OK, 2560 rows affected (0.17 sec)
Records: 2560  Duplicates: 0  Warnings: 0

mysql> show status like '%binlog_cache%';
+-----------------------+-------+
| Variable_name         | Value |
+-----------------------+-------+
| Binlog_cache_disk_use | 15    |
| Binlog_cache_use      | 110   |
+-----------------------+-------+
2 rows in set (0.00 sec)
```

图 3-56 binlog 的 cache 和使用情况（续）

本案例同时也说明，tmp 下面的空间不能随便乱用。当然最重要的还是避免，产生大事务，这才是根本。MySQL 不是用来处理大事务的。

3.3.7 一次两表关联导致的故障分析

在关系数据库中，两个表进行关联是很普遍的现象。但是如果SQL编写不当也会造成巨大的问题。多年前的一个早上，笔者接报一个数据库无法正常工作。立即登录进行处理，在后台看到一个SQL已经执行很久了。本着故障优先的原则先进行止损，结束该会话。然后在慢日志中找到了该SQL，SQL慢日志所示如图3-57，执行该操作的人想把两个表中ID不一样的值变成一样的值。这个操作由于仅有关联条件，没有过滤条件，所以是全表扫描全表锁定的。好在及时终止了事务，终止时该SQL执行了666s（11min）。

```
Query_time: 666.381028  Lock_time: 0.000248 Rows_sent: 0  Rows_examined: 758133810
SET timestamp=1513229142;
update sp
inner join  spd on sp.code = spd.code
set sp. ID = spd. ID
where  spd. ID<>sp. ID;
```

图 3-57 笛卡儿积 SQL 慢日志

这种两表关联 where 后谓词条件有且仅有不等于的 SQL 就造成笛卡儿积。笛卡儿积会在多表关联时将两个表的行数进行乘积的运算。例如一个有 1 万行数据的表 A，即 ID 从 1 到 1 万，拥有主键。进行 SELECT* FROM WHERE AWHERE ID!=1 时候，无论是否对 ID 建立索引都会执行全表扫描。返回数据量就是 9999 条。在这个基础上如何执行表 A 和表 B（表 B

和表 A 的表结构一模一样，但是数据仅有一条不一样）的关联，找出表 A 和表 B 不一样的 ID，SQL 语句如果写成 SELECT * FROM WHERE A,B WHEREA.ID!=B.ID，那么实际运算就是 1 万行乘以 1 万行，即 1 亿行的扫描。

所以该故障中的这个 SQL，在终止会话的时候已经扫描了 7 亿 5 千 8 百多万行。有关笛卡儿积的具体情况，在 7.2.3 节详细讲述。

本案例的结论是：在生产环境中 SQL 编写绝对要避免笛卡儿积的情况。

3.3.8 数据库连接数与连接复用不当的故障分析

数据库连接数是应用程序连接数据库所需要的通道数。如果通道全部被占用，则会造成数据库连接数满，这是一种常见的问题。因为关系数据库都有一个最大连接数限制，当连接数达到最大值后拒绝新的会话与数据库建立连接。如图 3-58 所示，MySQL 超过最大连接数，新的会话无法连接。在 MySQL 8 以后有一个管理员端口可以在数据库连接占满的情况下，通过管理员的专属端口登录进行抢救式维护。

```
[root@base ~]# mysql
ERROR 1040 (HY000): Too many connections
```

图 3-58 MySQL 超过最大连接数

不少公司会把出现连接数满定义为故障，所以运维人员会把这个值设置得很大，设置为几千甚至上万。笔者曾经见过有些人抱怨说为什么只给数据库设置了 1000 个连接的限制？在其他地方都是给到 10 万的。其实对于大部分数据库来说，连接数达到一定程度数据库已经不能正常工作了，根本达不到十万。尤其是活动连接数，一般的数据库（物理机）活动连接数达到几百的时候，说明 CPU 或 IO 已经大量排队了。如果是虚拟机的数据库，活动连接数在 20 个左右的时候，数据库基本会出现卡顿甚至无响应的情况。由于前序的 SQL 执行慢或无响应，后序 SQL 不能复用之前的会话，所以需要新创建一个会话，导致后序新的 SQL 会话不断叠加，非常容易造成连接数短时间陡然上升。

最容易出现的就是一个表中的一批记录被锁了（如忘记提交），甚至是整个表被锁了（例如在做影响写的 DDL，有些数据库的有些 DDL 不影响写），那么对这个表的写请求全部阻塞，用户端或接口等没有正常返回，就不断重试，这样短暂时间内连接数可能会上升到几十个甚至上百个，全部在等待，直到连接数用完。其场景非常类似于高速公路上发生了交通事故，后车连环发生故障，最终所有车道全部堵塞，后续车辆拥堵几公里。3.3.7 节的笛卡儿积锁表就是这样的场景。

多年来处理数据库连接数满基本分为两类：第一类大量慢 SQL，这些 SQL 未能及时返回结果，后续不断请求打开新会话。第二类就是锁，导致其他用到这个表的 SQL 未能及时返回结果，后续不断请求打开新会话。

不过也遇到过一次罕见的连接复用不当，引发了数据库锁，最终导致故障的案例。这是在生产中实际发生的事件，曾经一度找不到任何线索。通过本案例的分析希望能够启发读者，排查问题时应尝试从更多角度切入。

在一个交易场景的生产过程中，接到业务反馈，客户描述自己在平台上下单，点击生成订单成功，约 5 分钟后，该用户在平台的订单查询页面成功查询到该生成状态为成功的订单，后续再次进行查看时，却怎么都找不到该订单了，便联系客服进行查询。

客服通过后台系统也未能查询到该订单，基于平台订单号码结尾为自增整数的规则，发现用户描述的订单号码在系统中被跳过，前后序列号码的订单都正常生成并能够被正常查询到，便上报运维人员，联系相关开发人员进行排查。

开发人员通过应用日志检查，发现该订单确实曾经成功生成，也查询到该用户在订单查询页面查询到该订单的行为记录，该订单中的信息正确。

客户自述现象：用户在平台点击生成订单成功，过了 5 分钟在平台采购订单查询页面，能查询到该订单，但是过了一会当用户准备付款的时候这个订单不见了。

开发人员排查结果：

1）通过日志排查该订单生成成功。

2）数据库 binlog 没有该订单相关 insert 记录（没有 commit 的动作）。

最终通过假设和复现，认为是线程复用导致。线程复用流程如图 3-59 所示，Tomcat 线程池的连接复用了数据库的会话。

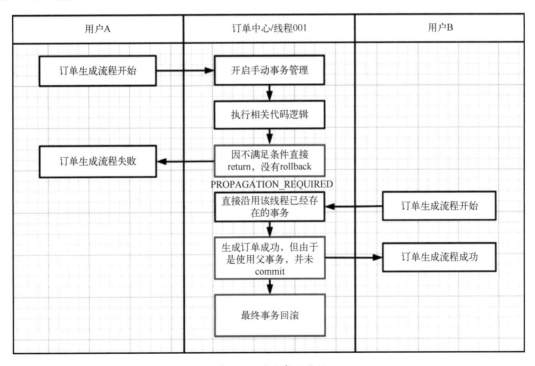

图 3-59　线程复用流程

本案例结论：避免连接数满的最好方式是事务完毕立即提交，以及 SQL 的效率尽可能达到极致。此外连接数不要设置过多，连接数不是越多越好。Redis 的单线程（读写数据库是单线程）效率是极高的这是一个很好的例证。

3.3.9 一次数据库 CPU 使用率 100%的故障分析

数据库的 CPU 满负荷几乎都是慢 SQL 造成的，多数集中在这两种场景下（也可能有其他场景，但是这两种概率较大）：
- 扫描大量的磁盘，出现 IO 等待；有时候还可以看到 CPU 内核态的使用率增加。
- 扫描大量内存而且还有排序等动作。

有一次看到监控报警，进入监控看到系统运行状态，SQL 全表扫描的监控图形如图 3-60 所示，几条图形右边快速升高的线条表示 IO 等待、CPU 内核态、CPU 使用率。此时数据库的负荷已经较高了，好在最后的 SQL 执行完毕后，整个数据库负荷下去了。但是这个过程中，或多或少影响了一些当时的业务，好在这些业务不是核心业务。

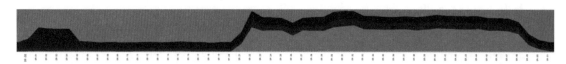

图 3-60 SQL 全表扫描的监控图形

如果有完善的数据库可观测性的软件，那么在监控图形上通过鼠标滑动可以直接展示出引发图形曲线变化的 SQL 语句，这一点 Oracle 在 2005 年就已经实现了。如果没有这类工具，要到后台日志中查找对应的 SQL。通过脱敏和美化后的 SQL，CPU 满负荷的 SQL 如图 3-61 所示，扫描了 882 万数据返回了 33 万 8000 行数据。执行了 1323s，和监控时间大致吻合。读取的数据量以及莫名其妙的分组排序也足以造成 CPU 的近似满负荷，以及大量的 IO 等。

```
# Query_time: 1323.091712   Lock_time: 0.020807  Rows_sent: 338639   Rows_examined: 8826251
SET timestamp=1686273997;
SELECT
        *,
        count(*) num,
        SUBSTR( created_date, 1, 10 ) cday
FROM
        log
GROUP BY
        id,
        no,
        cday,
        name,
        type;
```

图 3-61 CPU 满负荷的 SQL

从图 3-61 中可以看到，SQL 中 from 后面的表是 log，所以从表名就可以推断这个 log 表是一个日志表。而通常来说日志表的数据量一定很多。对于这句 SQL 所查询的 log 表来说，必须要关注，因为这种表不仅仅是行数多，可能还有二进制字段。这种表一旦发生全表扫描，会非常消耗服务器的 IO。而 CPU 在执行的过程中只能等待着 IO 的返回，所以 CPU 也有等待。通常所说的 IOWait（IO 等待）是一种空闲 CPU 时间。如果 CPU 核心因为没有工作而空闲，则该时间被计为空闲。但是，如果它因为某个进程在磁盘上等待而空闲，则 IO

时间将计入 IOWait。本案例就是 CPU 在等待 IO。

通常来说，分组一般最多三列进行统计，而且要有意义。就上述 SQL 而言，分组太多了，返回值也很多，所以造成了 CPU 较高的消耗。

本案例说明：对于在线系统的大表查询一定要慎重，大表查询一定要使用到索引，否则大表的全表扫描和排序会导致大量的磁盘 IO 操作，而如果 IO 能力不足则会导致查询缓慢和排序缓慢（大表排序无法在内存中完成，需要外存磁盘来完成，而磁盘 IO 能力不足则会加剧数据库负担），造成数据库的高负荷运作。

3.3.10 一次数据库索引不当引起的故障分析

数据库的表没有建立索引，那么查询时一定是全表扫描。建立了索引但是由于种种原因优化器没有用到索引，也会是全表扫描。那么一旦用到了索引是不是就可以了？也未必。曾经有这么一次故障就是有索引，而且 SQL 也使用到了索引，但是还是出了问题。其原因就是使用的索引效率不高。

该 SQL 经过脱敏后大致如此：SELECT NVL(SUM(t.MONEY), '0') FROM DETAIL t WHERE t.DETAIL_CODE = :1 AND t.STATUS = '10' AND t.TYPE = :2。这条语句在早上 9 点到 10 点半的 90 分钟内执行了 349 次，平均每次 118s，单次读 800MB，估算累计读数据达 272GB 以上。贡献了系统负荷的 40%。而这个数据库全库总数据大小约 100GB。其中 TYPE 为 3～5 种账本，DETAIL_CODE 是订单明细编码。真实数据是什么样呢？TYPE 有较大的倾斜，一个就占了 80% 的数据量。而其实表中的 DETAIL_CODE 几乎是唯一的。如果直接用 DETAIL_CODE 作为过滤条件就非常好。但是索引是基于 TYPE+STATUS+DETAIL_CODE 而建的，这里的索引采用了一般开发人员理解的传统索引方式，采用像先省、后市、再区县的从区分度小到大，范围逐步收敛的方式来建立，先找大范围再逐步缩小范围。但是这样的效率不高。知道问题的原因就有了对策：为 DETAIL_CODE 建立索引。通过调整索引后，数据库负荷马上就降了下来。经过统计，性能提升了约 2800 倍。故障就这样被解决了。

在交易业务场景中，有时候主要查询对象往往会是一对多的主项表与子项表关联查询。在生产中，是不是会经常有同时关联多列进行查询的语句呢？比如 SELECT * FROM A,B WHERE A.X=B.X AND A.Y=B.Y AND A.Z=B.Z...这样的同时关联多列进行查询？关联的几个列区分度由小往大，比如 X 是省、Y 是市、Z 是区这样的层级。然后在两个表上分别建立这些关联列的组合索引，用于提升关联效率，但是如此建立查询的组合索引，真的能够提升效率吗？以下通过实验来解答。如图 3-62 所示，建立主表（main 表）和明细表（detail 表）的模拟表，main 表和 detail 表是 1 对多的关系。

主项表与子项表间为一对多的关系，m_code 为主项号，在主项表中为唯一，子项表中每个主项号会出现 10 次，对应 10 个在子项表中唯一的子项号 d_code，buyer 字段模拟客户，会重复购买，而 flag 字段则可以理解为状态或标记，只有 0 和 1 两个值会出现，模拟数据中为随机出现，而根据前文的场景描述，日常开发习惯的关联 SQL 如图 3-63 所示。

```
mysql> desc main;
+--------+-------------+------+-----+---------+-------+
| Field  | Type        | Null | Key | Default | Extra |
+--------+-------------+------+-----+---------+-------+
| id     | int         | NO   | PRI | NULL    |       |
| m_code | int         | YES  | MUL | NULL    |       |
| buyer  | varchar(30) | YES  |     | NULL    |       |
| flag   | int         | YES  | MUL | NULL    |       |
+--------+-------------+------+-----+---------+-------+
4 rows in set (0.01 sec)

mysql> desc detail;
+--------+-------------+------+-----+---------+-------+
| Field  | Type        | Null | Key | Default | Extra |
+--------+-------------+------+-----+---------+-------+
| id     | int         | NO   | PRI | NULL    |       |
| m_code | int         | YES  | MUL | NULL    |       |
| d_code | int         | YES  |     | NULL    |       |
| buyer  | varchar(30) | YES  |     | NULL    |       |
| flag   | int         | YES  | MUL | NULL    |       |
+--------+-------------+------+-----+---------+-------+
5 rows in set (0.01 sec)
```

图 3-62　建立主表和明细表的模拟表

```
mysql> select * from main join detail on main.flag=detail.flag and main.buyer=detail.buyer and
main.m_code=detail.m_code where main.m_code=33;
+----+--------+-------+------+-----+--------+--------+-------+------+
| id | m_code | buyer | flag | id  | m_code | d_code | buyer | flag |
+----+--------+-------+------+-----+--------+--------+-------+------+
| 33 |     33 | A33   |    0 | 333 |     33 |    333 | A33   |    0 |
+----+--------+-------+------+-----+--------+--------+-------+------+
1 row in set (0.14 sec)
```

图 3-63　日常开发习惯的关联 SQL

后台可以看到该语句查询了 110000 条数据，也就是两张表的全表查询，如图 3-64 所示。

```
# Time: 2023-06-02T08:11:32.268451Z
# User@Host: root[root] @ localhost []  Id:     69
# Query_time: 0.138234  Lock_time: 0.000012 Rows_sent: 1  Rows_examined: 110000
SET timestamp=1685693492;
select * from main join detail on main.flag=detail.flag and main.buyer=detail.buyer and
main.m_code=detail.m_code where main.m_code=33;
```

图 3-64　两表关联的全表扫描

为了提升效率，建立组合索引，如图 3-65 所示，对主项表和子项表分别建立索引。

```
mysql> create index IDX_JOIN on main(flag,buyer,m_code);
Query OK, 0 rows affected (0.27 sec)
Records: 0  Duplicates: 0  Warnings: 0

mysql> create index IDX_JOIN on detail(flag,buyer,m_code);
Query OK, 0 rows affected (1.22 sec)
Records: 0  Duplicates: 0  Warnings: 0
```

图 3-65　建立组合索引

再进行查询，看到查询速度有所提升了，后台也能看到查询数量大幅减少。建立组合索引后的执行情况，如图 3-66 所示，扫描行数从 110000 减少到了 10001。

```
mysql> select * from main join detail on main.flag=detail.flag and main.buyer=detail.buyer and main.m_code=detail.m_code where main.m_code=33;
+----+--------+-------+------+-----+--------+--------+-------+------+
| id | m_code | buyer | flag | id  | m_code | d_code | buyer | flag |
+----+--------+-------+------+-----+--------+--------+-------+------+
| 33 |     33 | A33   |    0 | 333 |     33 |    333 | A33   |    0 |
+----+--------+-------+------+-----+--------+--------+-------+------+
1 row in set (0.01 sec)

# Time: 2023-06-02T08:15:47.098935Z
# User@Host: root[root] @ localhost []  Id:    69
# Query_time: 0.013756  Lock_time: 0.000012 Rows_sent: 1  Rows_examined: 10001
SET timestamp=1685693747;
select * from main join detail on main.flag=detail.flag and main.buyer=detail.buyer and main.m_code=detail.m_code where main.m_code=33;
```

图 3-66　建立组合索引后的执行情况

扫描了 10001 行得到了查询结果，这种为关联定制的组合索引在生产中常常遇到，但是这样的组合索引看似为关联查询定制，实际真的是效率最高的吗？先关掉这个组合索引，尝试建立一个简单的单列索引，如图 3-67 所示，为区分度高的列单独建立索引。

```
mysql> alter table main alter index IDX_JOIN invisible;
Query OK, 0 rows affected (0.03 sec)
Records: 0  Duplicates: 0  Warnings: 0

mysql> alter table detail alter index IDX_JOIN invisible;
Query OK, 0 rows affected (0.03 sec)
Records: 0  Duplicates: 0  Warnings: 0

mysql> create index IDX_CODE on main(m_code);
Query OK, 0 rows affected (0.11 sec)
Records: 0  Duplicates: 0  Warnings: 0

mysql> create index IDX_CODE on detail(m_code);
Query OK, 0 rows affected (0.87 sec)
Records: 0  Duplicates: 0  Warnings: 0
```

图 3-67　为区分度高的列单独建立索引

再看该 SQL 的查询效率，细粒度列索引执行效果如图 3-68 所示，扫描行数从 10001 减少到 2。

```
# Time: 2023-06-02T08:24:16.301508Z
# User@Host: root[root] @ localhost []  Id:    69
# Query_time: 0.002247  Lock_time: 0.000012 Rows_sent: 1  Rows_examined: 2
SET timestamp=1685694256;
select * from main join detail on main.flag=detail.flag and main.buyer=detail.buyer and main.m_code=detail.m_code where main.m_code=33;
```

图 3-68　细粒度列索引执行效果

从实验可见，只需要扫描两行，就得到了查询结果，是效率最优的查询方式。

由此可见，关联查询中，一般来说都是选择唯一列（或至少有一列是唯一列）进行关联。而且只要关联一个列即可，不用多个，让这个唯一列去发挥作用。如果无法做到关联唯一列，那么就要从设计上下手，变成理想的模型。多列分别关联的场景，不宜做成组合索引，更加不应该把区分度低的放在前面。关联列考虑最优的效率，往往也要考虑查询条件以及数据分布。

3.3.11 一次数据库主从延迟过大的故障分析

在实际工作中，发现 MySQL 会有主从延迟过大的故障。之所以会有这种现象，是因为 MySQL 的主从复制依靠 Binlog 日志，而这个日志是等事务结束后才开始写。所以如果是大事务则会导致写入时间长，而且一次性写入大量日志导致 Binlog 文件过大。而从库要从主库上获取 Binlog 日志，并且在从库上进行应用（解析日志在从库上再次执行），这也是需要时间的。

经历过多次 MySQL 出现主从延迟的情况，排查发现很多是由于主库上出现了大事务（这种最为普遍）。以下通过模拟复现一下当时发生延迟的场景。

两台 MySQL 服务器，建立了主从复制关系。主库地址最后一位 IP 为 138，从库地址最后一位 IP 为 91。主库 138 上有一张千万数据量的表 thousand，如图 3-69 所示。

```
mysql> select count(*) from thousand;
+----------+
| count(*) |
+----------+
| 10000001 |
+----------+
1 row in set (3.13 sec)
```

图 3-69　1000 万行的案例表

在主库上开启事务对千万级别的表进行全表更新，制造一个大事务。人为制造大事务，该事务执行了 19min，更新了将近 1000 万行数据，如图 3-70 所示。

```
mysql> begin;
Query OK, 0 rows affected (0.00 sec)

mysql> update thousand set company=name;
Query OK, 9999999 rows affected (19 min 2.36 sec)
Rows matched: 10000001  Changed: 9999999  Warnings: 0

mysql> commit;
Query OK, 0 rows affected (11.92 sec)
```

图 3-70　人为制造大事务

在事务提交后，在从库上观察主从状态，如图 3-71 所示，大事务导致主从状态延迟，延迟了 181s。

```
mysql> show slave status\G;
*************************** 1. row ***************************
               Slave_IO_State: Waiting for source to send event
                  Master_Host: 10.60.143.138
                  Master_User: repl
                  Master_Port: 3306
                Connect_Retry: 60
              Master_Log_File: binlog.000008
          Read_Master_Log_Pos: 197
               Relay_Log_File: mysql8-relay-bin.000004
                Relay_Log_Pos: 15086114
        Relay_Master_Log_File: binlog.000007
             Slave_IO_Running: Yes
            Slave_SQL_Running: Yes
              Replicate_Do_DB:
          Replicate_Ignore_DB:
           Replicate_Do_Table:
       Replicate_Ignore_Table:
      Replicate_Wild_Do_Table:
  Replicate_Wild_Ignore_Table:
                   Last_Errno: 0
                   Last_Error:
                 Skip_Counter: 0
          Exec_Master_Log_Pos: 961733916
              Relay_Log_Space: 708138297
              Until_Condition: None
               Until_Log_File:
                Until_Log_Pos: 0
           Master_SSL_Allowed: No
           Master_SSL_CA_File:
           Master_SSL_CA_Path:
              Master_SSL_Cert:
            Master_SSL_Cipher:
               Master_SSL_Key:
        Seconds_Behind_Master: 181
Master_SSL_Verify_Server_Cert: No
                Last_IO_Errno: 0
                Last_IO_Error:
               Last_SQL_Errno: 0
               Last_SQL_Error:
  Replicate_Ignore_Server_Ids:
             Master_Server_Id: 1
                  Master_UUID: e3be5a41-d73b-11ec-9b9d-0050569607b5
             Master_Info_File: mysql.slave_master_info
                    SQL_Delay: 0
          SQL_Remaining_Delay: NULL
          SQL_Remaining_Delay: NULL
      Slave_SQL_Running_State: Waiting for Replica Worker queue
           Master_Retry_Count: 86400
                  Master_Bind:
      Last_IO_Error_Timestamp:
     Last_SQL_Error_Timestamp:
               Master_SSL_Crl:
           Master_SSL_Crlpath:
           Retrieved_Gtid_Set: e3be5a41-d73b-11ec-9b9d-0050569607b5:49-108
            Executed_Gtid_Set: 3b64c057-ff8b-11ed-a325-005056968d7c:1-14,
e3be5a41-d73b-11ec-9b9d-0050569607b5:1-107
                Auto_Position: 1
         Replicate_Rewrite_DB:
                 Channel_Name:
           Master_TLS_Version:
       Master_public_key_path:
        Get_master_public_key: 0
            Network_Namespace:
1 row in set, 1 warning (0.00 sec)

ERROR:
No query specified
```

图 3-71 大事务导致主从状态延迟

从复现可以发现，当主库上出现大事务时，从库会出现一定程度的延迟情况。那么在这时候，如果主库上有其他事务接着执行，就会发生从库无法同步的现象，需要等待前一个大事务同步完成。大事务期间继续执行其他事务如图 3-72 所示，此时在主库上插入一条数据。

```
mysql> insert into thousand values(15000003, 'aaa', 10, 10, 10, current_timestamp, 'aaa');
Query OK, 1 row affected (0.09 sec)

mysql> select max(id) from thousand;
+-----------+
| max(id)   |
+-----------+
| 15000003  |
+-----------+
1 row in set (0.01 sec)
```

图 3-72　大事务期间继续执行其他事务

可以看到，主库上能够完成写入，并且可以查询。但是到从库上进行查询时，发现这一条新数据是还未同步完成的。大事务堵塞其他事务如图 3-73 所示，新增的 ID 还没有被从库获得。

```
mysql> select max(id) from thousand;
+-----------+
| max(id)   |
+-----------+
| 15000002  |
+-----------+
1 row in set (0.00 sec)
```

图 3-73　大事务堵塞其他事务

MySQL 的最佳实践是轻量型应用，不适用做大事务处理。这里的轻量型是指短平快的事务，快进快出，而不是 ERP 等复杂的应用场景。

3.3.12　一次数据库主从不一致的故障分析

对于 Oracle 数据库和 PostgreSQL 数据库来说，一般情况下，从库是只读不可写的。所以几乎不太可能出现主从不一致的情况。

而 MySQL 默认场景下从库是可写的，MySQL 只读参数如图 3-74 所示，read_only 和 super_read_only 两个参数默认是关闭的。

```
mysql> show variables like '%read%only';
+------------------------+-------+
| Variable_name          | Value |
+------------------------+-------+
| innodb_read_only       | OFF   |
| read_only              | OFF   |
| super_read_only        | OFF   |
| transaction_read_only  | OFF   |
+------------------------+-------+
4 rows in set (0.02 sec)
```

图 3-74　MySQL 只读参数

read_only 是指一般账户只读，但是 root 账户仍是可写的。super_read_only 是指连 root 账户也不可写。不过 MySQL 更多的是用在互联网场景，并没有强一致性的要求，所以也可以解释为什么从库默认可写。当然 MySQL 也可以设置成和 Oracle 数据库一样的从库不可写，MySQL 只读参数可以在线修改，如图 3-75 所示，不过这样就丧失了一定的灵活性。毕竟根据 CAP 理论，一致性和可用性在分布式中是矛盾的。CAP 理论会在 4.6 节具体阐述。

```
mysql> set global read_only=1;
Query OK, 0 rows affected (0.00 sec)

mysql> set global super_read_only=1;
Query OK, 0 rows affected (0.00 sec)

mysql> show variables like '%read%only';
+-----------------------+-------+
| Variable_name         | Value |
+-----------------------+-------+
| innodb_read_only      | OFF   |
| read_only             | ON    |
| super_read_only       | ON    |
| transaction_read_only | OFF   |
+-----------------------+-------+
4 rows in set (0.01 sec)
```

图 3-75　MySQL 只读参数在线修改

修改全局参数可以人工修改，无须重启数据库实例，但是对应的 my.cnf 中应该同步修改，防止数据库或操作系统重启而恢复默认值。

Oracle 数据库和 PostgreSQL 数据库选择从库只读是有道理的。但是时至今日，不少人仍然没有意识到从库可写带来的风险。接下来通过案例和案例的分析来讨论这个问题。

有一次接到朋友的电话，反馈他们的 MySQL 数据库经历了一次主从切换导致丢失了半个月的数据。初听觉得不可思议，随着对故障的分析，真实的原因逐渐明了。原来是因为从库不知道什么原因（现实中这种莫名其妙的原因很多）被写入了。这个写入有几种情况：

- 从库比主库多写了一条。
- 在从库上进行了一条数据的更新。
- 在从库上删除了一条数据。

以上情况都称为从库被写入，由于 MySQL 主从的机制是主库发送消息，从库到主库拿归档日志，然后将归档日志中的数据变化在本地重新执行一下，从而达到和主库一样的数据变化。所以从库的变化主库不会感知，这个数据流是单向的。

通常 MySQL 搭建完成主从高可用后，都会在从库上执行检查主从运行状态的命令。MySQL 主从检查如图 3-76 所示，其中 Slave_IO_Running、Slave_SQL_Running 是两个线程，一个 IO 线程的作用是从主库获取 Binlog 日志，另外一个 SQL 线程是应用日志。两个线程均为 Yes 表示主从数据库连接成功并且应用正常。而且图 3-76 中的 Last_IO_Errno、Last_IO_Error、Last_SQL_Errno、Last_SQL_Error 要么是 0 要么是空。

```
mysql> show slave status\G;
*************************** 1. row ***************************
               Slave_IO_State: Waiting for source to send event
                  Master_Host: 106.75.226.145
                  Master_User: repl
                  Master_Port: 3306
                Connect_Retry: 60
              Master_Log_File: binlog.000005
          Read_Master_Log_Pos: 1222
               Relay_Log_File: 10-24-1-231-relay-bin.000002
                Relay_Log_Pos: 411
        Relay_Master_Log_File: binlog.000005
             Slave_IO_Running: Yes
            Slave_SQL_Running: Yes
              Replicate_Do_DB:
          Replicate_Ignore_DB:
           Replicate_Do_Table:
       Replicate_Ignore_Table:
      Replicate_Wild_Do_Table:
  Replicate_Wild_Ignore_Table:
                   Last_Errno: 0
                   Last_Error:
                 Skip_Counter: 0
          Exec_Master_Log_Pos: 1222
              Relay_Log_Space: 627
              Until_Condition: None
               Until_Log_File:
                Until_Log_Pos: 0
           Master_SSL_Allowed: No
           Master_SSL_CA_File:
           Master_SSL_CA_Path:
              Master_SSL_Cert:
            Master_SSL_Cipher:
               Master_SSL_Key:
        Seconds_Behind_Master: 0
Master_SSL_Verify_Server_Cert: No
                Last_IO_Errno: 0
                Last_IO_Error:
               Last_SQL_Errno: 0
               Last_SQL_Error:
```

图 3-76　MySQL 主从检查

很遗憾，当出现上述的 MySQL 数据库从库被写入后，这两个线程并不能立即感知到问题，这两个线程的状态还是 Yes。只有当下次主库操作这些数据的时候，再从主库传过来数据变更（在行模式下变更包括变更前的行值和变更后的行值，而行模式是目前推荐的 binlog 模式）发现对不上，那么 MySQL 数据库的线程就感知到了数据的不一致。为了防止错误蔓延，数据库停止复制传输，这时线程的状态会变成 No。通过实验模拟 MySQL 两个线程是 Yes 的情况下，依然可以数据不一致。第一步对比现有主从数据如图 3-77 所示，root@ MySQL8M 代表主库（Master），root@MySQL8S 代表从库（Slave）。

```
[root@MySQL8M ~]# mysql -uroot -p1 -e "select * from xxg.xuexiaogang"
mysql: [Warning] Using a password on the command line interface can be insecure.
+----+---+---+---+---+---+
| id | a | b | c | d | e |
+----+---+---+---+---+---+
|  1 | 1 | 1 | 1 | 1 | 1 |
|  2 | 2 | 2 | 2 | 2 | 2 |
|  3 | 3 | 3 | 3 | 3 | 3 |
|  4 | 4 | 4 | 4 | 4 | 4 |
|  5 | 5 | 5 | 5 | 5 | 5 |
+----+---+---+---+---+---+

[root@MySQL8S ~]# mysql -uroot -p1 -e "select * from xxg.xuexiaogang"
mysql: [Warning] Using a password on the command line interface can be insecure.
+----+---+---+---+---+---+
| id | a | b | c | d | e |
+----+---+---+---+---+---+
|  1 | 1 | 1 | 1 | 1 | 1 |
|  2 | 2 | 2 | 2 | 2 | 2 |
|  3 | 3 | 3 | 3 | 3 | 3 |
|  4 | 4 | 4 | 4 | 4 | 4 |
|  5 | 5 | 5 | 5 | 5 | 5 |
+----+---+---+---+---+---+
```

图 3-77 第一步对比现有主从数据

目前主从数据库数据一致，检查从库线程状态如图 3-78 所示，两个线程都是 Yes，并且无 Error 的报错。

```
[root@MySQL8S ~]# mysql -uroot -p1 -e "show slave status\G"
mysql: [Warning] Using a password on the command line interface can be insecure.
*************************** 1. row ***************************
               Slave_IO_State: Waiting for source to send event
                  Master_Host: 10.60.143.32
                  Master_User: xuexiaogang
                  Master_Port: 3306
                Connect_Retry: 60
              Master_Log_File: binlog.000002
          Read_Master_Log_Pos: 5976760
               Relay_Log_File: MySQL8S-relay-bin.000014
                Relay_Log_Pos: 5971096
        Relay_Master_Log_File: binlog.000002
             Slave_IO_Running: Yes
            Slave_SQL_Running: Yes
              Replicate_Do_DB:
          Replicate_Ignore_DB:
           Replicate_Do_Table:
       Replicate_Ignore_Table:
      Replicate_Wild_Do_Table:
  Replicate_Wild_Ignore_Table:
                   Last_Errno: 0
                   Last_Error:
```

图 3-78 检查从库线程状态

模拟从库（MySQL8S）写入，从库被意外写入如图 3-79 所示，ID 为 1 的这行数据的 a 列被改成 11。

```
[root@MySQL8S ~]# mysql -uroot -p1 -e "update xxg.xuexiaogang set a=11 where id=1"
mysql: [Warning] Using a password on the command line interface can be insecure.

[root@MySQL8S ~]# mysql -uroot -p1 -e "select * from xxg.xuexiaogang"
mysql: [Warning] Using a password on the command line interface can be insecure.
+------+------+------+------+------+------+
| id   | a    | b    | c    | d    | e    |
+------+------+------+------+------+------+
|    1 |   11 |    1 |    1 |    1 |    1 |
|    2 |    2 |    2 |    2 |    2 |    2 |
|    3 |    3 |    3 |    3 |    3 |    3 |
|    4 |    4 |    4 |    4 |    4 |    4 |
|    5 |    5 |    5 |    5 |    5 |    5 |
+------+------+------+------+------+------+
```

图 3-79　从库被意外写入

此时主库（MySQL8M）不知道从库被写入，主库继续写入业务，主库写入了一行全部为 6 的数据，如图 3-80 所示。

```
[root@MySQL8M ~]# mysql -uroot -p1 -e "insert into xxg.xuexiaogang values (6,6,6,6,6,6)"
mysql: [Warning] Using a password on the command line interface can be insecure.

[root@MySQL8M ~]# mysql -uroot -p1 -e "select * from xxg.xuexiaogang"
mysql: [Warning] Using a password on the command line interface can be insecure.
+------+------+------+------+------+------+
| id   | a    | b    | c    | d    | e    |
+------+------+------+------+------+------+
|    1 |    1 |    1 |    1 |    1 |    1 |
|    2 |    2 |    2 |    2 |    2 |    2 |
|    3 |    3 |    3 |    3 |    3 |    3 |
|    4 |    4 |    4 |    4 |    4 |    4 |
|    5 |    5 |    5 |    5 |    5 |    5 |
|    6 |    6 |    6 |    6 |    6 |    6 |
+------+------+------+------+------+------+
```

图 3-80　主库继续写入业务

此时数据能不能到同步到从库呢？答案是可以。从库继续同步数据如图 3-81 所示，id 为 6 的这行数据已经到从库（MySQL8S），而且此时从库 id 为 1 的这行数据中 a 列与主库已经不一致了，但是主从状态是好的。

```
[root@MySQL8S ~]# mysql -uroot -p1 -e "select * from xxg.xuexiaogang"
mysql: [Warning] Using a password on the command line interface can be insecure.
+------+------+------+------+------+------+
| id   | a    | b    | c    | d    | e    |
+------+------+------+------+------+------+
|    1 |   11 |    1 |    1 |    1 |    1 |
|    2 |    2 |    2 |    2 |    2 |    2 |
|    3 |    3 |    3 |    3 |    3 |    3 |
|    4 |    4 |    4 |    4 |    4 |    4 |
|    5 |    5 |    5 |    5 |    5 |    5 |
|    6 |    6 |    6 |    6 |    6 |    6 |
+------+------+------+------+------+------+
```

图 3-81　从库继续同步数据

主库（MySQL8M）继续工作，如图 3-82 所示，此时主库也需要更新 id 为 1 的这行数据，但是更新的是 b 列而不是 a 列。更新完毕以后，又新增了一行 id 为 7 的数据。

```
[root@MySQL8M ~]# mysql -uroot -p1 -e "update  xxg.xuexiaogang set b=2 where id=1"
mysql: [Warning] Using a password on the command line interface can be insecure.

[root@MySQL8M ~]# mysql -uroot -p1 -e "insert into  xxg.xuexiaogang values (7,7,7,7,7,7)"
mysql: [Warning] Using a password on the command line interface can be insecure.

[root@MySQL8M ~]# mysql -uroot -p1 -e "select * from xxg.xuexiaogang"
mysql: [Warning] Using a password on the command line interface can be insecure.
+------+------+------+------+------+------+
| id   | a    | b    | c    | d    | e    |
+------+------+------+------+------+------+
|   1  |   1  |   2  |   1  |   1  |   1  |
|   2  |   2  |   2  |   2  |   2  |   2  |
|   3  |   3  |   3  |   3  |   3  |   3  |
|   4  |   4  |   4  |   4  |   4  |   4  |
|   5  |   5  |   5  |   5  |   5  |   5  |
|   6  |   6  |   6  |   6  |   6  |   6  |
|   7  |   7  |   7  |   7  |   7  |   7  |
+------+------+------+------+------+------+
```

图 3-82　主库继续工作

此时读者会不会想更新的不是同一列，不冲突？答案是否定的。主从中断如图 3-83 所示，在主库更新 id 为 1 的数据时，主从发生了错误，Slave_IO 线程状态为 Yes，进而影响了 id 为 7 的后续操作。

```
[root@MySQL8S ~]# mysql -uroot -p1 -e "select * from xxg.xuexiaogang"
mysql: [Warning] Using a password on the command line interface can be insecure.
+------+------+------+------+------+------+
| id   | a    | b    | c    | d    | e    |
+------+------+------+------+------+------+
|   1  |  11  |   1  |   1  |   1  |   1  |
|   2  |   2  |   2  |   2  |   2  |   2  |
|   3  |   3  |   3  |   3  |   3  |   3  |
|   4  |   4  |   4  |   4  |   4  |   4  |
|   5  |   5  |   5  |   5  |   5  |   5  |
|   6  |   6  |   6  |   6  |   6  |   6  |
+------+------+------+------+------+------+
[root@MySQL8S ~]# mysql -uroot -p1 -e "show slave status\G"
mysql: [Warning] Using a password on the command line interface can be insecure.
*************************** 1. row ***************************
               Slave_IO_State: Waiting for source to send event
                  Master_Host: 10.60.143.32
                  Master_User: xuexiaogang
                  Master_Port: 3306
                Connect_Retry: 60
              Master_Log_File: binlog.000002
          Read_Master_Log_Pos: 5977997
               Relay_Log_File: MySQL8S-relay-bin.000014
                Relay_Log_Pos: 5971692
        Relay_Master_Log_File: binlog.000002
             Slave_IO_Running: Yes
            Slave_SQL_Running: No
              Replicate_Do_DB:
          Replicate_Ignore_DB:
           Replicate_Do_Table:
       Replicate_Ignore_Table:
      Replicate_Wild_Do_Table:
  Replicate_Wild_Ignore_Table:
                   Last_Errno: 1032
                   Last_Error: Coordinator stopped because there were error(s) in the worker(s). The most recent failure being: Worker 1 failed executing transaction '790b8b3a-8b09-11ed-94e4-0050569684a9:20686' at master log binlog.000002, end_log_pos 5977659. See error log and/or performance_schema.replication_applier_status_by_worker table for more details about this failure or others, if any.
```

图 3-83　主从中断

跳过不一致的事务的方法如图 3-84 所示。停止复制，找到从库上接收的 GTID 和执行的 GITD。制造一个空事务，跳过去。开启复制，最终两个线程的状态都是 Yes。

```
                Retrieved_Gtid_Set: 790b8b3a-8b09-11ed-94e4-0050569684a9:2-20687
                 Executed_Gtid_Set: 4754d172-8b0f-11ed-a614-005056965d55:1-12,
790b8b3a-8b09-11ed-94e4-0050569684a9:1-20685
                     Auto_Position: 1
              Replicate_Rewrite_DB:
                      Channel_Name:
                Master_TLS_Version:
            Master_public_key_path:
             Get_master_public_key: 0
                 Network_Namespace:


mysql> stop slave;
Query OK, 0 rows affected, 1 warning (0.00 sec)

mysql> SET gtid_next = '790b8b3a-8b09-11ed-94e4-0050569684a9:20686';
Query OK, 0 rows affected (0.00 sec)

mysql> BEGIN;COMMIT;
Query OK, 0 rows affected (0.00 sec)

Query OK, 0 rows affected (0.00 sec)

mysql> SET gtid_next = 'AUTOMATIC';
Query OK, 0 rows affected (0.00 sec)

mysql> START SLAVE;
Query OK, 0 rows affected, 1 warning (0.01 sec)

mysql> show slave status\G;
*************************** 1. row ***************************
               Slave_IO_State: Waiting for source to send event
                  Master_Host: 10.60.143.32
                  Master_User: xuexiaogang
                  Master_Port: 3306
                Connect_Retry: 60
              Master_Log_File: binlog.000002
          Read_Master_Log_Pos: 5977997
               Relay_Log_File: MySQL8S-relay-bin.000015
                Relay_Log_Pos: 451
        Relay_Master_Log_File: binlog.000002
             Slave_IO_Running: Yes
            Slave_SQL_Running: Yes
```

图 3-84 跳过不一致的事务

对比主从数据库数据，如图 3-85 所示，可以发现此时部分数据是不一致的。但是主从状态恢复可以继续复制了。

以上是 MySQL 基于 GTID 模式下且 Binlog 是行模式的，GTID 全称 Global Transaction Identifiers，也称为全局事务 ID。是 MySQL 5.6 以后出现的特性，GTID 事务是全局唯一性的，并且一个事务对应一个 GTID 值。一个 GTID 值在同一个 MySQL 实例上只会执行一次。

```
[root@MySQL8S ~]# mysql -uroot -p1 -e "select * from xxg.xuexiaogang"
mysql: [Warning] Using a password on the command line interface can be insecure.
+------+------+------+------+------+------+
| id   | a    | b    | c    | d    | e    |
+------+------+------+------+------+------+
|    1 |   11 |    1 |    1 |    1 |    1 |
|    2 |    2 |    2 |    2 |    2 |    2 |
|    3 |    3 |    3 |    3 |    3 |    3 |
|    4 |    4 |    4 |    4 |    4 |    4 |
|    5 |    5 |    5 |    5 |    5 |    5 |
|    6 |    6 |    6 |    6 |    6 |    6 |
|    7 |    7 |    7 |    7 |    7 |    7 |
+------+------+------+------+------+------+
[root@MySQL8M ~]# mysql -uroot -p1 -e "select * from xxg.xuexiaogang"
mysql: [Warning] Using a password on the command line interface can be insecure.
+------+------+------+------+------+------+
| id   | a    | b    | c    | d    | e    |
+------+------+------+------+------+------+
|    1 |    1 |    2 |    1 |    1 |    1 |
|    2 |    2 |    2 |    2 |    2 |    2 |
|    3 |    3 |    3 |    3 |    3 |    3 |
|    4 |    4 |    4 |    4 |    4 |    4 |
|    5 |    5 |    5 |    5 |    5 |    5 |
|    6 |    6 |    6 |    6 |    6 |    6 |
|    7 |    7 |    7 |    7 |    7 |    7 |
+------+------+------+------+------+------+
```

图 3-85 对比主从数据库数据

GTID 相较传统复制（非 GTID）的优势如下。

1）主从搭建更加简便，不用手动特地指定 position 位置。
2）复制集群内有一个统一的标识，识别、管理上更方便。
3）故障转移更容易，不用像传统复制那样需要找 log_file 和 log_pos 的位置。
4）通常情况下，GTID 是连续没有空洞的，更能保证数据的一致性，零丢失。
5）比传统的复制更加安全，一个 GTID 在一个 MySQL 实例上只会执行一次，避免重复执行导致数据混乱或主从不一致。

MySQL 的主从也有非 GTID 模式，但是不推荐使用，因为从 MySQL 5.6 之后有了变革。此外，还有 Binlog 的语句模式，又可以搭配出不同的组合。感兴趣的读者朋友可以尝试一下。

通过上述实验表明，在对数据一致性有要求的场景下，从库一定要开启只读。否则即使两个线程都是 Yes 依然也无法保证一致性，而一旦发生数据库切换就带来了数据不一致的情况。

对于本案例中由于监控不完善导致中断半个月的数据，教训实在是惨痛。虽然这件事情发生在别的公司和企业，但是也为笔者自己的工作敲响了警钟。

当然如果想要一致性的高可用架构，Oracle 官方推荐的是基于 Paoxs 协议的 MGR。尽管对于 MGR 的评价不一，但是它是理论上可以做到一致性的高可用架构。笔者在工作中也使用过 Router+MGR 的架构，使用起来没有什么问题。这主要是遵循了 MySQL 官方使用的最佳实践：MySQL 是轻量型数据库，用于快进快出的短小事务，禁止大事务，也禁止从 Oracle 的系统迁移到 MySQL 上，因为 MySQL 不是替代 Oracle 的，原本的 Oracle 业务场景不适合 MySQL 数据库。MGR 也涉及网络，所以并没有做异地的机房部署。只是本地不同机房的部署,避免了网络上长距离传输影响 MGR 一致性。

3.3.13 一次 Redis 数据库无法启动的故障分析

本节的 Redis 故障案例超过一般的认知。这里有许多问题,之前的开发人员和运维人员不知道怎么衔接的,留下了较大的隐患。而这些都随着一次意外而浮现,通过对其体系的研究,抽丝剥茧之后还原了事件真相,并且重现。

某天有人找笔者求助,请求检查一个 Redis 数据库的连接情况。笔者用 Redis 的客户端工具 redis-cli 登录到 Redis 数据库,在进入数据后执行了 client list 的命令。Redis 客户端连接的检查如图 3-86 所示,从这个命令可以看到当前数据库的连接访问情况。

```
127.0.0.1:6379> client list
id=3 addr=127.0.0.1:6379 laddr=10.70.9.43:17237 fd=8 name= age=6762126 idle=1 flags=M db
events=r cmd=ping user=(superuser) redir=-1
id=3380480 addr=127.0.0.1:64719 laddr=127.0.0.1:6379 fd=7 name= age=8 idle=1 flags=N db=
61466 events=r cmd=client user=default redir=-1
127.0.0.1:6379>
```

图 3-86　Redis 客户端连接的检查

如果是主从架构,通常只看主库就可以。从库一般承担灾备的角色,无人连接。

当时的开发反馈的结果是主库能连接上,从库报没有权限连接。一般出现这种问题是没有输入密码造成的。但是主从架构的 Redis,要么主库从库都有密码,要么主库从库都没有密码。不太可能出现一个库设置了密码,而另外一个库没有密码。

从这个出发点考虑,建议这套主从的 Redis 最好统一设置密码。笔者让运维人员通知开发人员,准备统一重新设置密码。设置密码以后重启数据库,应用程序重连 Redis。之后在计划的时间点进行了 Redis 实例的重启。密码的问题很容易就解决了,然而更大的问题也来了。

Redis 数据库重启之后,开发人员反馈,应用程序报错,无法连接 Redis 实例。再次检查 Redis 数据库的主从,看到了在主从之外还有一个"哨兵",就是典型的哨兵模式。该模式笔者在工作中遇到过几次,也出过几次问题。其问题不是哨兵本身造成的,而是开发人员或开发框架造成的。

早在 2015 年,笔者经历过一次哨兵模式。当时开发人员说他的 jedis 要连接 Redis 的哨兵 IP 和 Redis 主库的 IP,共计两个 IP 地址,并且两个都要成功,如果只有一个成功是不行的。当时对于这个场景并不了解,笔者更加推崇 Redis 的 Cluster 模式。如果要求哨兵和主库两个 IP 都健康,那么主库发生问题岂不是就是全局问题吗?所以当年笔者建议开发人员采用 Cluster 模式,恰好笔者也做好了 Cluster 模式。只是开发说他的 jedis 连接不上笔者提供的 Redis 的 Cluster。无奈之下笔者用 C++做了一个 demo 连接上了 Redis 的 Cluster。

而在 2021 年再次出现了 6 年前的场景。开发那里的配置是要求连接哨兵 IP 和 Redis 主库 IP。

有了之前的经验,笔者请开发人员打开他的连接配置文件,检查配置连接字符串。发现连接字符串中配置连接了 Redis 的 6379 端口(默认端口没有问题)以及 26379 端口和 36379 端口,而 26379 端口和 36379 端口应该是哨兵的端口。但是这个问题和 6 年前的又有一点不一样，为什么这里会有两个哨兵的端口？理论上一个哨兵地址就行。这种设置没有相关资料可供参考，询问开发人员为什么这样配置？开发人员也不知道。

于是，考验笔者的时候到了。执行 cat /etc/sentinel.conf。看到这个文件只有一个配置端口的地方，配置的端口是 26379。启动哨兵进程，启动成功，但是开发人员反馈连接不上。然后把这个端口改成 36379，启动哨兵进程，启动成功，开发人员反馈还是连接不上。

笔者让开发人员在应用程序的连接字符串中把两个 IP 地址去掉一个，然后重启应用进行尝试。开发人员尝试修改应用程序的连接字符串，但是修改后的配置文件不生效。而且开发人员依然斩钉截铁地告诉笔者应用程序有且仅有这一个配置文件。

笔者建议他把应用程序配置文件的 IP 都清空，还是不起作用。此时开发也觉得可能配置文件不对，开发人员找到真正的配置文件去修改就行。但是他不知道还有什么地方可以修改配置。

这个时候笔者不能去改 Redis 的源码，支持哨兵多端口。笔者从原理进行推断想到了一个办法：先通过配置哨兵的配置文件/etc/sentinel.conf，配置端口 26379 启动哨兵进程。这个启动是成功的。然后修改/etc/sentinel.conf 把端口改成 36379，再次启动哨兵进程。等于启动了两遍哨兵，当然也就有了两个哨兵进程。由于 Redis 第一次依靠配置文件启动成功，已经驻留在系统中了。修改配置文件端口第一个哨兵进程并不能感知，再次启动新的端口哨兵进程，那么就有两个不同端口的哨兵进程了。启动效果如图 3-87 所示，出现了多哨兵进程，一个 Redis 的 6379 端口进程，一个 26379 的哨兵进程，一个 36379 的哨兵进程。

```
[root@localhost ~]# ps ef | grep redis root
root      4505    1   1  Apr01 ?   01:40:38 redis-server    *:6379
root      4510    1   0  Apr01 ?   00:13:32 redis- sentinel *:26379
root      4514    1   0  Apr01 ?   00:13:09 redis- sentinel *:36379
```

图 3-87 多哨兵进程

请开发人员再次启动应用进行连接，开发人员反馈应用程序连接 Redis 一切正常。

果然事情的真相就是这样的：之前有人启动了两个哨兵进程，然后告知开发人员这样连接，开发人员对此也是一知半解，盲目进行连接，而开发人员也不知道具体通过哪个连接字符串连接。在这里面还隐藏了当时主从库密码不一致的隐患。然后随着这次统一密码的重启，把多年前的隐患暴露了。

本案例是一次真实的事件，整个过程抽丝剥茧、层层深入、大胆假设，最后接近了事件的真相。

3.3.14 一次 MySQL 数据库数据类型不恰当导致的故障

有一次开发人员小 F 找到笔者反馈一个数据库故障，说是查询数据不对。了解了整个事情的来龙去脉，笔者找到了问题的原因：数据库设计是基于开发角度设计，而不是遵循客观的实际情况。下面对这个问题进行一下复现。

小 F 说的问题如图 3-88 所示，小 F 不解的是单独查询大于 6 月 1 日的数据有 166 条，单独查询小于 6 月 10 日的数据有 9834 条。为什么合在一起查询是 0？逻辑上说不过去。

```
mysql> select count(*) from car2 where time>'2023-06-01';
+----------+
| count(*) |
+----------+
|      166 |
+----------+
1 row in set (0.01 sec)
mysql> select count(*) from car2 where time<'2023-06-10';
+----------+
| count(*) |
+----------+
|     9834 |
+----------+
1 row in set (0.03 sec)
mysql> select * from car2 where time>'2023-06-01' and time<'2023-06-10';
Empty set (0.00 sec)
```

图 3-88 模拟范围查询是空

笔者的第一反应是可能在 6 月 1 日到 10 日无数据。而小 F 说在他们其他的团队涉及类似的业务表上是有数据的。范围查询符合预期如图 3-89 所示。

```
mysql> select count(*) from car where time>'2023-06-01';
+----------+
| count(*) |
+----------+
|       15 |
+----------+
1 row in set (0.00 sec)

mysql> select count(*) from car where time<'2023-06-10';
+----------+
| count(*) |
+----------+
|     9994 |
+----------+
1 row in set (0.03 sec)

mysql> select count(*) from car where time>'2023-06-01' and time<'2023-06-10';
+----------+
| count(*) |
+----------+
|        9 |
+----------+
1 row in set (0.00 sec)
```

图 3-89 范围查询符合预期

于是检查两个表的表结构，脱敏精简后的表结构如图 3-90 所示，两个表除了表名不一样，其他全部都一样。不过这里的时间字段的类型引起了笔者的注意，因为就是根据时间条件去检索。但是实际字段的类型不是 datetime 而是字符串。

时间属性的字段用字符串的数据类型，这在国内也不占少数。无论在 Oracle 上还是 MySQL 上都遇到过许多。其原因为表结构是开发人员自行设计，而在 Java 中处理时间要转换成字符串，所以为了简便就把这个字段类型定义为字符串。一旦定义成字符串各种问题就来了。于是有了解决问题的方向，打开两个表进行数据样本检查，如图 3-91 所示。

```
mysql> show create table car\G
*************************** 1. row ***************************
       Table: car
Create Table: CREATE TABLE `car` (
  `id` int DEFAULT NULL,
  `a` int DEFAULT NULL,
  `time` varchar(30) COLLATE utf8mb4_unicode_ci DEFAULT NULL,
  KEY `t1` (`time`)
) ENGINE=InnoDB DEFAULT CHARSET=utf8mb4 COLLATE=utf8mb4_unicode_ci
1 row in set (0.01 sec)

mysql> show create table car2\G
*************************** 1. row ***************************
       Table: car2
Create Table: CREATE TABLE `car2` (
  `id` int DEFAULT NULL,
  `a` int DEFAULT NULL,
  `time` varchar(30) COLLATE utf8mb4_unicode_ci DEFAULT NULL,
  KEY `t2` (`time`)
) ENGINE=InnoDB DEFAULT CHARSET=utf8mb4 COLLATE=utf8mb4_unicode_ci
1 row in set (0.00 sec)
```

图 3-90　脱敏精简后的表结构

```
mysql> select * from car limit 5;
+------+------+---------------------+
| id   | a    | time                |
+------+------+---------------------+
|    0 |    0 | 2023-06-15 00:00:00 |
|    1 |    1 | 2023-06-14 00:00:00 |
|    2 |    2 | 2023-06-13 00:00:00 |
|    3 |    3 | 2023-06-12 00:00:00 |
|    4 |    4 | 2023-06-11 00:00:00 |
+------+------+---------------------+
5 rows in set (0.00 sec)

mysql> select * from car2 limit 5;
+------+------+---------------------+
| id   | a    | time                |
+------+------+---------------------+
|    0 |    0 | 2023/06/15 00:00:00 |
|    1 |    1 | 2023/06/14 00:00:00 |
|    2 |    2 | 2023/06/13 00:00:00 |
|    3 |    3 | 2023/06/12 00:00:00 |
|    4 |    4 | 2023/06/11 00:00:00 |
+------+------+---------------------+
5 rows in set (0.00 sec)
```

图 3-91　两个表数据样本对比

通过对比读者可以发现时间字段的格式不一样。由于是字符串，那么就将字符串的每个字符转换成 ASCII 码去比较，结果"-"与"/"无法匹配。这应该是不同开发人员的读写习惯造成的，可能甲习惯 YYYY-MM-DD 这样的格式，而乙习惯 YYYY/MM/DD 这样的格式。其实如果把数据类型定义对了，如图 3-92 所示，时间字段采用时间类型属性定义。那么无论怎么查询，结果都是对的。不同格式化的时间查询比较如图 3-93 所示。

```
mysql> show create table car3\G
*************************** 1. row ***************************
       Table: car3
Create Table: CREATE TABLE `car3` (
  `id` int DEFAULT NULL,
  `a` int DEFAULT NULL,
  `time` datetime DEFAULT NULL,
  KEY `t3` (`time`)
) ENGINE=InnoDB DEFAULT CHARSET=utf8mb4 COLLATE=utf8mb4_unicode_ci
1 row in set (0.01 sec)

mysql> select * from car3 limit 5;
+------+------+---------------------+
| id   | a    | time                |
+------+------+---------------------+
|    0 |    0 | 2023-06-15 00:00:00 |
|    1 |    1 | 2023-06-14 00:00:00 |
|    2 |    2 | 2023-06-13 00:00:00 |
|    3 |    3 | 2023-06-12 00:00:00 |
|    4 |    4 | 2023-06-11 00:00:00 |
+------+------+---------------------+
5 rows in set (0.00 sec)
```

图 3-92　时间字段采用时间类型属性定义

```
mysql> select * from car3 where time>'2023/06/01' and time<'2023/06/10';
+------+------+---------------------+
| id   | a    | time                |
+------+------+---------------------+
|   13 |   13 | 2023-06-02 00:00:00 |
|   12 |   12 | 2023-06-03 00:00:00 |
|   11 |   11 | 2023-06-04 00:00:00 |
|   10 |   10 | 2023-06-05 00:00:00 |
|    9 |    9 | 2023-06-06 00:00:00 |
|    8 |    8 | 2023-06-07 00:00:00 |
|    7 |    7 | 2023-06-08 00:00:00 |
|    6 |    6 | 2023-06-09 00:00:00 |
+------+------+---------------------+
8 rows in set (0.01 sec)

mysql> select * from car3 where time>'2023-06-01' and time<'2023-06-10';
+------+------+---------------------+
| id   | a    | time                |
+------+------+---------------------+
|   13 |   13 | 2023-06-02 00:00:00 |
|   12 |   12 | 2023-06-03 00:00:00 |
|   11 |   11 | 2023-06-04 00:00:00 |
|   10 |   10 | 2023-06-05 00:00:00 |
```

图 3-93　不同格式化的时间查询比较

```
|   9  |   9  | 2023-06-06 00:00:00 |
|   8  |   8  | 2023-06-07 00:00:00 |
|   7  |   7  | 2023-06-08 00:00:00 |
|   6  |   6  | 2023-06-09 00:00:00 |
+------+------+---------------------+
8 rows in set (0.00 sec)
```

图 3-93 不同格式化的时间查询比较（续）

本案例虽然没有导致数据库崩溃，但是影响了业务逻辑。其实把时间定义成字符串的问题很多，不仅仅是两种不同格式的问题（YYYY-MM-DD 和 YYYY/MM/DD），还遇到过在时间中多一个 T，类似于 2023-06-02 00:00:00 和 2023-06-02T00:00:00，这些都给后续带来了很多问题。数据库提供的这些数据类型应该是结合实际设计设置，是围绕着数据库进行开发，而不是数据库围绕着应用程序去适配。

3.3.15 一次数据库全表查询的优化

多年前，每次出差后都需要填写报支单。当时公司的报支系统使用起来很慢。那时候不懂，在选择的时候仅仅选择笔者的工号，为什么报支单填写得这么慢？然后当笔者报支单出问题的时候，去财务部当面处理时，眼看着财务人员打开与笔者同样的界面，在选择笔者的工号后，还多选了公司的名字。单击查询，秒出。即使这个数据库不是笔者维护的，该报支系统也不是笔者开发的，但是瞬间明白其原因了。

首先构造一个 MySQL 的实验环境，如图 3-94 所示。g 表有 9 条数据，id 有一个单列索引，a 列和 b 列组成了复合索引。

```
mysql> select * from g;
+------+------+------+
| id   | a    | b    |
+------+------+------+
|   1  |   1  |   1  |
|   2  |   2  |   2  |
|   3  |   3  |   3  |
|   4  |   4  |   4  |
|   5  |   5  |   5  |
|   6  |   6  |   6  |
|   7  |   7  |   7  |
|   8  |   8  |   8  |
|   9  |   9  |   9  |
+------+------+------+
9 rows in set (0.00 sec)

mysql> show create table g\G
*************************** 1. row ***************************
       Table: g
Create Table: CREATE TABLE `g` (
  `id` int DEFAULT NULL,
  `a` int DEFAULT NULL,
  `b` int DEFAULT NULL,
  KEY `g1` (`id`),
  KEY `gab` (`a`,`b`)
) ENGINE=InnoDB DEFAULT CHARSET=utf8mb4 COLLATE=utf8mb4_unicode_ci
1 row in set (0.01 sec)
```

图 3-94 MySQL 复合索引环境

MySQL 复合索引执行计划如图 3-95 所示，分别执行三句 SQL：
SQL1：select * from g where a=1 and b=1;
SQL2：select * from g where a=1;
SQL3：select * from g where b=1;

SQL1 用到了复合索引 gab，而且 key_len 是 10，说明 a 列和 b 列都发生作用。而 SQL2 也用到了复合索引 gab，但是 key_len 是 5.说明只有 a 列起到了作用（因为 utf8mb4 编码中，一个字符占 4 个字节，可以为空加一个字节）。而 SQL3 没有用到任何索引。

```
mysql> explain select * from g where a=1 and b=1;
+----+-------------+-------+------------+------+---------------+------+---------+-------------+------+----------+-------+
| id | select_type | table | partitions | type | possible_keys | key  | key_len | ref         | rows | filtered | Extra |
+----+-------------+-------+------------+------+---------------+------+---------+-------------+------+----------+-------+
|  1 | SIMPLE      | g     | NULL       | ref  | gab           | gab  | 10      | const,const |    1 |   100.00 | NULL  |
+----+-------------+-------+------------+------+---------------+------+---------+-------------+------+----------+-------+
1 row in set, 1 warning (0.00 sec)

mysql> explain select * from g where a=1 ;
+----+-------------+-------+------------+------+---------------+------+---------+-------+------+----------+-------+
| id | select_type | table | partitions | type | possible_keys | key  | key_len | ref   | rows | filtered | Extra |
+----+-------------+-------+------------+------+---------------+------+---------+-------+------+----------+-------+
|  1 | SIMPLE      | g     | NULL       | ref  | gab           | gab  | 5       | const |    1 |   100.00 | NULL  |
+----+-------------+-------+------------+------+---------------+------+---------+-------+------+----------+-------+
1 row in set, 1 warning (0.00 sec)

mysql> explain select * from g where b=1 ;
+----+-------------+-------+------------+------+---------------+------+---------+------+------+----------+-------------+
| id | select_type | table | partitions | type | possible_keys | key  | key_len | ref  | rows | filtered | Extra       |
+----+-------------+-------+------------+------+---------------+------+---------+------+------+----------+-------------+
|  1 | SIMPLE      | g     | NULL       | ALL  | NULL          | NULL | NULL    | NULL |    9 |    11.11 | Using where |
+----+-------------+-------+------------+------+---------------+------+---------+------+------+----------+-------------+
1 row in set, 1 warning (0.00 sec)
```

图 3-95 MySQL 复合索引执行计划

这是因为复合索引要求第一列（也叫前导列或先导列）必须被用到。其原因还是索引的顺序问题。假设有两个维度年龄和身高，每个年龄都有自己年龄段的身高排序。查找一个年龄段的身高是已经排序好的，那么如果跳过年龄段直接查询这些结果呢？能不能保证不同年龄段混合后的顺序不改变？答案是否定的。

所以在工作中经常看到建立了复合索引，但是没有起到作用，就是因为 where 条件中没有第一列。

案例中的财务系统，它的索引建立一定是遵循类似省、市、区、县这种一个个级别。

此外还有种情况需要说明。生产环境中经常会出现一个状态或是标记，再带上常规数据进行精确查询的场景。如果这是一个高频度场景，开发人员通常会选择建立一个组合索引来提升效率，索引中包含 A（状态或标记列）以及 B（其他业务列）两列。那么，建立索引时 A、B 两列的先后顺序的不同是否会影响到查询效率？如何才能使组合索引达到预期？接下来通过不同数据库中的实验来说明。这个问题在国内还是比较普遍的。

首先在 Oracle 数据库中，建立 Oracle 组合索引实验表，如图 3-96 所示。

模拟写入 10 万行数据，状态数据 0 和 1 的总和大致相等，如图 3-97 所示，其中表数据 A 列由一位数字 0 或 1 构成，模拟生产中标记或状态等拥有大量相同数据的字段。

而 B 列为随机字符串，模拟常规数据。建立 B 列单列索引和 AB 列组合索引，如图 3-98 所示。

```
SQL> desc comb_a;
名称                                          是否为空？ 类型
 ----------------------------------------- -------- ----------------------------
 ID                                                 NUMBER(10)
 A                                                  NUMBER(5)
 B                                                  VARCHAR2(30)
```

图 3-96　建立 Oracle 组合索引实验表

```
SQL> select count(a),a from comb_a group by a;

  COUNT(A)          A
---------- ----------
     50070          1
     49930          0
```

图 3-97　状态数据 0 和 1 的总和大致相等

```
SQL> select i.INDEX_NAME,i.INDEX_TYPE,i.TABLE_NAME,c.column_name,i.NUM_ROWS,i.VISIBILITY from user_indexes i,
user_ind_columns c where i.index_name=c.index_name and i.table_name = 'COMB_A';

INDEX_NAME       INDEX_TYPE   TABLE_NAME   COLUMN_NAME            NUM_ROWS VISIBILIT
---------------- ------------ ------------ -------------------- ---------- ---------
IDX_COMB_AB      NORMAL       COMB_A       A                        100000 VISIBLE
IDX_COMB_AB      NORMAL       COMB_A       B                        100000 VISIBLE
IDX_COMB_B       NORMAL       COMB_A       B                        100000 VISIBLE
```

图 3-98　建立 B 列单列索引和 AB 列组合索引

虽然建立了两个索引，但是两个索引的效率是不一样的。B 列单列索引效率高，所以先屏蔽 B 列的单列索引。观察 AB 列组合索引的效果，Oracle 索引跳跃扫描如图 3-99 所示，在关闭 B 列索引后，使用到了 AB 组合索引。但是由于 where 后面谓词没有使用到组合索引的先导列，那么按照常规思路会产生全表扫描。但是 Oracle 用它独特的跳跃扫描避开了全表扫描。

```
SQL> alter index IDX_COMB_B invisible;

索引已更改。

SQL> select * from comb_A where b = 'lAT7x21Ngb';

        ID A B
---------- - ------------------------------
       649 0 lAT7x21Ngb


执行计划
----------------------------------------------------------
Plan hash value: 1050187020

--------------------------------------------------------------------------------------------
| Id  | Operation                            | Name        | Rows | Bytes | Cost (%CPU)| Time     |
--------------------------------------------------------------------------------------------
|   0 | SELECT STATEMENT                     |             |    1 |    18 |     5   (0)| 00:00:01 |
|   1 |  TABLE ACCESS BY INDEX ROWID BATCHED | COMB_A      |    1 |    18 |     5   (0)| 00:00:01 |
|*  2 |   INDEX SKIP SCAN                    | IDX_COMB_AB |    1 |       |     3   (0)| 00:00:01 |
--------------------------------------------------------------------------------------------
```

图 3-99　Oracle 索引跳跃扫描

```
Predicate Information (identified by operation id):
---------------------------------------------------

   2 - access("B"='lAT7x21Ngb')
       filter("B"='lAT7x21Ngb')
```

统计信息

```
          0  recursive calls
          0  db block gets
          9  consistent gets
          0  physical reads
          0  redo size
        712  bytes sent via SQL*Net to client
        405  bytes received via SQL*Net from client
          2  SQL*Net roundtrips to/from client
          0  sorts (memory)
          0  sorts (disk)
          1  rows processed
```

图 3-99 Oracle 索引跳跃扫描（续）

这个功能在 20 年前就有，不是新特性。本案例要讲述的是跳跃扫描和正确建立索引的差别对比，以及其他数据库在此场景下的效果对比。从上图可以得知，该 SQL 在索引跳跃扫描以后执行了 9 次的逻辑读。

接下来打开 B 列的单列索引，使得优化器可以识别到 B 列的单列索引。由于 B 列的单列索引对该 SQL 来说效率最高，所以优化器选择了单列索引。Oracle 选择单列索引如图 3-100 所示，命中最高效索引的情况下，发生了 4 次逻辑读。

```
SQL> alter index IDX_COMB_B visible;

索引已更改。

SQL> select * from comb_A where b = 'lAT7x21Ngb';

        ID  A B
---------- -- ----------
       649  0 lAT7x21Ngb
```

执行计划
```
----------------------------------------------------------
Plan hash value: 1066783570

------------------------------------------------------------------------------------------------
| Id  | Operation                            | Name      | Rows | Bytes | Cost (%CPU)| Time     |
------------------------------------------------------------------------------------------------
|   0 | SELECT STATEMENT                     |           |    1 |    18 |     3   (0)| 00:00:01 |
|   1 |  TABLE ACCESS BY INDEX ROWID BATCHED | COMB_A    |    1 |    18 |     3   (0)| 00:00:01 |
|*  2 |   INDEX RANGE SCAN                   | IDX_COMB_B|    1 |       |     1   (0)| 00:00:01 |
------------------------------------------------------------------------------------------------

Predicate Information (identified by operation id):
---------------------------------------------------

   2 - access("B"='lAT7x21Ngb')
```

图 3-100 Oracle 选择单列索引

统计信息
```
----------------------------------------------------------
          0  recursive calls
          0  db block gets
          4  consistent gets
          0  physical reads
          0  redo size
        712  bytes sent via SQL*Net to client
        405  bytes received via SQL*Net from client
          2  SQL*Net roundtrips to/from client
          0  sorts (memory)
          0  sorts (disk)
          1  rows processed
```

图 3-100　Oracle 选择单列索引（续）

由此可见，索引跳跃扫描并不是最优的。只是 Oracle 退而求其次的无奈之举，避免了全表扫描。用实验来说明一下没有索引跳跃扫描的后果。关闭跳跃扫描的索引，该 SQL 回到了理论上缺失前导列的场景，如图 3-101 所示。在该场景下，产生了 362 次逻辑读，比起之前只有个位数的逻辑读次数提高了 40 倍。

```
SQL> alter index IDX_COMB_AB invisible;
```

索引已更改。

```
SQL> select * from comb_A where b = 'lAT7x21Ngb';

        ID  A B
---------- -- ----------
       649  0 lAT7x21Ngb
```

执行计划
```
----------------------------------------------------------
Plan hash value: 941785496

---------------------------------------------------------------------------
| Id  | Operation         | Name   | Rows | Bytes | Cost (%CPU)| Time     |
---------------------------------------------------------------------------
|   0 | SELECT STATEMENT  |        |    1 |    18 |   103   (1)| 00:00:01 |
|*  1 |  TABLE ACCESS FULL| COMB_A |    1 |    18 |   103   (1)| 00:00:01 |
---------------------------------------------------------------------------

Predicate Information (identified by operation id):
---------------------------------------------------

   1 - filter("B"='lAT7x21Ngb')
```

统计信息
```
----------------------------------------------------------
          0  recursive calls
          0  db block gets
        362  consistent gets
```

图 3-101　关闭跳跃扫描的索引

```
        0  physical reads
        0  redo size
      708  bytes sent via SQL*Net to client
      405  bytes received via SQL*Net from client
        2  SQL*Net roundtrips to/from client
        0  sorts (memory)
        0  sorts (disk)
        1  rows processed
```

图 3-101　关闭跳跃扫描的索引（续）

接下来看看在 MySQL 中会如何？建立 MySQL 组合索引实验表如图 3-102 所示。

```
mysql> desc comb_a;
+-------+-------------+------+-----+---------+-------+
| Field | Type        | Null | Key | Default | Extra |
+-------+-------------+------+-----+---------+-------+
| id    | int         | NO   | PRI | NULL    |       |
| a     | int         | YES  | MUL | NULL    |       |
| b     | varchar(30) | YES  | MUL | NULL    |       |
| c     | varchar(30) | YES  |     | NULL    |       |
+-------+-------------+------+-----+---------+-------+
4 rows in set (0.01 sec)


mysql> alter table comb_a alter index idx_b invisible;
Query OK, 0 rows affected (0.03 sec)
Records: 0  Duplicates: 0  Warnings: 0


mysql> alter table comb_a alter index idx_ab invisible;
Query OK, 0 rows affected (0.02 sec)
Records: 0  Duplicates: 0  Warnings: 0
```

图 3-102　建立 MySQL 组合索引实验表

模拟写入数据库过程省略，数据分布和 Oracle 实验完全一样。建立一个组合索引，包含 A、B 两列，A 列在前以及 B 列的单独索引。实验开始先将两个索引全部关闭。

无索引查询之前的 MySQL 逻辑读如图 3-103 所示，当前已经发生的逻辑读请求次数是 23992。由于索引全部不可见，所以这个 SQL 的执行计划是全表扫描。

```
mysql> explain select * from comb_a where b='wqMeAIZRed';
+----+-------------+--------+------------+------+---------------+------+---------+------+--------+----------+-------------+
| id | select_type | table  | partitions | type | possible_keys | key  | key_len | ref  | rows   | filtered | Extra       |
+----+-------------+--------+------------+------+---------------+------+---------+------+--------+----------+-------------+
|  1 | SIMPLE      | comb_a | NULL       | ALL  | NULL          | NULL | NULL    | NULL | 100076 |    10.00 | Using where |
+----+-------------+--------+------------+------+---------------+------+---------+------+--------+----------+-------------+
1 row in set, 1 warning (0.01 sec)

mysql> show status like 'innodb_buffer_pool_read%s';
+---------------------------------------+-------+
| Variable_name                         | Value |
+---------------------------------------+-------+
| Innodb_buffer_pool_read_requests      | 23992 |
| Innodb_buffer_pool_reads              | 1435  |
+---------------------------------------+-------+
2 rows in set (0.01 sec)
```

图 3-103　无索引查询之前的 MySQL 逻辑读

将该 SQL 真实地执行一次，无索引查询之后的 MySQL 逻辑读如图 3-104 所示，可以看到逻辑读请求从 23992 次增加到了 25239 次，增加了 1247 次。

```
mysql> select * from comb_a where b='wqMeAIZRed';
+----+------+------------+------+
| id | a    | b          | c    |
+----+------+------------+------+
|  0 |    1 | wqMeAIZRed | NULL |
+----+------+------------+------+
1 row in set (0.15 sec)

mysql> show status like 'innodb_buffer_pool_read%s';
+---------------------------------+-------+
| Variable_name                   | Value |
+---------------------------------+-------+
| Innodb_buffer_pool_read_requests| 25239 |
| Innodb_buffer_pool_reads        | 1435  |
+---------------------------------+-------+
2 rows in set (0.01 sec)


# Query_time: 0.147735  Lock_time: 0.000011 Rows_sent: 1   Rows_examined: 100000
SET timestamp=1687957500;
select * from comb_a where b='wqMeAIZRed';
```

图 3-104　无索引查询之后的 MySQL 逻辑读

同时根据后台慢日志显示，的确发生了 10 万行的全表扫描，返回一行记录。

接下来打开 B 列的单列索引，有单列索引查询之前的 MySQL 逻辑读如图 3-105 所示，该 SQL 的执行计划是索引扫描，预估返回一行。当前逻辑读请求是 26602 次。

```
mysql> alter table comb_a alter index idx_b visible;
Query OK, 0 rows affected (0.02 sec)
Records: 0  Duplicates: 0  Warnings: 0

mysql> explain select * from comb_a where b='wqMeAIZRed';
+----+-------------+--------+------------+------+---------------+-------+---------+-------+------+----------+-------+
| id | select_type | table  | partitions | type | possible_keys | key   | key_len | ref   | rows | filtered | Extra |
+----+-------------+--------+------------+------+---------------+-------+---------+-------+------+----------+-------+
|  1 | SIMPLE      | comb_a | NULL       | ref  | IDX_B         | IDX_B | 123     | const |    1 |   100.00 | NULL  |
+----+-------------+--------+------------+------+---------------+-------+---------+-------+------+----------+-------+
1 row in set, 1 warning (0.00 sec)

mysql> show status like 'innodb_buffer_pool_read%s';
+---------------------------------+-------+
| Variable_name                   | Value |
+---------------------------------+-------+
| Innodb_buffer_pool_read_requests| 26602 |
| Innodb_buffer_pool_reads        | 1436  |
+---------------------------------+-------+
2 rows in set (0.01 sec)
```

图 3-105　有单列索引查询之前的 MySQL 逻辑读

将该 SQL 执行一次，有单列索引查询之后的 MySQL 逻辑读如图 3-106 所示，可以看到逻辑读请求从 26602 次增加到了 26612 次，只增加了 10 次。

同时根据后台慢日志显示，只发生了 1 行的扫描，返回一行记录。

接下来，关闭 B 列单列索引，打开 AB 两列复合索引。无前导列查询之前的 MySQL 逻辑读如图 3-107 所示，当前已经发生的逻辑读请求次数是 28470。由于 SQL 没有用到复合索引的前导列，所以这个 SQL 的执行计划是全表扫描。

```
mysql> select * from comb_a where b='wqMeAIZRed';
+----+------+------------+------+
| id | a    | b          | c    |
+----+------+------------+------+
|  0 |    1 | wqMeAIZRed | NULL |
+----+------+------------+------+
1 row in set (0.01 sec)

mysql> show status like 'innodb_buffer_pool_read%s';
+---------------------------------+-------+
| Variable_name                   | Value |
+---------------------------------+-------+
| Innodb_buffer_pool_read_requests | 26612 |
| Innodb_buffer_pool_reads        | 1436  |
+---------------------------------+-------+
2 rows in set (0.01 sec)

# Query_time: 0.001177  Lock_time: 0.000011 Rows_sent: 1  Rows_examined: 1
SET timestamp=1687957856;
select * from comb_a where b='wqMeAIZRed';
```

图 3-106　有单列索引查询之后的 MySQL 逻辑读

```
mysql> alter table comb_a alter index idx_b invisible;
Query OK, 0 rows affected (0.02 sec)
Records: 0  Duplicates: 0  Warnings: 0

mysql> alter table comb_a alter index idx_ab visible;
Query OK, 0 rows affected (0.01 sec)
Records: 0  Duplicates: 0  Warnings: 0

mysql> explain select * from comb_a where b='wqMeAIZRed';
+----+-------------+--------+------------+------+---------------+------+---------+------+--------+----------+-------------+
| id | select_type | table  | partitions | type | possible_keys | key  | key_len | ref  | rows   | filtered | Extra       |
+----+-------------+--------+------------+------+---------------+------+---------+------+--------+----------+-------------+
|  1 | SIMPLE      | comb_a | NULL       | ALL  | NULL          | NULL | NULL    | NULL | 100076 |    10.00 | Using where |
+----+-------------+--------+------------+------+---------------+------+---------+------+--------+----------+-------------+
1 row in set, 1 warning (0.00 sec)

mysql> show status like 'innodb_buffer_pool_read%s';
+---------------------------------+-------+
| Variable_name                   | Value |
+---------------------------------+-------+
| Innodb_buffer_pool_read_requests | 28470 |
| Innodb_buffer_pool_reads        | 1436  |
+---------------------------------+-------+
2 rows in set (0.01 sec)
```

图 3-107　无前导列查询之前的 MySQL 逻辑读

　　将该 SQL 执行一次，无前导列查询之后的 MySQL 逻辑读如图 3-108 所示，可以看到逻辑读请求从 28470 次增加到了 29717 次，同样增加了 1247 次。而且这个和 Oracle 的执行效果一样。

```
mysql> select * from comb_a where b='wqMeAIZRed';
+----+------+------------+------+
| id | a    | b          | c    |
+----+------+------------+------+
|  0 |    1 | wqMeAIZRed | NULL |
+----+------+------------+------+
1 row in set (0.15 sec)

mysql> show status like 'innodb_buffer_pool_read%s';
+---------------------------------+-------+
| Variable_name                   | Value |
+---------------------------------+-------+
```

图 3-108　无前导列查询之后的 MySQL 逻辑读

```
| Innodb_buffer_pool_read_requests | 29717 |
| Innodb_buffer_pool_reads         | 1436  |
+----------------------------------+-------+
2 rows in set (0.01 sec)

# Query_time: 0.150462  Lock_time: 0.000010 Rows_sent: 1  Rows_examined: 100000
SET timestamp=1687957920;
select * from comb_a where b='wqMeAIZRed';
```

图 3-108　无前导列查询之后的 MySQL 逻辑读（续）

同时根据后台慢日志显示，的确发生了 10 万行的全表扫描，返回一行记录。在这个基础上增加前导列条件，有前导列查询之前的 MySQL 逻辑读如图 3-109 所示，当前已经发生的逻辑读请求次数是 30112 次。该 SQL 的执行计划为使用复合索引，且预估返回 1 行。

```
mysql> explain select * from comb_a where b='wqMeAIZRed' and a=1;
+----+-------------+--------+------------+------+---------------+--------+---------+-------------+------+----------+-------+
| id | select_type | table  | partitions | type | possible_keys | key    | key_len | ref         | rows | filtered | Extra |
+----+-------------+--------+------------+------+---------------+--------+---------+-------------+------+----------+-------+
|  1 | SIMPLE      | comb_a | NULL       | ref  | IDX_AB        | IDX_AB | 128     | const,const |    1 |   100.00 | NULL  |
+----+-------------+--------+------------+------+---------------+--------+---------+-------------+------+----------+-------+
1 row in set, 1 warning (0.00 sec)

mysql> show status like 'innodb_buffer_pool_read%s';
+----------------------------------+-------+
| Variable_name                    | Value |
+----------------------------------+-------+
| Innodb_buffer_pool_read_requests | 30112 |
| Innodb_buffer_pool_reads         | 1438  |
+----------------------------------+-------+
2 rows in set (0.00 sec)
```

图 3-109　有前导列查询之前的 MySQL 逻辑读

将该 SQL 执行一次，有前导列查询之后的 MySQL 逻辑读如图 3-110 所示，可以看到逻辑读请求从 30112 次增加到了 30122 次，只增加了 10 次。其效果和 B 列单列索引一致。同时根据后台慢日志显示，只发生了 1 行的扫描，返回一行记录。

```
mysql> select * from comb_a where b='wqMeAIZRed' and a=1;
+----+---+------------+------+
| id | a | b          | c    |
+----+---+------------+------+
|  0 | 1 | wqMeAIZRed | NULL |
+----+---+------------+------+
1 row in set (0.00 sec)

mysql> show status like 'innodb_buffer_pool_read%s';
+----------------------------------+-------+
| Variable_name                    | Value |
+----------------------------------+-------+
| Innodb_buffer_pool_read_requests | 30122 |
| Innodb_buffer_pool_reads         | 1438  |
+----------------------------------+-------+
2 rows in set (0.01 sec)

# Query_time: 0.001157  Lock_time: 0.000013 Rows_sent: 1  Rows_examined: 1
SET timestamp=1687957976;
select * from comb_a where b='wqMeAIZRed' and a=1;
```

图 3-110　有前导列查询之后的 MySQL 逻辑读

这个实验结果展示了 MySQL 没有 Oracle 在执行计划中的跳跃扫描，但是效果和 Oracle 的跳跃扫描一样，都绕过了索引的第一列。区别是 MySQL 需要在 SQL 中写出低区分度的谓词，而 Oracle 是自动识别无视低区分度的第一列。不过如果这一列不是低区分度的，那是都无法实现的。

就以上实验得出的结果如图 3-111 所示，MySQL 也有跳跃索引，在只有 AB 两列复合索引的场景下，在 where 后面只有 B 列的谓词时，有时也能出现跳跃索引的效果（在 Extra 中的注释显示了 Using index for skip scan）。

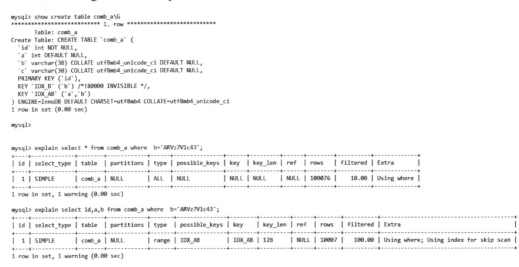

图 3-111　MySQL 也有跳跃索引

这里的区别是产生跳跃索引需要索引覆盖的条件，因为 AB 为复合索引，而 MySQL 中除了主键以外的索引都叫作二级索引或辅助索引，这些非主键索引后面都带着主键。所以 SELECT ID、A、B 列的时候不需要回表；而如果 SELECT* 的时候需要回表，那么就没有索引跳跃扫描了，而是全表查询。

所以由于索引设置不当或者 SQL 编写时谓词不当都会造成全表扫描，数据库有时候可能做出正确的判断，但是仅限特殊场景。更多场景的时候都需要精心设计，否则很容易在大表上引发灾难性的查询，导致故障。

第 4 章

如何进行数据库设计

　　DBA 的主要职责之一就是数据库设计，这个设计指的是数据架构的设计，需要根据实际场景选择适当的数据库架构、具体的数据库的选型进行数据库的拆分和合并。

　　不同行业不同系统，从技术层面来说，抽象到最高，总结成一句话就是：数据是架构的中心。仔细想想，其实 IT 做的一切工作都是围绕着数据开展的：数据的产生，数据的存储，数据的消费，数据的流动……只不过是根据不同的需求，变化数据的形态和服务方式。

　　从某种意义上说，系统=业务逻辑×数据。很多架构问题都出现在数据层。例如，常见的"烟囱式系统"带来的种种问题，特别是数据孤岛问题，其本质上的原因是没有将数据层打通，如果不从数据架构去思考，就可能"头疼医头、脚疼医脚"，治标不治本；如果将数据层（数据库）治理好，就像打通"任督二脉"一样，起到四两拨千斤的效果。

　　从某种意义上说，"单机"就是最好的架构。因为单机（单套）架构就是将数据层面打通了，很多问题得以解决，并且开发和运维的工作量大大降低，系统稳定性大大提高。这里单机的含义并不是否定高可用，高可用是必要的。单机是指不要分库分表，避免不必要的分布式架构。架构设计不好，就会导致开发和运维无穷无尽的问题。当然当业务达到一定的体量（比如日交易 1000 万笔以上）的时候可以考虑使用必要的分布式数据库的架构。

本章内容
数据库都有哪些架构
根据实际场景选择数据库架构
五个维度谈数据库选型
数据库拆分的利与弊
如何看待数据库的合并
CAP 理论与分布式数据库
如何看待数据库与中间件

4.1 数据库都有哪些架构

数据库架构按照部署方式来划分主要分为集中式架构与分布式架构。

集中式架构中，数据库可以是单实例的数据库，也可以是高可用模式的主从架构、主主架构。

分布式架构中，数据库可以是一主多从的全量分布式，也可以是数据分片的均匀分布式。

这些是对数据库狭义的架构定义，都是数据库的外部架构。其实广义上来说数据库还有内部架构，内部架构决定了数据库所能承载的业务场景。

有时候数据库选型很大程度上要适配业务场景。架构可以从技术视角观察，也可以从业务视角观察。既然和业务有适配关系，那么数据库的架构也会受到业务的影响。除了技术上的架构，还有业务上的规划导致的业务拆分而形成的分散性架构和汇聚型架构。

4.1.1 集中式架构

集中式数据库是指数据存储在一个中央服务器上的数据库系统。所有的数据都集中存储在中央服务器，并通过网络对该中央服务器进行访问和管理。在集中式数据库中，所有的数据操作和查询都需要通过中央服务器进行处理。单实例模式就是典型的集中式架构，是只有一个数据库实例可以提供读写服务的部署模式，这种架构在非正式环境中比较多见。

在正式环境中几乎都是要求高可用架构的。在 OLTP 数据库中，主要的高可用架构有主从、主主、多写等。

主从架构也叫作主备架构，是最常见的容灾架构，有一个主节点和至少一个从节点。主节点有读写功能，从节点只有读的功能，不支持写入。因为一旦写入数据，则造成了主从数据库的数据不一致。通常来说，从库会被强制设定为只读角色，当有数据被写入时则会被数据库拒绝。Oracle 数据库的主从容灾高可用架构叫作 ADG，此为数据库自带功能，主库将归档日志发送给备库。备库通过解析归档日志，重放数据库在主库上的变化从而达到镜像功能。数据流向为主库推向从库。ADG 架构下因为主库为单一源头，备库在 19c 以前不可写，所以一致性得到了保证。备库从 19c 开始，间接可写，Oracle 官方文档 DML 重定向如图 4-1 所示，备库接收了写请求后，并没有直接写入备库，而是发给主库，由主库完成后再次接收归档日志进行复制，从而保证一致性。此架构设计得很精妙，如果有备库少量写入的业务场景，则该业务场景可以实现双活。

MySQL 数据库的主从高可用架构有很多，但是大多数为非官方出品。官方推出的高可用架构现在称之为 MySQL InnoDB Cluster。这个架构是由 MGR+Router+Shell 三个部分组成的。

非官方高可用架构有 MHA（不过自从 MySQL 5.6 开始有 GTID 后就没有必要再使用这种架构了，而且 MHA 已经多年无人维护了）和 Keepalive+MySQL 的主主架构等。

PostgreSQL 数据库的容灾高可用也是主从架构，不过个人比较喜欢的是一个主库带两个从库。这两个从库之一与主库实时同步，另外一个从库为延迟从库与主库相差一定的时间，比如 4 个小时或者 8 个小时。这样做的好处是，一旦主库上发生一个 DELETE 的误删除操作，那么第一个从库上的数据也同步被误删除操作所删除。好在延迟从库上还有最近的数据，依靠延迟从库的表和归档日志可以最大限度恢复数据。

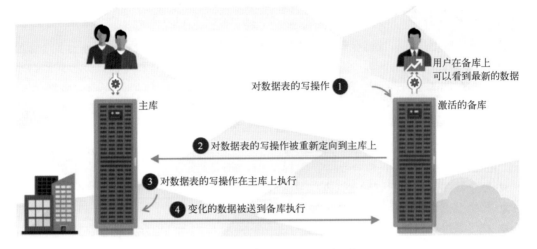

图 4-1　Oracle 官方文档 DML 重定向

通常 OLTP 的数据库都是集中式的：Oracle、MySQL、DB2、SQL Server、PostgreSQL 一般以单机集中式的形式被大家熟知。当然其中很多产品也可以做成分布式。Oracle 的 sharding（Oracle 的分库分表）如图 4-2 所示。

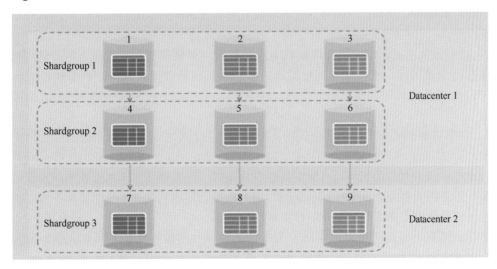

图 4-2　Oracle 的 sharding

其实 Oracle、MySQL、PostgreSQL 等数据库都有其对应的分布式架构。比如 Oracle 现在也支持分布式，只不过大家在认知领域还是觉得 Oracle 的集中式做得最好。当然 OLTP 的分布式数据库也有，如 TiDB 和 OceanBase。而 OceanBase 也推出了自己的单机版，称为分布式一体化数据库。所以集中式数据库和分布式数据库其实没有明显的界限，也没有绝对的优胜者。只是在大多数领域，一般的业务需求集中式数据库能满足。

4.1.2 分布式架构

分布式数据库是指将数据存储在多个服务器上的数据库系统。数据被划分为多个部分，并分布在不同的服务器上。每个服务器都可以独立地进行数据操作和查询。分布式数据库允许在多个服务器上同时进行数据处理，提供了更高的可靠性和性能。

分布式架构有多种实现方式，即分片集群模式、全量主从模式和共享存储模式。

1）分片集群模式是一种将多台（至少 3 台）数据库组合在一起提供服务的模式。集群节点通过共享数据或数据分片来提高负载均衡以提升性能和稳定性。如果采用分片，即数据按照一定的规律平均分散到每个数据节点上，优点如下。

- 每个数据库承载的数据量只占总数据量的 n 分之一，发生全量读的压力相应减小。
- 每个节点都可以读写，形成了多读多写。
- 利用最终一致性来保障数据完整。
- 数据库的集群通过分片可以进行水平扩展（水平扩展和扩展时影不影响业务是两回事）。
- 数据节点通常都有副本，一旦一个节点发生不可读，不影响整体集群服务，业务连续性可以保障。

分片集群需要极高扩展性并需要应用程序为此做出让步，需要应用程序根据键值将请求调度到特定数据库；所以在查询方面受到限制，只能做基于分片键的简单查询，而且不能跨分片进行查询，这也被称为水平分片。

2）全量主从模式，每个数据库承载了全量数据，单独存放在数据库所对应的存储上。写集中在一个数据库，变更到其他数据库，其他数据库承载读请求。这种架构模式与普通集中架构的主从架构的区别是有一个统一的入口，应用程序不用人为干预 SQL 的请求和切换。

3）共享存储模式与全量主从模式的主要区别是数据集中存储，每个数据库实例无须单独存储。

以上是主流关系型的架构，有集中式也有分布式。在当今环境下集中式依然占据很大比重，其原因是绝大多数企业（可能在 99%以上）的数据规模都是 TB 级别甚至不到 1TB，这种数据规模是集中式领域数据库非常容易处理的。

但是依然有一些互联网大厂以及一些行业龙头企业的数据量是非常惊人的，那么就需要分布式数据库的领域。分布式数据库有关系型为主的数据库，也有 NoSQL 的分布式数据库。

分布式数据库主要是分布式存储和分布式计算，拥有存算分离、多副本的特点。代表为关系型的 TiDB、OceanBase、TDSQL、PolarDB 以及 NoSQL 的 MongoDB、Redis、Elasticsearch 等。它们最少是 3 台机器组成集群，有的最少需要 10 台服务器组成集群。这些集群单一节点故障对用户无感知，部分数据库产品做到了任意的扩容与缩容（有的是无感的扩容和缩容）。

4.1.3　数据库内部的体系架构

数据库的内部体系架构涉及各个组件的工作原理，这个架构决定了数据库所对应的业务场景。通常来说，内部体系架构越简单的数据库，其处理的业务场景也越简单。如果选择体系架构复杂的数据库处理简单的场景，属于杀鸡用牛刀；如果选择体系简单的数据库处理复杂业务场景，属于小马拉大车。云和恩墨公司绘制的一张 Oracle18c 的架构图，如图 4-3 所示。该图详细描述了 Oracle 18c 的内部架构和每个进程的工作职责。可以看出，其内部架构十分复杂。Oracle 作为一个通用型数据库适配了绝大多数的业务场景，所以需要复杂的体系架构来支持通用场景的技术要求。

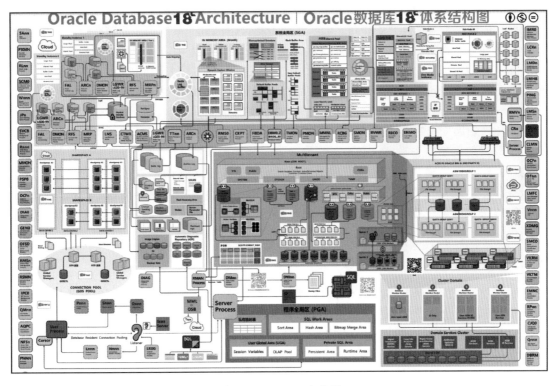

图 4-3　Oracle18c 架构图

4.1.4　"烟囱"式的数据库架构

数据库选型需要适配业务，业务有先建设与后建设之分，也有不同归属建设之分。"烟囱"式的数据库架构就是在这一背景下诞生的。传统信息化建设时，只按照当时局部的需求进行建设，没有全局规划。需要一个建立一个，久而久之，最后从全景图上看就是密密麻麻地建有各种数据库，而这些数据库之间既无数据交互，也毫无业务关联。

常见的"烟囱"式系统带来的种种问题，特别是数据孤岛问题，其本质上的原因就是没有将数据层打通。有些业务是紧耦合的，但是业务又分别建立了自己的数据库。造成这种问题的原因是业务团队各自为政，因为资源核算问题所以一般杜绝复用。而后建系统要和前期的系统做交互。交互的方式多种多样，有接口、数据同步、消息队列甚至中转数据库等。这些无形中又增加了整体的软硬件成本、开发复杂度和运维复杂度。除了个别架构师偏好这种模式以外，大部分开发团队和运维团队都对此持负面评价。

"烟囱"式的数据库架构资源利用率不高。每个数据库都是按照自己的峰值进行采购或申请的，而90%以上的时间是空闲的，所以造成了大量的资源浪费。

"烟囱"式的数据库架构开发效率也是较低的。曾经有一个公司，下属多套系统，有Oracle、MySQL、SQL Server等多种数据库，每种数据库是一个系统。而这些数据库是一个耦合度较高的整体，只是因为先后建设以及投资方不一样导致现在这个局面。每套系统都有自己的开发团队，所以每个系统都要做接口。源头系统变更，后续几个系统也要跟着变更。在信息化主管的力推下，将这些数据库全部统一到一个 Oracle 数据库中，甚至数据表都合并在一个 Schema 下，这样多个分散的系统合并成一个整体系统。所有接口全部取消，原本分系统之间的协同不需要了，效率大大提升，相关应用程序和数据库下线节约了大量的成本。开发团队合并统一安排，人手不足的情况也得到了解决。

避免"烟囱"式的数据库架构的核心思想是：架构层面不应该从技术架构角度去设计，而是从数据架构层面去考虑。必须意识到数据是架构的中心。一切工作都围绕着数据，数据的产生、存储、消费、流动都是根据不同的需求，变化数据的形态和服务方式。若为了微服务分库，或者在中台规划上划分不合理，会导致一个业务流程涉及多个数据库。而分库不能解决数据一致性问题，架构层面也不好控制，还会带来一系列纷乱繁杂的数据库同步需求。

4.1.5 独立业务线的数据库架构

独立业务线的数据库架构是笔者自定义的，不是单机架构，也不是"烟囱"式架构，是指针对不与其他业务线交互的单一业务线的数据库架构。

这种独立的数据库架构有一点类似运营商的架构模式，比如中国移动的业务是全国性的，但是其每个省市的分公司都有自己独立的数据库架构，甚至采用不同的数据库产品。这样即使遇到极端情况，如江苏移动出现问题，也不会影响到浙江移动等其他移动公司。

这种架构通常用于一个大型的系统中，为了避免单一链路上的问题影响到全局。这种架构的关键点在于，每条业务链路很少和其他业务链路相交互。本质上就是各自运行各自的，基本不会发生数据交互。这也就是说按照业务的上下游来划分数据库。

这里要注意的是，如果这个架构的前提不成立，即每条业务链路和其他业务链路频繁交互甚至是紧耦合,那么就不能采用该架构。因为该架构的主要价值是对抗各种风险,主要体现在单一链路不影响全局。如果是紧耦合,那么单一链路故障与全局故障没有明显区别,业务全流程无法走通,没有必要采用这种独立性很强的数据库架构。

4.2 根据实际场景选择数据库架构

同样的数据库在不同的场景中架构是不一样的,这里的场景是指不同的业务要求、不同的环境要求。

数据库领域最严苛的业务场景出现在金融领域,之所以被称为最严苛,是因为金融领域的数据库对强一致性有硬性的规定。例如,转账的场景中数据不允许丢失、不允许重复、不允许错。而强一致性的约束条件在某些互联网场景中没有强制要求,而网络不好情况下导致连发了两条内容一样的微博是勉强能接受的,至少不算故障或问题。基于以上的业务要求,即使同样的 MySQL 架构,在强一致环境下就要使用 MGR 的架构来保证强一致的场景,因为从 8.0.22 的版本后 MySQL 的 MGR 是强一致的;而在一致性要求不那么严格的场景下就可以用 MySQL 的主从架构。

如果有的企业业务量大、并发量大,那么企业级的数据库应用通常采用物理机的部署模式。而如果业务量小,使用的用户少,则可以采用虚拟机的部署模式,甚至可以采用容器化的部署模式。

如果有本地容灾需求,需要做到单节点故障对应用无感知,则需要类似 Oracle 的 RAC 这样的架构。而如果有异地容灾需求,需要主从库或者实时同步的异地库。

如果有人为误操作的可能性但是不可控,则数据库在有从库的基础上还需要一个延迟数据库的架构,以避免误删除的可能性。

当然以上的诉求可能未必是单一来源的,可能是叠加的,既要高可用,又要有一致性和异地容灾等。那么这个数据库架构就会复杂很多,数据库产品很多,不同产品的解决方案不一样。这里仅说明 Oracle 最大可用性体系结构(Maximum Availability Architecture,MAA),此架构目前属于满足金融级别需求的架构。Oracle 的 MAA 架构为高可用性、数据保护和灾难恢复提供了一系列标准——MAA 参考体系结构。每个 MAA 参考体系结构或高可用性层都使用一组最佳的 Oracle 功能,如 Oracle 的 RAC、Oracle 的 ASM、Oracle 的 ADG 等,当这些功能在一起部署时,可以可靠地实现计划外停机和计划内维护事件的目标服务级别。

4.3 五个维度谈数据库选型

数据库种类很多,如何对数据库选型?这是一个老生常谈的话题,毕竟数据库选型确定并部署后很少会改变。因为应用开发几乎都是围绕着数据库进行的,无论使用的开发语言是 Java、Python、PHP 还是其他开发语言,其主体还是 SQL。因为无论使用 Oracle、MySQL、PostgreSQL、SQL Server 还是 DB2 数据库,这些数据库都不能直接识别开发语言中的代码,数据库只能识别 SQL（包括 NoSQL 数据库对应的语法，它可能不叫 SQL。现在越来越多的 NoSQL 都开始支持 SQL，甚至像 Pulsar 这样的中间件都开始支持 SQL 了）。

以建立一个电商网站为例，在线浏览商品，这些商品的信息都是从数据库中检索出来的，那么必须写一句 Select。如果下了一个订单，一定会在订单所在的数据库最终执行一条 Insert。如果进行了一次支付，一定会将账户的余额字段进行一次 Update。

但是不同数据库的 SQL 语法并不是全都一样的，所以在数据库移植的过程中应用程序要做相应的改变，这个改变可能是微小的，也可能是巨大的。所以数据库一旦选定便很难更换，因为更换的代价太大。

每种数据库的体系和使用场景不一样，开发风格和运维风格也大不相同。所以一旦选定了数据库，也间接决定了技术团队的技术路线。

4.3.1 从业务场景特征维度

在 4.1.3 节中提到了数据库内部的体系架构。体系架构越复杂，就可以承载操作数据逻辑更复杂的业务场景；反之体系结构简单的数据库更加合适操作数据逻辑简单的业务场景。通常说的轻量型数据库就是指的体系结构相对简单的数据库，而轻量操作就是处理数据逻辑简单的操作。轻量操作并没有一个统一的标准，但是可以设定一个标准：经过极致优化后单一事务涉及的表为 1～2 个，SQL 语句为 1-3 个，事务时长控制在 100ms 以内为轻量操作。对于数据库来说，一个合格的 SQL 100ms 已经是很宽松的要求了。支付宝的付款、微信的发朋友圈、微博的博文都是简单的数据读写场景，难度在于业务场景的高并发。重量操作也没有统一的标准，也可以设定一个：如果经过极致优化后的主要场景事务要操作 3 个以上的表，事务中的 SQL 大于 3 个甚至几十个，需要 1s 以上甚至多秒的操作都属于重量操作。典型代表为财务结算、企业级应用等。

除了提到的支付、社交场景外，几乎所有互联网场景都属于典型的轻量操作为主的场景，所以该业务场景比较适合使用 MySQL。加上 MySQL 的开源属性，使得在互联网浪潮下非常流行。MySQL 主打的是轻量、快进快出的场景。而企业级的复杂流程、ToB 业务场景、ERP 等都是与 MySQL 最佳实践相背离的，非 MySQL 所擅长。

互联网公司也使用 PostgreSQL，比如去哪儿既使用 MySQL，也使用 PostgreSQL，有两个数据库团队。其实阿里最初也是用 Oracle 的，只是后来因为费用问题以及阿里云的战略提出了"去 O"化。

有些数据库简单场景和复杂场景都适配，这种称为通用型数据库，如 Oracle、DB2、SQL Server，属于全能型选手。有些数据库在逻辑简单的场景下发挥出色，如 PostgreSQL 和 MySQL。还有一些数据库在细分领域表现突出，如各种 NoSQL。所以选型首先看自己的业务逻辑是简单还是复杂，根据场景选择适合的数据库。

4.3.2 从数据规模大小维度

不同数据库对于数据存储的最佳实践是不一样的。有的数据库可以处理 GB 级别的数据，有的数据库可以处理 TB 级别的数据，还有的数据库可以处理 PB 级别的数据。不同企业的数据规模大小不一，国内绝大多数企业的核心数据都是 GB 级别到 TB 级别。而互联网企业的数据量是非常大的，这些大的数据更多是非结构化数据累积的。

在当今的技术成熟度下，互联网大厂、通信运营商、大型银行等体量非常大的企业要么有自己的数据库，要么使用分布式数据库，如 TiDB、OceanBase 等，选择依据之一就是数据规模。而绝大多数企业的数据量都不大，选择集中式数据库甚至是单体数据库也能发挥较好的作用。

单从数据库存储来说，Oracle、MySQL、PostgreSQL 等都可以做得很大。但是从个人最佳实践来说，推荐 MySQL 的单库不要超过 1TB，PostgreSQL 的单库不要超过 5TB，Oracle 的单库不要超过 100TB。所以业务上可以估算数据规模，选择适合自己体量的数据库，过高地估计了数据量会造成浪费，过少地估计了数据量会造成数据库后期维护的成本高。实际上许多公司的数据库规模都超过了以上建议的规模，运行也没有出现过大问题，比如笔者就听说过有的公司 Oracle 数据量超过 500TB。

4.3.3 从用户自身开发团队能力维度

如果用户自身团队的 DBA 能力较强，可以主动分析需求并进行数据库的业务逻辑设计，充分了解数据库开发技术，那么可以考虑选用开源数据库，如 MySQL 和 PostgreSQL 等。当然这里优先考虑的还是业务场景，即使最优秀的数据库设计人员也无法让一个不适用该场景的数据库满足业务场景。在这个前提下，DBA 需要能意识到，开发是把业务诉求合理地转换成计算机语言。这里的合理性很重要，如果业务没说清楚需求，DBA 应该多问几句，直到需求明确。如果需求不合理，那么要加以拒绝。拒绝不合理的需求非常重要。如果用户自身的应用开发人员、集成商应用开发人员或外包应用开发人员不了解数据库，编写 SQL 时没有按照数据库的开发规范去编写 SQL，对于索引、执行计划等数据库基础知识比较薄弱，也不能听从 DBA 的建议和意见，那么需要考虑使用商用数据库，如 Oracle、SQL Server 等。这类数据库有很强的抗冲击能力和自动化运维能力。对于不良的 SQL 容忍度要比开源数据库好一些。

如果是个初创的互联网公司（这里说的是消费互联网而不是产业互联网），在这个公司文化中有着追求技术提升的理念，可以提供较高的薪资待遇招聘有经验的 DBA，并且可以授权 DBA 控制好应用开发的应用程序质量，那么可以考虑开源数据库，如 MySQL 和 PostgreSQL 等。反之如果 DBA 没有话语权，开发人员也对于 DBA 的意见不太重视，那么需要考虑的是商用数据库，如 Oracle、SQL Server 等。

综上所述，数据库选型的一个参考维度是公司和企业的开发人员技术能力。而结合这些因素由 DBA 来选型，当然这里 DBA 要熟悉多种数据库，而不是只熟悉其中一种。

4.3.4 从用户自身运维团队能力维度

运维通常包含主机运维、网络运维和数据库运维等。数据库运维团队包含 DBA，而数据库运维不可避免地涉及主机和网络等。所以数据库运维不单单是 DBA 的事情，也会和其他岗位的工作有交叉。运维团队（含 DBA）水平的高低以及对应用开发人员的管控能力也是数据库选型的一个维度。运维团队（以下主要指运维 DBA），的能力高低也是数据库选型的一个维度。如果运维团队能力强，能够协调控制开发质量，那么企业就可以选用一些维护难度高的数据库及其相关技术栈，比如各类开源组件和分布式数据库。反之，如果运维团队人手不足或者能力不足，无法控制开发质量，那么应该选择一些稳定性较高的数据库，比如使用成熟的商业数据库，避免因为使用过多的技术栈，导致系统复杂度上升，稳定性下降。

许多人对于 DBA 的工作有一些误解。其原因主要如下。
- 有些公司的运维人员（含 DBA）属于传统的基础型运维，没有可以让应用开发人员信服的能力。
- 有些公司运维人员（含 DBA）的岗位和地位低于研发人员。
- 有些公司技术负责人是应用开发出身，对运维人员（含 DBA）不太重视。
- 有些公司的运维（含 DBA）不了解开发工作，没有开发技能。

其实 DBA 的职责之一就是开发管控。在实际生产环境中几乎所有的数据库问题和大部分的中间件问题都来自 SQL 质量。此问题涉及稳定性和业务连续性。SQL 质量带来了稳定性和拆分数据库的问题，进而使得架构再次走向复杂，成本再次上升。拆分数据库后资源不复用，最重要的是无论先建单位还是后建单位都对对方的开发质量没有信心，要求分库来规避应用程序对数据库的冲击。SQL 质量管控是一个需要长期大力度支持的工作，而且前置工作成本越低效果越好。所以 DBA 涉及从需求把握到数据库设计、再到代码实现全流程控制。

Google 的负责运维的 SRE（站点可靠性工程师）在公司内部拥有绝对的话语权，当 VP 面对开发团队和运维团队争论时，只会说："不用跟我说那么多，听 SRE 的就完了！"理由只有一个：你今天不听运维的，明天还得听运维的。所以，SRE 在 Google 中是非常有话语权的。当然，SRE 能够有这样的重要性，是因为他们能够非常详细地帮助开发团队分析开发存在的问题，会造成什么危害，能不能做得更好。

当研发与运维（包括 DBA）意见向左时候，以运维（DBA）为准，研发服从运维。当然前提是运维能力足够强。

4.3.5 从公司管理能力维度

这里的企业管理能力是指企业对信息化系统稳定性的管控能力，体现在对于即使很小的故障也不允许一而再、再而三地发生，对于故障能够深挖问题原因，能够做到举一反三、杜绝同类故障等。如果企业的管理能力足够，那么可以选择的数据库范围就大，商用的和开源的都可以选择。如果这些方面重视程度不够、管控能力较低，那么一定要选择数据库稳定程度高的产品。

4.4 数据库拆分的利与弊

将原本的一个数据库拆开成多个数据库的情况并不少见。随着海量数据的累积，有的企业的数据库在存储和性能上出现瓶颈，或者是在读写分离、微服务以及中台建设上需要重新规划数据库的布局。这些都会导致将原来的一个数据库拆分成多个数据库。拆分数据库可以在一定程度上缓解当下遇到的困难，但是也会带来新的问题。

4.4.1 数据库拆分的背景

在业务数据量累计达到一定的数据规模后，可能就需要考虑对数据库进行拆分和归档。主要原因是可能受制于硬件存储的能力，由于硬件无法扩容或达到上限，需要分散存储压力或者释放存储空间。

数据库的演进就是分分合合的过程。数据库拆分如图 4-4 所示，企业积累了大量数据，单机存储达到极限，那么就要进行数据库的拆分。

1）在线数据库拆分。拆分的原因是业务产生的数据过多，还没到达到归档的时间，但是单一数据库的容量已经很庞大了。这里的庞大是相对的，笔者经历过单表容量达 100TB 的情况，这个容量其实并不大，但是有些使用者可能认为 500GB 的容量就很大了。数据量大小取决于企业的规模，可能有的企业硬件资源少，单机存储和计算能力都不足，所以在达到单机数据库的处理上限后，就需要将数据库拆分成多个小数据库。

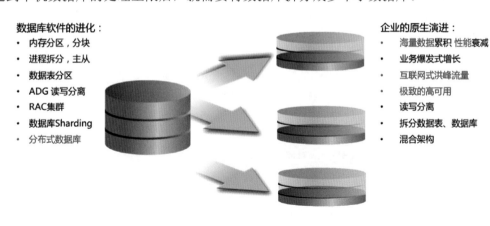

图 4-4　数据库拆分

2）在线数据库归档。归档是将在线数据库中已经不需要存储的数据，归档到一个历史数据库中，以备不时之需，从而减少在线数据库的存储容量。通过碎片整理等方式释放在线数据库的磁盘空间。

3）在线数据库性能不足。刚才提到的数据量的庞大是非常主观的。而客观上可能是达到性能瓶颈时会考虑拆分数据库。性能瓶颈虽然客观但是也涉及诸多因素，最主要的是 SQL 效率。如果数据库设计得好，SQL 质量高，100TB 的容量也不会出现性能问题。反之如果数据库设计得糟糕，SQL 质量低下，那么 100GB 的容量也可能出现性能问题。

以上是数据库拆分最常见的 3 个原因。在当今的主流数据库版本和硬件的环境中，数据库的容量和处理能力上的问题已经基本解决了，而性能不足的问题在开发质量可以掌控的前提下，也是可以解决的。

4.4.2 数据库拆分的三大问题：一致性、数据关联、数据同步

当数据库从一个紧密的整体数据库拆分成若干个个体数据库后，数据库就分成了若干个独立的数据库。彼此之间再无关系数据库的一致性、原子性的保障。拆分虽然缓解了单机存储和计算能力不足的问题，也变相缓解了不良设计和应用程序效率低的问题，但是仅仅是缓解，拆分数据库后带来的副作用也是很大的。

1）数据一致性问题。由于在一个关系数据库中，遵循着 ACID 的中的一致性，但是多个数据库之间的一致性很难甚至无法保证，即使引入分布式锁和在架构上控制两阶段提交，但是对于链路很长的事务来说依然不能 100%保证。在数据的一致性都无法保证的情况下，正常的业务逻辑都很难走下去。

2）跨库数据关联问题。数据库中纯数据的拆分（不是拆分表）难度不大。但是拆分数据库以后，还是无法避免多个数据库之间的关联查询。比如会员数据库、订单数据库和物流数据库要进行关联查询才能得到需要的数据，而此时由于数据库的拆分则需要跨网络进行跨库关联。这样的架构性能低下，具体实验数据会在 5.3.1 节中给出。

3）数据同步问题。当数据库集中在一起时，支付订单费用使得订单生效以及扣除用户资金是在一个事务中完成的。而现在需要在接口完成，效率比起原有的效率要低很多。所以拆分数据库后交互的性能会下降。

4）数据汇聚问题。数据同步首先要保证实时性和一致性。一致性是一个无法绕开的问题。如果要保证实时性和一致性，那么只有通过日志级别的同步技术才可以达到。这时候需要引入更多的技术栈，如 OGG 和 Flink CDC 等技术。而这些技术是需要大量人力来保障的，一般来说要一个小规模的团队。但是即使使用这些较为成熟的技术依然会带来一些小概率的问题，如上游数据库增加默认值字段带来的旧数据无法更新问题，上游数据库大事务带来的通道堵塞的问题等。导致开发人员在设计实现的时候考虑了太多问题，以至于单凭借开发人员是无法完成的，需要运维和架构一起完成。其代价得高以至于在一些企业中无法实现。

正是因为以上的问题，导致业务开始抱怨数据不准、查询数据效率低下，以及需求周期长和代价大，以至于许多开发要么寻求分布式数据库的解决方案，要么考虑把数据库重新合在一起。

4.4.3 分表带来的问题：一致性、聚合、排序、扩缩容

分库是将数据库按照功能分开，即上面提到的数据库拆分。而分表一般是在分库的基础上将数据库再次按照一定的规则分解形成分表。这个带来的好处就是通过水平扩展降低低效 SQL 对数据库的冲击。但是带来的问题也是显而易见的，会在分库的基础上继续增加查询难度。单库单表订单查询如图 4-5 所示，单库单表的查询就是一个简单的 SQL。

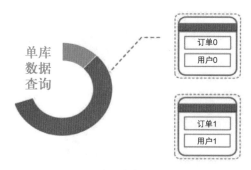

图 4-5 单库单表订单查询

而如果是分表查询，则跨库多表订单查询如图 4-6 所示。如果分库的维度是用户维度，而查询的维度是时间维度，那么一定时间内的用户数据必然是跨所有数据库和相关的数据表的。

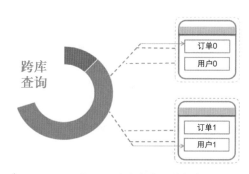

图 4-6 跨库多表订单查询

在真实的环境中由于分库分表，客户端和应用程序要么通过一个中间件（包含数据库和数据表的路由规则）连接数据库，要么在应用程序中写定（数据库和数据表的路由规则）。分别从各个数据库的各个分表中提取结果集。跨库汇聚集合如图 4-7 所示。

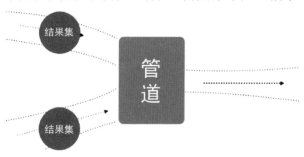

图 4-7 跨库汇聚集合

在结果集汇聚以后需要再次做分组聚合以及排序，如图 4-8 所示。因为每个结果集的分组和排序与全局的分组和排序无法保证是一致的。

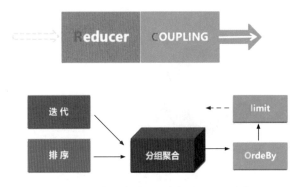

图 4-8　跨库汇聚再次分组或再次重新排序

从这个流程来说，开发人员的工作量是上升的。而且跨库查询的一致性依然没有办法保证。

此外，分表一般采用的都是按照一个维度平均拆分或采用哈希拆分，其目的是平均存储数据以便分摊压力。这样做的目的也是当性能（SQL 全表扫描导致）再次达到瓶颈或存储达到瓶颈时可以水平扩展。

但是在做水平扩展的时候请注意一点。比如一个表有 1000 万数据，平均分布在 10 台机器上，每台机器上的数据表是 100 万。现在需要扩展为 20 台，为了数据均匀，可能的做法是将每台旧机器上的 50 万数据迁移到新机器上，而且要对迁移的 50 万数据进行删除。这个过程为了保证数据的一致性，最好是停止外部写入，也就是停止对外服务进行迁移。这个代价是很大的，所以支持水平扩展和不停机水平扩展是两个概念。

数据库拆分带来的是弊大于利，其开发成本和运维成本比较高。只有深切经历过集中式数据库和分库后开发运维工作的人才能深刻体会到个中滋味。

4.5　如何看待数据库的合并

数据库架构在分分合合中不断演进，每次变革都是针对当下的问题采取了符合当前实际的方案，用来解决当前的主要矛盾。而在经历了大量数据库拆分后，当下数据库合并的想法受到了越来越多的关注。

4.5.1　为何要做数据库合并

2010 年左右，由于数据量的增加和硬件条件的制约，拆分数据库逐渐流行。而在 2020 年左右，由于拆分数据库给应用开发人员和架构师带来了很多副作用，加上硬件技术的进步，现在越来越多的应用开发人员也在思考过度拆分数据库对开发效率和稳定性都有较大的影响，开始考虑数据库的合并。

这是因为当数据库进行了合并后，对应用开发人员来说降低了工作量，对运维（含 DBA）人员来说降低了工作难度，对企业来说降低了成本。

4.5.2 数据库合并的意义：降成本、提升稳定性

数据库合并有着诸多收益。这些收益在 IT 支出成本上和系统稳定性两个大方向上有明显的体现。

1）一致性：单机关系数据库的基本功能，对于分库来说是无法 100%保证的巨大鸿沟。在分库的业务系统中，几乎每天都为了数据的准确性而动用大量的人力物力去弥补一致性不足带来的遗留问题。没有一个关键业务可以容忍数据是不准确的或者是错误的。数据要再三确认以及多方核实才能使用，对业务方来说体验不好。而将耦合度较高的数据库合并，轻而易举解决了一致性问题。

2）业务需求：当使用数据时需要去不同数据库上获取数据，再自行组合。而采用 Hadoop 技术又只能解决 T+1 的离线场景，业务对于实时的跨库关联数据有迫切的需求。而将业务上关联度较高的数据库合并，解决了数据关联的问题。

3）架构瘦身：合并数据库后负责接口和转发的应用服务器不被需要了，负责异步和解耦的消息队列也不被需要了。系统架构可以成倍缩小。

4）降低开发复杂度：分库后开发需要在不同数据库直接做接口。而合并数据库后只要通过授权就可以直接获得其他的不同 Schema（原本属于不同数据库实例）的数据。开发难度和工作量都大幅下降。

5）性能提升：同库的直接访问比接口调用要快 10 倍到 10 万倍（有些接口写得差可能是分钟级的调用）。

6）缓解运维压力：维护一个数据库实例和两个 Tomcat 的工作量比起维护 10 个数据库实例以及几十个 Tomcat 还有各式各样的消息队列来说，难度不是线性的而是指数级别的上升。

7）降本增效：减少了大量的数据库和中间件，这些资源可以释放或退租。甚至可以缩减开发团队的规模。

以上 1~4 条属于 IT 的成本支出，5~7 条属于系统稳定性的提升。

4.5.3 数据库合并带来的问题：鸡蛋放在一个篮子里

数据库合并有没有风险？答案是有。若干个数据库合并在一起存在一定的风险。如果一个数据库损坏则意味着全部数据丢失，或者一个数据库发生性能问题，则全部数据库都可能因为个体性能而受到影响。

要解决以上问题首先要说明几个概念。

1）拆分不一定风险小。因为微服务等原因将一个整体系统拆分成 n 个数据库。虽然任意数据库出现性能问题导致短暂宕机不影响全体数据库，但是由于是一个整体，最终流程依然无法走通。只是略微缓解了问题同时发生的时间点，而没有从根本上解决全局的问题。由于单一数据库问题导致的全局数据的回退或者纠偏工作量不比合并数据库的代价小。

2）合并不一定风险大。对于数据库来说，只要遵守数据库的开发规范（前提是制定了），则不用担心合并数据库后带来的性能问题。

3）可以选择独立业务线的数据架构。在4.1.4节提过，可以按照几乎不交互的业务线来减少数据库，也是一种折中的架构方式。既做到了一定的拆分，也把需要关联的数据库有所合并。

4.5.4　数据库合并的前提：高质量 SQL、硬件的进步、稳定的基础环境

高质量 SQL 是数据库合并的前提保障，合并的关键是能不能遵守开发规范。因为遵守了开发规范的 SQL 一定是高效的 SQL。曾经有开发团队问过笔者，新建项目是否可以复用之前的数据库。其实原有数据库负荷不大，于是笔者就问："你们能遵守开发规范吗？"开发团队想了想说："那我们不复用数据库了。"言下之意就是做不到遵守规范，无法避免低效 SQL，造成的数据库问题。

高性能的硬件是数据库合并的物理基础，10 年前的硬件和现在的硬件已经不可同日而语。四分之一的 Exadata 一体机硬件性能指标如图 4-9 所示，仅仅凭借其强大的 IO 吞吐和存储就已经可以满足数据库合并的要求了。

型号	Exadata X9M-2 HC 1/4配
Exadata 数据库服务器	2台
数据库服务器总CPU Cores	128个
数据库服务器总内存容量	2TB （最大扩展到4TB）
内部高速互联交换机（36端口100GB/s RoCE网络交换机）	2台
管理交换机（Management Switch）	1台
Exadata 分布式智能存储服务器（HC高容量配置）	3台
存储服务器CPU总核数	96个用于SQL卸载处理
持久化内存（TB）	4.5TB
NVMe PCIe闪存卡（TB）	76.8 TB
SAS-3磁盘容量（TB）	648 TB
两副本冗余后可用容量(TB)	245.4 TB
三副本冗余后可用容量(TB)	192.4 TB
最高SQL闪存带宽	135GB/s
最高SQL PMEM 读取IOPS	560万
最高SQL 闪存 写入 IOPS	184.2万

图 4-9　四分之一的 Exadata 一体机硬件性能指标

稳定的基础环境是数据库合并的基础。数据库需要与之配套的硬件冗余和集群使得基础环境在部分损坏的极端情况下依然可以对外提供服务，如 Oracle 的 RAC 和 MySQL 的 MGR。尤其是 Oracle 的一体机自带 RAC，存储多副本，最大限度上实现了软硬一体的基础环境和物理环境，属于性价比极高的解决方案。

其实数据库合并只要做好一件事，那就是编写高质量的 SQL，在此基础之上，其他分库分表导致的问题都可因为数据库的合并而自然而然地解决了。

4.6　CAP 理论与分布式数据库

CAP 理论主要是针对分布式数据库而言的，在集中式数据库中不涉及 CAP 理论。因为分布式使得数据库围绕着一系列的问题展开了复杂的理论论证，而这一切的根源是一致性与

分布式的扩展性在一定程度上是对立的。

4.6.1 CAP 理论概述

CAP 理论是指计算机分布式系统的三个核心特性：一致性（Consistency）、可用性（Availability）和分区容错性（Partition Tolerance）。

1）C：Consistency 即一致性，访问所有的节点得到的数据应该是一样的。注意，这里的一致性指的是强一致性，也就是数据更新完，访问任何节点看到的数据完全一致，要和弱一致性、最终一致性区分开来。

2）A：Availability 即可用性，所有的节点都保持高可用性。注意，这里的高可用性还包括不能出现延迟，比如节点 B 由于等待数据同步而阻塞请求，那么节点 B 就不满足高可用性。也就是说，任何没有发生故障的服务必须在有限的时间内返回合理的结果集。

3）P：Partition Tolerance 即分区容错性，这里的分区是指网络意义上的分区。由于网络是不可靠的，所有节点之间很可能出现无法通信的情况，在节点不能通信时（通常也可以理解为节点故障或损坏），要保证系统可以继续正常服务。一般分布式数据库都要满足这个条件，即数据分开分散不是集中在一个数据节点或单一数据节点的。在一定程度上也可以说数据库的主从也是满足这个条件。

在 CAP 理论中，一致性指的是多个节点上的数据副本必须保持一致；可用性指的是系统必须在任何时候都能够响应客户端请求；而分区容错性指的是系统必须能够容忍分布式系统中的某些节点或网络分区出现故障或延迟。CAP 理论认为，分布式系统最多只能同时满足其中的两个特性，而无法同时满足全部三个特性。这是因为在分布式系统中，网络分区和节点故障是不可避免的，而保证一致性和可用性需要跨节点协调，这会增加网络延迟和系统复杂度。

4.6.2 CAP 理论的延展

分布式数据库必然受到 CAP 理论的约束。如果选择了 P 为基准条件或前提条件，那么 A 和 C 只能满足一个，舍去一个，如果重视一致性，那么就是 CP 类型，这类架构代表的数据库产品有 MongoDB、HBase；如果重视可用性，那么就是 AP 类型，这类架构代表的数据库产品有 Cassandra、CouchDB。同理，如果选择了 C 为基准条件或前提条件，那么 P 和 A 只能满足一个，舍去一个，如果重视分布式，那么就是 CP 类型，这个组合已经出现过；如果重视一致性，那么就是 CA 类型，即集中式部署非分布式，这也就是传统的关系数据库。

而通常说的分布式数据库一定要选择 P 这个要素，否则就不满足分布式这个特性。而作为典型分布式架构的分片式（Sharding）就是受到了这个约束，要么为了一致性牺牲一部分性能，要么为了性能牺牲一部分一致性，实现最终一致性，而不是强一致性。最终一致性是指大多数数据节点完成，即可进入后续数据处理之中，少数节点后续慢慢同步数据。区块链应用场景中的比特币就是采用这种方式。

Sharding 通常指的是将数据打散分布在不同的数据节点上。比如 Redis 的 Cluster 就是典型的 Sharding 架构，属于 AP 类型。也有基于第三方或中间件的架构和解决方案，比如基于 MySQL 数据库做分库分表，也属于 AP 类型。

数据库技术在不断进步，CAP 理论还没有办法突破，但是都在不断地提升边界范围。CAP 范围的延展如图 4-10 所示。在传统关系数据库（Traditional RDBMS）上增加一定的扩展性，而降低一定的一致性，就产生了云数据库（Cloud DBMS）。在 NoSQL 数据库的基础上增加一些一致性，而降低一些扩展性就形成了 NewSQL 数据库。这一切背后的原因都是 CAP 理论无法同时满足。

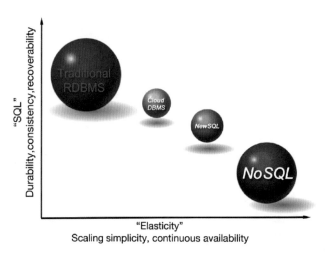

图 4-10　数据库 CAP 范围的延展

4.6.3　分库分表不是分布式数据库

分库分表不是分布式数据库。因为分库导致各个数据库物理隔离，原子性和一致性没有保障。而分表使得跨分片的查询变得非常困难。分布式数据库对外是一个整体，有全局事务控制，是满足一致性甚至是强一致性的数据库。

在 CDH（Cloudera Hadoop）的全家桶中的数据库，虽然不拥有强一致性，但是拥有最终一致性，所以也算分布式数据库。

而如果由若干个独立的数据库为基础，依靠应用开发人员在应用程序控制全局事务、分布式锁、两阶段提交的数据库组合，不能称之为分布式数据库。

分库分表和分布式数据库最主要的区别是有没有全局一致性。

分库分表在国内提及较多，在国外提及较少，并不是说明国内的数据量比国外的数据量多。而是在国外很多开发人员对数据库比较了解，其开发的应用程序 SQL 质量较高。不需要分库分表这种解决方案。

能不拆分的数据库最好不要拆分，而分表的问题比分库更加多。如果了解数据库性能，那么对于绝大多数企业来说，集中式数据库的单机架构都可以应付业务场景，而不需要分库分表。如果个别企业超出了单机的上限，那么可以考虑分布式数据库。

4.7 如何看待数据库与中间件

数据库是 IT 需要中的一个重要环节，但是不是唯一环节。数据库在中间件、操作系统、存储设备这些环节中起到承上启下的作用。而操作系统和存储设备是不需要直接和应用程序打交道的，而且稳定性极高。只有数据库和中间件是直接与应用程序交互，而且是紧密交互。两者中任意一个出现问题都会导致全链路受到影响。

4.7.1 数据库与中间件：上下游的"难兄难弟"

这里中间件一般指的是 Tomcat 和消息队列。

Tomcat 常见的问题是频繁 FullGC 和 OOM。通常来说，出现这两种问题极有可能是从数据库中提取了大量的数据，可能一次是几十万，也可能是几百万，甚至是上亿行的数据。或者数据量不是非常大，但是数据的对象非常大，比如有大的二进制字段存放的大容量数据。

以上具体达到什么阈值会导致发生频繁 FullGC 和 OOM 具体要视 Tomcat 所在服务器的配置以及 JVM 的情况而定。有些配置较低的 Tomcat 可能在提取几万数据的时候就会遇到这两种情况。

而消息队列作为中间传输和解耦的环节，如果从数据库中获取大量数据向消息队列中写入，消息队列来不及写入或消费（读取消息队列数据写入另外一个数据）速度远远低于写入速度，都会导致消息队列堆积，通信中断。

有时候从 Tomcat 上发出了一个上亿表的 COUNT 统计导致数据库不能服务。也有时候数据库上返回了全字段的几百万数据导致 Tomcat 不能服务。所以说数据库和中间件是紧密的伙伴和相互影响的上下游。

4.7.2 中间件内存溢出的原因：源头还是数据库

中间件通常由运维来维护，但是由于如今数据库重要程度越来越高，所以中间件和数据库可以视作一个整体来维护。

案例 1：用户登录读取一个规则表做判定。这个规则表的行数只有个位数。但是有人在维护规则表时不小心写入了 1 万多行规则数据。JVM 随即由每次获取几条数据，而变成每次获取 1 万多行数据。很快多个负载均衡的 Tomcat 都出现了频繁 FullGC，最终导致了 Tomcat 不可用，酿成了故障。

案例 2：从一个几百万的业务表中根据单号取一条数据。结果开发页面新功能时，漏传了单号条件。该 SQL 就变成了 SELECT*FROM 业务表。导致直接返回了几百万行数据。只要该页面打开，JVM 就从数据库直接提取几百万行数据，即使不停重启 Tomcat 也无法解决，酿成了故障。

可以看出两个案例中 Tomcat 出故障的原因不一样，但是根源都是取数据过多。而取数据的是 SQL，所以其源头还是和数据库有关。希望这两个案例对大家有所帮助。

通常每次发生频繁 FullGC 和 OOM，可以检查一下数据库当时有没有大量提取数据的

SQL，大概率是找得到的。定位问题也就容易了。

4.7.3 适当减少数据库与中间件的数量：避免不必要的故障节点

曾经有人问笔者，如何处理一个扣款的场景。笔者不假思索地回答，将余额这个字段设置为非负数。每一笔交易都去减余额，遇到最后一次如果为负数，数据库自动校验不通过就可以了。提问者觉得就这么简单吗？没有其他架构层面的支持吗？这里没有考虑到消息队列、异步处理等环节，是不是草率了？

其实这个问题很容易回答，数据库是物理学和数学的结合，最终都会落实到这两个层面上。以上问题涉及数学运算和物理时间。如果余额为 10 元，那么无论怎么优化都无法同时购买一件 5 元和一件 6 元的商品，这是基本常识。

所以既然最终都会面对一个成功、一个失败的结果，那么为什么中间还要去异步处理？或者增加消息队列等画蛇添足的环节？两个扣款的事务理论上都不超过 50ms，经过异步设计，进队列、出队列两个步骤可能花费几秒，不仅效率大大降低，而且先后顺序不能保证。而在一个数据库下，一切会变得简单。不仅仅是逻辑简单了，而且架构也简单了。

当单机写入性能不足时可以考虑消息队列来削峰填谷，但是绝大多数业务场景都没有达到单机的写入上限。对外进行异步解耦可能会需要消息队列，但是尽可能不要滥用，否则对资源来说是浪费，对稳定性来说多了一个故障的环节。

DBA 不能只管数据库。许多中间件故障和问题之根源或定位也需要 DBA 参与。这也对 DBA 的技能提出了要求，需要 DBA 掌握更多的技能。

第 5 章

如何做好数据库之间的数据同步

在日常工作环境中,为了满足跨库查询的需求,数据库之间经常需要进行数据传输。这种数据传输过程需要将多个数据库(无论它们是同构还是异构的)的数据汇聚到一起,并确保多个数据库之间数据一致,这个过程被称为数据同步。而在数据同步的过程中往往会面临各种问题。作为 DBA,解决这些问题、实现良好的数据同步是必修功课。

本章内容
数据同步的作用
数据库同步的分类
同构数据库数据同步的示例
异构数据库同步的实例——基于 **OGG**

5.1 数据同步的作用

数据同步在数据库管理中扮演着重要的角色，涵盖了数据传输、数据汇聚和数据迁移三个方面。它的目标是确保数据库间的一致性，并满足业务需求。

5.1.1 实现数据传输

由于各种原因，相关的业务数据可能分散在不同数据库中，即数据层未打通。而业务上需要这些数据来支持，所以数据就需要进行传输。数据传输流向如图 5-1 所示，这种传输可能是单向的，也可能是双向的，可能是单向数据同步的实时报表场景，也可能是双向数据同步的数据高可用场景。其中，单向实时报表是指将数据从一个或多个源数据库传输到报表数据库，以生成实时报表和分析数据。这种传输是单向的，即数据只从源数据库流向报表数据库，不进行反向同步。而双向高可用是指通过实现数据库之间的双向同步，实现数据的互写和故障切换。这种架构旨在提供高可用性和容错性，以确保在一个数据库发生故障时能够无缝切换到另一个数据库。

图 5-1　数据传输流向

5.1.2 实现数据汇聚

数据汇聚是典型的分库后导致数据无法关联查询的补救措施。它通常是指将多个数据库中的数据合并到一个数据库中，以便实现数据的关联查询和综合分析。如图 5-2 所示，更多真实业务场景的需求是图右侧的将多个数据库的数据汇聚到一个数据仓库，以支持跨数据源的分析操作。

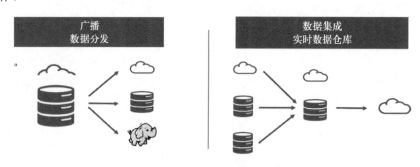

图 5-2　数据分发和数据汇聚

5.1.3 实现数据迁移

数据同步还有一个重要作用就是数据迁移。数据迁移是特殊的数据同步，特殊的点是它只有很短暂的过程，而不是日常业务。

数据库迁移需要数据同步。当数据库运行的硬件环境开始老化时，采购新的硬件服务器到位后需要数据库进行一次迁移。这个迁移只有很短的停机窗口，甚至可能没有停机窗口。

数据库升级需求数据同步。当数据库运行的版本已经没有支持，而新版本特性又能解决当下业务痛点时，需要对数据库进行升级。数据库升级类似于迁移，不过可以争取到短暂的停机窗口。

5.2 数据库同步的分类

5.2.1 同构数据库的同步

同构数据库的同步是指同数据库产品同版本或跨版本的数据传输。典型的是数据库主从数据同步，Oracle 的 ADG、PostgreSQL 的流复制以及 MySQL 的主从都属于这种。或者是通过第三方数据工具实时捕获解析上游数据库的日志进而将数据变化传递到下游数据库，Oracle 的 OGG 和 FlinkCDC 属于这种。

5.2.2 异构数据库的同步

异构数据库的同步是指不同数据库产品的数据传输，比如 MySQL 到 Oracle，即使是同一个公司的产品也是异构数据库。甚至是同版本的数据库但是由于是异构操作系统，而导致无法建立主从关系，如 Windows 操作系统下的 Oracle 和 Linux 操作系统下的 Oracle。

而现在许多国产数据库上线涉及从 Oracle、MySQL、SQL Server、DB2 这些数据库迁移到国产数据库，需要长期共存涉及诸多异构数据库的同步。

5.2.3 数据库到消息队列的同步

关系数据库与关系数据库同步是直接写入的。而如果是涉及接口或解耦的情况下进行数据同步，可能数据不是直接放在关系数据库中，而是放在消息队列中。比如将 Oracle 和 MySQL 的数据变化放到 Kafka 和 RabbitMQ 之中，后面由其他系统结合自己实际需要进行订阅读取。

5.2.4 数据库到 Hadoop 的数据同步

虽然大多数系统中，只要符合软件工程的管理，基本上单机数据库都能解决自身业务场景。但是这种理想情况极少，许多系统中都涵盖了 Hadoop 这类大数据系统。需要将关系数据库中的数据实时传输到 Hadoop 中处理。通常有三种方案：

- 将数据送到消息队列中，由开发人员编写订阅程序，从消息队列取数据进行加工处理。
- 将数据送到 HBase 中，由开发人员完成后续数据处理。

- 将数据送到 Hive 中，由开发人员完成后续数据处理。由于 Hive 处理效率低，通常会采用 Impala 来处理，而 Impala 是映射 Hive 的数据。低版本的 Hive 不支持写，开发人员加工和处理的工作量还是不小的。造成这个问题的原因主要是早期 Hive 的不成熟。

5.3 同构数据库数据同步的示例

每种数据库同步方式都有其自身的特点和适用性，用对了可以解决许多问题；反之用错了就会带来很大的问题。而同步方式本身并没有问题，存在的就是合理的，每种方式必定是对应最初产生的需求而开发的。

5.3.1 使用 dblink 实现 Oracle 数据库之间的数据同步

当需要在不同的 Oracle 数据库之间进行数据同步时，dblink 是一个非常实用的工具。不过在现实工作中，dblink 始终承受着不同程度的诟病。之所以出现这种毁誉参半的现象，主要是对其定位不清和使用不当。本节会讲述 dblink 的使用场景和注意事项，正是这些使用场景的不当和注意事项的疏忽才造成了它在使用中的问题。而如果正确使用，它可以发挥出很大的作用。

首先 dblink 可以帮助用户在不同的数据库之间建立连接，实现数据的快速传输和访问。这是 dblink 的主要功能之一，在本节中介绍如何使用 dblink 来实现 Oracle 数据库之间的数据同步。

1. 使用 dblink 实现数据同步

本例用两台同网段 IP 结尾分别为 101 和 103 的服务器（下文分别用 101 和 103 代指，其中 103 是上游数据库，101 是下游数据库）进行数据同步，两台服务器能够使用各自的管理员账号登录。

在 101 数据库上登录，如图 5-3 所示，下游数据库的用户名和密码都是 tang。

```
[oracle@dev-dbteam-143101 ~]$ sqlplus tang/tang@t

SQL*Plus: Release 19.0.0.0.0 - Production on 星期二 5月 16 16:10:00 2023
Version 19.3.0.0.0

Copyright (c) 1982, 2019, Oracle.  All rights reserved.

上次成功登录时间: 星期二  5月   16 2023 15:32:47 +08:00

连接到:
Oracle Database 19c Enterprise Edition Release 19.0.0.0.0 - Production
Version 19.3.0.0.0
```

图 5-3　登录下游数据库

在 103 数据库上，也进行用户登录。如图 5-4 所示，用户名是 tang2，密码是 tang。这里特意用了与下游数据库不同的用户名以示区别。

```
[oracle@adg1 ~]$ sqlplus tang2/tang@t

SQL*Plus: Release 19.0.0.0.0 - Production on 星期二 5月 16 16:09:12 2023
Version 19.3.0.0.0

Copyright (c) 1982, 2019, Oracle.  All rights reserved.

上次成功登录时间: 星期二 5月   16 2023 15:45:14 +08:00

连接到:
Oracle Database 19c Enterprise Edition Release 19.0.0.0.0 - Production
Version 19.3.0.0.0
```

图 5-4　上游数据库登录

下面演示从 101 数据库连接到 103 数据库。首先在 101 数据库上配置能够登录 103 数据库的连接信息。该文件是下游数据库主动发起连接上游数据库的连接信息，如图 5-5 所示。

```
[oracle@dev-dbteam-143101 ~]$ vim /u01/app/oracle/product/19.3.0/db/network/admin/tnsnames.ora
```

在配置文件中插入：

```
t3 =
  (DESCRIPTION =
    (ADDRESS = (PROTOCOL = TCP)(HOST = 10.60.143.103)(PORT = 1521))
    (CONNECT_DATA =
      (SERVER = DEDICATED)
      (SERVICE_NAME = tang)
```

图 5-5　修改 tnsnames 文件

注意：需要修改目标地址以及目标服务名。HOST 的地址是上游数据库的地址，PORT 是上游数据库的监听端口。SERVICE_NAME 是上游数据库的数据库名（恰巧和用户名同名而已）。t3 是个别名，可以起任何符合规定的名字，a1、hhhh 等都可以。

配置完毕后，尝试从 101 数据库服务器远程登录 103 数据库上验证连接是否正常。
注意：登录的用户名密码需要使用 103 数据库上的用户名及密码，如图 5-6 所示。

```
[oracle@dev-dbteam-143101 ~]$ sqlplus tang2/tang@t3

SQL*Plus: Release 19.0.0.0.0 - Production on 星期二 5月 16 16:18:05 2023
Version 19.3.0.0.0

Copyright (c) 1982, 2019, Oracle.  All rights reserved.

上次成功登录时间: 星期二 5月   16 2023 16:09:14 +08:00

连接到:
Oracle Database 19c Enterprise Edition Release 19.0.0.0.0 - Production
Version 19.3.0.0.0
```

图 5-6　在下游数据库远程登录上游数据库

看到连接成功表示 dblink 的基础条件都已经具备。首先在上游的 103 数据库上建立一张目标表，并写入一些测试数据。如图 5-7 所示。

```
SQL> create table t103(id int, a int, b varchar2(30), primary key (id));
```

表已创建。

图 5-7　上游数据库建表

然后在下游的 101 数据库服务器上登录下游本地会话，并建立与 103 数据库的远程连接，即 dblink 连接。如图 5-8 所示，x4 是 dblink 的名字，可以是任意的合法名字。

```
SQL> create database link x4 connect to tang2 identified by tang using 't3';
```

数据库链接已创建。

图 5-8　建立 dblink

由于是下游数据库连接上游数据库，所以这里 connect to 后面的用户名是上游数据库的名字 tang2，密码也是上游数据库的密码 tang。其实这很好理解：要去访问别人的数据，就要以别人的身份去连接。

至此 dblink 就已经建立完成。在下游数据库上尝试查询上游 103 数据库上的数据。如图 5-9 所示。@后面表示查询的是远端上游数据库。

```
SQL> select * from t103@x4;

        ID          A B
---------- ---------- ------------------------------
         0          1 fKA2sExjB0
         1          1 ngbpq5zxfY
         2          1 T7Ktvutktj
         3          0 8wpm4dz7Cc
         4          1 TrUULW8N3L
         5          1 ZrUtkuPEf8
         6          0 5sC3piVdRX
         7          0 omB7t0l7Ki
         8          1 FLBHWF77j7
         9          1 1REduxKjho
```

已选择 10 行。

图 5-9　通过 dblink 查询

可以看到已经能够成功查询，那么就实现了 dblink 的第一个功能远程访问。此功能其实并不推荐大家使用，甚至是极力反对的。下文中会介绍为什么不建议直接使用 dblink 读取数据。

接下来尝试第二个功能：数据迁移。将上游数据库数据迁移到下游数据库，如图 5-10 所示。

既然可以远程查询和迁移，那么同样也可以关联两台不同数据库服务器上的数据进行关联查询，如图 5-11 所示。

第 5 章 如何做好数据库之间的数据同步 223

```
SQL> create table t101(id int, a int, b varchar2(30), primary key (id));

表已创建。

SQL> insert into t101 select * from t103@x4;

已创建 10 行。

SQL> commit;

提交完成。

SQL> select * from t101;

        ID          A B
---------- ---------- ------------------------------
         0          1 fKA2sExjB0
         1          1 ngbpq5zxfY
         2          1 T7Ktvutktj
         3          0 8wpm4dz7Cc
         4          1 TrUULW8N3L
         5          1 ZrUtkuPEf8
         6          0 5sC3piVdRX
         7          0 omB7t0l7Ki
         8          1 FLBHWF77j7
         9          1 1REduxKjho

已选择 10 行。
```

图 5-10　在下游数据库建立空表进行迁移

```
SQL> select t101.id,t101.b b101,t103.b b103 from t101 join t103@x4 on t101.id=t103.id;

        ID B101                           B103
---------- ------------------------------ ------------------------------
         0 fKA2sExjB0                     fKA2sExjB0
         1 ngbpq5zxfY                     ngbpq5zxfY
         2 T7Ktvutktj                     T7Ktvutktj
         3 8wpm4dz7Cc                     8wpm4dz7Cc
         4 TrUULW8N3L                     TrUULW8N3L
         5 ZrUtkuPEf8                     ZrUtkuPEf8
         6 5sC3piVdRX                     5sC3piVdRX
         7 omB7t0l7Ki                     omB7t0l7Ki
         8 FLBHWF77j7                     FLBHWF77j7
         9 1REduxKjho                     1REduxKjho

已选择 10 行。
```

图 5-11　本地数据关联远程数据

2. dblink 的使用局限性

至此，Oracle 中 dblink 的初级使用方式就已经涵盖到了。但是以上的操作除了数据迁移，都不建议广大读者在真实环境中使用。这主要是基于 dblink 的跨库查询存在性能上的问题，对 dblink 的衍生做一个简单的效率测试，看看同样的查询，跨库与不跨库的效率差距。本地关联数据与远程关联比较如图 5-12 所示。

```
SQL> select count(*) from t join t101 on t.id=t101.id;

  COUNT(*)
----------
    100000

已用时间：  00: 00: 00.42
SQL> select count(*) from t join t103@x4 on t.id=t103.id;

  COUNT(*)
----------
    100000

已用时间：  00: 00: 00.92
```

图 5-12 本地关联数据与远程关联比较

从图中可以看出，在数据基本相同的情况下，用时有着近一倍的差距。那么在生产环境中，表宽度更大、返回数据量更大的情况下，dblink 所带来的效率影响就会更大。所以这里不仅仅是不推荐，甚至是反对这样使用 dblink。这不是 dblink 适合使用的场景。

此外 dblink 有一个 SCN 的问题存在于老版本中。如图 5-13 所示，云和恩墨公司在 2018 年时提到过该问题的预警。

图 5-13 dblink 的 SCN 问题

其实质是关于低版本数据库在 2019 年 6 月份以后通过 dblink 访问高版本数据库可能会遇到连接失败、退出、拒绝连接等问题（高版本访问低版本没有问题）。这里说的低版本都是 12c 之前的版本，18c 和 19c 目前没有这个问题。

在两个库通过 dblink 进行分布式事务时，假设 B 库的 SCN 值要高于 A 库的 SCN，因此要将 B 库的 SCN 值同步到 A 库，但是如果 B 库的 SCN 过高，这样同步到 A 库之后，使得 A 库面临 SCN 耗尽的风险，那么 A 库会拒绝同步 SCN，这个时候就会报 ORA-19706: Invalid SCN 错误。

SCN 是 System Change Number 的缩写，是事务改变的序号。由于现在许多系统还存在着低版本的数据库，所以在遇到这些数据库的时候操作要注意这些。

其实重点就是 dblink 会产生分布式事务（因为跨数据库实例），即使是通过 dblink 的查询操作。这就使得有时候在数据库监控上看到大事务，而没有查到锁。这其实就是 dblink 在进行大范围查询的操作。

所以在做数据迁移的时候，最好是同版本或是高版本访问低版本。多年以前在公安系统工作时候，就用此方法进行过 TB 级别的数据迁移，整个过程停机时间大约为 1min。

如图 5-14 所示，有一个老旧的数据库系统，图 5-14 左边内网地址为 192.168.0.1 的数据库。硬件过旧，数据库版本也需要升级，数据库大约有 5TB 的数据，其中部分表还有 BLOB 的图片数据。需要将数据库迁移到图 5-14 右边内网地址为 192.168.0.2 的数据库。

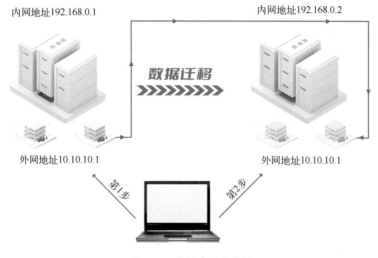

图 5-14 数据库迁移升级

一般服务器都有两个网卡，这里就利用双网卡的环境进行迁移。外网地址即对外服务的地址，业务网段。而内网地址是指仅供机房内部使用的地址。为了便于阅读对图 5-14 左边数据库称为 A 数据库，即旧数据库，图 5-14 右边数据库称为 B 数据库，即新数据库。将 A 数据库上的表结构、视图、注释、存储过程、索引、函数等所有静态的元数据导出到 B 数据库。A 数据库已有外网地址 10.10.10.1 对外服务，B 数据库也配置与 A 数据库一样的外网地址 10.10.10.1，但是网线不接入交换机。A、B 两个数据库通过网线直连构成了 192.168.0.1 与 192.168.0.2 直接通信。

A、B 两个数据库通过内网地址建立 dblink，由 B 数据库向 A 数据库发起 dblink 连接。将 A 数据库上当前时间（如当前是 2020 年 3 月 1 日上午 10 时）之前的主要业务数据通过 insertintotableselect*fromtable@dblink 的方式分批迁移到 B 数据库上。通常主要业务表在一个数据库中只有 1~5 张大的业务流水表，这里的业务流水是每秒几百条以上的流水，一个小时几条的可以当作静态数据来处理。

分批需要结合实际硬件吞吐能力进行估算，可能一批是 10 万行，也可能一批是 20 万行。这个过程可能要做 1~2 天。这个过程中 A 数据库对外服务不受影响。

待 dblink 迁移工作完成，此时可能是 2020 年 3 月 3 日上午 10 时。则 B 数据库已经拥有了 A 数据库除了 2020 年 3 月 1 日上午 10 时到 2020 年 3 月 3 日上午 10 时这 48 个小时以外的所有数据。此时将 A 数据库对应网卡地址是 10.10.1.1 的网线拔下，把 B 数据库网卡配置好 IP 为 10.10.10.1 的网卡对应的网线接入交换机。交换机在 30s 后完成了通信，应用程序通过断网重连机制也恢复了读写数据库的功能。

此时再用同样的方法，将 48 小时的增量数据分批导出 B 数据库。自此完成了数据库迁移。此举适合在业务低峰时期操作，能接受有短暂（几十秒）的停止服务的场景。因为这个方法没有做到无缝，技术含量也较低。但是如果使用 OGG 等技术，需要的商业费用和支持相对本方法要高。综合来说，本方案性价比最高，这种使用场景是推荐的，只要注意高版本和低版本的 SCN 即可。

5.3.2 使用物化视图实现远程 Oracle 数据库之间的数据同步

通过 dblink 连接上远程数据库，可以实现简单的查询以及数据同步，但这种方式在网络上会出现较大的性能影响，网络上的消耗是业内基本禁止直接使用 dblink 查询远程数据库的主要原因，尤其是多表复杂查询时。现实中，如果真的需要查询，则需要将远程的数据映射到本地，这样进行关联的时候就是读取本地的物理 IO 或逻辑 IO，而减少了网络上的消耗。

将数据映射到本地的做法是建立物化视图。dblink 可以用于物化视图。但是物化视图使用也有其合适的场景，使用好了节约成本，使用不好造成数据库的负担。以下通过具体案例加以说明。

1. 使用物化视图实现数据同步

首先，在数据源所在的上游数据库服务器 103 上创建源表的物化视图日志，如图 5-15 所示。

```
[oracle@adg1 ~]$ sqlplus tang2/tang@t

SQL*Plus: Release 19.0.0.0.0 - Production on 星期三 5月 17 14:57:42 2023
Version 19.3.0.0.0

Copyright (c) 1982, 2019, Oracle.  All rights reserved.

上次成功登录时间: 星期二 5月   16 2023 17:17:55 +08:00

连接到:
Oracle Database 19c Enterprise Edition Release 19.0.0.0.0 - Production
Version 19.3.0.0.0

SQL> create materialized view log on t103;

实体化视图日志已创建。
```

图 5-15 创建源表的物化视图日志

之所以要创建物化视图日志，是因为物化视图在下游数据库是一张实际存在的表，它是上下游数据库中对应表的实体映射，这个过程是数据实时传输的。最终在下游数据库中占有磁盘空间。这个日志的目的是记录上游数据变化的情况，下游通过访问这些日志捕获数据变化，达到数据同步的效果，初始物化视图日志内容如图 5-16 所示。

```
SQL> select * from xxg;

    ID NAME          AGE
---------- ---------- ----------
     1 a              1
     2 b              2
     3 c              3

SQL> desc  xxg;
 名称                                      是否为空? 类型
 ----------------------------------------- -------- ----------------------------
 ID                                        NOT NULL NUMBER(38)
 NAME                                               VARCHAR2(10)
 AGE                                                NUMBER(38)

SQL> CREATE MATERIALIZED VIEW LOG ON xxg;

实体化视图日志已创建。


SQL> select * from MLOG$_xxg;

未选定行
```

图 5-16 初始物化视图日志内容

在新增一条 ID 为 4 的数据，而且将 ID 是 2 的这行数据的 AGE 字段改为 30 以后。xxg 表就变成了如图 5-17 所示。

```
SQL> select * from xxg;

                                    ID NAME                              AGE
-------------------------------------- ---------- -------------------------
                                     1 a                                  1
                                     2 b                                 30
                                     3 c                                  3
                                     4 d                                 41
```

图 5-17 操作物化视图基表数据

物化视图日志如图 5-18 所示，该日志表记录了变更数据行的 ID，以及操作类型 DMLTYPE$$是新增的 Insert(I)还是更新的 Update(U)。

可以看到在物化视图日志中记录了新增数据、更新数据等变化。正因为有了这些日志，所以物化视图可以进行增量刷新和断点续传。

```
SQL> select * from MLOG$_xxg;

    ID SNAPTIME$$   DMLTYPE$$ OLD_NEW$$ CHANGE_VECTOR$$              XID$$
---------- ---------- --------- --------- -------------------------- ----------
     4 4000/1/1     I         N         FE                         5629542483
     2 4000/1/1     U         U         08                         5629542483
```

图 5-18 查看物化视图日志

在下游目标数据库服务器 101 上建立远程连接 x4，如图 5-19 所示。

```
[oracle@dev-dbteam-143101 ~]$ sqlplus tang/tang@t

SQL*Plus: Release 19.0.0.0.0 - Production on 星期三 5月 17 15:10:29 2023
Version 19.3.0.0.0

Copyright (c) 1982, 2019, Oracle.  All rights reserved.

上次成功登录时间：星期二 5月   16 2023 17:21:24 +08:00

连接到：
Oracle Database 19c Enterprise Edition Release 19.0.0.0.0 - Production
Version 19.3.0.0.0

SQL> create database link x4 connect to tang2 identified by tang using 't3';

数据库链接已创建。
```

图 5-19 建立物化视图的 dblink

建立 dblink 后，如图 5-20 所示，检查连接情况。

```
SQL> select count(*) from t103@x4;

  COUNT(*)
----------
    100000
```

图 5-20 进行 dlink 的验证

从验证结果看到在下游数据库 101 中可以查询到上游数据库 103 上的表以及数据了。下面在下游数据库 101 上创建一张相同结构的物化视图，如图 5-21 所示。不过这种创建方式无法将上游数据库的注释、索引等对象带过来。如果需要这些对象，则手工建立。

```
SQL> create table t101 as select * from t103@x4 where 1=2;

表已创建。

SQL> select count(*) from t101;

  COUNT(*)
----------
         0
```

图 5-21 创建物化视图对应表

通过简单查询，可以看到表已经存在，不过其中还没有数据。接下来建立一个简单的物化视图，用于执行同步动作，如图 5-22 所示。

```
SQL> create materialized view t101 on prebuilt table as select * from t103@x4;

实体化视图已创建。
```

图 5-22 建立一个简单的物化视图

对该物化视图执行手动刷新。如图 5-23 所示。

```
SQL> exec dbms_mview.refresh('t101','C');

PL/SQL 过程已成功完成。

SQL> select count(*) from t101;

  COUNT(*)
----------
    100000
```

图 5-23　刷新物化视图

图中第一个参数是物化视图的视图名，本实验是 t101。第二个参数是刷新方式 C，代表全量。

从上图中，能够看到数据已经成功同步过来了。至此一个简单的物化视图远程同步数据的动作就已经完成了。

再深入讲解其中的一些常用参数，使物化视图满足实际工作需求。在建立物化视图的语句中，refresh 后面可以有三个不同的选项 complete/fast/force，如果不主动选择 refresh 参数，默认选择 force 选项，也就是 Oracle 会自行判断能否进行增量刷新，如果能就增量，不能就全量，而 fast 就是指定使用增量刷新，complete 对应的为全量刷新。要注意如果使用 complete 参数，在源表数据量大的情况下，刷新动作会占用大量时间。

而 on demand/on commit 参数则可以指定物化视图刷新的触发方式，on commit 参数可以使物化视图在源数据提交时自动刷新目标数据，on demand 选项则可以通过设定时间，进行定时刷新数据的动作，若不为其设定时间则不会自动刷新，需要手动进行触发。

接下来简单演示一下使用 on demand 做定时自动刷新物化视图的方式，同时目标库中没有指定的对应目标表，使用视图直接建立。而这种方式是普遍流行的做法。

当下数据库服务器 103 中有源表 t3，如图 5-24 所示。注意：用做数据源的表需要有主键列。

```
SQL> select * from t3;

        ID          A
---------- ----------
         0          0


SQL> create materialized view log on t3;

实体化视图日志已创建。
```

图 5-24　为 t3 建立物化视图日志

在下游数据库服务器 101 上，在没有目标表的情况下直接建立物化视图，如图 5-25 所示。

```
SQL>select * from t1;
select * from t1;

第 1 行出现错误:
```

图 5-25　直接建立物化视图并且指定刷新频率

ORA-00942：表或视图不存在

```
SQL>create materialized view t1 refresh on demand
  2   start with sysdate next sysdate+5/86400
  3   as select * from t3@x4;
```

图 5-25　直接建立物化视图并且指定刷新频率（续）

可以看到建立完物化视图后，源表的表结构以及数据都已经同步到了目标位置，建立物化视图的参数中，start with sysdate nest sysdate+5/86400 表示每 5s 进行一次自动刷新，refresh fast 参数则表示刷新方式为增量。

在上游数据库中插入一条新数据，如图 5-26 所示。

```
SQL> insert into t3 values (1,1);

已创建 1 行。

SQL> commit;

提交完成。
```

图 5-26　上游数据库写入数据

如图 5-27 所示，可以看到数据已经成功同步过来了。现在再进行跨库的查询动作已经没有网络问题了，全部是本地 IO 操作。

```
SQL> select * from t1;

        ID          A
---------- ----------
         0          0
         1          1
```

图 5-27　下游数据库中查询数据

以上是通过物化视图方式实现 Oracle 数据库之间数据同步的过程。可以说，dblink 的物化视图解决了多个 Oracle 数据库直接的跨库查询问题。但是也来了一些问题和思考，即物化视图的优劣。

2．物化视图的优势与劣势

物化视图的优势包括以下几方面。

优势 1：不需要额外的产品支持，只要几个命令完成。

优势 2：日志级别增量同步，同步效率高。

优势 3：刷新频率可以自由调控。

优势 4：可以针对表级同步。

优势 5：可以结合实际情况，部分列传输到下游数据库，而不用全部传输（敏感数据可以不传输）。

优势 6：断网恢复后可以续传。

优势 7：上游数据库增加字段，如果下游数据库不需要这个字段，不影响物化视图同步数据。

优势 8：下游数据库可以结合自己的查询条件建立与上游数据库不同的索引。

物化视图的劣势包括以下三方面。

劣势 1：上游数据库增加字段，如果下游数据库也要看到这个字段，则物化视图需要重新建立，表数据量多，则重建间隔长，影响下游数据库读取。

劣势 2：无论数据库在刷新周期内是不是有数据变化，下游数据库都定期主动拉取上游数据库的数据。可能产生无用功，消耗上下游两端的数据库的资源。

劣势 3：如果物化视图过多，而且刷新频率较高，对数据库消耗会非常大。笔者曾经处理过一个系统，上下游数据库有 50 多张表，每隔 5s 去刷新一下物化视图，使得数据库负荷居高不下，最终采用了 OGG 的替代方案，替代后上下游数据库各减轻了 40%左右的负荷。这个劣势并不是说明物化视图不好，而是说明使用的方法出现了问题。

从上述案例可以看出，无论是 dblink 还是物化视图都是要解决跨库查询的问题，那么为什么要跨库查询？这里主要是规划问题，其实这个问题和第 4 章的数据库分与合的最底层的问题是一致的，缺乏规划或者说管理上存在壁垒。

数据库同步方式有很多，但是最好的恰恰是不同步。如果需要跨库查询说明这两个数据库的业务数据耦合度比较紧密，那么建立数据库的时候就应该建立在一起，而不是分开建立。

5.3.3 使用插件实现 MySQL 数据库之间的数据同步

在 Oracle 中，通过 dblink 来实现远程数据查询的目的。那么在使用 MySQL 的项目中，如何实现类似的功能？答案是通过插件来实现。下面通过具体案例来说明如何使用插件在 MySQL 数据库之间进行数据同步。

准备两台在同一网段下的服务器，分别以 134 和 138 为 IP 结尾（后文用 134 和 138 表示）。以 138 作为数据源，134 作为要进行跨库查询的目标库。

首先在 134 上做环境的准备，如图 5-28 所示。

```
mysql> show engines ;
| Engine             | Support | Comment                                                        | Transactions | XA   | Savepoints |
| FEDERATED          | NO      | Federated MySQL storage engine                                 | NULL         | NULL | NULL       |
| MEMORY             | YES     | Hash based, stored in memory, useful for temporary tables      | NO           | NO   | NO         |
| InnoDB             | DEFAULT | Supports transactions, row-level locking, and foreign keys     | YES          | YES  | YES        |
| PERFORMANCE_SCHEMA | YES     | Performance Schema                                             | NO           | NO   | NO         |
| MyISAM             | YES     | MyISAM storage engine                                          | NO           | NO   | NO         |
| MRG_MYISAM         | YES     | Collection of identical MyISAM tables                          | NO           | NO   | NO         |
| BLACKHOLE          | YES     | /dev/null storage engine (anything you write to it disappears) | NO           | NO   | NO         |
| CSV                | YES     | CSV storage engine                                             | NO           | NO   | NO         |
| ARCHIVE            | YES     | Archive storage engine                                         | NO           | NO   | NO         |
9 rows in set (0.00 sec)
```

图 5-28 MySQL 插件检查

使用 Federated 插件来进行远程连接，该插件是一个独立的存储引擎，其功能就是实现 dblink 的功能。所以检查 Federated 是否开启，从上图中可见插件已经默认安装，但是未启用。

若插件未安装，可以使用管理员 root 登录到 MySQL 数据库中，在命令行中提示符下执行 install plugin federated soname'ha_federated.so';语句进行安装。

安装后需要进行启用的动作。在操作系统提示符下编辑 MySQL 的配置文件，如图 5-29 所示，向 my.cnf 文件中插入 federated。

```
[root@base ~]# vim /etc/my.cnf
[mysqld]
federated
```

图 5-29 编辑 MySQL 配置文件

修改配置文件后，需要重启数据库生效，如图 5-30 所示。

```
[root@base etc]# service mysqld restart
Stopping mysqld:                                           [  OK  ]
Starting mysqld:                                           [  OK  ]
```

图 5-30 重启数据库使配置生效

重启完成后，再检查 Federated 是否开启，如图 5-31 所示。可以看到 Federated 对应的 Support 列已经是 YES 了，代表已经开启成功。

```
mysql> show engines ;
+--------------------+---------+----------------------------------------------------------------+--------------+------+------------+
| Engine             | Support | Comment                                                        | Transactions | XA   | Savepoints |
+--------------------+---------+----------------------------------------------------------------+--------------+------+------------+
| FEDERATED          | YES     | Federated MySQL storage engine                                 | NO           | NO   | NO         |
| MEMORY             | YES     | Hash based, stored in memory, useful for temporary tables      | NO           | NO   | NO         |
| InnoDB             | DEFAULT | Supports transactions, row-level locking, and foreign keys     | YES          | YES  | YES        |
| PERFORMANCE_SCHEMA | YES     | Performance Schema                                             | NO           | NO   | NO         |
| MyISAM             | YES     | MyISAM storage engine                                          | NO           | NO   | NO         |
| MRG_MYISAM         | YES     | Collection of identical MyISAM tables                          | NO           | NO   | NO         |
| BLACKHOLE          | YES     | /dev/null storage engine (anything you write to it disappears) | NO           | NO   | NO         |
| CSV                | YES     | CSV storage engine                                             | NO           | NO   | NO         |
| ARCHIVE            | YES     | Archive storage engine                                         | NO           | NO   | NO         |
+--------------------+---------+----------------------------------------------------------------+--------------+------+------------+
9 rows in set (0.01 sec)
```

图 5-31 Federated 插件已开启

至此，Federated 所需环境就已经准备完成了。来到作为数据源的 138 数据库上，建立一张表，并写入几条数据，如图 5-32 所示。

```
mysql> select * from t138;
+----+---+---+
| id | a | b |
+----+---+---+
|  0 | 0 | a |
|  1 | 1 | b |
|  2 | 2 | c |
+----+---+---+
3 rows in set (0.01 sec)
```

图 5-32 在 MySQL 源端建表写入数据

随后来到作为目标端的 134 数据库上，用远程联机的方式建立一张同名表，如图 5-33 所示。

```
mysql> create table t138 (id int, a int ,b varchar(20)) engine=federated connection='mysql://root:1@10.60.143.138:3306/x/t138';
Query OK, 0 rows affected (0.02 sec)

mysql> select * from t138;
+------+------+------+
| id   | a    | b    |
+------+------+------+
|    0 |    0 | a    |
|    1 |    1 | b    |
|    2 |    2 | c    |
+------+------+------+
3 rows in set (0.13 sec)
```

图 5-33 建立远程映射表

从图 5-33 可见，表的存储引擎指定为 federated。建表语句中的 connect 后面跟着的是远程数据库的连接字符串：

用户名 root
密码 1
IP 地址 10.60.143.138
端口号 3306
database 名字 x
表名 t138

表建立完成后，在 134 数据库上直接可以查询 138 数据库中指定的表的数据了。同 dblink 一样的，查询效率比在数据源直接查询会有所降低。如果在 138 数据库上再插入一条数据，如图 5-34 所示。在 134 数据库上也能够实时查询到，如图 5-35 所示。

```
mysql> insert into t138 values(3,3,'d');
Query OK, 1 row affected (0.00 sec)
```

图 5-34 MySQL 源端新增数据

```
mysql> select * from t138;
+------+------+------+
| id   | a    | b    |
+------+------+------+
|    0 |    0 | a    |
|    1 |    1 | b    |
|    2 |    2 | c    |
|    3 |    3 | d    |
+------+------+------+
4 rows in set (0.01 sec)
```

图 5-35 MySQL 目标端查询数据

如此，便也能够在 134 数据库上进行跨库的关联查询，如图 5-36 所示。

```
mysql> select * from t134 join t138 where t134.id=t138.id;
+------+------+------+------+------+
| id   | a    | id   | a    | b    |
+------+------+------+------+------+
|    0 |    0 |    0 |    0 | a    |
+------+------+------+------+------+
1 row in set (0.12 sec)
```

图 5-36 MySQL 跨库关联查询

注意：虽然这样可以实现远程查询，但是在实际工作中不推荐大数据表远程关联。这种场景仅限于读取远程的一个数据量不大的配置表。

5.3.4 利用数据同步实现 MySQL 的版本升级

一般来说，数据库的主从高可用架构要求主库与从库的操作系统和版本要一致，不过 MySQL 可以不一致。这就减少了一个约束条件，该约束的释放也带来了一个新的解决方案。这个方案是让 MySQL 的从库版本高于主库，通过主从切换来实现数据库的版本升级。这种方法可以最大限度地减少停机时间，同时确保数据库服务的连续性。注意：MySQL 的 MGR 要求版本一致，MySQL 的物理克隆从库也需要版本一致。这里说的不一致是通过逻辑备份建立的主从数据库。

本案例要将 MySQL 5.7 升级到 MySQL 8.0.33。实验环境说明如下：

IP 为 10.60.148.154 主库（MySQL 5.7）和 10.60.148.155（MySQL 8.0.33）。下文用 IP 位数 154 和 155 代指两台数据库。为数据库添加 GTID 属性，如图 5-37 所示。

```
root@xue-test1-148154:[/root]mysql -uroot -p1 -e "select version()"
mysql: [Warning] Using a password on the command line interface can be insecure.
+------------+
| version()  |
+------------+
| 5.7.25-log |
+------------+


root@xue-test1-148154:[/root]cat /etc/my.cnf
# For advice on how to change settings please see
# http://dev.mysql.com/doc/refman/5.7/en/server-configuration-defaults.html

[mysqld]
datadir=/var/lib/mysql
socket=/var/lib/mysql/mysql.sock
symbolic-links=0
log-error=/var/log/mysqld.log
pid-file=/var/run/mysqld/mysqld.pid
######################################################
gtid_mode=ON
log-slave-updates=ON
enforce-gtid-consistency=ON
server-id=1
log-bin=binlog
binlog_format=row
```

图 5-37 MySQL 主库准备工作

设置好 GTID 模式与 server-id（主从数据库的 ID 必须不一致）和开启 binlog 以及 binlog 的模式后，对数据库进行重启，如图 5-38 所示。

将数据库中非系统自带的 schema 导出，并且传输到从库服务器上，如图 5-39 所示，至此 154 主库设置基本完毕。

```
root@xue-test1-148154:[/root]service mysqld restart
Redirecting to /bin/systemctl restart  mysqld.service

root@xue-test1-148154:[/root]mysql -uroot -p1 -e "show databases"
mysql: [Warning] Using a password on the command line interface can be insecure.
+--------------------+
| Database           |
+--------------------+
| information_schema |
| maxwell            |
| mysql              |
| new                |
| performance_schema |
| sys                |
+--------------------+
```

图 5-38　对数据库进行重启

```
root@xue-test1-148154:[/root]mysqldump -uroot -p1  --set-gtid-purged=ON --databases maxwell new  >all.sql
mysqldump: [Warning] Using a password on the command line interface can be insecure.
Warning: A partial dump from a server that has GTIDs will by default include the GTIDs of all transactions, even those that changed suppressed parts of the database.
root@xue-test1-148154:[/root]
root@xue-test1-148154:[/root]scp all.sql root@10.60.148.155:/setup/
root@10.60.148.155's password:
all.sql                                                                                          100%   32KB
root@xue-test1-148154:[/root]
```

图 5-39　备份主库传输到源端

从 Oracle 官网上下载 MySQL 8.0.33 的安装介质并且上传到 155 的从库上。如图 5-40 所示，使用 rpm -ihv 命令将图中涉及的 rpm 包一次性全部用空格连接进行安装，这样解决了相互依赖的问题。

root@xue-test2-148155:[/setup]tar -xvf mysql-8.0.29-1.el7.x86_64.rpm-bundle.tar
mysql-community-client-8.0.29-1.el7.x86_64.rpm
mysql-community-client-plugins-8.0.29-1.el7.x86_64.rpm
mysql-community-common-8.0.29-1.el7.x86_64.rpm
mysql-community-debuginfo-8.0.29-1.el7.x86_64.rpm
mysql-community-devel-8.0.29-1.el7.x86_64.rpm
mysql-community-embedded-compat-8.0.29-1.el7.x86_64.rpm
mysql-community-icu-data-files-8.0.29-1.el7.x86_64.rpm
mysql-community-libs-8.0.29-1.el7.x86_64.rpm
mysql-community-libs-compat-8.0.29-1.el7.x86_64.rpm
mysql-community-server-8.0.29-1.el7.x86_64.rpm
mysql-community-server-debug-8.0.29-1.el7.x86_64.rpm
mysql-community-test-8.0.29-1.el7.x86_64.rpm
root@xue-test2-148155:[/setup]

root@xue-test2-148155:[/setup]rpm -ihv rpm -ihv mysql-community-*

图 5-40　解压 rpm 安装

```
warning: mysql-community-client-8.0.29-1.el7.x86_64.rpm: Header    V4   RSA/SHA256
Signature, key ID 3a79bd29: NOKEY
Preparing...                          ################################# [100%]
Updating / installing...
   1:mysql-community-common-8.0.29-1.e################################# [ 14%]
   2:mysql-community-client-plugins-8.################################# [ 29%]
   3:mysql-community-libs-8.0.29-1.el7################################# [ 43%]
   4:mysql-community-client-8.0.29-1.e################################# [ 57%]
   5:mysql-community-icu-data-files-8.################################# [ 71%]
   6:mysql-community-server-8.0.29-1.e################################# [ 86%]
   7:mysql-community-libs-compat-8.0.2################################# [100%]
```

图 5-40　解压 rpm 安装（续）

安装好数据库软件，需要启动数据库完成数据库的初始化。根据生成的随机密码进行登录。然后设置一个满足复杂度要求的密码，再去掉复杂度限制，最终将密码设置为 1，如图 5-41 所示。

```
root@xue-test2-148155:[/setup]systemctl start mysqld
root@xue-test2-148155:[/setup]cat /var/log/mysqld.log | grep "password"
2023-07-01T07:13:57.244769Z 6 [Note] [MY-010454] [Server] A temporary password is generated for root@localhost: NENjlm.SK2q8

root@xue-test2-148155:[/setup]mysql -uroot -p
Enter password:
Welcome to the MySQL monitor.  Commands end with ; or \g.
Your MySQL connection id is 8
Server version: 8.0.29

Copyright (c) 2000, 2022, Oracle and/or its affiliates.

Oracle is a registered trademark of Oracle Corporation and/or its
affiliates. Other names may be trademarks of their respective
owners.

Type 'help;' or '\h' for help. Type '\c' to clear the current input statement.

mysql> set password='12345#Qweasdzxc';
Query OK, 0 rows affected (0.02 sec)

mysql> set global validate_password.mixed_case_count=0;
Query OK, 0 rows affected (0.00 sec)

mysql> set global validate_password.number_count=0;
Query OK, 0 rows affected (0.00 sec)

mysql> set global validate_password.length=0;
Query OK, 0 rows affected (0.01 sec)

mysql> set global validate_password.policy=0;
Query OK, 0 rows affected (0.00 sec)

mysql> set global validate_password.special_char_count=0;
Query OK, 0 rows affected (0.00 sec)

mysql> set password='1';
Query OK, 0 rows affected (0.01 sec)
```

图 5-41　数据库初始化并且修改管理员密码

对照主库一样设置好 GTID 模式与 server-id（主从数据库的 ID 必须不一致），开启 binlog 以及 binlog 的模式后，对数据库进行重启。如图 5-42 所示。

```
root@xue-test2-148155:[/root]vim /etc/my.cnf
root@xue-test2-148155:[/root]cat /etc/my.cnf
# For advice on how to change settings please see
# http://dev.mysql.com/doc/refman/8.0/en/server-configuration-defaults.html

[mysqld]

datadir=/var/lib/mysql
socket=/var/lib/mysql/mysql.sock
log-error=/var/log/mysqld.log
pid-file=/var/run/mysqld/mysqld.pid
##########################################
gtid_mode=ON
log-slave-updates=ON
enforce-gtid-consistency=ON
server-id=2
log-bin=binlog
binlog_format=row
root@xue-test2-148155:[/root]systemctl restart mysqld
```

图 5-42 从库设置参数并且重启

重启从库并导入主库备份数据后，检查从库可见只有系统自带的 schema，如图 5-43 所示。

```
root@xue-test2-148155:[/root]mysql -uroot -p1 -e "show databases;"
mysql: [Warning] Using a password on the command line interface can be insecure.
+--------------------+
| Database           |
+--------------------+
| information_schema |
| mysql              |
| performance_schema |
| sys                |
+--------------------+
root@xue-test2-148155:[/root]

root@xue-test2-148155:[/root]mysql -uroot -p1 -e "reset master"
mysql: [Warning] Using a password on the command line interface can be insecure.

root@xue-test2-148155:[/root]mysql -uroot -p1< /setup/all.sql
mysql: [Warning] Using a password on the command line interface can be insecure.
```

图 5-43 重置从库 GTID 并导入主库备份数据

在从库上设定主从数据库的连接关系，并且启动主从复制线程，如图 5-44 所示。

```
root@xue-test2-148155:[/root]mysql -uroot -p1 -e "change master to master_host='10.60.148.154',master_port=3306,master_user='root',master_password='1',master_auto_position=1;"

mysql: [Warning] Using a password on the command line interface can be insecure.
root@xue-test2-148155:[/root]mysql -uroot -p1 -e "start slave"
```

图 5-44 建立 MySQL 主从关系

检查主从关系,跨版本主从数据库主从线程工作正常如图 5-45 所示。

```
mysql: [Warning] Using a password on the command line interface can be insecure.
*************************** 1. row ***************************
               Slave_IO_State: Waiting for source to send event
                  Master_Host: 10.60.148.154
                  Master_User: root
                  Master_Port: 3306
                Connect_Retry: 60
              Master_Log_File: binlog.000001
          Read_Master_Log_Pos: 154
               Relay_Log_File: xue-test2-148155-relay-bin.000002
                Relay_Log_Pos: 364
        Relay_Master_Log_File: binlog.000001
             Slave_IO_Running: Yes
            Slave_SQL_Running: Yes
              Replicate_Do_DB:
          Replicate_Ignore_DB:
           Replicate_Do_Table:
       Replicate_Ignore_Table:
      Replicate_Wild_Do_Table:
  Replicate_Wild_Ignore_Table:
                   Last_Errno: 0
                   Last_Error:
```

图 5-45　跨版本主从数据库主从线程工作正常

在主库上写入数据,如图 5-46 所示。

```
root@xue-test1-148154:[/root]mysql -uroot -p1 -h 10.60.148.154 -e " select version() "
mysql: [Warning] Using a password on the command line interface can be insecure.
+-----------+
| version() |
+-----------+
| 5.7.25-log |
+-----------+
root@xue-test1-148154:[/root]mysql -uroot -p1 -h 10.60.148.154 -e " create database x"
mysql: [Warning] Using a password on the command line interface can be insecure.
root@xue-test1-148154:[/root]mysql -uroot -p1 -h 10.60.148.154 -e "create table x.t (id int,a int)"
mysql: [Warning] Using a password on the command line interface can be insecure.
root@xue-test1-148154:[/root]mysql -uroot -p1 -h 10.60.148.154 -e " insert into x.t values (1,1),(2,2)"
mysql: [Warning] Using a password on the command line interface can be insecure.
root@xue-test1-148154:[/root]mysql -uroot -p1 -h 10.60.148.154 -e " select * from  x.t "
mysql: [Warning] Using a password on the command line interface can be insecure.
+----+---+
| id | a |
+----+---+
|  1 | 1 |
|  2 | 2 |
+----+---+
```

图 5-46　主库建表写入数据

在从库上读取数据,如图 5-47 所示。

至此数据库跨版本同步完成,随时可以切换数据库以达成数据库升级的目的。注意:跨版本升级只能跨一个大版本,如 5.6 到 5.7 和 5.7 到 8.0,5.6 是不能直接切换到 8.0 的。

```
root@xue-test2-148155:[/root]mysql -uroot -p1 -h 10.60.148.155 -e " select version() "
mysql: [Warning] Using a password on the command line interface can be insecure.
+-----------+
| version() |
+-----------+
| 8.0.29    |
+-----------+

root@xue-test2-148155:[/root]mysql -uroot -p1 -h 10.60.148.155 -e " select * from  x.t "
mysql: [Warning] Using a password on the command line interface can be insecure.
+------+------+
| id   | a    |
+------+------+
|    1 |    1 |
|    2 |    2 |
+------+------+
root@xue-test2-148155:[/root]
```

图 5-47　在从库上读取数据

5.3.5　采用多源复制的功能实现 MySQL 数据库之间的数据同步

在实际工作中，由于种种原因对 MySQL 进行了分库处理。但是正是由于这种分库导致了数据无法关联查询，分库的弊端就体现了出来。如果需要跨库查询，或是需要从不同的库中获取数据进行汇聚，此时不需要额外的工具和组件，仅依靠 MySQL 自带的功能即可完成。这里不需要开发增加工作量做数据转换或双写，也不需要大数据技术栈。

采用 MySQL 多源复制的功能即可达到多主一从的汇聚。本节就演示多源复制的搭建，在实际工作中曾经用这个解决方案替代了 ETL 抽取到 Hadoop 的方案，解决了实时性和准确性的问题。MySQL 多源复制是 MySQL 主从复制的衍生技术，基于上一小节的主从复制的知识进行拓展。

本次演示实验数据库共计 3 台：10.60.143.134（作为源数据库 1，以下简称 134）、10.60.143.138（作为源数据库 2，以下简称 138）和 10.60.143.91（作为目标汇聚数据库，以下简称 91）。首先在每台参与的节点库上都建立用于复制的账户（仅展示了一台，每台都进行相同的操作，测试环境为了方便，都使用 1 作为密码），如图 5-48 所示。

```
mysql> create user 'repl'@'%' identified by '1';
Query OK, 0 rows affected (0.04 sec)

mysql> grant replication slave on *.* to 'repl'@'%' ;
Query OK, 0 rows affected (0.01 sec)

mysql> ALTER USER 'repl'@'%' IDENTIFIED WITH mysql_native_password BY '1';
Query OK, 0 rows affected (0.01 sec)
```

图 5-48　建立复制账户

在目标库上修改一下配置文件 my.cnf，如图 5-49 所示，并且在修改完成后重启数据库。

从两个源数据库 134 和 138 上分别进行数据库的导出，如图 5-50 所示，134 数据库上备份导出的数据库叫作 x134，导出的文件叫作 x134.sql；138 数据库上备份导出的数据库叫作 x138，导出的文件叫作 x138.sql。分别把 x134.sql 和 x138.sql 使用 scp 命令传输到目标数据库 91 上。

```
[root@mysql8 ~]# vim /etc/my.cnf
gtid_mode=ON
log-slave-updates=ON
enforce-gtid-consistency=ON

[root@mysql8 ~]# service mysqld restart
Stopping mysqld:                                          [  OK  ]
Starting mysqld:                                          [  OK  ]
```

图 5-49 修改配置文件

```
[root@base ~]# mysqldump -uroot -p1 --set-gtid-purged=on --databases  x134  > x134.sql
mysqldump: [Warning] Using a password on the command line interface can be insecure.
Warning: A partial dump from a server that has GTIDs will by default include the GTIDs of all transactions, even those
that changed suppressed parts of the database. If you don't want to restore GTIDs, pass --set-gtid-purged=OFF. To make a
complete dump, pass --all-databases --triggers --routines --events.
[root@base ~]# vi x134.sql
[root@base ~]# scp x134.sql root@10.60.143.91:/root/
x134.sql                                                                                        100%
2448       2.4KB/s   00:00

[root@mysql8 ~]# mysqldump -uroot -p1 --set-gtid-purged=on --databases x138  > x138.sql
mysqldump: [Warning] Using a password on the command line interface can be insecure.
Warning: A partial dump from a server that has GTIDs will by default include the GTIDs of all transactions, even those
that changed suppressed parts of the database. If you don't want to restore GTIDs, pass --set-gtid-purged=OFF. To make a
complete dump, pass --all-databases --triggers --routines --events.
[root@mysql8 ~]#  scp x138.sql root@10.60.143.91:/root/
x138.sql                                                                                        100%
2283       2.2KB/s   00:00
```

图 5-50 源数据库导出

两个数据库源库上都是分别挑选了其中一个库作为同步对象的场景，如果要做全库同步，可以带上--all-databases 的参数。在 91 的目标数据库上找到传输过来的备份文件，在目标库所在的服务器上执行导入，如图 5-51 所示。

```
[root@mysql8 ~]# mysql -u root -p   < x134.sql
[root@mysql8 ~]# mysql -u root -p   < x138.sql
```

图 5-51 导入目标数据库

导入完成后登录数据库，已经成功导入到目标数据库中，如图 5-52 所示。

```
mysql> select * from x134.t134;
+------+
| id   |
+------+
|    0 |
+------+
1 row in set (0.00 sec)

mysql> select * from x138.t138;
+------+
| id   |
+------+
|    0 |
+------+
1 row in set (0.00 sec)
```

图 5-52 目标数据库验证导入数据

在目标数据库 91 上建立多个主从复制通道，如图 5-53 所示。

```
mysql> change replication source to
    -> source_host='10.60.143.134',source_user='repl',
    -> source_password='1',source_auto_position=1
    -> for channel 'C1';


mysql> change replication source to
    -> source_host='10.60.143.138',source_user='repl',
    -> source_password='1',source_auto_position=1
    -> for channel 'C2';
```

图 5-53　建立多个主从复制通道

在目标数据库 91 上开启复制线程，如图 5-54 所示。

```
mysql> START SLAVE;
Query OK, 0 rows affected, 1 warning (0.15 sec)
```

图 5-54　开启复制线程

在源端两个主库上分别写入数据进行验证，如图 5-55 所示。

```
mysql> insert into t134 values(134);
Query OK, 1 row affected (0.01 sec)


mysql> insert into t138 values(138);
Query OK, 1 row affected (0.02 sec)
```

图 5-55　源端分别写入验证数据

在目标端 91 的数据库上，可以获得同步的数据，如图 5-56 所示。

```
mysql> select * from x138.t138;
+-----+
| id  |
+-----+
|   0 |
| 138 |
+-----+
2 rows in set (0.00 sec)

mysql> select * from x134.t134;
+-----+
| id  |
+-----+
|   0 |
| 134 |
+-----+
2 rows in set (0.00 sec)
```

图 5-56　验证多源复制数据

在目标端 91 的数据库上可以进行关联查询，如图 5-57 所示。

```
mysql> select * from x138.t138 join x134.t134 on t138.id=t134.id;
+------+------+
| id   | id   |
+------+------+
|    0 |    0 |
+------+------+
1 row in set (0.00 sec)
```

图 5-57 多源复制解决跨库关联查询

多源复制在 MySQL 5.7 后期版本出现，从架构上解决了分库带来的汇聚问题。由于是 MySQL 原生自带的，所以比起其他第三方工具等有着兼容性好以及稳定性高的优势。尤其对于 DDL 的变更是其他技术栈无法比拟的。

MySQL 虽然不擅长大量数据的分析，但是对于使用到索引情况下的多表关联和数据分析还是有很大作用的。

5.3.6 采用主从模式实现 PostgreSQL 主从数据同步

本小节将探讨如何搭建基于主从复制的 PostgreSQL 环境，将详细介绍设置主数据库和从数据库的步骤，以及配置它们之间的复制关系。

从官网 https://www.postgresql.org/ftp/source/v15.3/ 上下载源码，本案例以较新的 15 版本作为演示，如图 5-58 所示。

```
[root@dev-dbteam-148180 setup]# tar -zxvf postgresql-15.3.tar.gz
.
.|
.
postgresql-15.3/Makefile
postgresql-15.3/README
postgresql-15.3/COPYRIGHT
postgresql-15.3/GNUmakefile.in
postgresql-15.3/.gitattributes
postgresql-15.3/aclocal.m4
postgresql-15.3/INSTALL
[root@dev-dbteam-148180 setup]# ll
total 29252
drwxrwxrwx 6 1107 1107     4096 May  9 05:26 postgresql-15.3
-rw-r--r-- 1 root root 29946539 Jul  1 16:50 postgresql-15.3.tar.gz
```

图 5-58 下载源码压缩包上传服务器并且解压

解压后，如图 5-59 所示，进入解压目录对编译程序授予可执行权限。这个命令将使得 configure 文件可以被执行。

```
[root@dev-dbteam-148180 setup]# cd postgresql-15.3

[root@dev-dbteam-148180 postgresql-15.3]#chmod +x ./configure
```

图 5-59 授权执行

配置安装 PostgreSQL 数据库软件，并指定安装路径为/opt/pgsql_15.3，如图 5-60 所示，这个命令告诉编译器在安装软件时将文件安装到指定的目录中。

```
[root@dev-dbteam-148180 postgresql-15.3]# ./configure --prefix=/opt/pgsql_15.3 --with-pgport=1922 --with-wal-blocksize=16
.
.
.
checking for dbtoepub... no
checking whether gcc -std=gnu99 supports -Wl,--as-needed... yes
configure: using compiler=gcc (GCC) 4.8.5 20150623 (Red Hat 4.8.5-44)
configure: using CFLAGS=-Wall -Wmissing-prototypes -Wpointer-arith -Wdeclaration-after-statement -Werror=vla -Wendif-labels -
configure: using CPPFLAGS= -D_GNU_SOURCE
configure: using LDFLAGS=  -Wl,--as-needed
configure: creating ./config.status
config.status: creating GNUmakefile
config.status: creating src/Makefile.global
config.status: creating src/include/pg_config.h
config.status: creating src/include/pg_config_ext.h
config.status: creating src/interfaces/ecpg/include/ecpg_config.h
config.status: linking src/backend/port/tas/dummy.s to src/backend/port/tas.s
config.status: linking src/backend/port/posix_sema.c to src/backend/port/pg_sema.c
config.status: linking src/backend/port/sysv_shmem.c to src/backend/port/pg_shmem.c
config.status: linking src/include/port/linux.h to src/include/pg_config_os.h
config.status: linking src/makefiles/Makefile.linux to src/Makefile.port
```

图 5-60 指定安装路径

依次执行 gmake 命令，如图 5-61 所示，这两个命令刷屏信息太多，不详细展示。

```
# gmake world
# gmake install-world
```

图 5-61 依次执行 gmake 命令

为软件创建一个软连接，如图 5-62 所示，并且创建密码。

```
[root@dev-dbteam-148180 ~]# ln -s /opt/pgsql_15.3 /opt/pgsql


[root@dev-dbteam-148180 setup]# groupadd pg15
[root@dev-dbteam-148180 setup]# useradd pg15 -g pg15
[root@dev-dbteam-148180 setup]# passwd pg15
Changing password for user pg15.
New password:
BAD PASSWORD: The password contains less than 1 uppercase letters
Retype new password:
passwd: all authentication tokens updated successfully.
```

图 5-62 创建软连接和创建 pg 用户

如图 5-63 所示，数据库目录为/pgdata/pg15/pg_root。

```
[root@dev-dbteam-148180 ~]# mkdir -p /pgdata/pg15/pg_root
[root@dev-dbteam-148180 ~]# chown -R pg15:pg15 /pgdata/pg15
```

图 5-63 创建 PostgreSQL 的数据目录并且授权

切换到创建好的 pg15 的用户下，并且编辑环境变量，如图 5-64 所示。

```
[root@dev-dbteam-148180 ~]# su - pg15
[pg15@dev-dbteam-148180 ~]$ vim .bash_profile
```

图 5-64 编辑环境变量

在 .bash_profile 文件内添加如下配置信息：

```
export PGPORT=1922
export PGUSER=postgres
export PGDATA=/pgdata/pg15/pg_root
export LANG=en_US.utf8

export PGHOME=/opt/pgsql
export LD_LIBRARY_PATH=$PGHOME/lib:/lib64:/usr/lib64:/usr/local/lib64:/lib:/usr/lib:/usr/local/lib
export DATE=`date +"%Y%m%d%H%M"`
export PATH=$PGHOME/bin:$PATH:.
export MANPATH=$PGHOME/share/man:$MANPATH
alias rm='rm -i'
alias ll='ls -lh'
```

修改环境变量文件完毕，保存退出。执行 source 命令使环境变量生效，如图 5-65 所示。

```
[pg15@dev-dbteam-148180 ~]$ source .bash_profile
```

图 5-65　使环境变量生效

以上的步骤操作在从库上再来一遍。

切换到 pg15 的用户下，如图 5-66 所示，执行 initdb -E UTF-8 --locale=C -D /pgdata/pg15/pg_root -U postgres -W 命令完成数据库初始化。

```
[root@dev-dbteam-148180 ~]# su - pg15
Last login: Sat Jul  1 17:51:05 CST 2023 on pts/0

[pg15@dev-dbteam-148180 ~]$ initdb -E UTF-8 --locale=C -D /pgdata/pg15/pg_root -U postgres -W
The files belonging to this database system will be owned by user "pg15".
This user must also own the server process.

The database cluster will be initialized with locale "C".
The default text search configuration will be set to "english".

Data page checksums are disabled.

Enter new superuser password:
Enter it again:

fixing permissions on existing directory /pgdata/pg15/pg_root ... ok
creating subdirectories ... ok
selecting dynamic shared memory implementation ... posix
selecting default max_connections ... 100
selecting default shared_buffers ... 128MB
selecting default time zone ... Asia/Shanghai
creating configuration files ... ok
running bootstrap script ... ok
performing post-bootstrap initialization ... ok
syncing data to disk ... ok

initdb: warning: enabling "trust" authentication for local connections
initdb: hint: You can change this by editing pg_hba.conf or using the option -A, or --auth-local and --auth-host, the next time you run initdb.

Success. You can now start the database server using:

    pg_ctl -D /pgdata/pg15/pg_root -l logfile start
```

图 5-66　初始化 PostgreSQL 主数据库

对 PostgreSQL 的配置文件进行编辑，如图 5-67 所示，在数据库主目录下打开 postgresql.conf 进行必要的参数修改。

```
[pg15@dev-dbteam-148180 ~]$ cd /pgdata/pg15/pg_root/
[pg15@dev-dbteam-148180 pg_root]$ vim postgresql.conf
```

图 5-67　编辑 postgresql.conf 参数

打开配置文件添加如下信息：

```
listen_addresses = '*'              # 指定服务器用于侦听来自客户端应用程序的连
                                      接的 TCP/IP 地址
max_connections = 1000              # 确定与数据库服务器的最大并发连接数，如果改
                                      变需要重启
shared_buffers = 128MB              # 共享缓存区大小不应超过总内存 1/4，一般为 1/8
wal_level = logical                 # 服务器记录事务日志的级别有 minimal、replica、
                                      logical 三种
synchronous_commit = off            # 何时向客户端确认事务提交成功
checkpoint_timeout = 30min          # 执行自动检查点的时间间隔
archive_mode = on                   # 归档模式开关参数
archive_command = '/bin/date'       # 配置归档命令
max_wal_senders = 10                # 最大的传输归档进程
wal_keep_size = 128                 # 指定保存在 pg_wal 目录中的过去日志文件段的
                                      最小大小

log_destination = 'csvlog'          # 记录服务器消息的方法
logging_collector = on              # 控制日志收集方式的配置参数
log_directory = 'log'               # 创建日志文件的目录
log_file_mode = 0600                # 创建日志文件的文件名的模式
wal_log_hints = on                  # 记录特定提示位（hint-bit）的变化
```

修改配置文件完毕，保存退出。如图 5-68 所示，运行 pg_ctl start 启动数据库。看到 server started 表示启动成功。

```
[pg15@dev-dbteam-148180 ~]$ pg_ctl start
waiting for server to start....2023-07-01 18:01:20.305 CST [19809] LOG:  redirecting log output to logging collector process
2023-07-01 18:01:20.305 CST [19809] HINT:  Future log output will appear in directory "log".
 done
server started

[pg15@dev-dbteam-148180 ~]$ psql
psql (15.3)
Type "help" for help.

postgres=#
```

图 5-68　启动 PostgreSQL 数据库

为了使从库可以连接到主库，需要修改 pg_hba.conf 文件，如图 5-69 所示，这里运行所有连接（正式环境结合实际而定），在文件最后添加 host replication all 0.0.0.0/0 md5。

```
[pg15@dev-dbteam-148180 pg_root]$ vim pg_hba.conf

[pg15@dev-dbteam-148180 pg_root]$ cat pg_hba.conf |grep "all"
# This file controls: which hosts are allowed to connect, how clients
# DATABASE can be "all", "sameuser", "samerole", "replication", a
# database name, or a comma-separated list thereof. The "all"
# USER can be "all", a user name, a group name prefixed with "+", or a
# "all", "sameuser", "samerole" or "replication" makes the name lose
# If you want to allow non-local connections, you need to add more
# allows any local user to connect as any PostgreSQL user, including
# the database superuser.  If you do not trust all your local users,
local   all             all                                     trust
host    all             all             127.0.0.1/32            trust
host    all             all             ::1/128                 trust
local   replication     all                                     trust
host    replication     all             127.0.0.1/32            trust
host    replication     all             ::1/128                 trust
host    replication     all             0.0.0.0/0               md5
```

图 5-69　修改 pg_hba.conf 文件

重新加载配置文件，如图 5-70 所示，使用 pg_ctl 命令重新加载配置文件的变化。PostgreSQL 的很多参数改变可以先修改底层配置文件后用 reload 重新加载，而无须重启数据库实例。

```
[pg15@dev-dbteam-148180 pg_root]$ pg_ctl reload
server signaled
```

图 5-70　重新加载配置文件

登录主数据库创建实验数据库和表，在主库上模拟建表写入数据，如图 5-71 所示。

```
[pg15@dev-dbteam-148180 pg_root]$ psql
psql (15.3)
Type "help" for help.

postgres=# create database xxg;
CREATE DATABASE

postgres=# \c xxg
You are now connected to database "xxg" as user "postgres".

xxg=# create table xuexiaogang (id int ,n int);
CREATE TABLE
xxg=# insert into xuexiaogang values (1,1);
INSERT 0 1
xxg=# insert into xuexiaogang values (2,2);
INSERT 0 1
xxg=# select * from xuexiaogang;
 id | n
----+---
  1 | 1
  2 | 2
(2 rows)
```

图 5-71　在主库上模拟建表写入数据

创建一个复制账户用来执行主从复制,如图 5-72 所示。到此为止,主库上的工作基本完成。

```
postgres=# CREATE USER repuser REPLICATION LOGIN CONNECTION LIMIT 10 ENCRYPTED PASSWORD '1';
CREATE ROLE
```

图 5-72 创建复制账户

在从库上执行 pg_basebackup -R -D /pgdata/pg15/pg_root -Fp -Xs -v -P -h 10.60.148.180 -p 1922 -U repuser。如图 5-73 所示,从库将主库进行了全库备份并且在从库上做了备份的恢复还原,从而达到了主库和从库的基本一致。

```
[root@dev-dbteam-148181 ~]# su - pg15

[pg15@dev-dbteam-148181 ~]$
[pg15@dev-dbteam-148181 ~]$ pg_basebackup -R -D /pgdata/pg15/pg_root -Fp -Xs -v -P -h 10.60.148.180 -p 1922 -U repuser
Password:
pg_basebackup: initiating base backup, waiting for checkpoint to complete
pg_basebackup: checkpoint completed
pg_basebackup: write-ahead log start point: 0/2000028 on timeline 1
pg_basebackup: starting background WAL receiver
pg_basebackup: created temporary replication slot "pg_basebackup_23072"
30402/30402 kB (100%), 1/1 tablespace
pg_basebackup: write-ahead log end point: 0/2000100
pg_basebackup: waiting for background process to finish streaming ...
pg_basebackup: syncing data to disk ...
pg_basebackup: renaming backup_manifest.tmp to backup_manifest
pg_basebackup: base backup completed
[pg15@dev-dbteam-148181 ~]$
```

图 5-73 在从库上执行复制

执行 pg_ctl start 启动从库,如图 5-74 所示,如果遇到权限问题,使用 root 对数据目录进行赋权(结合报错实际情况)。

```
[pg15@dev-dbteam-148181 ~]$ pg_ctl start
waiting for server to start....2023-07-02 11:19:59.960 CST [31530] FATAL:  data directory "/pgdata/pg15/pg_root" has invalid permissions
2023-07-02 11:19:59.960 CST [31530] DETAIL:  Permissions should be u=rwx (0700) or u=rwx,g=rx (0750).
 stopped waiting
pg_ctl: could not start server
Examine the log output.
[pg15@dev-dbteam-148181 ~]$ exit
logout

[root@dev-dbteam-148181 ~]# chmod 0700 /pgdata/pg15/pg_root/
[root@dev-dbteam-148181 ~]# su - pg15
Last login: Sun Jul  2 11:19:54 CST 2023 on pts/0
[pg15@dev-dbteam-148181 ~]$ pg_ctl start
waiting for server to start....2023-07-02 11:21:14.913 CST [31559] LOG:  redirecting log output to logging collector process
2023-07-02 11:21:14.913 CST [31559] HINT:  Future log output will appear in directory "log".
 done
server started
```

图 5-74 启动从库

在主库上继续写入数据,在从库上检查数据是否同步正常。如图 5-75 所示,在 148.180 上写入数据后,在 148.181 上可以看到数据同步。至此,PostgreSQL 的主从高可用搭建验证完毕。

```
[pg15@dev-dbteam-148180 pg_root]$ psql
psql (15.3)
Type "help" for help.

postgres=# \c xxg;
```

图 5-75 主从数据同步验证

```
You are now connected to database "xxg" as user "postgres".
xxg=# select * from xuexiaogang;
 id | n
----+---
  1 | 1
  2 | 2
(2 rows)

xxg=# insert into xuexiaogang values (3,3);
INSERT 0 1
xxg=# select * from xuexiaogang;
 id | n
----+---
  1 | 1
  2 | 2
  3 | 3
(3 rows)


[pg15@dev-dbteam 148181 pg_root]$ psql
psql (15.3)
Type "help" for help.

postgres=# \c xxg;
You are now connected to database "xxg" as user "postgres".
xxg=# select * from xuexiaogang;
 id | n
----+---
  1 | 1
  2 | 2
  3 | 3
(3 rows)
```

图 5-75 主从数据同步验证（续）

5.4 异构数据库同步的实例——基于 OGG

一个 IT 系统中只有一种数据库是非常理想化的，但是这种场景不多。很多公司都会有不止一种数据库，甚至有些公司有几十种数据库技术栈。当这些系统的数据需要交互的时候，就涉及异构数据库的数据同步。

解决这种问题主要使用的技术栈是 CDC（Change Data Capture），其核心是解析上下游数据库的日志，捕获变化量从而进行数据变更，使得下游数据实时跟随上游数据库的改变。比较有代表性的商用产品是 OGG（Oracle GoldenGate）和 FlinkCDC。前者属于商用，后者属于开源。本节内容主要介绍使用 OGG 实现异构数据库的数据同步。

5.4.1 CDC 简介

OGG（Oracle GoldenGate）是 CDC 技术的典型代表。因此，在介绍 OGG 之前，需要先了解 CDC 技术的基本概念。

CDC 的具体定义是指变化数据捕获（Change Data Capture），通过识别和捕获对数据库中的数据所做的更改（包括数据或数据表的插入、更新、删除等），然后将这些更改按发生的顺序完整记录下来，并实时通过消息中间件传送到下游流程或系统。与之前通过建立主从关系进行数据库同步不同，CDC 技术作为第三方中间件提供数据同步服务。这种独立性使得 CDC 不仅适用于同构数据库，而且也支持异构数据库。

5.4.2 OGG 概述

对于 OGG，可能广大 DBA 并不陌生。OGG 最初是一家从事 Oracle 数据库与 Oracle 数据库同步的公司，当时叫作 GoldenGate，后来被 Oracle 收购，改名为 Oracle GoldenGate，简称 OGG。如今发展到可以同步几乎所有的数据库以及打通数据库到消息队列和大数据的中间件解决方案。

对于 OGG 如何设定从非 Oracle 的数据库与 Oracle 数据库进行同步，以及 Oracle 和大数据（Hadoop）的数据同步，其实并没有太多的资料。而熟悉 OGG 的读者可能对于 OGG 的命令行实施和维护体验不佳。但是 OGG 有它自己的图形化界面安装和维护模式，本节所编写的是比较新的实现方式，相关资料甚少。希望可以帮助大家普及这些技术，以及对技术进行推广。

5.4.3 OGG 微服务版的安装部署

微服务版的特色就是图形化，相比较之前采用命令行搭建 OGG 来说，实施、监控和维护这些方面会大大简便。但是如果不熟悉 OGG 的原理以及没有做过命令行模式的搭建，会依然觉得很难。因为网上并没有很多参考资料，即使笔者以前写的资料和本节比起来也是很小的一部分，更重要的是官方文档并没有本节记录得详细。可以说这是第一份最为详细的微服务图形化界面的手册。

OGG 可以做 Oracle、MySQL 和大数据之间的各种数据传输，但是它其实是三个软件。即 OGG for Oracle、OGG for MySQL、OGG for BigData，只要将它们部署在同一台机器上就可以实现相互之间的数据同步。

OGG 的软件需要单独部署，而无须在数据库服务器上进行安装，这样做的好处是不对数据库进行改造。本次选用了 143.108 这个 IP 的服务器作为 OGG 的服务器。在 143.108 的服务器上建立安装目录，如图 5-76 所示。

```
[root@dev-dbteam-143108 ~]# mkdir /u01
[root@dev-dbteam-143108 ~]# mkdir /u01/stage
[root@dev-dbteam-143108 ~]# mkdir /u01/stage/oggsc
```

图 5-76 建立 OGG for Oracle 的软件目录

将从官网下载的安装介质进行解压，如图 5-77 所示，解压后的文件夹为 fbo_ggs_Linux_x64_Oracle_services_shiphome。这是 3 个 OGG 软件之一。

```
[root@dev-dbteam-143108 ~]# cd /tmp/
[root@dev-dbteam-143108 tmp]# unzip 213000_fbo_ggs_Linux_x64_Oracle_services_shiphome.zip
[root@dev-dbteam-143108 tmp]# ll
total 372392
-rw-rw-r-- 1 otcadmin otcadmin 381014026 Jul  3 09:49 213000_fbo_ggs_Linux_x64_Oracle_services_shiphome.zip
drwxr-xr-x 3 root     root            19 Jul 29  2021 fbo_ggs_Linux_x64_Oracle_services_shiphome
-rw-r--r-- 1 root     root          2409 Aug 11  2021 OGG-21.3.0.0-README.txt
-rw-r--r-- 1 root     root        306495 Aug 11  2021 oracle-goldengate-release-notes_21.3.pdf
```

图 5-77 解压安装包

软件统一安装到 Oracle 的操作系统用户下,所以需要用 root 用户解压后对 u01 向下的目录级联授权。执行 chmod 777 -R /u01,然后切换到 Oracle 账户进行安装。本次安装采用图形化安装,所以 CentOS 7 的操作系统上需要安装图形化界面的相关系统软件以及 VNC 软件。yum install 必需的软件包如图 5-78 所示。

```
[root@dev-dbteam-143108 home]# yum whatprovides "*/xhost"

[root@dev-dbteam-143108 home]# yum -y install xorg-x11-utils*

[root@dev-dbteam-143108 home]# yum -y install *vnc*

[root@dev-dbteam-143108 ~]# yum grouplist

[root@dev-dbteam-143108 ~]# yum groupinstall -y "GNOME Desktop"
```

图 5-78 yum install 必需的软件包

yum install 所需的 yum 源在这里不介绍了,请读者自行查询。安装完毕所有依赖软件,配置 VNC 服务,如图 5-79 所示,也可以自行搜索对照。

```
[root@dev-dbteam-143108 home]# vim /etc/systemd/system/vncserver@:1.service

[Unit]
Description=Remote desktop service (VNC)
After=syslog.target network.target

[Service]
Type=forking
User=root
Type=simple
ExecStartPre=-/usr/bin/vncserver -kill %i
ExecStart=/sbin/runuser -l root -c "/usr/bin/vncserver %i"
PIDFile=/root/.vnc/%H%i.pid
ExecStop=-/usr/bin/vncserver -kill %i

[Install]
WantedBy=multi-user.target
```

图 5-79 编辑 VNC 配置文件

VNC 服务在启动之前需要设置密码,如图 5-80 所示,看到 active 的标识表示服务启动成功。

如果需要调整 VNC 的端口,如图 5-81 所示,修改端口号。修改完毕保存,重新加载配置文件并且重启服务。

```
[root@dev-dbteam-143108 .vnc]# vncpasswd
Password:
Verify:
Would you like to enter a view-only password (y/n)? n
A view-only password is not used
[root@dev-dbteam-143108 .vnc]# systemctl restart vncserver@:1.service
[root@dev-dbteam-143108 .vnc]# systemctl status vncserver@:1.service
● vncserver@:1.service - Remote desktop service (VNC)
   Loaded: loaded (/etc/systemd/system/vncserver@:1.service; enabled; vendor preset: disabled)
   Active: active (running) since Mon 2023-07-03 13:50:56 CST; 6s ago
  Process: 72154 ExecStartPre=/usr/bin/vncserver -kill %i (code=exited, status=2)
 Main PID: 72212 (Xvnc)
   CGroup: /system.slice/system-vncserver.slice/vncserver@:1.service
           ▶ 72212 /usr/bin/Xvnc :1 -auth /root/.Xauthority -desktop dev-dbteam-143108:1 (root) -fp catalogue:/etc/X11/fontpath.d
                   -geometry 102...

Jul 03 13:50:56 dev-dbteam-143108 systemd[1]: Starting Remote desktop service (VNC)...
Jul 03 13:50:56 dev-dbteam-143108 vncserver[72154]: Can't find file /root/.vnc/dev-dbteam-143108:1.pid
Jul 03 13:50:56 dev-dbteam-143108 vncserver[72154]: You'll have to kill the Xvnc process manually
Jul 03 13:50:56 dev-dbteam-143108 systemd[1]: Started Remote desktop service (VNC).
```

图 5-80　为 VNC 设置密码并启动服务

```
[root@dev-dbteam-143108 .vnc]# vim /usr/bin/vncserver
.
.
.
$vncPort = 8080 + $displayNumber;
.
.
.
[root@dev-dbteam-143108 .vnc]# systemctl daemon-reload
[root@dev-dbteam-143108 .vnc]# systemctl restart vncserver@:1.service
```

图 5-81　VNC 端口修改

设置一下窗口，如图 5-82 所示。至此，OGG 安装完毕。

```
[root@dev-dbteam-143108 .vnc]# xhost +
access control disabled, clients can connect from any host
[root@dev-dbteam-143108 .vnc]# su - oracle
Last login: Mon Jul  3 13:08:12 CST 2023 on pts/3
[oracle@dev-dbteam-143108 Disk1]$ DISPLAY=:1.0
[oracle@dev-dbteam-143108 Disk1]$ export DISPLAY
[oracle@dev-dbteam-143108 Disk1]$ echo $DISPLAY
:1.0
```

图 5-82　设置窗口

5.4.4　使用 OGG for Oracle 实现 Oracle 数据库之间的数据同步

前一小节的内容详细介绍了 OGG 微服务的基本配置。接下来，将继续介绍 OGG for Oracle 的配置和基本使用。

1. OGG for Oracle 的软件安装

使用 VNC 的客户端程序连接服务器进入图形化界面，如图 5-83 所示，运行 ./runInstaller。出现了安装界面第一步。

单击"Next"按钮，来到了第二步选择软件安装目录，如图 5-84 所示。

```
[oracle@dev-dbteam-143108 Disk1]$ pwd
/tmp/fbo_ggs_Linux_x64_Oracle_services_shiphome/Disk1
[oracle@dev-dbteam-143108 Disk1]$ ./runInstaller
Starting Oracle Universal Installer...

Checking Temp space: must be greater than 120 MB.    Actual 84687 MB    Passed
Checking swap space: must be greater than 150 MB.    Actual 4095 MB    Passed
Checking monitor: must be configured to display at least 256 colors.    Actual 16777216    Passed
Preparing to launch Oracle Universal Installer from /tmp/OraInstall2023-07-03_02-07-36PM. Please wait ...
```

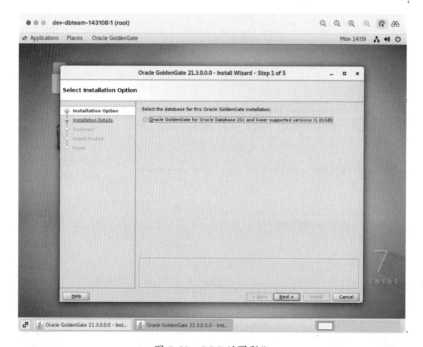

图 5-83 OGG 的图形化

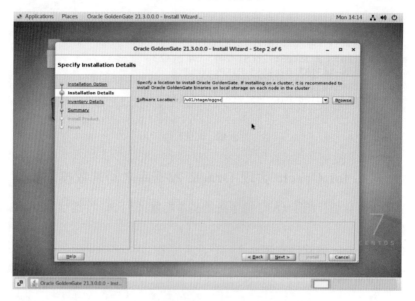

图 5-84 指定 OGG for Oracle 的软件目录

第三步，选择默认。如图 5-85 所示，该目录之前已经建立。

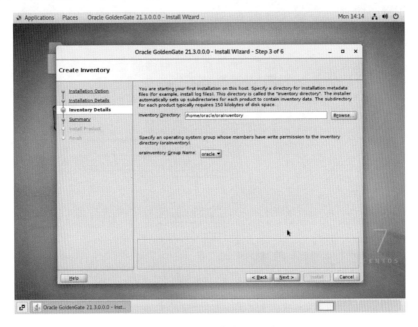

图 5-85　Oracle 清单目录选择

第四步，单击"Install"按钮。如图 5-86 所示，最后确认设置。

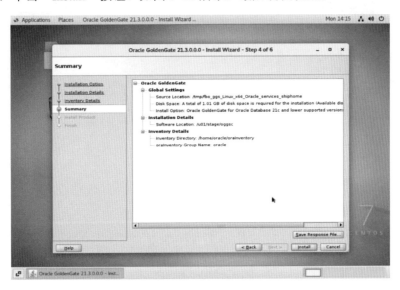

图 5-86　安装总览

第五步，等待安装结束。如图 5-87 所示，在安装完成后，最后需要用操作系统 root 用户执行配置脚本，设置环境变量。

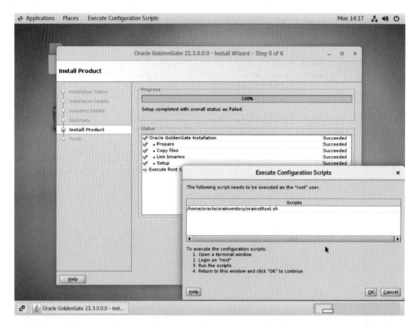

图 5-87 执行配置脚本

软件安装完成，如图 5-88 所示，单击"Close"按钮关闭对话框。

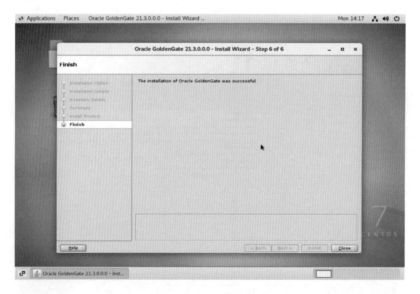

图 5-88 OGG 安装完成界面

以上仅仅是安装软件，随后启动图形化配置。在指定路径，如图 5-89 所示，这里填写本机地址以及一个未被使用的端口，并新建一个用于 Service Manager 的目录。然后单击"Next"按钮。

第 5 章　如何做好数据库之间的数据同步

```
[oracle@dev-dbteam-143108 Disk1]$ cd /u01/stage/oggsc/bin/
[oracle@dev-dbteam-143108 bin]$ ./oggca.sh
```

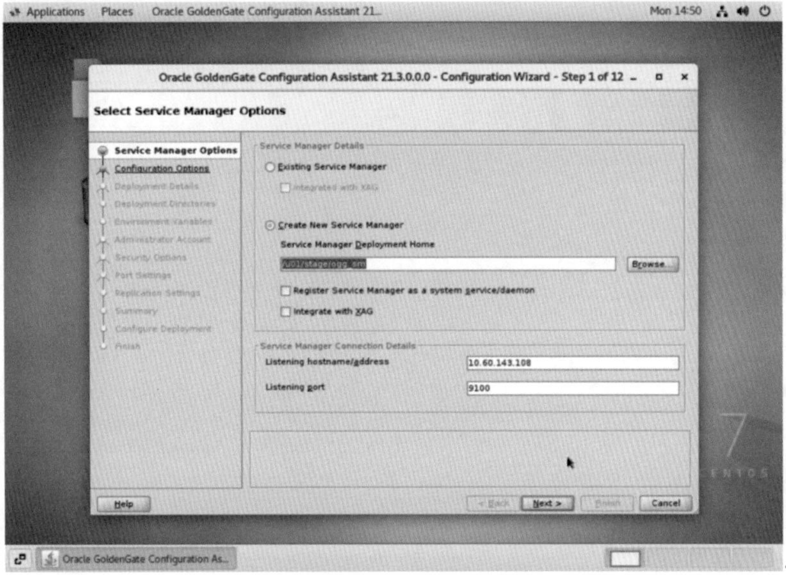

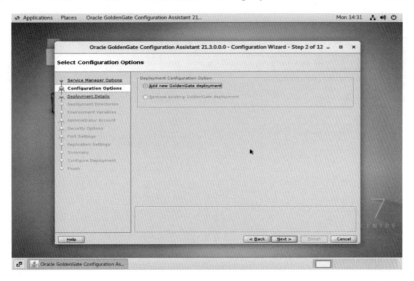

图 5-89　启动 OGG 配置界面

进入第二步，如图 5-90 所示，选择新建一个 deployment，单击"Next"按钮。

图 5-90　新建 deployment

进入第三步，如图 5-91 所示，给 deployment 起一个名字，软件所在目录是刚才步骤中已经定义过的，无须改变，单击"Next"按钮。

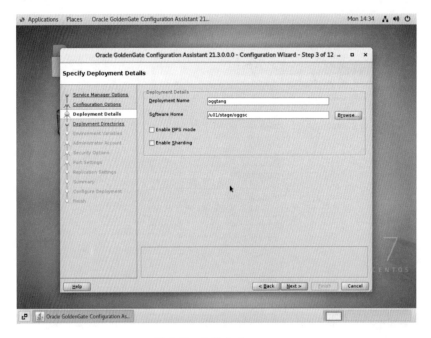

图 5-91 配置 deployment

第四步,选择 deployment 的路径。如图 5-92 所示,基于软件目录自适应,单击"Next"按钮。

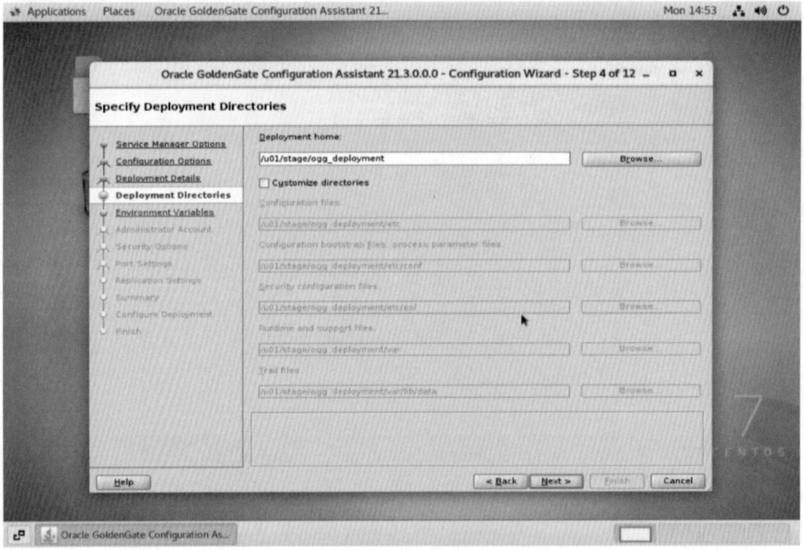

图 5-92 选择 deployment 路径

第五步,基于前面步骤的设置,保持默认,如图 5-93 所示,单击"Next"按钮。

第 5 章　如何做好数据库之间的数据同步　257

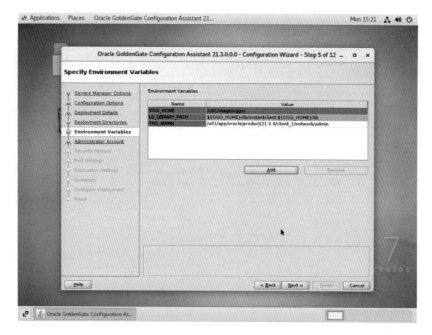

图 5-93　确认环境变量

第六步，如图 5-94 所示，设置管理员用户名和密码，此用户为后续 Web 访问界面的登录用户，单击"Next"按钮。

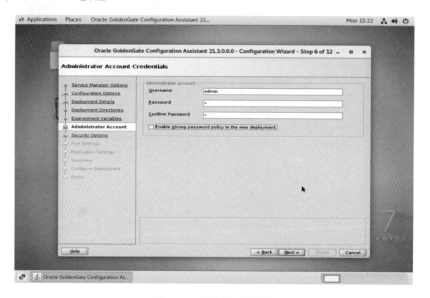

图 5-94　设置管理员账户

第七步跳过，第八步非常重要。预设 Web 界面端口，如图 5-95 所示，这些端口都要牢记，不可冲突。第一个 Administration Service port 端口填写一个未被使用的端口，后面会顺延这个端口号自动填写，单击"Next"按钮。

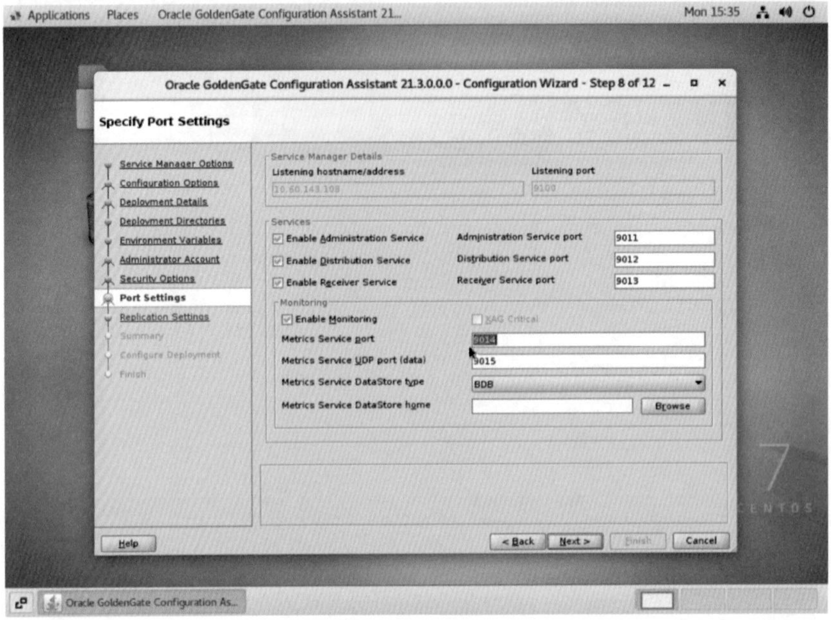

图 5-95 预设 Web 界面端口

第九步，如图 5-96 所示，设定一个自定义的 Schema，单击"Next"按钮。

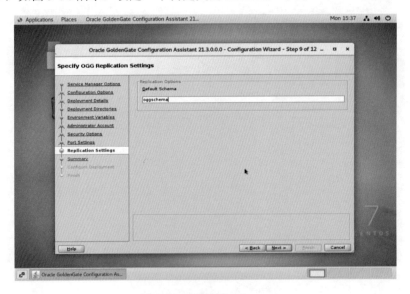

图 5-96 设定默认 Schema

第十步，如图 5-97 所示，最后确认前面设置的信息，单击"Next"按钮。

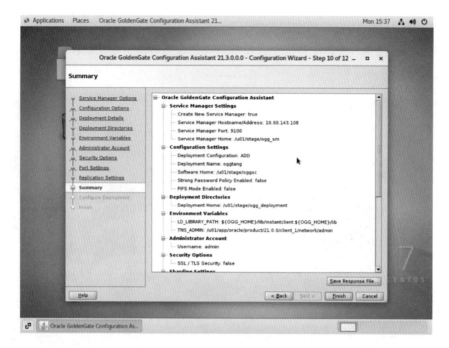

图 5-97　Web 配置总览

第十一步省略，最终如图 5-98 所示，Web 页面配置完毕。

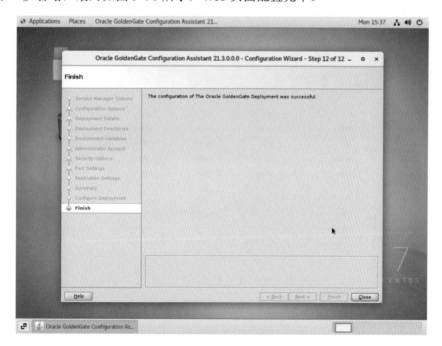

图 5-98　Web 页面配置完毕

2. 启动 OGG for Oracle 的准备工作

安装完成后使用第一步中配置的地址及端口登录服务管理页面，如图 5-99 所示。

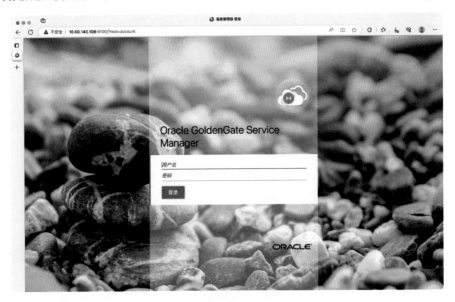

图 5-99 登录服务管理页面

使用第六步中配置的 admin 账号密码登录，管理界面如图 5-100 所示。

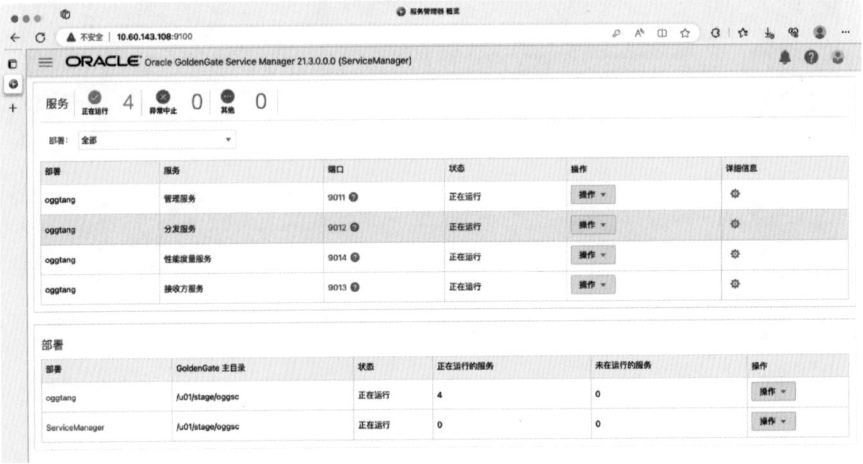

图 5-100 登录后进入管理界面

演示从 Oracle 到 Oracle 的数据同步。首先登录作为数据源的上游数据库所在 CDB 做环境以及账号的准备。如图 5-101 所示，这个步骤和传统 OGG 实施以及 ADG 的搭建步骤一致，如果已经做过可以忽略。

```
SQL> archive log list;
数据库日志模式              非存档模式
自动存档                   禁用
存档终点                   USE_DB_RECOVERY_FILE_DEST
最早的联机日志序列           6
当前日志序列                8

SQL> shutdown immediate ;
数据库已经关闭。
已经卸载数据库。
ORACLE 例程已经关闭。

SQL> startup mount
ORACLE 例程已经启动。

Total System Global Area 2516582192 bytes
Fixed Size                  9137968 bytes
Variable Size             570425344 bytes
Database Buffers         1929379840 bytes
Redo Buffers                7639040 bytes
数据库装载完毕。
SQL> alter database archivelog;

数据库已更改。

SQL> alter database open;

数据库已更改。

SQL> alter pluggable database all open;

插接式数据库已变更。
```

图 5-101 开启上游数据库归档

由于不是全部数据库都作为传输对象,到指定的 PDB 中打开附加日志,如图 5-102 所示。

```
SQL> ALTER DATABASE FORCE LOGGING;

数据库已更改。

SQL> ALTER DATABASE ADD SUPPLEMENTAL LOG DATA;

数据库已更改。

SQL> alter session set container=oggpdb;

会话已更改。

SQL> ALTER DATABASE ADD SUPPLEMENTAL LOG DATA;

数据库已更改。
```

图 5-102 PDB 级别打开附加日志

在 CDB 以及 PDB 下建立表空间给数据抽取账号做准备,如图 5-103 所示。

```
SQL> create tablespace ogg datafile'/u01/app/oracle/oradata/XXG/ogg.dbf'
  2 SIZE 100M AUTOEXTEND ON NEXT 100M MAXSIZE UNLIMITED;
```

表空间已创建。

```
SQL> alter session set container=oggpdb;
```

会话已更改。

```
SQL> create tablespace ogg datafile'/u01/app/oracle/oradata/XXG/oggpdb/ogg.dbf'
  2 SIZE 100M AUTOEXTEND ON NEXT 100M MAXSIZE UNLIMITED;
```

表空间已创建。

```
SQL> alter session set container=ORCLPDB;
```

会话已更改。

```
SQL> create tablespace ogg datafile'/u01/app/oracle/oradata/XXG/orclpdb/ogg.dbf'
  2 SIZE 100M AUTOEXTEND ON NEXT 100M MAXSIZE UNLIMITED;
```

表空间已创建。

图 5-103 增加 OGG 单独的数据文件

在 CDB 下创建账号并赋权，在每个 PDB 中将其默认表空间指定为前面建立的表空间，如图 5-104 所示。

```
SQL> create user c##ogg identified by ogg default tablespace ogg;
```

用户已创建。

```
SQL> begin
  2 DBMS_GOLDENGATE_AUTH.GRANT_ADMIN_PRIVILEGE
  3(GRANTEE => 'c##ogg', PRIVILEGE_TYPE => 'CAPTURE', GRANT_SELECT_PRIVILEGES => TRUE,
  4 DO_GRANTS => TRUE, CONTAINER => 'ALL' );
  5 end;
  6 /
```

PL/SQL 过程已成功完成。

```
SQL> grant dba to c##ogg;
```

授权成功。

图 5-104 设置 PDB 中 c##ogg 的默认表空间

第 5 章　如何做好数据库之间的数据同步　263

```
SQL> alter session set container=oggpdb;
```

会话已更改。

```
SQL> alter user c##ogg default tablespace ogg;
```

用户已更改。

```
SQL> grant dba to c##ogg;
```

授权成功。

```
SQL> alter session set container=orclpdb;
```

会话已更改。

```
SQL> alter user c##ogg default tablespace ogg;
```

用户已更改。

图 5-104　设置 PDB 中 c##ogg 的默认表空间（续）

在 CDB 下启用参数，如图 5-105 所示。

```
SQL> show parameters enable_goldengate_replication;

NAME                                 TYPE        VALUE
------------------------------------ ----------- ------------------------------
enable_goldengate_replication        boolean     FALSE

SQL> alter system set enable_goldengate_replication=TRUE scope=BOTH;

系统已更改。

SQL> show parameters enable_goldengate_replication;

NAME                                 TYPE        VALUE
------------------------------------ ----------- ------------------------------
enable_goldengate_replication        boolean     TRUE
```

图 5-105　enable_goldengate_replication 参数启用

3．OGG for Oracle 的初始连接通道配置

打开 Web 浏览器，登录图形化管理的界面，如图 5-106 所示。单击箭头所示的 9011 的超链接跳转管理服务页面（手动输入地址和端口号也可跳转）。

图 5-106　进行抽取进程的配置

进入 9011 页面后（注意和上图中的 9100 是不一样的），如图 5-107 所示，单击箭头所指的按钮，切换到配置页面。

图 5-107　进入配置页面

单击左侧边框的配置选型，如图 5-108 所示，即将进入最终的配置进程页面。

图 5-108　配置进程

单击"身份证明"旁边的"+"，依次填写用户 ID 和密码，如图 5-109 所示。这里的用户 ID 是由用户名+数据库 IP 地址+端口+数据库服务名组成的。用户名是之前建立的 CDB 下的用户 c##ogg，10.60.143.150 是数据库地址，oggpdb 是上游数据库的数据库服务名，端口是默认的，如果自定义需要在 IP 后面加":端口号"。

图 5-109　添加连接信息

添加成功后，如图 5-110 所示，单击箭头所指按钮。

图 5-110　测试连接上游 Oracle

如果有问题则会报错，如果成功则出现如图 5-111 所示界面，显示检查点信息。

图 5-111　连接上游数据库成功的界面

如图 5-112 所示,单击"TRANDATA 信息"一栏右边的加号,新增 Schema。

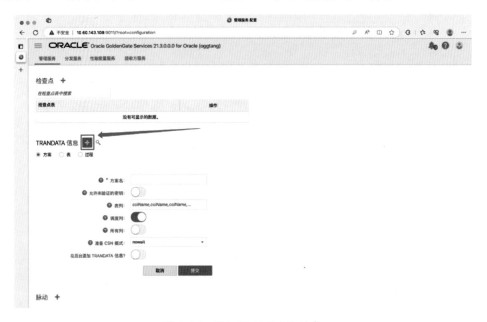

图 5-112　增加 TRANDATA 信息

方案名使用需要进行同步的 Schema 名,如图 5-113 所示。

图 5-113　添加 TRANDATA 的方案名

添加完毕即可显示如图 5-114 所示界面。单击图中箭头所指,可以查询到添加成功的

Schema，并能看到右边准备好实例化的表的数量，表示能够正确读取到该 Schema 的信息。

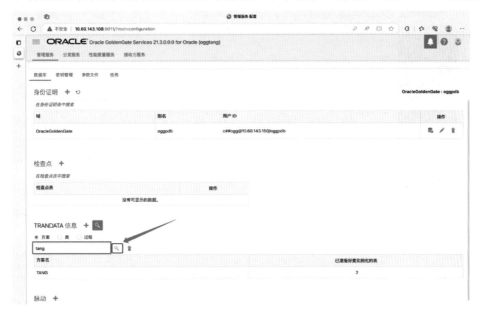

图 5-114 添加 TRANDATA 方案成功

然后需要用同样的方式添加上游 PDB 所在的 CDB 连接，如图 5-115 所示（实测不添加 CDB 不行），但是不需要为 CDB 添加 TRANDATA 信息。添加完后如下，有两个身份证明，上面是 PDB，下面是 CDB。

图 5-115 添加 CDB 连接信息

用类似的方式配置下游数据库，用与上游相同的方式建立 OGG 专用账号，并赋权（详情参考前文）。配置完毕，添加下游 PDB 节点，下游只需要配置 PDB 节点的连接，不需要 CDB 的相关配置，如图 5-116 所示，第二行为下游目标 PDB 数据库的连接。

图 5-116　添加下游数据库

注意：虽然下游数据库无须配置 CDB，但是 PDB 中需要添加 TRANDATA 信息，即 Schema 信息，如图 5-117 所示。

图 5-117　配置下游数据库的 TRANDATA 信息

上下游的连接都配置完毕后，单击左侧切换到概览页面，如图 5-118 所示，准备添加抽取进程以及投递进程。图中提取和复制这一行中两个框的加号按钮即为分别添加进程的入口。提取是 OGG 从上游数据库拿数据，复制是 OGG 把拿到的数据送到下游。

4．OGG for Oracle 的提取进程配置

首先单击"提取"右侧的加号，配置提取进程，如图 5-119 所示，一般情况选择集成的提取（Change Synchronization/Integrated Extract），单击"下一步"按钮。

第 5 章 如何做好数据库之间的数据同步 269

图 5-118 单击左侧边栏概览操作

图 5-119 配置提取进程类型

在配置提取选型页面中，需要输入进程名称和线索名称，如图 5-120 所示。进程名称不能超过 8 个字符，可以自定义（最好是自己便于理解和记忆的）。填写一个线索名称（线索名称很重要，用于给下游进程匹配相应的上游进程）。

图 5-120 配置提取进程图 1

往下翻看页面，如图 5-121 所示，先选第二个框中身份证明信息，这里选择 Oracle GoldenGate。然后选身份证明别名，这里在下拉列表中选 CDB 对应的名字。完成该步骤的选择，才会出现图中的第一个框的隐藏信息。在"注册到 PDB"的下拉列表中，选择上游数据库所在的 PDB，单击"下一步"按钮。

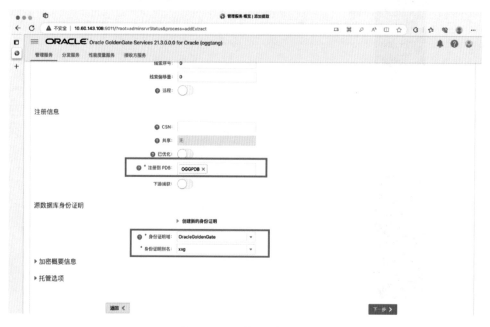

图 5-121　配置提取进程图 2

进入如图 5-122 所示界面，该界面默认只有前两行数据，后面的所有信息全部要求用户手工输入。

图 5-122　编辑提取参数文件所示的界面

在这个参数中新增几行数据：EXTTRAIL ex.DDL INCLUDE ALL（用于同步 DDL 语句）和 TABLE oggpdb.tang.*;（用于指定范围，这里是 PDB.SCHEMA.TABLE 格式）。编辑完成后单击"创建并运行"按钮。

随后就能看到在页面上有一个提取进程被创建出来，如图 5-123 所示。并且是绿色的监控运行状态。

图 5-123　成功创建提取进程

5．OGG for Oracle 的复制进程配置

按照同样的步骤新增复制进程，如图 5-124 所示。一般选择集成的复制，单击"下一步"按钮。

图 5-124　选择复制进程类型

在复制选型页面，如图 5-125 所示，进程名称可以自定义，身份证明域选择下拉列表中的 Oracle GoldenGate，身份证明别名在下拉列表中选择下游目标数据库。

图 5-125 复制选型参数设置

重点提示，图中的线索名称要填写和提取进程相同的线索名称。前文提到写的是 ex，所以这里也要写 ex，才能接收到对应抽取进程的数据。然后单击"下一步"按钮。

在复制进程参数文件页面，如图 5-126 所示，初始也只有前两行。后面两行需要用户自行添加。添加 ddl include all（同步 ddl）以及添加 MAP oggpdb.feng.*, TARGET pdbtang.tang.*;（指定上下游精确同步位置）。

图 5-126 设置复制进程参数文件

创建并运行后，看到右边复制一栏中有进程成功启动，如图 5-127 所示。

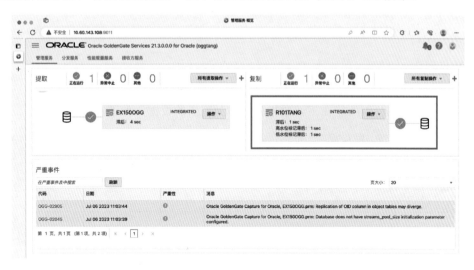

图 5-127 创建复制进程成功

重点提示：所有进程创建成功并不能代表运行成功，需要进行验证。在 IP 为 150 的上游源库上建表，如图 5-128 所示。

SQL> create table test_ogg(id int, primary key(id));

表已创建。

图 5-128 上游数据库建立测试同步表

在下游目标库中，能看到表结构的同步（需要几秒的时间，视系统繁忙程度和配置而定，1～20s 都有可能，正常都是 1s），如图 5-129 所示。

```
SQL> desc test_ogg;
ERROR:
ORA-04043: 对象 test_ogg 不存在


SQL> desc test_ogg;
ERROR:
ORA-04043: 对象 test_ogg 不存在


SQL> desc test_ogg;
 名称                                      是否为空? 类型
 ----------------------------------------- -------- ----------------------------
 ID                                        NOT NULL NUMBER(38)
```

图 5-129 下游数据库得到同步表结构

在上游数据库写入数据，如图 5-130 所示。

```
SQL> insert into test_ogg values(3);
```

已创建 1 行。

```
SQL> commit;
```

提交完成。

图 5-130 上游数据库写入数据

同样，在 1 到几秒，就能在下游数据库查询到同步的数据，如图 5-131 所示。

```
SQL> select * from test_ogg;
```

未选定行

```
SQL> select * from test_ogg;
```

未选定行

```
SQL> select * from test_ogg;

        ID
----------
         3
```

图 5-131 下游数据库同步数据

实测 Oracle 对 Oracle 数据库的 DDL 和 DML 都能成功同步。这是基于上下游数据库都是空表的情况，不过如果一开始上游数据库已经存在数据怎么办？下面演示通过 Initial Load 提取功能实现历史数据的初始化转移并持续同步。

6．OGG for Oracle 的初始化追平配置

首先建立初始化投递，传输历史数据。如图 5-132 所示，在图中选择"初始加载提取"（Initial Load 提取），而不是"集成的提取"。单击"下一步"按钮。

图 5-132 初始化提取类型设置

在提取选项中自定义提取进程名字，设定初始化提取进程名称，如图 5-133 所示。单击"下一步"按钮。

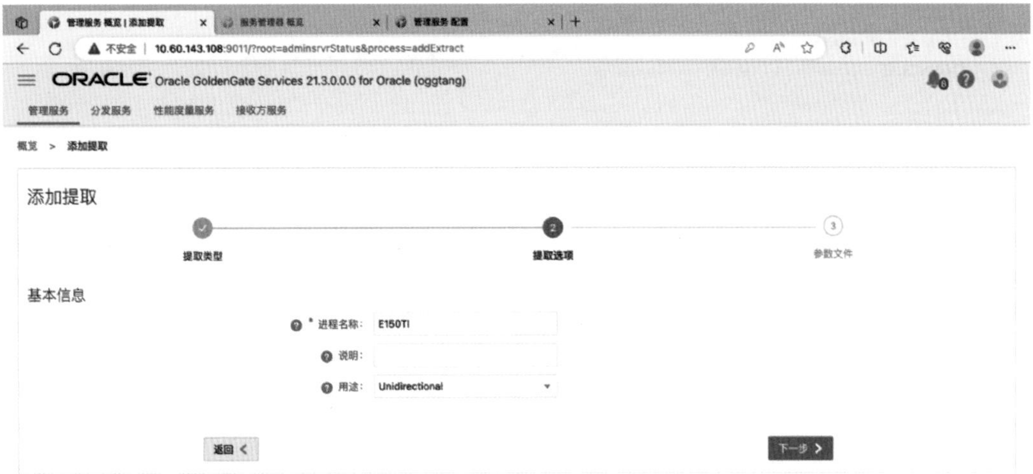

图 5-133　设定初始化提取进程名称

在图 5-134 中，新增参数：

userid c##ogg@10.60.143.150/xxg, password ogg（这里是填写 CDB 信息）
extfile ee purge（ee 为线索，可以自己命名）
TABLE oggpdb.tang.*;（精确数据源的位置 TABLE 后面第一个是 PDB 名字，第二个是 Schema 名字，第三个是表名，*代表所有表）

然后单击"创建并运行"按钮。

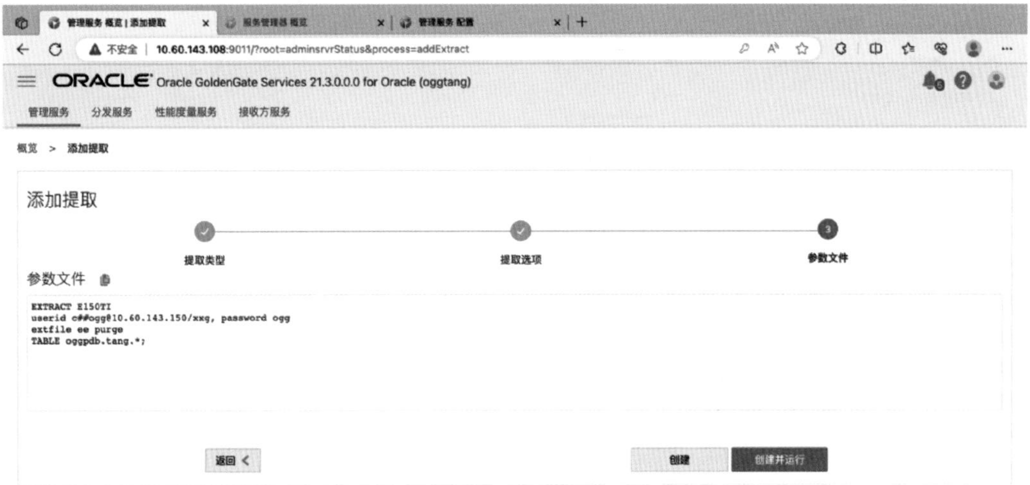

图 5-134　新增参数

建立完成后可能会看到节点状态变为绿色一次，然后就停止了。如图 5-135 所示，单击"操作"菜单的"详细信息"可以查询运行状态，类似命令行模式的 viewreport。

图 5-135 通过后台日志查看进程信息

在如图 5-136 所示的页面中可以查看参数和运行状态。

图 5-136 进程信息 1

在如图 5-137 所示的界面中，可以看到初始化的数据有多少。在报告的最后看到两张表的共计 9 条数据已经被导出了。

第 5 章　如何做好数据库之间的数据同步　277

图 5-137　进程信息 2

下面需要建立对应的初始化复制（投递）进程，如图 5-138 所示，这里一定要选择"非集成的复制"（也就是 Classic Replicat）。这点和常规复制的选择是不一样的。单击"下一步"按钮。

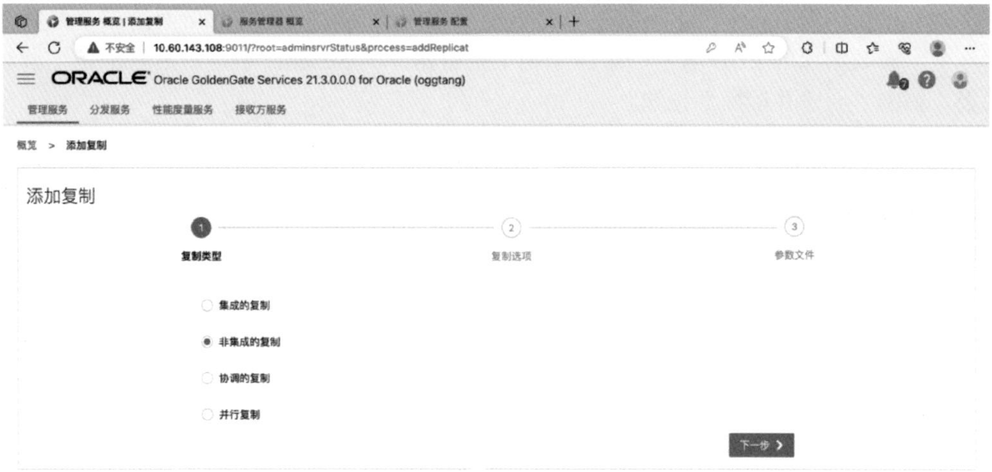

图 5-138　初始化复制类型选取

在初始化复制选项页面进行勾选，如图 5-139 所示，进程名称自定义。"身份证明域"选择 Oracle GoldenGate，"身份证明别名"是下游目标 PDB 的数据库，注意"线索名称"对应前面初始化投递进程配置中手动写的配置名称，前文用的是 ee，这里也用 ee。

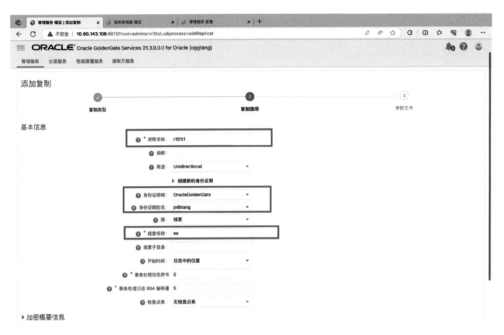

图 5-139　初始化复制选项设置

最后配置初始化参数，如图 5-140 所示，前两行为自动生成，使用者需要自己配置第三行。第三行为精确匹配。MAP 后面的三个参数分别是第一个上游数据库的 PDB 名字，第二个是上游数据库的 Schema 名字，第三个是上游数据库的表。TARGET 后面的三个参数分别是第一个下游数据库的 PDB 名字，第二个是下游数据库的 Schema 名字，第三个是下游数据库的表。单击"创建并运行"按钮。

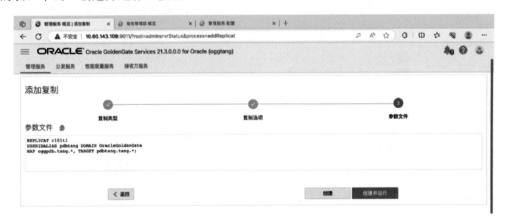

图 5-140　初始化复制参数设置

单击之后会出现启动又停止的情况，说明数据已经传输完成。直接到目标库中查询，可以看到数据已经同步完成了。

最终 OGG 微服务监控页面如图 5-141 所示，有一个初始化提取进程（黄色），一个初始化复制进程（红色），还有一对提取和复制进程（一对绿色）。进程名称虽然是自定义的，但是为了维护还是应该有一定的规则。这里两个 TI 是用于投递历史数据的，两个 TC 是数据的持续同步，保证历史数据同步后，新数据也能及时同步。

图 5-141　OGG 微服务监控页面

以上是 Oracle 对 Oracle 数据库的同步，下面介绍 MySQL 到 Oracle 数据库同步的设置。

5.4.5　使用 OGG for MySQL 实现 MySQL 与异构数据库的同步

OGG for MySQL 的图形化安装与 OGG for Oracle 的步骤基本一致，不占用篇幅赘述。注意：在同一台机器上想要同时安装 OGG for MySQL 与其他 OGG 数据源的版本，需要建立不同的目录安装，防止覆盖安装。在初始化配置的过程中也要注意，不要不同版本使用相同端口。如图 5-142 所示，使用了 8001 的管理端口（图 5-142 的地址栏中是 10.60.143.108:8001）。这里显示了在同一台机器上安装部署不同数据源的 OGG 版本采用了不同的版本和端口。

首先假设在上游源端 MySQL 上已经有存量数据，现在需要将这些数据同步给其他异构数据库。本次演示同步到 Oracle 场景。

在上游 MySQL 数据库上建立账号用于数据抽取，如图 5-143 所示。

同 OGG for Oracle 的操作相同，单击管理服务右边的 8001 端口号，跳转至 OGG for MySQL 的管理界面。

如图 5-144 所示，输入上游 MySQL 数据库的 IP 地址、端口、刚才已经建立的管理员（ogg）和对应的密码。注意：端口下面的"数据库名称"是指 MySQL 中的 Database 的名字。

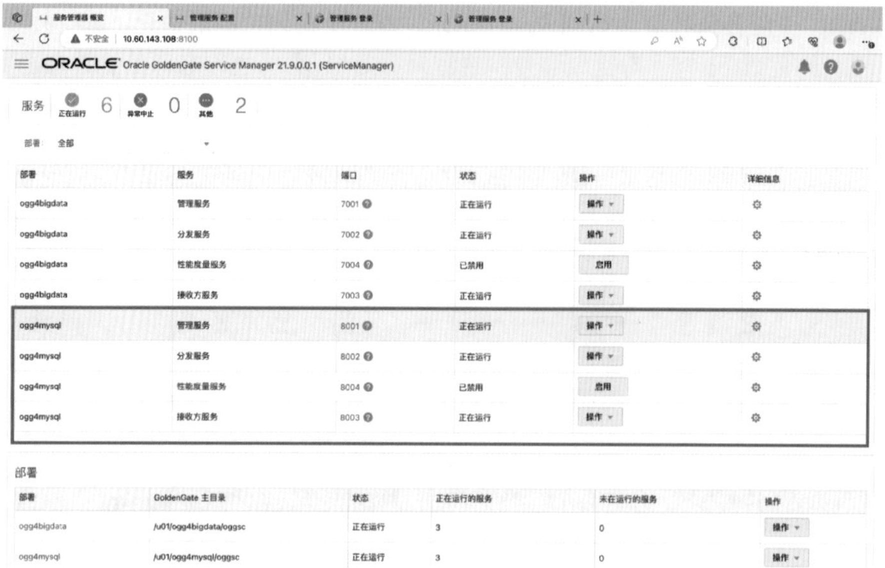

图 5-142 OGG 的管理端口

```
mysql> create user 'ogg'@'%' identified by 'ogg';
Query OK, 0 rows affected (0.01 sec)

mysql> grant all on *.* to 'ogg'@'%';
Query OK, 0 rows affected (0.01 sec)
```

图 5-143 上游 MySQL 建立提取数据的管理员账户

图 5-144 配置上下游的数据库连接

身份证明别名填写 2 个及 2 个以上的字符，完成填写后，单击"提交"按钮。在弹出的身份证明的页面中，如图 5-145 所示，单击箭头所指按钮进行连接测试。

图 5-145 MySQL 连接测试

连接配置完成后，在"概览"的标签页中，单击添加提取进程。如图 5-146 所示，填写"进程名称"为 e32c（名字可以自定义），"线索名称"为 mm（名字可以自定义），选择前面新增的数据库连接配置，单击"下一步"按钮。

图 5-146 填写提取进程选项

在参数文件页面中，如图 5-147 所示，第三行的 EXTFILE 配置为绝对路径。其原因是需要将提取的 MySQL 数据主动送到 Oracle 的复制进程目录下，所以直接将文件指向目标的绝对路径，可以方便 Oracle 复制进程读取数据，需要填写绝对路径。

图 5-147　MySQL 的 OGG 初始化参数配置

本次演示为 MySQL 的 c 库下的数据库中的所有表，所以 TABLE 选项后为 c.*，如果要指定自定义的表，可以自行修改。单击"创建并运行"按钮。

在管理服务页面看到建立的初始化（INITIAL LOAD）已经执行完毕。如图 5-148 所示，观测到状态变为绿色后快速变为黄色，数据抽取时状态为绿色，抽取完成进程会结束变为黄色。

图 5-148　观测 MySQL 提取进程初始化

单击该初始化进程的"操作"按钮，可以在其报告页中检查日志。如图 5-149 所示，指定的上游数据库的数据都已经抽取完成。

第 5 章 如何做好数据库之间的数据同步 283

图 5-149 MySQL 初始化提取报告详情

抽取完成后，转到复制进程的页面（注意 Web 页面：从 OGG for MySQL 的页面到 OGG for Oracle 的页面）。在 OGG for Oracle 的页面中配置投递进程，利用与前文中 Oracle 到 Oracle 的初始化投递相同的步骤建立投递进程（R101TM）并启动。如图 5-150 所示，MySQL 的 c 库下的 d 表的 1000 条数据已经传输到了 PDBTANG 的数据库的 TANG 的 Schema 下的 D 表。

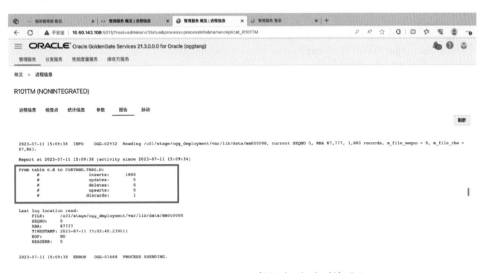

图 5-150 OGG for Oracle 的复制进程报告详情页面

总结规律：初始化投递只会投递数据，并不会投递表结构，数据只会投递到对应表名下。所以下游数据库必须建立和上游数据库相同的表结构。此外，由于 Oracle 要求使用 OGG 服务的 Oracle 数据库需要有主键，而 MySQL 没有这个要求。所以在复制投递时若出现上游 MySQL 源表中对应下游 Oracle 表中主键字段存在重复值的，只会写入一条以保证下游表的主键不重复。

5.4.6 使用 OGG for BigData 实现大数据组件与异构数据库的数据同步

下面介绍从数据库到大数据系统（Hadoop 组件）的数据同步。本节仅介绍到 Kafka 的数据同步，到 HBase 和 Hive 的就不做展示了。使用的是 OGG 21c for Oracle 微服务版以及 OGG 21c for BigData 微服务版进行配合完成。

基于前文中安装 OGG 21c for Oracle 的基础，在同一台机器上安装 OGG 21c for BigData 微服务版。

安装 Kafka 集群，现在最新的 Kafka 集群包含了 ZooKeeper，不再需要单独搭建。如果还是用旧版的 Kafka，则还是需要 ZooKeeper 支持的（中间件的技术栈也是数据库从业者应该掌握的）。

下面演示 ZooKeeper 的安装及搭建方式。如图 5-151 所示，编辑 zoo.cfg 指定 datadir 和 datalogdir 路径，并且将三个 ZooKeeper 的地址全部写入每个节点的 ZooKeeper 的 zoo.cfg 文件。注意最后新增 server 地址及端口的配置，这里演示 3 节点集群的配置。

```
root@xue-test1-148154:[/root]tar -zxvf apache-zookeeper-3.7.1-bin.tar.gz
.
.
.
root@xue-test1-148154:[/root]vim /zookeeper/apache-zookeeper-3.7.1-bin/conf/zoo.cfg
tickTime=2000
initLimit=10
syncLimit=5
dataDir=/zookeeper/data/zookeeper
dataLogDir=/zookeeper/data/logs
clientPort=2181
maxClientCnxns=60
autopurge.snapRetainCount=3
autopurge.purgeInterval=1
server.1=10.60.148.154:2888:3888
server.2=10.60.148.155:2888:3888
server.3=10.60.148.156:2888:3888
```

图 5-151 下载解压 ZooKeeper 的二进制文件

ZooKeeper 集群的重点是每个节点需要一个标识文件 myid，如图 5-152 所示。在其中一个节点 10.60.148.155 上设定 myid，3 个节点分别是 1、2、3。

```
root@xue-test1-148154:[/zookeeper/data]mkdir -p /zookeeper/data/zookeeper
root@xue-test1-148154:[/zookeeper/data]mkdir -p /zookeeper/data/log
root@xue-test1-148154:[/zookeeper/data/zookeeper]pwd
/zookeeper/data/zookeeper
root@xue-test1-148154:[/zookeeper/data/zookeeper]echo 1 > myid
```

图 5-152 生成标识文件

在指定目录下执行 zkServer.sh 启动 ZooKeeper 进程，如图 5-153 所示。启动完成后，检查 ZooKeeper 的角色。

```
root@xue-test1-148154:[/root]cd /zookeeper/apache-zookeeper-3.7.1-bin/bin/
root@xue-test1-148154:[/zookeeper/apache-zookeeper-3.7.1-bin/bin]./zkServer.sh start
/usr/bin/java
ZooKeeper JMX enabled by default
Using config: /zookeeper/apache-zookeeper-3.7.1-bin/bin/../conf/zoo.cfg
Starting zookeeper ... STARTED
root@xue-test3-148156:[/zookeeper/apache-zookeeper-3.7.1-bin/bin]./zkServer.sh status
/usr/bin/java
ZooKeeper JMX enabled by default
Using config: /zookeeper/apache-zookeeper-3.7.1-bin/bin/../conf/zoo.cfg
Client port found: 2181. Client address: localhost. Client SSL: false.
Mode: leader
```

图 5-153　启动 ZooKeeper 进程

一般 3 个 ZooKeeper 组成的集群一个是 leader，其余两个是 follower（也可以将 ZooKeeper 的 bin 加入到环境变量中，就不用输入绝对路径了），如图 5-154 所示。

```
root@xue-test3-148156:[/zookeeper/apache-zookeeper-3.7.1-bin/bin]./zkServer.sh status
/usr/bin/java
ZooKeeper JMX enabled by default
Using config: /zookeeper/apache-zookeeper-3.7.1-bin/bin/../conf/zoo.cfg
Client port found: 2181. Client address: localhost. Client SSL: false.
Mode: follower
```

图 5-154　ZooKeeper 从节点角色

至此，ZooKeeper 集群的安装搭建就已经完成。然后进行 Kafka 的安装，如图 5-155 所示，编辑 Kafka 的配置文件 server.properties。其重点在于 broker.id 与 ZooKeeper 的 myid 要对应起来。

```
root@xue-test1-148154:[/root]tar -zxvf kafka_2.12-3.5.0.tgz
root@xue-test1-148154:[/root]vim /kafka/config/server.properties
############################# Server Basics #############################

# The id of the broker. This must be set to a unique integer for each broker.
broker.id=1
```

图 5-155　编辑 Kafka 的配置文件

使用命令后台启动 Kafka，如图 5-156 所示，最后加上 & 变量。这样 Kafka 进程不会因为当前窗口的关闭而关闭。

```
root@xue-test1-148154:[/kafka/bin]./kafka-server-start.sh /kafka/config/server.properties &
```

图 5-156　使用命令后台启动 Kafka

其他两个节点依次启动，三个节点的 Kafka 启动完成后，开始创建主题。如图 5-157 所示，创建了一个 xxg 的主题。

```
root@xue-test1-148154:[/root/CMAK-master]kafka-topics.sh --create --bootstrap-server
10.60.148.154:9092,10.60.148.155:9092,10.60.148.156:9092 --replication-factor 2 --partitions 2 --topic xxg
 [2023-07-14 16:15:46,867] INFO [ReplicaFetcherManager on broker 1] Removed fetcher for partitions Set(xxg-0) (
kafka.server.ReplicaFetcherManager)
 [2023-07-14 16:15:46,898] INFO [LogLoader partition=xxg-0, dir=/tmp/kafka-logs] Loading producer state till
offset 0 with message format version 2 (kafka.log.UnifiedLog$)
 [2023-07-14 16:15:46,900] INFO Created log for partition xxg-0 in /tmp/kafka-logs/xxg-0 with properties {} (
kafka.log.LogManager)
 [2023-07-14 16:15:46,901] INFO [Partition xxg-0 broker=1] No checkpointed highwatermark is found for partition
xxg-0 (kafka.cluster.Partition)
 [2023-07-14 16:15:46,902] INFO [Partition xxg-0 broker=1] Log loaded for partition xxg-0 with initial high
watermark 0 (kafka.cluster.Partition)
Created topic xxg.
```

图 5-157 新建主题

利用 Kafka 原始工具和命令观测生产消息和消费消息。如图 5-158 所示，在 154 的机器上发送了一个 Hello World，在 156 的机器上可以观察到消息送达。

```
root@xue-test1-148154:[/root]kafka-console-producer.sh --broker-list 10.60.148.154:9092 --topic xxg
>Hello World

root@xue-test3-148156:[/kafka/bin]/kafka/bin/kafka-console-consumer.sh --bootstrap-server 10.60.148.156:9092
--topic xxg --from-beginning
Hello World
```

图 5-158 对主题进行消息收发

能够正常发送、接收表示集群已经搭建完成，且工作正常。下面利用 OGG 实现从 Oracle（也可以是其他数据库，这里仅仅演示 Oracle）中抽取数据，并将 trail 文件投递到 OGG 21c for BigData 的目录下提供给 Kafka 读取。首先，用图形化界面安装 OGG for BigData，此过程和安装 OGG for Oracle 一致，参考前文。

登录对应的 Web 管理页面登录，建立提取进程。如图 5-159 所示，填写参数时，第四行的提取文件路径需要填写绝对路径，可以将 trail 文件投递到指定的路径下。

图 5-159 建立从 Oracle 到 BigData 的提取进程

配置并运行后，管理界面进程出现报错，如图 5-160 所示。该报错官方文档并没有说要怎么解决。

经过各种实践，得知需要在命令行界面进行手工配置（官方文档目前还没有告知需要手工执行），需要执行的命令是图中框内的命令。如果有读者使用过传统的 OGG 软件，一定知道在软件目录下有个程序叫作 ggsci。这里其实就是要在 ggsci 中去执行，但是官网没有告知。而且这个软件在 OGG for BigData 的软件中也改了名字，不再叫作 ggsci 而是叫作 adminclient。

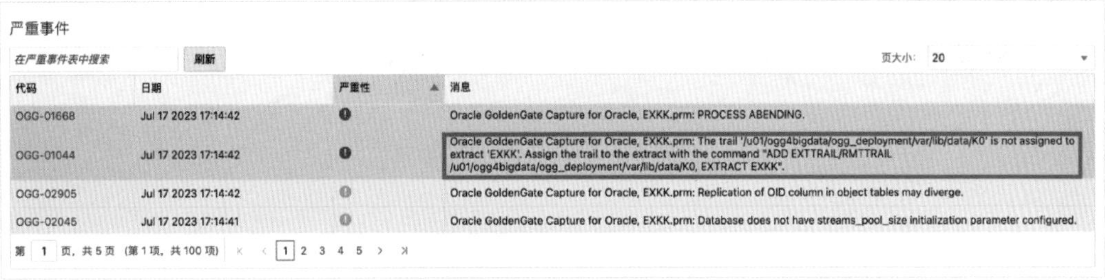

图 5-160　管理界面进程出现报错

按照上图中框内的命令，在 adminclient 命令行中进行添加提取进程。如图 5-161 所示。图中有两个 Trail，第一个是 k0，第二个是/u01/ogg4bigdata/ogg_deployment/var/lib/data/k0。第一个是 Web 界面代入的，不过无法使用。第二个是人工添加，属于正确配置。但是若两个同时存在，则不能正常工作，抽取进程不支持同时给两个 Trail 文件写入。所以只能用它的工具进行配置后再手工进行删除。这是官方文档中没有的部分。

```
OGG (http://localhost:9100 oggtang) 2> ADD EXTTRAIL /u01/ogg4bigdata/ogg_deployment/var/lib/data/k0,
EXTRACT E150K

OGG (http://localhost:9100 oggtang) 3> INFO EXTTRAIL *

       Local Trail: k0
      Seqno Length: 6
 Flip Seqno Length: yes
           Extract: E150K
             Seqno: 0
               RBA: 0
         File Size: 500M

       Local Trail: /u01/ogg4bigdata/ogg_deployment/var/lib/data/k0
      Seqno Length: 6
 Flip Seqno Length: yes
           Extract: E150K
             Seqno: 0
               RBA: 0
         File Size: 500M

OGG (http://localhost:9100 oggtang) 9> DELETE EXTTRAIL k0

OGG (http://localhost:9100 oggtang) 10> INFO EXTTRAIL *

       Local Trail: /u01/ogg4bigdata/ogg_deployment/var/lib/data/k0
      Seqno Length: 9
 Flip Seqno Length: no
           Extract: EXKK
             Seqno: 0
               RBA: 1326
         File Size: 500M
```

图 5-161　手工添加 forBigData 的提取进程

提示：删除的是 k0，保留的是/u01/ogg4bigdata/ogg_deployment/var/lib/data/k0。如果删

除后者，那么就是 DELETEEXTTRAIL /u01/ogg4bigdata/ogg_deployment/var/lib/data/k0。

完成操作后，回到 Web 配置页重新启动提取进程，如图 5-162 所示。进程图标显示绿色，表示成功。

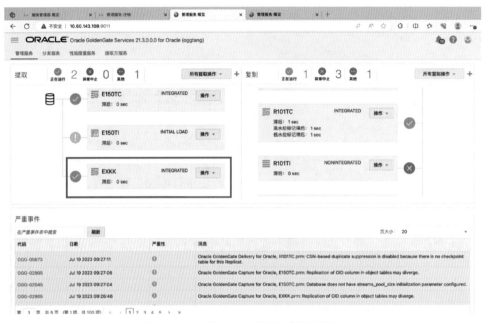

图 5-162　回到 Web 配置页重启提取进程

提取进程完成后需要配置到 Kafka 的复制进程。第一步要安装 Kafka 所需要的依赖，可以使用官方推荐的联网获取的方式，如图 5-163 所示。

```
[root@dev-dbteam-143108 ~]# cd /u01/ogg4bigdata/oggsc/opt/DependencyDownloader/
[root@dev-dbteam-143108 DependencyDownloader]# ./kafka.sh 3.5.0
java version "1.8.0_181"
Java is installed.
Apache Maven 3.8.2 (ea98e05a04480131370aa0c110b8c54cf726c06f)
Maven is accessible.
Root Configuration Script
INFO: This is the Maven binary [../../ggjava/maven/bin/mvn].
INFO: This is the location of the settings.xml file [./docs/settings_np.xml].
INFO: This is the location of the toolchains.xml file [./docs/toolchains.xml].
INFO: The dependencies will be written to the following directory[../dependencies/kafka_3.5.0].
INFO: The Maven coordinates are the following:
INFO: Dependency 1
INFO: Group ID [org.apache.kafka].
INFO: Artifact ID [kafka-clients].
INFO: Version [3.5.0]
INFO: Dependency 2
INFO: Group ID [org.apache.kafka].
INFO: Artifact ID [connect-json].
INFO: Version [3.5.0]
[INFO] Scanning for projects...
[INFO]
[INFO] ---------------< oracle.goldengate:dependencyDownloader >---------------
[INFO] Building dependencyDownloader 1.0
[INFO] --------------------------------[ pom ]---------------------------------
Downloading from central: https://repo.maven.apache.org/maven2/org/apache/maven/plugins/maven-clean-plugin/2.5/maven-clean-plugin-2.5.pom
```

图 5-163　在联网环境下安装 Kafka 依赖

由于源在外网，如果没有网络条件，可以从 Kafka 所在的服务器上将依赖直接传输到 OGG 所在的服务器中。在 OGG 服务器上建立对应的目录，如图 5-164 所示，用来放置依赖文件。

```
[root@dev-dbteam-143108 otcadmin]# mkdir -p  /var/lib/kafka/libs
```

图 5-164　在 OGG 服务器上建立对应的目录

从 Kafka 所在的服务器上将 Kafka 自身所用的依赖传输过来，如图 5-165 所示。

```
root@xue-test1-148154:[/kafka/bin]scp /kafka/libs/ root@10.60.143.108 /var/lib/kafka/libs/
```

图 5-165　将 Kafka 自身所用的依赖传输过来

编辑 Kafka 的配置文件，如图 5-166 所示，bootstrap.servers 配置地址填写上面搭建的集群地址，将三个 Kafka 节点的 IP 和端口写入，用于连接 Kafka 服务，端口默认 9092。

```
[root@dev-dbteam-143108 ogg]# cd /u01/ogg4bigdata/ogg_deployment/etc/conf/ogg/
[root@dev-dbteam-143108 ogg]# vim kafka.properties
bootstrap.servers=10.60.148.154:9092,10.60.148.155:9092,10.60.148.156:9092
acks = 1
compression.type = gzip
reconnect.backoff.ms = 1000

value.serializer = org.apache.kafka.common.serialization.ByteArraySerializer
key.serializer = org.apache.kafka.common.serialization.ByteArraySerializer
```

图 5-166　编辑 kafka 配置文件

回到 OGG for BigData 的 Web 配置界面操作。与 OGG 21c for Oracle 不同，OGG for BigData 仅将数据投递给 Kafka，不需要添加连接配置，直接添加复制进程即可。如图 5-167 所示，选择"经典复制"。

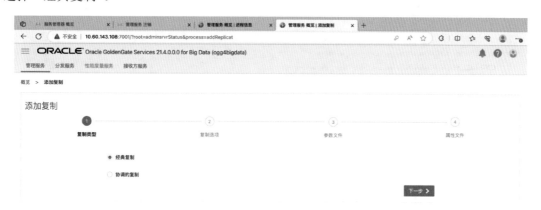

图 5-167　添加复制进程

在 OGG for BigData 复制选项界面，如图 5-168 所示，在线索名称中填写前面 Oracle 中提取进程相应的名称（删除一个简称的 k0，留下了一个绝对路径的 k0），最后在"目标"下拉列表中选择 Kafka。单击"下一步"按钮。

图 5-168　配置复制选项

在 OGG for BigData 的参数文件界面，如图 5-169 所示，需要明确上游的 MAP 的信息，这里的三个参数分别是数据库名 OGGPDB、Schema 名称 TANG 和表名 KK，以及下游的 Target 信息。这里的 xxg 是随意编写的，*也是无所谓的。因为 Kafka 属于消息中间件，对数据会照单全收。单击"下一步"按钮。

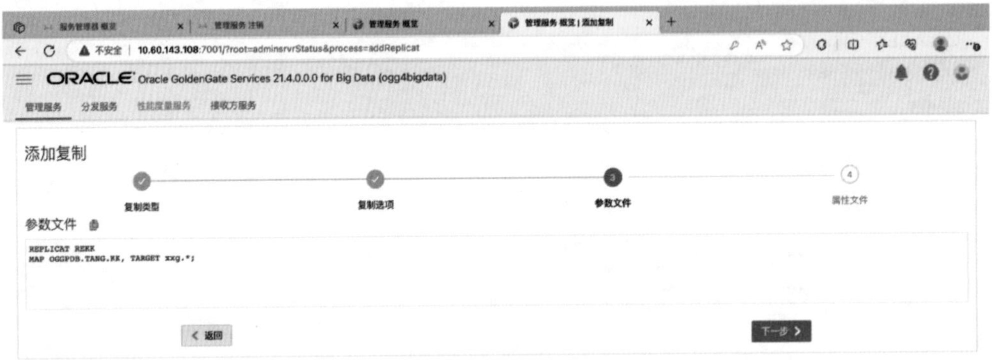

图 5-169　配置参数

最后在 OGG for BigData 的属性文件配置页面中，进行 Kafka 属性文件的配置。如图 5-170 所示，要写明 Kafka 的属性文件名字、上游对应的表以及依赖的 Kafka 相关 lib 包的路径。具体说明如下：gg.handler.kafkahandler.kafkaProducerConfigFile 选项填写前面编写的 kafka.properties 文件名，gg.handler.kafkahandler.topicMappingTemplate 选项填写目标 topic 名，可以填写已经创建的 topic，OGG 会向该 topic 中投递；也可以填写不存在的 topic，那么 OGG 会建立一个新的 topic 并向其投递。

配置后单击"创建并运行"按钮。

图 5-170 配置 Kafka 属性文件

最终在管理界面看到了复制进程正常启动并且运行，如图 5-171 所示。

图 5-171 OGG for Big Data 复制进程启动成功

在上游 Oracle 数据库中进行数据操作，如图 5-172 所示，进行了 DML 和 DDL 的动作。

```
SQL> insert into kk values(888);

已创建 1 行。

SQL> commit;

提交完成。

SQL> update kk set id=777 where id=888;

已更新 1 行。

SQL> commit;

提交完成。

SQL> alter table kk add a int;

表已更改。

SQL> commit;

提交完成。

SQL> insert into kk values(666,666);

已创建 1 行。

SQL> commit;

提交完成。

SQL> alter table kk add b int default 0;

表已更改。

SQL> delete from kk where id=666;

已删除 1 行。

SQL> commit;

提交完成。
```

图 5-172 模拟上游数据库操作数据

随后在 Kafka 所在服务器上开启接收指定的 topic 的命令。如图 5-173 所示，可以观察到数据库中 insert、update、delete 操作，在 Kafka 中对应的 op_type 值为 I/U/D，而更新数据则会用 before 和 after 表示。

```
root@xue-test2-148155:[/kafka/bin]./kafka-console-consumer.sh --bootstrap-server 10.60.148.155:7092 --topic KK --from-beginning
{"table":"XXG.KK","op_type":"I","op_ts":"2023-07-18 15:11:08.000000","current_ts":"2023-07-18 15:11:11.895000","pos":"00000000010000003367","after":{"ID":888}}
{"table":"XXG.KK","op_type":"U","op_ts":"2023-07-18 15:15:42.000000","current_ts":"2023-07-18 15:15:46.344000","pos":"00000000010000003504","before":{"ID":888},"after":{"ID":777}}
{"table":"XXG.KK","op_type":"I","op_ts":"2023-07-18 15:21:33.000000","current_ts":"2023-07-18 15:21:37.689000","pos":"00000000010000005282","after":{"ID":666,"A":666}}
{"table":"XXG.KK","op_type":"D","op_ts":"2023-07-18 15:24:10.000000","current_ts":"2023-07-18 15:24:13.850000","pos":"00000000010000007214","before":{"ID":666,"A":666,"B":0}}
```

图 5-173 命令行显示 Kafka 消费情况

关于 HBase 和 Hive 的演示类似于 Kafka，不过由于以前的 HBase 不方便关联，而 Hive 不支持更改，所以即使数据同步到两个组件中意义也不大。而投递到消息队列中，对于数据的结构和对接还是有很大的意义的。

至此使用 OGG 同步各种主流数据库的讲解完成了。CDC 技术作为最轻量级（指的是对上下游的影响）的数据同步解决方案可以适配不同数据库和中间件。但是无论哪种 CDC 技术都有其实施和维护成本，这是对于数据库规划和定位的补充。如果可以让相关联的数据库在一个数据库实例中，那么就可以最大限度上避免数据同步的成本。

第 6 章

认识 HTAP 技术

HTAP 又名混合事务/分析处理技术,是当前大家普遍关注的热门技术,如今已发展得比较成熟。如何认识这项技术?这项技术有怎样的意义和价值?实现方式又是如何的?本章将进行详细介绍。

本章内容
HTAP 的概念及其价值
HTAP 的几种实现方式

6.1 HTAP 的概念及其价值

6.1.1 HTAP：混合事物/分析处理

HTAP（Hybrid Transactional/Analytical Processing）的概念于 2016 年由 Gartner 提出，其标准定义是将事务处理（Transactional Processing）和分析处理（Analytical Processing）相结合的一种数据处理模式，它的目标是在同一数据平台上支持实时事务处理和复杂分析查询，将原本需要不同系统和架构来处理的事务和分析任务整合到一个统一的解决方案中。有趣的是，Oracle 数据库从未自称为 HTAP 数据库，但大多数使用 Oracle 数据库的用户都是将事务处理和分析查询放在一起进行。这种使用方式也能应对大规模数据的大量事务处理，很多场景下的分析性能也能满足需求。

在 1.1.3 节中，专门介绍了列式数据库，这一技术被广泛应用于分析查询领域。但单一的列式数据库作为一个产品，市场潜力较小。因此，目前许多列式存储与传统的用于事务处理的行式存储融合在一起，这种融合要达成的最终目标就是 HTAP。此外，基于融合思路，目前数据库领域还有融合 JSON、图数据库和时序数据库的多模数据库，以及软硬融合的一体机。本节主要介绍行列存储融合发展出的 HTAP 数据库。

6.1.2 HTAP 的价值

HTAP 并非适用于所有企业，个别企业的数据规模庞大，只有量身定做数据库才能满足自身要求。而大部分中小企业没有动辄 PB 级别的增量数据，数据量还是 TB 级别的规模，自身也没有庞大的技术团队支持 TP 交易的同时又能满足好实时的 AP 分析。在这种环境下，基于 HTAP 的解决方案对于中小企业来说契合度比较高。

1. HTAP 解决了异构数据库的数据结构问题

通常 OLTP 数据库和 OLAP 数据库都不是同构数据库。而异构数据库在字段定义的类型和精度都可能不一样，这就涉及数据映射的问题。而且当上游 OLTP 数据库变更（增加表或字段）的时候，必须对下游数据结构人工介入进行变更，且上游数据库的变更需要重新初始化该表。

然而，在 HTAP 数据库中，事务处理和分析处理都在同一个数据库中完成，不需要区分 OLTP 和 OLAP，也不存在异构数据库的情况。HTAP 数据库利用自身的机制，无须人工介入，可以自动完成这些操作。因此，HTAP 数据库在处理数据结构的变化时更加灵活和高效。

2. HTAP 解决了跨库查询的数据延迟问题

在数据库中交易与分析的数据是在一个事务中完成的，是强一致的，即使不在一个事务中也是毫秒级的，这个毫秒级别的延迟对于分析场景来说可以忽略不计，实时性得到了极大的保证。

3. HTAP 满足了查询分析的实时性需求

HTAP 的列式存储对聚合分析非常友好，可以高效地进行数据统计分析乃至下钻。由于实时性极高,用户对于当下的场景可以快速、便捷地查看热点交易数据,而不用去关注离线的历史数据。

4. HTAP 消除或减少了数据的冗余

由于 HTAP 数据库的数据是在线数据,所以 OLAP 时不会有多次抽取数据的版本对比,只保留最新的数据。这极大地节约了存储成本。数据不再需要从交易处理数据库复制到分析数据库,消除或减少了数据的冗余,这非常重要。一般从一个或多个 OLTP 数据库复制数据到一个 OLAP 数据库的传输环节,稳定性很难保障。

5. HTAP 解决了数据一致性的问题

除了 DDL 变更带来的初始化,基于 ETL 抽取数据还需要严格基于时间戳来进行比较。但是对于没有变更时间戳的数据,甚至不支持或没有时间戳的数据来说,ETL 会出现漏数据或者不得不每次采用全量抽取的方式,对上游数据库造成了不必要的压力。而如果上游进行了 Delete 这样的物理删除,ETL 无法感知数据的减少,造成了数据的不一致。

HTAP 数据库既支持交易又支持分析,不需要 ETL 这个环节,也不需要数据同步与搬迁。从根本上来说就是一份数据,不存在数据不一致的可能性。

6. HTAP 解决了数据传输高成本的问题

上游的 OLTP 可能有几种,下游的 OLAP 也可能有几种,中间的数据同步技术也可能有几种,Oracle 官方 OGG 宣传图如图 6-1 所示。这些需要分别熟悉这些技术栈的专业人员来完成,所以可能是人员要求多,或者是单一人员复合程度高。无论哪种组合,一个数据同步团队的成本是非常高的,每年花费一百万以上。以最为成熟和稳定的 Oracle GoldenGate 来说,如果企业上下游涉及多种数据源,其实施和维护难度依然很大。

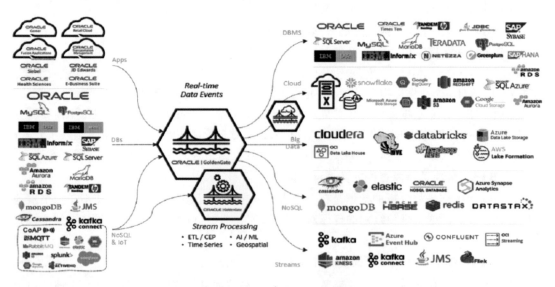

图 6-1 Oracle 官方 OGG 宣传图

而 HTAP 数据库无论采用垂直还是水平或者共享存储都不再需要第三方数据传输工具，这就从根本上解决了中小企业无专业从事数据同步团队的问题。

7. HTAP 解决了数据传输稳定性的问题

在一致性和稳定性约束下，ETL 都不会被纳入可选的技术方案，只能采用日志级别的同步技术。在这个场景下，通常采用 CDC（Change Data Capture，变化数据捕获）技术，通过识别和捕获对数据库中的数据所做的更改（包括数据或数据表的插入、更新、删除等），然后将这些更改按发生的顺序完整记录下来，并实时通过消息中间件传送到下游流程或系统的过程。无论是商用的 Oracle GoldenGate 还是开源的 FlinkCDC 技术都属于中间件范畴，也都较为稳定，但是在上游数据库出现大事务的场景下，对于日志级同步来说是不稳定的，甚至会引起传输中断。而 HTAP 只会有一定的延迟，最终不会引起传输故障，数据传输稳定性很高。

8. HTAP 实现了一定程度大数据离线计算的替代

由于 HTAP 拥有了实时分析的能力，用户几乎不需要再去使用 Hadoop 等离线分析系统，大部分场景都会在具有 HTAP 功能的数据库中完成。

6.2 HTAP 的几种实现方式

本节将介绍 HTAP（混合事务/分析处理）的几种实现方式，具体包括垂直方向的实现（以 Oracle 为代表的实现方式）以及水平方向的实现（以 MySQL 为代表的实现方式）。具体来说，垂直实现方式指的是在单台机器上采用不同的数据存储策略，以满足不同类型的数据库场景需求。相对应的，水平实现方式则是在多台机器上使用不同的存储策略，依靠机器之间的数据同步来保证数据一致性。

6.2.1 垂直方向的实现：以 Oracle 数据库为例

Oracle 采用了垂直实现方式，在磁盘上按照行模式进行数据存储，用以实现 OLTP 的数据库场景；在内存上按照列式进行数据存储，用于实现 OLAP 的数据库场景。若一个事务中的数据发生变化，在更新磁盘的行数据的同时，同步更新内存中的列数据，以达到行列混存且保障数据一致性。行存储的磁盘和列存储的内存的对应的使用场景以及它们之间的数据流动关系，如图 6-2 所示。

图 6-2 来自 Oracle 的行列混合存储

该垂直模式比起水平的实现方式来说最大的优势比较明显：就是 HTAP 的实现过程中没有延迟。其最大劣势是对资源要求高，不过随着硬件的进步，TB 级别的内存已经比较常见了。除了 Oracle 使用这种方式，阿里的 PolarDB 也采用这种方式。

6.2.2 水平方向的实现：以 MySQL、TiDB 数据库为例

MySQL 采用了水平实现方式，在主库上按照行进行数据存储，用以实现 OLTP 的数据库场景。在从库上按照列进行数据存储，用于实现 OLAP 的数据库场景。使用 HeatWave 插件实现 MySQL 数据库的 HTAP 的一般架构如图 6-3 所示。当 HeatWave 加载一个表时，数据被共享并分布在 HeatWave 节点之间。加载表后，表上的 DML 操作将自动传播到 HeatWave 节点。同步数据不需要用户操作。这一实现通过 HeatWave 的插件来完成，该插件属于 MySQL 企业版，而且要在 Oracle 的云上才能使用。

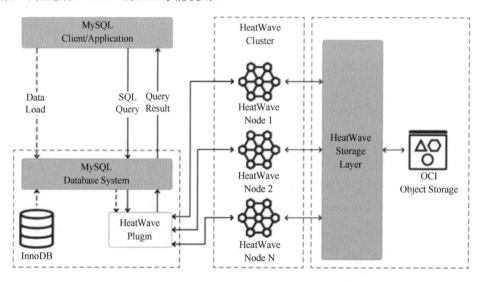

图 6-3 MySQL-HeatWave

TiDB 的实现方式：TiDB 也是采用了水平实现方式，在它的存储节点 TiKV 上按照行进行数据存储，用以实现 OLTP 的数据库场景。在分析节点 TiFlash 库上按照列进行数据存储，用于实现 OLAP 的数据库场景。TiKV 与 TiFlash 进行行列转换存储，并且 TiKV 与 TiFlash 的数据是强一致性的。当有写操作时，SQL 在 TiKV 库运行；当有聚合分析时，SQL 在 TiFlash 库运行，TiDB-TiFlash 架构如图 6-4 所示。

6.2.3 其他类型数据库的实现方式

国内其他类型数据库的实现方式还包括非行列存储，单一依靠读写分离方式。有些产品依靠自己提供的读写分离的特点，也能够实现在同一套数据库中的交易与分析。可以通过扩展读节点进行扩展。但是这种架构不是强一致性的，可能存在微小的延迟。

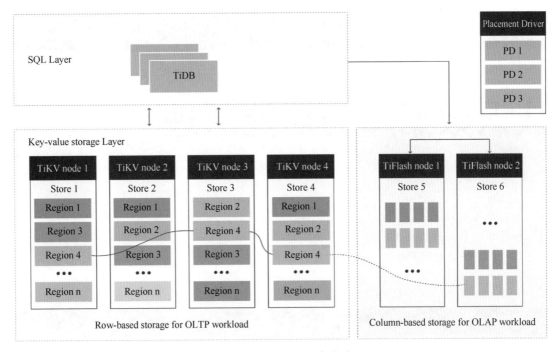

图 6-4　TiDB-TiFlash 架构图

国内一些云厂商采用共享存储模式，OLTP 的节点和 OLAP 的节点共享同一套数据文件。可以想象为 Oracle 的 RAC 双节点，只不过 RAC 的两个节点是双活的，而共享存储的 HTAP（见图 6-5）是一个节点负责写，一个节点（或多个节点）负责读。由于共享存储，解决了数据库一致性的问题。而且可以通过共享存储扩展读节点，使得 HTAP 的读节点可以扩展。

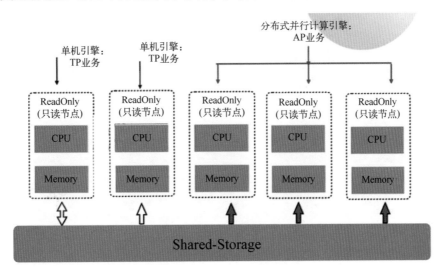

图 6-5　云厂商共享存储 HTAP 架构

第 7 章

认识数据库的功能原理

数据库是数据写入、存储和运算的载体,这些是数据库的物理属性。而 SQL 是将用户的逻辑和物理对象打通的桥梁。数据库是物理学和数学的高度结合。作为 DBA,应该对数据库的工作原理有一定的了解,对于内存、外存以及数据如何存取的物理操作非常清楚。

本章内容
优化器——基于统计学原理
数据库的查询——火山模型
数据库 AI——向量化
编译执行

7.1 优化器——基于统计学原理

数据库中最复杂的一个组成部分是优化器，而优化器就和数学以及统计学有着不可分割的关系。数学和统计学密切相关，两者有着很多的交叉点。首先，数学是统计学的基础。统计学是一门应用数学的学科。其次，统计学反过来也影响了数学的发展。

不同数据库的优化器是不一样的。而优化器的基础就是来自于数据表的统计信息。数据库中的统计信息是指关于数据库中各个表和索引的统计数据，这些统计信息包括表的行数、列的基数、列的数据分布情况，以及索引的选择性和统计信息更新的时间等。通过统计信息，数据库可以更加有效地优化查询计划，提高查询性能。统计信息对于数据库系统的性能非常重要，它可以帮助数据库系统了解表中数据的分布情况，从而选择最优的查询执行计划。例如，当我们使用 WHERE 子句进行条件查询时，数据库可以利用统计信息来判断选择使用索引还是全表扫描，以及使用哪个索引。此外，统计信息还可以用于查询优化器对连接操作、聚合操作等进行优化。

不同数据库的优化器有强有弱，不过即使优化器再强，如果缺乏统计信息或统计信息错误，也无法得到正确的判断。业内普遍认为 Oracle 的优化器是数据库领域最强大的优化器。之所以强大是其算法强大，有人曾经说过 Oracle 的优化器是很多白发苍苍的数学系教授在孜孜不断地进行攻坚克难。这些算法必须依靠全面和精准的统计信息，算法和统计这些名词都是数学学科领域的专业术语，所以数据库的底层基本是物理学和数学的结合。

数据库的统计信息分为六大类。
- 表的统计信息。
- 列的统计信息。
- 索引的统计信息。
- 系统的统计信息。
- 数据字典的统计信息。
- 内部对象的统计信息。

本节主要介绍前 3 种，即表、列和索引的统计信息。建立统计信息实验表如图 7-1 所示。

```
SQL> create table xxgtjxx (id int,a int,b varchar2(10),t date);

表已创建。

  create or replace procedure tj10
  as
  begin
    for i in 1..10 loop
      insert into xxgtjxx values (i,1,'abcdef'||i,sysdate-i);
    end loop;
    commit;
2    3    4    5    6    7    8      end;
9  /

过程已创建。
```

图 7-1　建立统计信息实验表

写入 10 条数据后,统计信息表如图 7-2 所示。dba_tab_statistics 是表的统计信息的元数据,USER_TAB_COL_STATISTICS 是列的统计信息的元数据,DBA_IND_STATISTICS 是索引的统计信息的元数据。

```
SQL>exec tj10;

PL/SQL 过程已成功完成。


SQL>select count(*) from XXGTJXX;

  COUNT(*)
----------
        10


SQL>select table_name,num_rows,blocks from dba_tab_statistics where table_name='XXGTJXX';

TABLE_NAME      NUM_ROWS     BLOCKS
-----------     ---------    -------
XXGTJXX




SQL> select table_name,column_name,num_distinct,
  2  low_value,high_value,notes from user_tab_col_statistics
  3  where table_name='XXGTJXX';

未选定行



SQL>  select index_name,table_name,distinct_keys,
  2  num_rows,blevel,leaf_blocks from dba_ind_statistics
  3  where table_name='XXGTJXX';

未选定行
```

图 7-2　所有统计信息都为空

此时执行一个查找 ID 的 SQL,如图 7-3 所示,数据库执行 select * from xxgtjxx where id=1 时提示,使用到了动态采样。

选择执行计划的对象统计信息缺失、过期或不足时,CBO 优化器选择的执行计划就有可能不是最优的甚至是最差的。所以为了保证优化器在 SQL 硬解析时选择到最优的执行计划,需要根据数据库动态统计的设置级别,动态收集 Object 的统计信息。图 7-3 中的 level=2 是指默认对没有统计信息的表进行的采样级别,即 dba_tab_statistics 中该表的统计信息为空。可见没有统计信息一个简单的 SQL 都执行不了,数据库的优化器只能在执行之前强制去动态采样,然后再执行 SQL。接下来对实验表收集统计信息,如图 7-4 所示。

```
SQL> set autotrace on;
SQL> select  * from xxgtjxx   where id=1;

        ID         A B         T
---------- ---------- --------------
         1          1 abcdef1    08-7月 -23


执行计划
----------------------------------------------------------
Plan hash value: 2858722449

--------------------------------------------------------------------------
| Id  | Operation         | Name    | Rows  | Bytes | Cost (%CPU)| Time     |
--------------------------------------------------------------------------
|   0 | SELECT STATEMENT  |         |     1 |    42 |     3   (0)| 00:00:01 |
|*  1 |  TABLE ACCESS FULL| XXGTJXX |     1 |    42 |     3   (0)| 00:00:01 |
--------------------------------------------------------------------------

Predicate Information (identified by operation id):
---------------------------------------------------

   1 - filter("ID"=1)

Note
-----
   - dynamic statistics used: dynamic sampling (level=2)

统计信息
----------------------------------------------------------
         16  recursive calls
          0  db block gets
         28  consistent gets
          1  physical reads
          0  redo size
        784  bytes sent via SQL*Net to client
        617  bytes received via SQL*Net from client
          2  SQL*Net roundtrips to/from client
          0  sorts (memory)
          0  sorts (disk)
          1  rows processed
```

图 7-3 动态采样

```
SQL> begin
  2  dbms_stats.gather_table_stats
  3  ('xxg','xxgtjxx',estimate_percent =>100,method_opt => 'for table');
  4  end;
  5  /
PL/SQL 过程已成功完成。


SQL>select table_name,num_rows,blocks from dba_tab_statistics
  2  where table_name='XXGTJXX';

TABLE_NAME    NUM_ROWS     BLOCKS
-----------   ---------    -------
XXGTJXX             10          5


SQL> select table_name,column_name,num_distinct,
  2  low_value,high_value,notes from user_tab_col_statistics
  3  where table_name='XXGTJXX';

未选定行


SQL>  select index_name,table_name,distinct_keys,
  2  num_rows,blevel,leaf_blocks from dba_ind_statistics
  3  where table_name='XXGTJXX';

未选定行
```

图 7-4 收集实验表的统计信息

由于只对表收集统计信息，所以列和索引的统计信息还是为空。再次执行 select * from xxgtjxx where id=1，如图 7-5 所示，不再有动态采样的过程。不过因为没有索引，该 SQL 还是全表扫描。只是预估 ID=1 的行数是 1。

为 ID 列建立索引，建立索引的同时也是收集索引统计信息的过程，如图 7-6 所示。

```
SQL> explain plan for select * from xxgtjxx where id=1;
已解释。
SQL> select * from table(dbms_xplan.display);
PLAN_TABLE_OUTPUT
--------------------------------------------------------------------------------
Plan hash value: 1920144158

--------------------------------------------------------------------------
| Id  | Operation         | Name  | Rows  | Bytes | Cost (%CPU)| Time     |
--------------------------------------------------------------------------
|   0 | SELECT STATEMENT  |       |     1 |     5 |   450   (2)| 00:00:01 |
|*  1 |  TABLE ACCESS FULL| TJ_A  |     1 |     5 |   450   (2)| 00:00:01 |
--------------------------------------------------------------------------

Predicate Information (identified by operation id):
---------------------------------------------------

PLAN_TABLE_OUTPUT
--------------------------------------------------------------------------------

   1 - filter("ID"=1)
```

图 7-5 无索引 SQL 执行计划

```
SQL> create index tjxx_id on xxgtjxx (id);

索引已创建

SQL>  select index_name,table_name,distinct_keys,
  2   num_rows from dba_ind_statistics
  3   where table_name='XXGTJXX';

INDEX_NAME  TABLE_NAME  DISTINCT_KEYS    NUM_ROWS
----------  ----------  -------------    ----------
TJXX_ID     XXGTJXX                10           10
```

图 7-6 为 ID 列建立索引

此时利用存储过程对实验表新增 10 万条数据，如图 7-7 所示。

检查统计信息，发现有两条信息如图 7-8 所示。STATS_ON_CONVENTIONAL_DML 提示出现使用了 19c 的实时统计信息。

```
create or replace procedure tj10w
as
a varchar2(10);
begin
  for i in 11..100010 loop
       SELECT DBMS_RANDOM.STRING ('A', 10) into a FROM DUAL;
    insert into xxgtjxx values (i,1,a,sysdate-300+i/1440);
    end loop;
    commit;
    end;
  7    8    9    10    11    /
```

过程已创建。

```
SQL> set timing on;
SQL> exec tj10w;

PL/SQL 过程已成功完成。

已用时间:  00: 00: 22.95
```

图 7-7　实验表新增 10 万条数据

```
SQL> select table_name,num_rows,blocks,notes from dba_tab_statistics where table_name='XXGTJXX';

TABLE_NAME              NUM_ROWS   BLOCKS NOTES
-------------------    ---------- --------- -----------------------------
XXGTJXX                       10        5
XXGTJXX                    10009       43 STATS_ON_CONVENTIONAL_DML
```

图 7-8　出现了两条统计信息

Oracle 早期版本是当数据变化达到一定程度后才触发统计信息改变，而且还伴有定时任务辅助。如果不及时更新就会导致信息过久，从而优化器不能及时感知到变化。所以 12c 以后引入了实时统计信息，但是任何新功能都是一把双刃剑，有优点也有缺点，对于实时统计也一样。它最大的优点是，在优化器选择执行计划时，对统计信息缺失或统计不够准确的对象，能够动态地收集统计信息，从而获得相对好的执行计划。并且在 12c 中，优化器还能够自行决定统计收集级别和在内存中保存统计结果，供其他操作共享。然而，实时统计也存在一些缺点。首先，由于在硬解析过程需要收集和处理更多的统计信息，解析时间可能变得更长。其次，如果大量的动态收集操作发生，可能影响到数据库全体性能。因此，在实际中还需结合自己情况评估。

动态采样的参数如图 7-9 所示。在 19c 中默认是打开的，可以将其关闭。

```
SQL> show parameter OPTIMIZER_DYNAMIC_SAMPLING;

NAME                                 TYPE        VALUE
------------------------------------ ----------- ------------------------------
optimizer_dynamic_sampling           integer     2
```

图 7-9　动态采样参数

既然数据库已经动态感知到统计信息的变化，那么就再次收集全表的统计信息，如图 7-10 所示。

```
SQL> ANALYZE TABLE xxgtjxx compute STATISTICS;
```

表已分析。

```
SQL>select table_name,num_rows,blocks from dba_tab_statistics
  2 where table_name='XXGTJXX';

TABLE_NAME      NUM_ROWS    BLOCKS
------------    ---------   ---------
XXGTJXX           100010       496


SQL>  select index_name,table_name,distinct_keys,
  2 num_rows,blevel from dba_ind_statistics
  3 where table_name='XXGTJXX';

INDEX_NAME TABLE_NAME DISTINCT_KEYS   NUM_ROWS   BLEVEL
---------- ---------- -------------  ---------- ---------
TJXX_ID    XXGTJXX        100010       100010       1
```

图 7-10 以计算方式收集统计信息

收集过后表和索引的统计信息都更新了。再检查列的统计信息，如图 7-11 所示。

```
SQL> select column_name,num_distinct,
  2 low_value,high_value from user_tab_col_statistics
  3 where table_name='XXGTJXX';


COLUMN_NAME NUM_DISTINCT LOW_VALUE                  HIGH_VALUE
----------- ------------ -------------------------- ------------------
ID              100010   C102                       C30B010B
A                    1   C102                       C102
B               100010   4141414253706C756D57       7A7A794252514B444867
T               100010   787A090C151930             787B0708143538
```

图 7-11 列的统计信息

列的统计信息通常比表要重要，这里反映了列的数据分布情况（如空值、不同值、最大值和最小值等），这里不得不说，有些系统设计会把业务的默认时间设置到 1900 年或 2099 年，这就严重影响整体的统计信息的估算。在上图中，最小值和最大值看起来不能让人一眼识别，需要自定义一个函数进行转换，如图 7-12 所示。

```
create or replace function display_raw (rawval raw, type varchar2)
  return varchar2
  is
    cn      number;
    cv      varchar2(32);
    cd      date;
    cnv     nvarchar2(32);
    cr      rowid;
    cc      char(32);
  begin
    if (type = 'NUMBER') then
        dbms_stats.convert_raw_value(rawval, cn);
        return to_char(cn);
    elsif (type = 'VARCHAR2') then
        dbms_stats.convert_raw_value(rawval, cv);
        return to_char(cv);
    elsif (type = 'DATE') then
        dbms_stats.convert_raw_value(rawval, cd);
        return to_char(cd);
    elsif (type = 'NVARCHAR2') then
        dbms_stats.convert_raw_value(rawval, cnv);
        return to_char(cnv);
    elsif (type = 'ROWID') then
        dbms_stats.convert_raw_value(rawval, cr);
        return to_char(cnv);
    elsif (type = 'CHAR') then
        dbms_stats.convert_raw_value(rawval, cc);
        return to_char(cc);
    else
        return 'UNKNOWN DATATYPE';
    end if;
end;
/
```

图 7-12　自定义转换统计信息极值函数

再次检查列统计信息元数据表，如图 7-13 所示，现在可以清晰地看到每个列的最大值和最小值。

```
SQL> select a.column_name, a.num_distinct,
  2 display_raw(a.low_value,b.data_type) as low_val,
  3 display_raw(a.high_value,b.data_type) as high_val, b.data_type
  4 from dba_tab_col_statistics a, dba_tab_cols b
  5  where a.table_name='XXGTJXX' and a.table_name=b.table_name
  6 and a.column_name=b.column_name;


COLUMN_NAM NUM_DISTINCT LOW_VAL        HIGH_VAL        DATA_TYPE
---------- ------------ ----------     ----------      ----------
ID               100010 1              100010          NUMBER
A                     1 1              1               NUMBER
B                100010 AAABSplumW     zzyBRQKDHg      VARCHAR2
T                100010 12-9月 -22     08-7月 -23      DATE
```

图 7-13　检查列的统计信息元数据表

对有索引和无索引的列分别执行 SQL，观察 ID 列和 B 列的执行计划，如图 7-14 所示，虽然 ID 列有索引，B 列无索引，但是优化器预估出两个 SQL 的行数都是一行。因为结合列的统计信息，B 的区分度和 ID 的区分度是一样的。

```
SQL> select  id from xxgtjxx  where id=10;

       ID
----------
       10
执行计划
----------------------------------------------------------
Plan hash value: 4090917750

--------------------------------------------------------------------
| Id  | Operation         | Name    | Rows  | Bytes | Cost (%CPU)| Time     |
--------------------------------------------------------------------
|   0 | SELECT STATEMENT  |         |     1 |     4 |     1   (0)| 00:00:01 |
|*  1 |  INDEX RANGE SCAN | TJXX_ID |     1 |     4 |     1   (0)| 00:00:01 |
--------------------------------------------------------------------

Predicate Information (identified by operation id):
---------------------------------------------------

   1 - access("ID"=10)


统计信息
----------------------------------------------------------
          1  recursive calls
          0  db block gets
          3  consistent gets
          0  physical reads
          0  redo size
        552  bytes sent via SQL*Net to client
        619  bytes received via SQL*Net from client
          2  SQL*Net roundtrips to/from client
          0  sorts (memory)
          0  sorts (disk)
          1  rows processed


SQL> select  b from xxgtjxx   where b='DsltXkEKqr';

B
----------
DsltXkEKqr
执行计划
----------------------------------------------------------
Plan hash value: 2858722449

-----------------------------------------------------------------------
| Id  | Operation         | Name    | Rows  | Bytes | Cost (%CPU)| Time     |
-----------------------------------------------------------------------
|   0 | SELECT STATEMENT  |         |     1 |    10 |   137   (1)| 00:00:01 |
|*  1 |  TABLE ACCESS FULL| XXGTJXX |     1 |    10 |   137   (1)| 00:00:01 |
-----------------------------------------------------------------------
```

图 7-14　对比 ID 列和 B 列的执行计划

```
Predicate Information (identified by operation id):
---------------------------------------------------

   1 - filter("B"='DsltXkEKqr')


统计信息
----------------------------------------------------------
          1  recursive calls
          0  db block gets
        475  consistent gets
          0  physical reads
          0  redo size
        559  bytes sent via SQL*Net to client
        407  bytes received via SQL*Net from client
          2  SQL*Net roundtrips to/from client
          0  sorts (memory)
          0  sorts (disk)
          1  rows processed
```

图 7-14　对比 ID 列和 B 列的执行计划（续）

通过数据库的统计信息，数据库的优化器可以得知在一个 SQL 中涉及的表的体量是多大，以及列上数据的区分度是不是可以使用索引。但是现实中仅有区分度是不够的，因为数据不一定总是平均分布的。有很多场景下数据是倾斜的甚至是严重倾斜的。比如我国人口数据中，汉族总人口占比约为 91%，数据占比极高。利用已有的 xxgtjxx 表的 10 万样本数据，建立直方图（ZFT）表，如图 7-15 所示，在执行建表的同时，列的统计信息也被收集了（即使没有建立索引）。

```
SQL> create table zft as select * from xxgtjxx;

表已创建。

SQL> SELECT column_name,num_distinct FROM DBA_TAB_COL_STATISTICS WHERE TABLE_NAME = 'ZFT';

COLUMN_NAME               NUM_DISTINCT
--------------------      ------------
ID                              100010
A                                    2
B                               100010
T                                99168
```

图 7-15　创建直方图表

在建表的同时也生成了一个粗略的直方图（这个数据其实并不准确），如图 7-16 所示，数据库认为每个列的数据分布比较平均。

随后创建索引并收集索引的统计信息，如图 7-17 所示，ZFT（直方图）表上有两个不同的值，总数是 100010 行。那么按照平均的思想来说这两个值分别占 50%，也就是 50005 行。

```
SQL> select TABLE_NAME,COLUMN_NAME,ENDPOINT_NUMBER,ENDPOINT_VALUE from dba_histograms
  2  where TABLE_NAME='ZFT' order by COLUMN_NAME;

TABLE_NAME      COLUMN_NAME     ENDPOINT_NUMBER ENDPOINT_VALUE
--------------- --------------- --------------- ---------------
ZFT             A                             0              0
ZFT             A                             1              1
ZFT             B                             1      6.3594E+35
ZFT             B                             0      3.3882E+35
ZFT             ID                            0              1
ZFT             ID                            1         100010
ZFT             T                             1     2460134.83
ZFT             T                             0     2459835.85
```

已选择 8 行。

图 7-16 建表同时生成的粗略直方图

```
SQL> create index zft_a on zft (a);
```

索引已创建。

```
SQL> select index_name,table_name,distinct_keys,num_rows,blevel from DBA_IND_STATISTICS where table_name='ZFT';

INDEX_NAME          TABLE_NAME          DISTINCT_KEYS   NUM_ROWS   BLEVEL
------------------- ------------------- -------------   --------   ------
ZFT_A               ZFT                             2     100010        1
```

图 7-17 创建索引收集索引的统计信息

结合列的统计信息元数据表和表的对象元数据表联合查询。HISTOGRAM 是 NONE，表示没有做过详细的直方图，如图 7-18 所示。

```
SQL> select a.column_name,
  2         b.num_rows,
  3         a.num_distinct Cardinality,
  4         round(a.num_distinct / b.num_rows * 100, 2) selectivity,
  5         a.histogram,
  6         a.num_buckets
  7  from dba_tab_col_statistics a, dba_tables b
  8  where a.owner=b.owner
  9  and a.table_name=b.table_name
 10    and a.table_name='ZFT';

COLUMN_NAME          NUM_ROWS CARDINALITY SELECTIVITY HISTOGRAM       NUM_BUCKETS
-------------------- -------- ----------- ----------- --------------- -----------
ID                     100010      100010         100 NONE                      1
A                      100010           2           0 NONE                      1
B                      100010      100010         100 NONE                      1
T                      100010       99168       99.16 NONE                      1
```

图 7-18 没有做过详细的直方图

设置数据库不显示返回结果,只显示真实执行计划。在 A 列数据分布极端倾斜的情况下,如图 7-19 所示,执行的是全表扫描,预计行数是 50005(和之前推论一致),但是实际返回的是 10001 行,可见偏差较大。

```
SQL> set autotrace traceonly;
SQL> select  * from zft  where a=1;

已选择 100001 行。

执行计划
----------------------------------------------------------
Plan hash value: 2423196980

--------------------------------------------------------------------------
| Id  | Operation         | Name | Rows  | Bytes | Cost (%CPU)| Time     |
--------------------------------------------------------------------------
|   0 | SELECT STATEMENT  |      | 50005 | 1318K|   127   (1)| 00:00:01 |
|*  1 |  TABLE ACCESS FULL| ZFT  | 50005 | 1318K|   127   (1)| 00:00:01 |
--------------------------------------------------------------------------

Predicate Information (identified by operation id):
---------------------------------------------------

   1 - filter("A"=1)


统计信息
----------------------------------------------------------
          1  recursive calls
          0  db block gets
       7083  consistent gets
          0  physical reads
          0  redo size
    4162952  bytes sent via SQL*Net to client
      73717  bytes received via SQL*Net from client
       6668  SQL*Net roundtrips to/from client
          0  sorts (memory)
          0  sorts (disk)
     100001  rows processed
```

图 7-19 A 列数据分布极端倾斜

而在 A 列数据分布倾斜度较低的情况下,如图 7-20 所示,执行的也是全表扫描,预计行数是 50005(和之前推论一致)。但是实际返回的是 9 行,不仅仅偏差较大而且本应该使用索引的变成了全表扫描,这显然是不合理的。

```
SQL> select * from zft where a=0;

已选择 9 行。

执行计划
----------------------------------------------------------
Plan hash value: 2423196980

--------------------------------------------------------------------------
| Id  | Operation         | Name | Rows  | Bytes | Cost (%CPU)| Time     |
--------------------------------------------------------------------------
|   0 | SELECT STATEMENT  |      | 50005 | 1318K |   127   (1)| 00:00:01 |
|*  1 |  TABLE ACCESS FULL| ZFT  | 50005 | 1318K |   127   (1)| 00:00:01 |
--------------------------------------------------------------------------

Predicate Information (identified by operation id):
---------------------------------------------------

   1 - filter("A"=0)


统计信息
----------------------------------------------------------
         85  recursive calls
         19  db block gets
        599  consistent gets
        445  physical reads
       3488  redo size
       1023  bytes sent via SQL*Net to client
        391  bytes received via SQL*Net from client
          2  SQL*Net roundtrips to/from client
         12  sorts (memory)
          0  sorts (disk)
          9  rows processed
```

图 7-20 A 列数据分布倾斜度较低

直方图就是解决这种数据不均衡的场景的，通过查看数据的分布来得出最佳的执行效果。执行 dbms_stats.gather_table_stats 包来完成收集直方图的工作，如图 7-21 所示。第一个参数是 schema 名字，这里是 xxg。第二个参数是表名，这里是 zft。第三个参数是采样百分比，这里是 100%。第四个参数是要对哪一列做直方图，这里是所有列。第五个参数是做列的直方图时收集索引统计信息，这里是 true。最后一个参数是并行度，这里是开了 16 个并行。

```
SQL> exec dbms_stats.gather_table_stats('xxg','zft',estimate_percent => 100, method_opt => 'for all columns',cascade => true,degree => 16);
PL/SQL 过程已成功完成。
```

图 7-21 收集直方图

直方图完成后，检查直方图的统计结果。如图 7-22 所示，A 列有只有两个值，区分度很低，而且数据倾斜严重，值为 0 的有 9 个，值为 1 的有 100010 个。B 列、ID 列和 T 列由于区分度较高，所以分布得比较均匀。

```
SQL>    select   TABLE_NAME,COLUMN_NAME,ENDPOINT_NUMBER,ENDPOINT_VALUE    from dba_histograms
  2    where TABLE_NAME='ZFT' order by COLUMN_NAME,ENDPOINT_NUMBER;

TABLE_NAME           COLUMN_NAME          ENDPOINT_NUMBER ENDPOINT_VALUE
-------------------- -------------------- --------------- --------------
ZFT                  A                                  9              0
ZFT                  A                             100010              1
ZFT                  B                                  0     3.3882E+35
 :
ZFT                  B                                 73     6.3036E+35
ZFT                  B                                 74     6.3511E+35


TABLE_NAME           COLUMN_NAME          ENDPOINT_NUMBER ENDPOINT_VALUE
-------------------- -------------------- --------------- --------------
ZFT                  B                                 75     6.3594E+35
ZFT                  ID                                 0              1
ZFT                  ID                                 1           1334
 :
ZFT                  ID                                73          97344
ZFT                  ID                                74          98677
ZFT                  ID                                75         100010


TABLE_NAME           COLUMN_NAME          ENDPOINT_NUMBER ENDPOINT_VALUE
-------------------- -------------------- --------------- --------------
ZFT                  T                                  0     2459835.85
ZFT                  T                                  1     2459836.78
 :
ZFT                  T                                 74     2459904.38
ZFT                  T                                 75     2460134.83

已选择 230 行。
```

图 7-22 直方图的统计结果

再次结合列的统计信息元数据表和表的对象元数据表联合查询，表上做过详细的直方图如图 7-23 所示。HISTOGRAM 不再是 NONE，而是 FREQUENCY 和 HEIGHT BALANCED。FREQUENCY 表示该列的区分度较低，只用了两个桶就装载了数据分布。而 HEIGHT BALANCED 表示该列的区分度较高，用了 75 个桶来平均装载数据分布。

再次验证直方图效果，由于图 7-22 中已知 A 列值为 1 的有 100010 行，所以如图 7-24 所示，有直方图条件下执行查询倾斜度较高的情况，执行的是全表扫描，预计行数是 100000（和直方图显示基本一致），实际返回 100001 行。

```
SQL>
select a.column_name,
       b.num_rows,
       a.num_distinct Cardinality,
       round(a.num_distinct / b.num_rows * 100, 2) selectivity,
       a.histogram,
       a.num_buckets
from dba_tab_col_statistics a, dba_tables b
where a.owner=b.owner
and a.table_name=b.table_name
  and a.table_name='ZFT';

COLUMN_NAME      NUM_ROWS   CARDINALITY  SELECTIVITY  HISTOGRAM         NUM_BUCKETS
-------------    --------   -----------  -----------  ---------------   -----------
ID                 100010        100010          100  HEIGHT BALANCED            75
A                  100010             2            0  FREQUENCY                   2
B                  100010        100010          100  HEIGHT BALANCED            75
T                  100010        100010          100  HEIGHT BALANCED            75
```

图 7-23　表上做过详细的直方图

```
SQL> select  * from zft  where a=1;

已选择 100001 行。

执行计划
----------------------------------------------------------
Plan hash value: 2423196980

--------------------------------------------------------------------------
| Id  | Operation         | Name | Rows  | Bytes | Cost (%CPU)| Time     |
--------------------------------------------------------------------------
|   0 | SELECT STATEMENT  |      |  100K |  2636K|   127   (1)| 00:00:01 |
|*  1 |  TABLE ACCESS FULL| ZFT  |  100K |  2636K|   127   (1)| 00:00:01 |
--------------------------------------------------------------------------

Predicate Information (identified by operation id):
---------------------------------------------------

   1 - filter("A"=1)

统计信息
----------------------------------------------------------
          0  recursive calls
          0  db block gets
       7083  consistent gets
          0  physical reads
          0  redo size
    4162952  bytes sent via SQL*Net to client
      73717  bytes received via SQL*Net from client
       6668  SQL*Net roundtrips to/from client
          0  sorts (memory)
          0  sorts (disk)
     100001  rows processed
```

图 7-24　有直方图条件下执行查询倾斜度较高的情况

再次执行 A 列的谓词条件等于 0 的查询，由于图 7-22 中已知 A 列值为 0 的有 9 行，所以如图 7-25 所示，有直方图条件下执行查询倾斜度较低的情况执行的是索引扫描，不再是错误的全表扫描，预计行数是 9（和直方图显示完全一致），实际返回 9 行。

MySQL 的直方图和 Oracle 类似（MySQL8 版本才有直方图功能）。

Oracle 直方图语法：exec dbms_stats.gather_table_stats('xxg','zft',estimate_percent => 100, method_opt => 'for all columns',cascade =>true,degree => 16);

MySQL 直方图语法：Analyze table zft update histogram on column with 1024 buckets;

```
SQL> select a from zft where a=0;

已选择 9 行。


执行计划
----------------------------------------------------------
Plan hash value: 260406893

--------------------------------------------------------------------------
| Id  | Operation        | Name  | Rows | Bytes | Cost (%CPU)| Time     |
--------------------------------------------------------------------------
|   0 | SELECT STATEMENT |       |    9 |    27 |     1   (0)| 00:00:01 |
|*  1 |  INDEX RANGE SCAN| ZFT_A |    9 |    27 |     1   (0)| 00:00:01 |
--------------------------------------------------------------------------

Predicate Information (identified by operation id):
---------------------------------------------------

   1 - access("A"=0)


统计信息
----------------------------------------------------------
          1  recursive calls
          0  db block gets
          3  consistent gets
          0  physical reads
          0  redo size
        638  bytes sent via SQL*Net to client
        392  bytes received via SQL*Net from client
          2  SQL*Net roundtrips to/from client
          0  sorts (memory)
          0  sorts (disk)
          9  rows processed
```

图 7-25　有直方图条件下执行查询倾斜度较低的情况

两者仅仅是语法的不同，思路都一样。其实每个数据库都有类似的解决思路，关键是对于数据分布的巧妙应用。

7.2　数据库的查询——火山模型

本节将重点介绍查询执行过程中的一个重要模型——火山模型（Volcano Model）。火山模型是著名的查询执行模型，我们熟知的关系数据库大都采用了该模型，如 Oracle、SQL Server、MySQL 等。火山模型基本是事实上的关系数据库查询执行模型设计标准，在数据库领域中得到了广泛应用，理解火山模型有助于 DBA 理解数据库内部查询执行的工作原理。

1. 火山模型的基础概念和原理

火山模型最早出现在 1994 年 Goetz Graefe 的论文 "Volcano, An Extensible and Parallel Query Evaluation System"，目前已经成为实现关系数据库查询执行引擎的关键技术。火山模型的基本原理是将 SQL 查询中的每个操作抽象为操作符（Operator），每个操作符需要实现自己的 Next 方法。在执行查询时，整个 SQL 会构建成一个操作符树，从根节点到叶子节点自上而下地递归调用 Next 方法，每次调用 Next 方法，就会有一行数据从叶子节点传递到根节点，最终得到一条查询结果。

操作符树是火山模型中执行查询的一个重要概念，可以将其比作一个"计算流程图"。当我们在执行一个查询时，如通过 SQL 语句查询数据库中的数据，这个查询会被分解成一系列操作符，每个操作符都会执行一个特定的操作，如过滤数据、排序数据、聚合数据等。操作符树就是将这些操作符按照一定的顺序连接起来，形成一个"树状结构"。在这个树状结构中，每个操作符都会接收数据流，并对数据进行处理，然后将处理后的结果输出到下一个操作符。这个过程就像是数据在操作符之间"流动"。例如，可以将一个查询分解成以下几个操作符。

- 选择操作符：根据查询条件过滤数据。
- 排序操作符：按照指定的字段对数据进行排序。
- 分组操作符：按照指定的字段对数据进行分组。
- 聚合操作符：对每个分组进行聚合操作，如求和、求平均值等。

这些操作符可以被连接起来，形成一个操作符树。在这个树中，选择操作符接收原始数据流，然后将过滤后的数据流输出给排序操作符，排序操作符将排序后的数据流输出给分组操作符，分组操作符将分组后的数据流输出给聚合操作符，最终输出查询结果。

2. 从设计模式的角度理解火山模型

从设计模式的角度看，火山模型还被叫作迭代器模型，这又是为什么？要弄清楚这个问题首先需要简单了解下迭代器模式。迭代器模式是一种设计模式，它要求在不暴露底层数据结构的情况下，通过统一的迭代器接口遍历一个聚合对象的元素。

对应到火山模型的设计，操作符就是一个迭代器，多个迭代器连接成一个迭代器树，数据库就是聚合对象，每条数据记录是一个元素。每个操作符都实现了迭代器接口，也就是 Next 方法，该方法执行时先调用子节点的 Next 方法，获取子节点的数据，进行本节点特有的处理（如根据查询条件过滤）后，向上返回给父节点。基于火山模型，整个 SQL 查询过程也就变成了根节点不断调用 Next 方法，一条条数据记录经历多个操作符处理，最后生成查询结果的过程。

3. 火山模型在 SQL 执行中的应用

举例说明一下火山模型的执行过程。假设有一个简单的 SQL 查询语句：SELECT column3

FROM T1 WHERE column1 = 'value1' AND column2 = 'value2'，其中 T1 是待查询的数据表。在火山模型中，该查询语句将被转换为一个操作符树，一个简化的操作符树结构如图 7-26 所示。

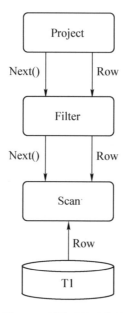

图 7-26 简单的操作符树

在这个操作符树中，Scan 操作符用于扫描整个表，Filter 操作符用于过滤符合条件的行，Project 操作符用于选择需要的列。在执行查询时，从根节点开始递归调用每个操作符的 Next 方法，每次从磁盘中读取一行数据，经过多个运算符处理最终将结果返回给用户。

4．火山模型的优点和局限性

经过上述讲解，相信大家对于火山模型的原理有了基本的认识。那么这时我们也会产生一些疑问，为什么火山模型会被运用到数据库查询执行引擎？是为了解决什么问题？其实，火山模型作为 20 世纪 90 年代提出的模型，想解决的问题也比较有年代感，那就是计算机内存资源十分昂贵和稀缺，如何在完成数据库查询操作的同时尽可能节省内存资源。在火山模型的设计里，整个查询处理过程数据是一条一条如同流水线一样在运算符之间流动的，只要没有 sort 之类破坏流水性的运算符出现，每个运算符只需要很少的缓存就可以运行起来，非常节省内存资源。此外，火山模型的设计还有以下优点。

- 独立优化：火山模型中每个操作符不需要考虑自己的父节点和子节点的类型和实现，只需要提供自己相应的接口和调用即可。这样每个操作符都是单独抽象，具体实现时不用考虑其他操作符，可以独立地对特定操作符做优化。
- 灵活性：不同运算符节点可以灵活组合，运算符树的结构可以千变万化，有限的运算符也可以通过组合不同的树结构来支持复杂的查询计划。后期想要增加新的运算符也只需要实现该运算符相应接口，加入到查询计划中构建即可。

- 并行性：运算符树中兄弟节点之间不相互依赖，可以支持并行处理，进而提升查询性能。
- 通用性：火山模型是一个独立于数据库底层数据模型的查询执行模型，可以应用于不同种类数据模型的数据库，包括行存数据库、列存数据库以及对象存储数据库等。

介绍完火山模型的优点，接下来也介绍一下它的缺点。火山模型的设计中整个 Next 调用链只会处理一行数据，完全无法发挥现代 CPU 的特性（如 SIMD 指令、多核架构），处理效率低下，而且每行数据处理都要调用多次 Next 方法，在大量数据处理下，函数调用的开销也非常大。针对火山模型的这些缺点，业内引入了向量化执行模型。

7.3 数据库 AI——向量化

向量化是一种将 SQL 查询语句转换成向量操作的优化技术。传统的 SQL 查询语句是基于行的操作，即一次只能处理一行数据，而向量化则是将多行数据组织成向量进行操作，一次可以处理多行数据，从而大幅提高查询性能。向量化的原理是基于 SIMD（Single Instruction Multiple Data）指令集的并行计算，即通过一条指令同时对多个数据进行计算，从而实现高效的向量计算。

1. 向量化在 SQL 优化中的应用

向量化在 SQL 优化中的应用非常广泛，可以用于优化各种类型的查询语句，包括聚合查询、连接查询、排序查询等。在聚合查询中，向量化可以将多个行数据转换成向量进行计算，从而大幅提高计算效率；在连接查询中，向量化可以将两个表中的多行数据转换成向量进行连接操作，从而减少磁盘 IO 和 CPU 计算等开销；在排序查询中，向量化可以将多个行数据转换成向量进行排序操作，从而大幅提高排序效率，被广泛应用于对大规模数据集的高效处理。具体来说，向量化在以下方面有着广泛的应用。

- 聚合查询：聚合查询是 SQL 中常见的查询之一，如 SUM、AVG、MAX、MIN 等。传统的聚合查询是基于迭代的方式实现的，需要对每一行数据进行单独处理，计算出聚合结果。通过向量化技术，可以将多行数据组织成一个向量进行计算，从而大幅提高聚合查询的效率。
- 连接查询：连接查询是 SQL 中常见的查询之一，如 INNER JOIN、LEFT JOIN、RIGHT JOIN 等。传统的连接查询也是基于迭代的方式实现的，需要对每一行数据进行单独处理，计算出连接结果。通过向量化技术，可以将多个数据集组织成向量进行计算，从而大幅提高连接查询的效率。
- 排序查询：排序查询是 SQL 中常见的操作之一，如 ORDER BY、GROUP BY 等。传统的排序查询是基于比较的方式实现的，需要对每一行数据进行单独比较，然后排序。通过向量化技术，可以将多行数据组织成一个向量进行排序，从而大幅提高排序查询的效率。
- 窗口函数：窗口函数包括 ROW_NUMBER、RANK、DENSE_RANK 等。传统的窗口函数也是基于迭代的方式实现的，需要对每一行数据进行单独处理，计算出窗口函数结果。通过向量化技术，可以将多行数据组织成一个向量进行计算，从而大幅提高窗口函数的效率。
- 子查询：子查询是 SQL 中常见的操作之一，可以嵌套在主查询中使用。传统的子查询也是基于迭代的方式实现的，需要对每一行数据进行单独处理，计算出子查询结果。但是，通过向量化技术，可以将多行数据组织成一个向量进行计算，从而大幅提高子查询的效率。

应用向量化技术时需要考虑局限性，包括数据类型限制、内存消耗、实现难度和数据分布等因素。因此，SQL 优化需要根据具体场景和需求，灵活选择使用向量化技术或其他优化方法。针对向量化技术的优化，SQL 引擎的开发者可以采取以下技术和方法来优化实现。

- 批量处理：批量处理可以将多个操作组合成一个操作，从而提高向量化的效率。例如，在进行聚合操作时，可以将多个值组合为一个向量，然后进行计算。
- SIMD 指令集：SIMD 指令集可以在一条指令中同时处理多个数据。SQL 引擎可以使用 SIMD 指令集来实现向量化计算，从而提高计算效率。
- 内存管理：合理的内存管理可以避免内存泄露和内存消耗过大等问题。SQL 引擎可以采用内存池等技术来优化内存管理。
- 并行计算：并行计算可以将任务分配给多个线程进行处理，从而提高计算效率。SQL 引擎可以采用多线程技术来实现并行计算。
- 数据预处理：数据预处理可以将常用的数据操作提前计算并缓存，从而减少重复计算的情况，提高计算效率。

在实现向量化优化时，SQL 引擎的开发者需要综合考虑各种因素，如数据类型、计算效率、内存消耗等，从而选择最优的实现方案。同时，应用向量化技术的 SQL 语句也需要进行优化，如使用合适的索引和优化查询语句等，从而更好地发挥向量化技术的优势。

2．向量化的优点和局限性

向量化是一种优化数据库 SQL 执行效率的方法，其主要优点如下。

- 高效：通过向量化技术，可以将多行数据组织成一个向量进行计算，从而大幅提高 SQL 查询的效率。相比传统的迭代方式，向量化可以在减少 CPU 指令的情况下完成更多的计算，从而大幅降低了查询的响应时间。
- 灵活：向量化技术可以应用于不同类型的操作，如聚合、连接、排序、窗口函数和子查询等。这使得向量化技术在 SQL 优化中有很大的灵活性，可以满足不同类型的查询需求。
- 可扩展：向量化技术可以在支持 SIMD 指令的 CPU 上实现，这种 CPU 越来越普及。此外，向量化技术可以通过多线程等方式进一步提高计算效率，使得它在大规模数据处理场景下也能够发挥出优势。

然而，向量化技术也存在一些局限性，主要如下。

- 数据类型限制：向量化技术只适用于数值型数据类型，如整型和浮点型等。对于字符串、日期等非数值型数据类型，向量化技术的应用受到限制。
- 内存消耗：向量化技术需要将多行数据组织成一个向量进行计算，这要占用一定的内存。如果数据集较大，向量化技术可能会导致内存消耗过大，甚至导致内存不足。
- 实现难度：向量化技术的实现涉及底层的 CPU 指令集和编程技巧，这对于 SQL 引擎的开发者来说可能是一个挑战。此外，不同的 CPU 架构和指令集可能需要不同的向量化实现方式，这也增加了向量化技术实现的难度。

最后，向量化技术的实现方法包括硬件向量化和软件向量化两种方式。硬件向量化需要硬件支持，如 Intel 的 SSE、AVX 指令集等，而软件向量化可以通过编写高效的代码实现，如使用 OpenMP、CUDA 等并行计算框架。同时，还可以采用一些开源的向量化库，如 SIMDjson、Arrow 等。

3．向量化的实现技术和方法

向量化是通过 CPU 在一个时钟周期内同时执行多个操作，从而提高计算效率的一种技术。实现向量化有硬件向量化和软件向量化两种方式。

（1）硬件向量化

硬件向量化是指利用专门的硬件支持来实现向量化技术，从而提高计算效率。常见的硬件向量化支持包括 SIMD 指令集和特殊的向量处理器等。

SIMD 指令集可以让处理器在一条指令中同时处理多个数据，从而实现并行计算。例如，Intel 的 SSE（Streaming SIMD Extensions）指令集提供了 128 位和 256 位的向量寄存器，可以同时处理多个数据，包括整数、浮点数、双精度浮点数等。AVX（Advanced Vector Extensions）是 SSE 的升级版，支持更长的向量寄存器，可以同时处理更多的数据。

SIMD 指令是向量化实现的基础，它可以在一条指令中同时处理多个数据元素。在处理多个相同数据类型的元素时，SIMD 指令可以大大提高处理效率，因为它可以将多个数据元素打包在一起，同时执行相同的操作。

例如，假设需要对一个包含 100 个整数的数组进行加法操作。在不使用 SIMD 指令的情况下，需要使用循环来遍历数组中的每个元素并进行加法操作。而使用 SIMD 指令，则可以将多个整数打包在一起，在一条指令中同时执行加法操作，从而大大提高处理速度。

通常，向量化实现使用 SIMD 指令的方式包括以下几个步骤。

1）将数据元素打包到向量寄存器中。
2）执行相应的操作，如加法、减法、乘法等。
3）将结果从向量寄存器中解包并存储回内存中。

需要注意的是，为了有效地使用 SIMD 指令，数据需要被适当地打包到向量寄存器中。如果数据未对齐或长度不是向量大小的整数倍，向量化效果可能会降低。此外，一些现代处理器还提供了特殊的 SIMD 指令，如 AVX 和 SSE 指令集。这些指令集提供了更多的向量化指令和更大的向量寄存器，可以进一步提高向量化的效率。

除了 SIMD 指令集之外，还有一些专门的向量处理器可以实现更高效的向量化计算。例如，Intel 的 Xeon Phi 是一款针对高性能计算的加速卡，具有超过 50 个处理核心和 512 位向量寄存器，可以实现高效的向量化计算。IBM 的 Power Systems 也提供了类似的加速卡，可以实现向量化计算加速。

在数据库优化中，硬件向量化可以加速查询的执行，提高数据库的整体性能。例如，在数据仓库等大数据量的场景中，使用向量化查询可以减少 CPU 占用率，从而加速查询的执行。此外，在某些场景下，向量化技术还可以减少内存带宽的占用，提高数据处理效率。

需要注意的是，硬件向量化通常需要特殊的硬件支持，因此在使用之前需要评估硬件的性能和适用性。同时，由于向量化计算是一种并行计算方式，因此在编程时需要考虑数据的并发访问和同步问题，以避免出现数据竞争和死锁等问题。

（2）软件向量化

软件向量化是指通过对算法和代码进行优化，使得 CPU 能够高效地执行向量化操作。下面介绍一些常见的软件向量化技术和方法。

1）循环展开是一种常见的向量化技术，可以将循环中的多个操作合并为一个向量操作，从而提高处理速度。其基本思想是将循环中的多次计算合并为一次，从而减少循环迭代次数，提高计算效率。循环展开可以通过手动编写代码实现，也可以由编译器自动实现。这样可以充分利用向量寄存器的并行性，避免了数据读取和指令调度的开销，从而大大提高了程序的运行效率。例如，GCC 编译器提供了 O3 优化选项，能够自动进行循环展开。循环展开的实现需要满足以下几个条件。

- 连续的内存布局：循环展开需要将多个操作合并到向量中进行，因此需要保证这些数据在内存中是连续的，这样才能一次读取到向量寄存器中。
- 对齐要求：向量寄存器的大小一般为 128 位或 256 位，因此循环展开需要将每次迭代的操作数个数设置为向量寄存器大小的整数倍，这样才能保证向量寄存器被充分利用。
- 循环结构：循环展开需要在循环结构中实现，因此需要满足循环结构的特点。

循环展开的缺点是会占用更多的内存空间，因为需要将多个操作合并为一个较大的操作，需要分配更多的内存空间。此外，循环展开需要满足一定的条件，才能达到最优的效果，因此需要仔细地设计循环展开算法，以满足各种条件的要求。

2）数据结构优化是软件向量化实现的另一个重要方面。通过合理的数据结构设计，可以进一步提高向量化的效率。在 SQL 查询中，常用的数据结构包括数组、哈希表和树等。这些数据结构都可以进行优化，以利用向量化指令的优势。

- 数组优化：数组优化数组是最简单的数据结构之一，也是最容易进行向量化的数据结构。通过对数组进行连续存储和对齐，可以大大提高向量化效率。此外，可以采用分块存储、分组存储等方式，进一步提高向量化效率。
- 哈希表优化：哈希表是一种常见的查询优化数据结构。在哈希表中，元素的排列是随机的，难以进行向量化。为了提高向量化效率，可以使用开放地址哈希表（Open Addressing Hash Table）。开放地址哈希表采用连续存储方式，元素的排列是有序的，可以更好地利用向量化指令。
- 树优化：树是一种重要的数据结构，广泛应用于各种查询操作。由于树的数据排列是随机的，难以进行向量化。为了提高向量化效率，可以采用预排序遍历（Pre-Order Traversal）或后序遍历（Post-Order Traversal）等方式，将树的数据排列转化为有序的，从而更好地利用向量化指令。

需要注意的是，软件向量化技术和方法的实现需要考虑算法的特点和硬件的性能。同时，软件向量化也需要考虑代码的可读性和维护性，避免过度优化导致代码难以理解和修改。

7.4 编译执行

1. 编译执行的基本概念和原理

编译执行是一种将程序代码转化为机器代码并且直接执行的技术。在编译执行中，程序代码被先编译成可执行代码，然后在执行时直接运行这些指令，而不是先将代码解释成机器指令，再执行机器指令。编译执行可以提高程序的性能，因为它避免了解释器的性能开销，并且可以利用各种编译器优化来提高执行效率。

编译执行的基本原理是将源代码转化为机器代码，然后直接执行。这个过程通常分为以下几个步骤。

1）词法分析：将源代码分解成词法单元，并且将它们组成语法树。

2）语法分析：根据语法规则对语法树进行分析，确定程序结构和语义。这一步通常会把语法树转化为抽象语法树（AST）。

3）语义分析：对 AST 进行分析，包括类型检查、常量折叠、死代码删除等，以消除不必要的操作并且提高程序性能。

4）中间代码生成：将 AST 转化为中间代码，通常是一种机器无关的代码表示形式。

5）优化：对中间代码进行各种优化，包括控制流分析、数据流分析、常量传播、表达式优化、循环展开等。

6）代码生成：将中间代码转化为目标机器代码，通常是汇编代码或机器码。

7）链接：将生成的目标代码和库文件链接在一起，生成最终可执行文件。

2. 编译执行在 SQL 优化中的应用

编译执行在 SQL 优化中的应用是一个很重要的话题，因为它可以帮助提高 SQL 查询的性能和效率。在 SQL 查询中，编译执行的基本原理是将 SQL 查询语句转换为可执行代码，以便在执行时能够更快地运行。

在 SQL 查询中，编译执行包括两个主要阶段：编译阶段和执行阶段。编译阶段是指将 SQL 查询语句转换为可执行代码的过程，执行阶段是指实际执行查询并返回结果的过程。

编译阶段的主要任务是将 SQL 查询语句转换为查询计划（Query Plan），即确定查询的执行顺序和算法。查询计划是一个树形结构，它描述了查询语句的执行顺序和每个操作所使用的算法。查询计划由优化器（Optimizer）生成，优化器负责根据查询的结构和查询优化规则，选择最优的查询计划。优化器使用了各种优化算法和技术，如基于成本的优化（Cost-Based Optimization）、规则优化（Rule-Based Optimization）、统计信息优化等。

执行阶段的主要任务是将查询计划转换为可执行代码，并执行这些代码以获得查询结果。在执行阶段，编译执行可以采取减少不必要指令的方式进行优化，具体来说，编译执行技术通过以下几种方式来减少不必要的指令。

1）常量折叠：常量折叠是一种编译执行中常用的优化技术。它的原理是在编译阶段对常量表达式进行求值，将常量表达式计算为单个常量值，从而减少计算量。在执行阶段，将使用已经计算好的常量值来代替原来的常量表达式，从而减少了运算次数和内存占用。这种优化技术适用于重复使用相同的常量表达式的情况，如在 WHERE 子句中使用的常量表达式。

2）死代码删除：死代码删除是指在编译阶段将不会被执行的代码从查询计划中删除，从而减少运行时间和内存占用。在 SQL 优化中，死代码删除是一种非常重要的优化技术，因为它可以避免执行不必要的操作，从而提高查询性能。例如，在查询语句中存在一个分支语句，但是分支条件永远不会为真，那么这个分支语句就是死代码，可以从查询计划中删除。

3）表达式优化：表达式优化是一种编译执行中常见的优化技术。它的原理是通过重新组合表达式来减少运算次数和内存占用，从而提高查询性能。例如，在 WHERE 子句中使用多个相同的表达式进行比较，可以通过将这些表达式合并为一个表达式来减少计算量。

4）循环展开：循环展开是指将循环中的迭代次数替换为具体的常数，从而减少循环开销。在 SQL 优化中，循环展开通常应用于查询语句中的嵌套循环，如在 JOIN 操作中使用的循环。通过将循环展开，可以减少循环次数，从而提高查询性能。

常量折叠、死代码删除、表达式优化和循环展开等技术都是编译执行中常用的优化技术，它们可以帮助减少计算量、减少循环次数，从而提高查询性能。

即时编译（Just-In-Time Compilation，JIT）也是编译执行的一种优化技术。在 SQL 优化中，JIT 可以帮助减少不必要的指令，从而提高查询性能。

JIT 的基本原理是将代码编译成本地机器代码，并在程序运行时立即执行。这种编译方式可以帮助克服传统解释执行中的一些性能瓶颈，如指令翻译和解释器开销。因为 JIT 可以针对特定的硬件和操作系统优化代码生成，所以它通常比传统解释执行更快。

在 SQL 优化中，JIT 可以帮助减少不必要的指令，从而提高查询性能。具体来说，JIT 可以在执行阶段对查询计划进行优化，以减少不必要的计算和内存访问。例如，JIT 可以对查询计划中的算术运算和逻辑运算进行优化，从而减少运算次数和内存占用。此外，JIT 还可以对查询计划中的条件语句和循环语句进行优化，以减少不必要的分支和循环开销。

在使用 JIT 进行优化时，需要注意以下几点。

1）JIT 需要足够的内存和 CPU 资源来进行编译和执行。因此，如果系统资源受限，JIT 可能会影响查询性能。

2）JIT 需要对查询计划进行动态分析和优化，因此它可能会增加查询计划的生成时间和执行时间。

3）JIT 的优化效果取决于查询的结构和数据分布情况。因此，不同的查询可能需要不同的 JIT 优化策略。

4）JIT 需要进行优化的代码必须满足一定的编码规范和语法规则，否则可能会出现编译错误或性能下降的问题。

JIT 可以在 SQL 优化中帮助减少不必要的指令，从而提高查询性能。通过对查询计划中的算术运算、逻辑运算、条件语句和循环语句进行优化，JIT 可以减少运算次数、内存占用、分支开销和循环开销,从而提高查询性能。但是,使用 JIT 进行优化时需要注意系统资源、编码规范和查询结构等因素,以确保优化效果和稳定性。

编译执行在 SQL 优化中的应用还包括使用预编译查询(Prepared Statement)来提高查询性能。预编译查询是一种将查询计划缓存起来并在后续查询中重用的技术,它可以减少查询计划的生成时间,并提高查询的性能和可重复性。此外,编译执行还可以结合批处理(Batch Processing)来提高查询性能,批处理是一种将多个查询请求一次性发送到数据库中执行的技术,它可以减少通信和开销,提高查询效率。

总的来说,编译执行在 SQL 优化中的应用可以帮助提高 SQL 查询的性能和效率,通过使用优化器和各种编译执行技术,可以减少查询计划的生成时间,并提高查询的性能和可重复性。

3. 编译执行的优点和局限性

编译执行的优点如下。

1)更快的执行速度:编译执行的执行速度比解释执行快得多,因为编译器将程序代码编译成可执行代码,以便在执行时能够更快地运行。

2)更高的执行效率:编译器可以对代码进行各种优化,如死代码删除、表达式优化和循环展开等,以减少运行时间和内存占用。

3)更好的可移植性:由于编译器将程序代码编译成可执行代码,因此可以轻松地在不同平台上运行,而不需要重新编写代码。

4)更好的安全性:编译执行可以检测代码中的错误和漏洞,并在编译过程中消除这些问题,从而提高程序的安全性。

虽然编译执行有很多优点,但也存在局限性。

1)编译开销:编译执行的开销相对较高,因为编译器需要将程序代码编译成可执行代码,这需要一定的时间和计算资源。

2)动态生成代码:对于需要动态生成代码的程序,编译执行不适用,因为编译器只有在编译时才能对代码进行优化。

3)平台依赖性:由于编译器生成的代码是针对特定机器的,因此需要为不同的平台各编译一份代码。

4)程序规模较小:在程序规模较小的情况下,编译执行可能比解释执行更慢,因为编译执行需要的时间比解释执行更长。

编译执行是一种非常有用的技术,可以提高程序的执行效率和可移植性,并增强程序的安全性。然而,它也存在一些限制和局限性,需要根据具体情况进行评估和选择。

第 8 章

认识数据库中的数学
（逻辑与算法）

如何操作数据库中的数据以及数据库中的相关运算包含着算法。比如最简单的二分查找法就是一种朴素的算法。而实际工作和工程中遇到的算法远不止于此。数据结构+算法构成了程序，作为 DBA，如何高效、巧妙地设计、优化这些存取操作也是非常重要的。设计和优化数据的存取就需要了解相关的数学知识和算法知识。

本章内容
数据库中的典型逻辑与算法
动态规划法在数据库中的应用
数据库开发中的逻辑思维

8.1 数据库中的典型逻辑与算法

8.1.1 从斐波那契数列到数据分析

有一个著名的关于兔子的数学问题：假设有一对兔子，长两个月就算长大成年了。之后的每个月都会生出 1 对兔子，生下来的兔子也都是长两个月就算成年，然后每个月也都会生出 1 对兔子。这里假设兔子不会死，每次都是只生 1 对兔子。

第一个月，只有 1 对小兔子。

第二个月，小兔子还没长成年，还是只有 1 对兔子。

第三个月，兔子长成年了，同时生了 1 对小兔子，因此有两对兔子。

第四个月，成年兔子又生了 1 对兔子，加上自己及上月生的小兔子，共有 3 对兔子。

第五个月，成年兔子又生了 1 对兔子，第三月生的小兔子现在已经成年且生了 1 对小兔子，加上本身两只成年兔子及上月生的小兔子，共 5 对兔子。

这样过了一年之后，会有多少对兔子？

把这些兔子的数量以对为单位列出数字就能得到一组数字：1、1、2、3、5、8、13、21、34、55、89、144、233。所以，过了一年之后，总共会有 233 对兔子。

这组数字可以形成一个有规律的数列，叫作"斐波那契数列"。在这个数列中的数字，被称为"斐波那契数"。这个数列是意大利数学家斐波那契在 1228 年首先提出的。这个数列的规律是这样的：它的第一项、第二项是 1，而从第三项起每一项都等于它的前两项之和。所以，如果不考虑兔子的死亡问题，我们还可以继续往下算出 n 年后兔子繁殖的数量。

斐波那契数列是一个满足以下特性的数列：

$f(1)=f(2)=1$

$f(n)=f(n-1)+f(n-2)$　　$(n\geqslant 3)$

通过递归法（递归法的效率不高）实现斐波那契数列的计算如图 8-1 所示。

```java
private static long fibonacciRecursion(int n){
    if (n <= 2){
        return 1;
    }else{
        return fibonacciRecursion( n: n-1) + fibonacciRecursion( n: n-2);
    }
}
```

图 8-1　Java 递归实现斐波那契数列

得到正确计算结果的耗时如图 8-2 所示。由于笔者笔记本计算机的运算能力所限，只能写到 50，后续无法支持。

```
第10个月，我们拥有了55只兔子！   计算耗时：0ms
第20个月，我们拥有了6765只兔子！   计算耗时：0ms
第30个月，我们拥有了832040只兔子！   计算耗时：2ms
第40个月，我们拥有了102334155只兔子！   计算耗时：194ms
第50个月，我们拥有了12586269025只兔子！   计算耗时：23057ms

Process finished with exit code 0
```

图 8-2　递归实现斐波那契数列耗时

以上算法实现的就是递归查询。递归查询是从结果上看，后期执行得很慢，其原因是递归。要算 f（10）就要知道 f（9）和 f（8），而算 f（9）要知道 f（8）和 f（7），以此类推，其运算比阶乘还复杂，并且运算量大。

但是如果转变思路，利用空间换取时间，用正推法计算，代码如图 8-3 所示。

```java
private static long fibonacciIterative(int n) {
    if (n <= 1){
        return 1;
    }else{
        long fib = 1;
        long prevFib = 1;
        for (int i = 2; i < n; i++){
            long temp = fib;
            fib += prevFib;
            prevFib = temp;
        }
        return fib;
    }
}
```

图 8-3　正推法计算结果计算斐波那契数列

其得到结果的计算耗时如图 8-4 所示。

```
第10个月，我们拥有了55只兔子！ 计算耗时：0ms
第20个月，我们拥有了6765只兔子！ 计算耗时：0ms
第30个月，我们拥有了832040只兔子！ 计算耗时：0ms
第40个月，我们拥有了102334155只兔子！ 计算耗时：0ms
第50个月，我们拥有了12586269025只兔子！ 计算耗时：0ms
第100个月，我们拥有了3736710778780434371只兔子！ 计算耗时：0ms
第1000个月，我们拥有了8177703259943977771只兔子！ 计算耗时：0ms

Process finished with exit code 0
```

图 8-4　正推法计算结果计算斐波那契数列耗时

由于计算不到 1ms，所以显示 0ms。其原理是每次都记录需要知道的上一节点数据，只要访问 2 个静态数据就行。在当今的笔记本计算机环境下也可以轻松应对最简单的四则运算。一个 CPU 一秒可以计算一亿次，而以上只是 1000 多个数字相加而已。

实时计算结果与计算结果单独存储在数据表中是两种思路。实时计算结果的这种现象也不少见，在实际的工作中笔者就处理过很多企业针对这些场景的优化。发现这些企业的财务系统和仓储系统，并没有账户余额和用户库存的表和字段。每次访问账户余额以及库存量的时候是采用 SUM 等方式，将用户从开户至今的所有流水进行运算得出余额，将仓库的所有进出库进行运算得出库存数据和出库数据。等于将用户多年的操作进行了一次快速回放。这种算法设计随着数据增加，执行效率线性降低，用户体验越来越差。如果涉及数据归档，则整个数据都无法保证准确性。

这种做法虽然比起递归计算斐波那契数列的复杂度降低了很多（递归计算斐波那契数列的算法复杂度是几何倍的，而 SUM 计算余额和库存的算法复杂度是线性正比例的），但是依然不可取。随着数据的积累，最终会导致平台上用户的请求越来越慢。一旦面临并发访问，其 SQL 造成的 CPU 消耗和 IO 读取压力比计算斐波那契数列的 CPU 资源消耗（递归运算斐波那契数列不产生 IO 压力）有过之而无不及。

不幸的是，时至今日许多公司依然采用 SUM 来计算余额和库存数据。而正确的做法可以参考银行的交易场景。用户的银行卡都绑定了短信提醒，每月发工资和奖金时会提醒，本月入账 X 元，当前账户余额 Y 元。银行的余额设计非常合理，读取余额仅仅是一行数据，效率很高。

作为技术人员，应该学习的是一种编程思想，而不仅仅是技术，应该思考如何设计实现逻辑，最优地实现需求。比如在处理不少业务场景的时候，只要设计得当，即使是传统的关系数据库也可以做到快速出结果，耗时仅为秒级甚至是毫秒级。这样会使得系统架构比较精简。反之如果是采用流式计算或者数据同步等技术将数据搬迁到大数据平台或者其他异构数据库中做处理，是将简单问题复杂化。设计实现逻辑的目的一定是复杂问题简单化。

正推法和递归法合在一起的完整 Java 代码如图 8-5 所示。

```java
package db;

import java.util.Date;

public class fibonacci {
    public static void main(String[] args) {
        System.out.println("递归法: ");
        timeResult_recursion(10);
        timeResult_recursion(20);
        timeResult_recursion(30);
        timeResult_recursion(40);
        timeResult_recursion(50);

        System.out.println("正推法: ");
        timeResult_iterative(10);
        timeResult_iterative(20);
        timeResult_iterative(30);
        timeResult_iterative(40);
        timeResult_iterative(50);
        timeResult_iterative(100);
        timeResult_iterative(1000);

    }

    //递归法
    private static long fibonacciRecursion(int n){
        if (n <= 2){
            return 1;
        }else{
            return fibonacciRecursion(n-1) + fibonacciRecursion(n-2);
        }
    }

    //正推法
    private static long fibonacciIterative(int n) {
        if (n <= 1){
            return 1;
        }else{
            long fib = 1;
            long prevFib = 1;
            for (int i = 2; i < n; i++){
                long temp = fib;
                fib += prevFib;
                prevFib = temp;
            }
            return fib;
        }
    }

    private static void timeResult_recursion(int n){
        Date startTime1 = new Date();
        long numRecursion = fibonacciRecursion(n);
        Date endTime1 = new Date();
        System.out.println("第" + n + "个月, 我们拥有了" + numRecursion + "只兔子! 计算耗时: " + (endTime1.getTime()-startTime1.get
    }

    private static void timeResult_iterative(int n){
        Date startTime = new Date();
        long numIterative = fibonacciIterative(n);
        Date endTime = new Date();
        System.out.println("第" + n + "个月, 我们拥有了" + numIterative + "只兔子! 计算耗时: " + (endTime.getTime()-startTime.getTi
    }

}
```

图 8-5　Java 实现斐波那契数列

图 8-5　Java 实现斐波那契数列（续）

具体的 Java 代码如下：

```java
package db;
import java.util.Date;
public class fibonacci {
    public static void main(String[] args) {
        System.out.println("递归法：");
        timeResult_recursion(10);
        timeResult_recursion(20);
        timeResult_recursion(30);
        timeResult_recursion(40);
        timeResult_recursion(50);
        System.out.println("正推法：");
        timeResult_iterative(10);
        timeResult_iterative(20);
        timeResult_iterative(30);
        timeResult_iterative(40);
        timeResult_iterative(50);
        timeResult_iterative(100);
        timeResult_iterative(1000);
    }
    //递归法
    private static long fibonacciRecursion(int n){
        if (n <= 2){
            return 1;
        }else{
            return fibonacciRecursion(n-1) + fibonacciRecursion(n-2);
        }
    }
```

```java
    }
    //正推法
    private static long fibonacciIterative(int n) {
        if (n <= 1){
            return 1;
        }else{
            long fib = 1;
            long prevFib = 1;
            for  (int i = 2; i< n; i++){
                long temp = fib;
                fib += prevFib;
                prevFib = temp;
            }
            return fib;
        }
    }

    private static void timeResult_recursion(int n){
        Date startTime1 = new Date();
        long numRecursion = fibonacciRecursion(n);
        Date endTime1 = new Date();
        System.out.println("第" + n + "个月，我们拥有了" + numRecursion + "只兔子！计算耗时： " + (endTime1.getTime()-startTime1.getTime()) + "ms");
    }

    private static void timeResult_iterative(int n){
        Date startTime = new Date();
        long numIterative = fibonacciIterative(n);
        Date endTime = new Date();
        System.out.println("第" + n + "个月，我们拥有了" + numIterative + "只兔子！计算耗时： " + (endTime.getTime()-startTime.getTime()) + "ms");
    }
}
```

这里是一种简单的算法，在 8.2 节中会涉及更加复杂的。这些思维如果用得好，那么在数据库的数据分析中会发挥非常大的作用。

8.1.2　增加数据维度处理——减少关联

有一次开发人员小 E 求助一个问题，应用场景是一个消息读取场景。开发人员设计了一个消息表，系统给用户发的消息或者通知都放在这里。当用户读取消息时，开发人员设计的是将已读的消息放入另外一个已读消息表。这样的设计在用户维度查询已读消息是没有问题的，可以直接读取已读消息表。但是要查询未读消息时问题就来了。下面对这个场景进行一下模拟。模拟写入 5 条消息数据，查询 message 表显示当前所有消息，如图 8-6 所示。

```
create table message (id int,sentid varchar2(10),

TXT VARCHAR2(100),receiveid varchar2(10),time date);

create table message_read (id int,sentid varchar2(10),

TXT VARCHAR2(100),receiveid varchar2(10),time date);


insert into message values (1,'T1','你好1','T2',sysdate-30+1);
insert into message values (2,'T1','你好2','T3',sysdate-30+1);
insert into message values (3,'T4','你好3','T5',sysdate-30+1);
insert into message values (4,'T6','你好4','T7',sysdate-30+1);
insert into message values (5,'T3','你好5','T1',sysdate-30+1);




SQL> select id,sentid,receiveid,txt from message;

ID SENTID     RECEIVEID  TXT
-- ---------- ---------- --------------
 1 T1         T2         你好1
 2 T1         T3         你好2
 3 T4         T5         你好3
 4 T6         T7         你好4
 5 T3         T1         你好5
```

图 8-6　创建两个消息表

当接收 ID 为 T3 和 T7 的两个用户发生读取消息的行为，则进行已读消息的处理。将两条数据复制到 message_read 表，如图 8-7 所示。

此时如果 T3 再次接收到新的未读消息，则对于 T3 来说既有已读的又有未读的。如图 8-8 所示，ID 为 2 和 6 的记录都是 T3 收到的消息，ID 为 2 的记录为已读（已经进入 message_read 表），ID 为 6 的记录还未读（仅存在于 message 表）。

```
SQL> insert into message_read select * from message where receiveid='T3';
1 row inserted

SQL> insert into message_read select * from message where receiveid='T7';
1 row inserted

SQL> commit;
Commit complete

SQL> select id,sentid,receiveid from message_read;

        ID SENTID     RECEIVEID
---------- ---------- ----------
         2 T1         T3
         4 T6         T7
```

图 8-7 写入已读消息表

```
SQL> insert into message values (6,'T2','你好6','T3',sysdate-30+1);

SQL> commit;
```

提交完成

```
SQL> select id,sentid,receiveid from message;

        ID SENTID     RECEIVEID
---------- ---------- ----------
         6 T2         T3
         1 T1         T2
         2 T1         T3
         3 T4         T5
         4 T6         T7
         5 T3         T1
```

已选择 6 行。

图 8-8 新增消息后的消息详情

在目前这种设计下，读取 T3 的已读消息还是简单的。直接查询 message_read 表即可，如图 8-9 所示。

```
SQL> select id,sentid,receiveid from message_read where receiveid='T3';

        ID SENTID     RECEIVEID
---------- ---------- ----------
         2 T1         T3
```

图 8-9 直接读取已读消息表

但是这种数据库设计下，如果要查询 T3 为未读消息就比较麻烦，一般开发人员会写成如图 8-10 所示的样式，来进行反连接的关联。

第 8 章　认识数据库中的数学（逻辑与算法）　335

```
SQL> select id,sentid,receiveid from message m where
  2  not exists (select * from message_read d where m.id=d.id) ;

    ID         SENTID     RECEIVEID
---------- ---------- ----------
     6         T2         T3
     1         T1         T2
     5         T3         T1
     3         T4         T5


SQL> select id,sentid,receiveid from message m where
  2  m.id not in (select id from message_read d where m.id=d.id) ;

    ID         SENTID     RECEIVEID
---------- ---------- ----------
     6         T2         T3
     1         T1         T2
     3         T4         T5
     5         T3         T1
```

图 8-10　关联表排除

出现以上这种问题的场景还常见于一些错误日志的场景，也是将错误的数据另存一份。当要检查没有错误的数据时，就做了两个表的排除。

通常对于这种写法，有改写的方法。但是这个问题主要归于数据库设计，推荐从设计上规避。具体如何做？需要增加一个维度。举例来说，如果用 0 代表否，1 代表是，那么 0 和 1 只能代表 2 个信息，如果要代表 4 个信息，那么就需要多一位数字来记录，如 00、01、10、11。增加了一个维度，增加了一倍的信息承载。

在信息表上增加一个已读标记位，默认是 0 表示未读，1 表示已读，如图 8-11 所示。

```
SQL> alter table message add read_flag char(1)  default '0' not null;

表已更改。

SQL> update message set read_flag =1 where RECEIVEID='T3';

已更新 1 行。

SQL> update message set read_flag =1 where RECEIVEID='T7';

已更新 1 行。

SQL> commit;

提交完成。
```

图 8-11　增加消息表字段

模拟在 5 条数据的初始情况下，T3 和 T7 将自己的消息设为已读。然后 T3 再次收到消息。那么如图 8-12 所示，查询全部消息和查询未读消息都变得非常简单，也便于后续维护。

```
SQL> select id,sentid,receiveid,read_flag from message where
  2 RECEIVEID='T3';

        ID SENTID     RECEIVEID  R
---------- ---------- ---------- -
         6 T2         T3         0
         2 T1         T3         1

SQL> select id,sentid,receiveid,read_flag from message where
  2 RECEIVEID='T3' and read_flag='0';

        ID SENTID     RECEIVEID  R
---------- ---------- ---------- -
         6 T2         T3         0
```

图 8-12　查询 T3 的消息

小结：这种增加标记的做法使得后续维护成本大大降低，而且相比较之前的做法存储数据量上也有较大的节约。很多场景中设计改进一小步，优化得到一大步。设计的收益有时候远大于改写和优化带来的改善。

8.1.3　多表关联算法——一对多与多对多

多表关联是一个最朴素的算法逻辑。当执行 A 表和 B 表进行关联时，通常会有 A.ID=B.ID 或者是其他的 A.关联列=B.关联列的关联条件。假设 A 表为学生学籍表，包含学号、姓名、班级等信息，有 1 万条数据。B 表为学生详情表，包含学号、身份证号、电话、邮箱等信息。从数据库建模的角度出发，A、B 两个表是一对一的关系。查询时通过学号进行关联（毕竟学生姓名有可能同名，但学号是唯一的）。如图 8-13 所示，两个表通过学号列进行关联。如果不带任何条件，那么根据下图的关系就是 A 表查询 1 万，B 表也查询 1 万。实际数据扫描了两万行。

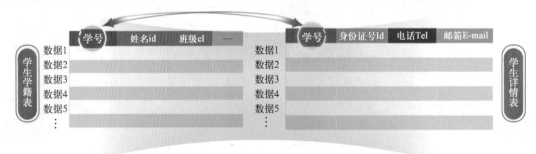

图 8-13　学号关联

从全局来看，如图 8-14 所示，所列都围绕着学号进行展开，整个外圈的数据列可以依次读取。

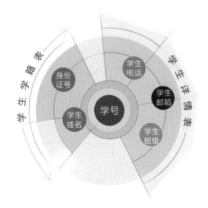

图 8-14　学号关联展开

当 A、B 两个表为一对多时，如图 8-15 所示，一个合同可能包含了 10 张发票。这种就是一对多的关系。

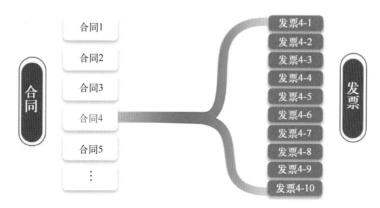

图 8-15　合同与发票对应关系

如果查询了 10 个合同，同时要显示发票表上的字段，那么就要关联发票表。实际扫描行数就是主表（合同表）扫描 10 行，子表（发票表）扫描 100 行（真实环境中可能是 1∶2 的比例也可能 1∶10000 的比例，本示意图是 1∶10）。

如果是多表关联，如图 8-16 所示，要进行综合查询时就需要多表的多倍计算。

笔者曾经遇到过这样一个问题：

在 MySQL 数据库的 SLOWLOG 中看到：

Query_time: 255.707167 Lock_time: 0.002991Rows_sent: 10000 Rows_examined: 8217255 解读这句话就是查询了约 821 万条数据，返回 1 万条数据，耗时 255s。开发人员反馈自己的程序不可能查询这么多数据，最大的表才 140 万行左右，不可能发生读取 821 万行的操作，请笔者帮他澄清或者解惑。

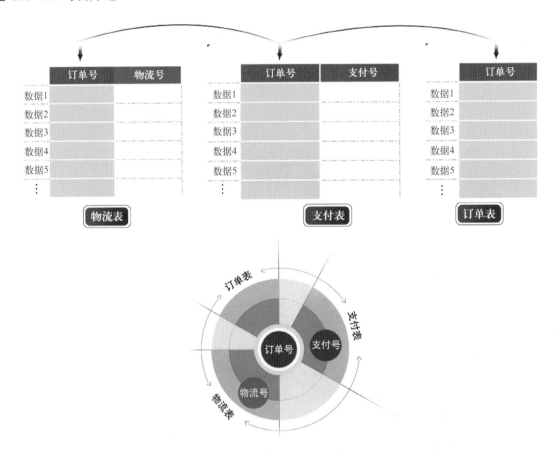

图 8-16　通过订单多表关联

其实这个问题就是通常讲的多表关联,也就是 DBA 经常对开发人员传达的禁止多表关联的原因,这里做实验来说明,构造三个有 10 万条数据的表,如图 8-17 所示。

```
-- 创建xuexiaogang1表
CREATE TABLE xuexiaogang1 (
  id INT(11) NOT NULL AUTO_INCREMENT,
  PRIMARY KEY (id)
);

-- 向xuexiaogang1表中插入1到1000000的整数值
INSERT INTO xuexiaogang1 (id)
SELECT n
FROM (
  SELECT @row := @row + 1 AS n
  FROM (SELECT @row := 0) r, INFORMATION_SCHEMA.COLUMNS c1, INFORMATION_SCHEMA.COLUMNS c2
) numbers
WHERE n <= 1000000;
```

图 8-17　构造三个有 10 万数据的表

```
-- 创建xuexiaogang2表
CREATE TABLE xuexiaogang2 (
  id INT(11) NOT NULL AUTO_INCREMENT,
  PRIMARY KEY (id)
);

-- 向xuexiaogang2表中插入1到1000000的整数值
INSERT INTO xuexiaogang2 (id)
SELECT n
FROM (
  SELECT @row := @row + 1 AS n
  FROM (SELECT @row := 0) r, INFORMATION_SCHEMA.COLUMNS c1, INFORMATION_SCHEMA.COLUMNS c2
) numbers
WHERE n <= 1000000;

-- 创建xuexiaogang3表
CREATE TABLE xuexiaogang3 (
  id INT(11) NOT NULL AUTO_INCREMENT,
  PRIMARY KEY (id)
);

-- 向xuexiaogang3表中插入1到1000000的整数值
INSERT INTO xuexiaogang3 (id)
SELECT n
FROM (
  SELECT @row := @row + 1 AS n
  FROM (SELECT @row := 0) r, INFORMATION_SCHEMA.COLUMNS c1, INFORMATION_SCHEMA.COLUMNS c2
) numbers
WHERE n <= 1000000;
```

图 8-17　构造三个有 10 万条数据的表（续）

确认数据写入成功，如图 8-18 所示。

```
mysql> select count(*) from xuexiaogang1;
+----------+
| count(*) |
+----------+
|  1000000 |
+----------+
1 row in set (0.04 sec)

mysql> select count(*) from xuexiaogang2;
+----------+
| count(*) |
+----------+
|  1000000 |
+----------+
1 row in set (0.06 sec)

mysql> select count(*) from xuexiaogang3;
+----------+
| count(*) |
+----------+
|  1000000 |
+----------+
1 row in set (0.03 sec)
```

图 8-18　查询三张 10 万的数据表

模拟查询，检查该 SQL 的执行计划，如图 8-19 所示。

```
mysql> Explain select * from xuexiaogang1,xuexiaogang2,xuexiaogang3 where xuexiaogang1.id=xuexiaogang2.id and xuexiaogang1.id=xuexiaogang3.id limit 100000,10;
+----+-------------+--------------+------------+--------+---------------+---------+---------+---------------------------+--------+----------+-------------+
| Id | select_type | table        | partitions | type   | possible_keys | key     | key_len | ref                       | rows   | filtered | Extra       |
+----+-------------+--------------+------------+--------+---------------+---------+---------+---------------------------+--------+----------+-------------+
|  1 | SIMPLE      | xuexiaogang1 | NULL       | index  | PRIMARY       | PRIMARY | 4       | NULL                      | 999000 |   100.00 | Using index |
|  1 | SIMPLE      | xuexiaogang2 | NULL       | eq_ref | PRIMARY       | PRIMARY | 4       | xuexiaogang.xuexiaogang1.id |      1 |   100.00 | Using index |
|  1 | SIMPLE      | xuexiaogang3 | NULL       | eq_ref | PRIMARY       | PRIMARY | 4       | xuexiaogang.xuexiaogang1.id |      1 |   100.00 | Using index |
+----+-------------+--------------+------------+--------+---------------+---------+---------+---------------------------+--------+----------+-------------+
3 rows in set, 1 warning (0.00 sec)
```

图 8-19　MySQL 多表关联执行计划

模拟查询，分页查询如图 8-20 所示。

```
mysql>select * from xuexiaogang1,xuexiaogang2,xuexiaogang3 where
    -> xuexiaogang1.id=xuexiaogang2.id and xuexiaogang1.id=xuexiaogang3.id
    -> limit 100000,10;
+--------+--------+--------+
| id     | id     | id     |
+--------+--------+--------+
| 100001 | 100001 | 100001 |
| 100002 | 100002 | 100002 |
| 100003 | 100003 | 100003 |
| 100004 | 100004 | 100004 |
| 100005 | 100005 | 100005 |
| 100006 | 100006 | 100006 |
| 100007 | 100007 | 100007 |
| 100008 | 100008 | 100008 |
| 100009 | 100009 | 100009 |
| 100010 | 100010 | 100010 |
+--------+--------+--------+
10 rows in set (0.22 sec)
```

图 8-20　MySQL 数据分页查询

最终的结果是查询了 100000+100000+100000=30 万行数据，再加上每个表 10 行，3 个表都一对一地返回也需要 10+10+10=30 行。共计 300030 行，如图 8-21 所示。

```
# Time: 2023-04-21T11:19:14.836066Z
# User@Host: root[root] @ localhost []  Id:    38
# Query_time: 0.214561  Lock_time: 0.000006 Rows_sent: 10  Rows_examined: 300030
SET timestamp=1682075954;
select * from xuexiaogang1,xuexiaogang2,xuexiaogang3 where xuexiaogang1.id=xuexiaogang2.id and xuexiaogang1.id=xuexiaogang3.id limit 100000,10;
```

图 8-21　MySQL 多表慢查询结果

以上是 3 个表关联，如果是 10 个表关联，那就是 100010×10。所以通过这个案例和实验，开发人员也知道了背后的原因。对于广大开发和设计人员来说，千万要控制表的关联数量，这是根本。如果因为历史原因已经很多表了，那么 where 后面的谓词条件一定要"控制"住，避免全表扫描，否则多表关联的结果就是 N 倍的累积。如果无法控制表的设计，那后期 SQL 上线就可能要出问题，这是最朴素的数据乘积算法。

数据库表设计把控难度相对大，根据笔者多年的经验得出的结论就是最好一开始就介入，结合需求设计。

多表关联一般指的是超过两张表的关联。数据库虽然没有限制表关联的数量，但是从本案例可以得出，表的关联越多运算量就越可能成倍地增加。

8.1.4 排除处理算法——笛卡儿积

生产中常常要面对同时查询多表的情况，这种称之为多表关联查询。其中最关键的两字便为"关联"，意为多个查询对象间要有一定程度的关联。而未使用关联或是不恰当的关联就会导致笛卡儿积的产生。笛卡儿积是指在数据库查询中，查询结果或是查询过程中的查询数量相当于与两个表各自数据量的笛卡儿乘积，这种现象会导致查询效率低下、结果数据量庞大或是结果不准确。

工作多年，时至今日笔者依然能时不时遇到笛卡儿积的现象。许多开发人员对此没有概念。本节通过实验演示笛卡儿积的模拟，分析其危害以及如何避免关联查询中笛卡儿积的产生。

数据库是物理学和数学的结合，这两大基础自然科学不因数据库产品的不同而不同，笛卡儿积在每个数据库上原理都一样。本次以 MySQL 为例。

首先如图 8-22 所示，在 MySQL 中建立两张简单的实验表。

```
mysql> select * from name;              mysql> select * from city;
+----+--------+                         +----+-------+
| id | name   |                         | id | city  |
+----+--------+                         +----+-------+
|  0 | J3EPDo |                         |  0 | ZEFbW |
|  1 | cwgTCW |                         |  1 | 0NgAe |
|  2 | vkJWfV |                         |  2 | 2jv8L |
|  3 | WlNBzR |                         |  3 | VBDFG |
|  4 | OoJuyz |                         |  4 | VPC2p |
|  5 | spA0jL |                         |  5 | aZv2V |
|  6 | SZwEK7 |                         |  6 | epEK9 |
|  7 | lMVJE7 |                         |  7 | KIaxI |
|  8 | eaBRL1 |                         |  8 | b03BR |
|  9 | 4MACDq |                         |  9 | JfwL9 |
+----+--------+                         +----+-------+
10 rows in set (0.01 sec)               10 rows in set (0.00 sec)
```

图 8-22 构造关联实验表

笛卡儿积查询如图 8-23 所示，两张表中各有 10 条数据，做一个最简单查询。进行两表关联查询。

```
mysql> select name,city from name,city;
+--------+-------+
| name   | city  |
+--------+-------+
| 4MACDq | ZEFbW |
| eaBRL1 | ZEFbW |
| lMVJE7 | ZEFbW |
         .
         .
         .
| vkJWfV | JfwL9 |
| cwgTCW | JfwL9 |
| J3EPDo | JfwL9 |
+--------+-------+
100 rows in set (0.00 sec)
```

图 8-23 笛卡儿积查询

该查询没有 where 条件做关联条件，得到的即为笛卡儿积的结果，两张 10 条数据的表得到了 100 条查询结果。从后台的慢日志中也可以得到相同的佐证，如图 8-24 所示。

```
# Time: 2023-05-09T01:28:42.379562Z
# User@Host: root[root] @ localhost []  Id:    38
# Query_time: 0.001022  Lock_time: 0.000013 Rows_sent: 100  Rows_examined: 20
SET timestamp=1683595722;
select name,city from name,city;
```

图 8-24 MySQL 笛卡儿积慢日志

非特殊场景下，这样的结果往往不是我们所需要的，在数据量增长或是表数量增加的情况下也极易产生巨量数据的查询。

以上场景在生产中较为少见，也较易避免，通过查看查询结果的数量便能够分辨。但是在某些场景下，多表关联查询的关联条件为不等于，这种查询同样会带来近似于笛卡儿积的查询结果，如图 8-25 所示。

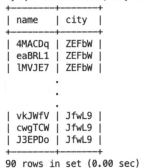

图 8-25 不等于关联笛卡儿积

查询得到的结果为 90 条，仅仅比上一次查询得到完整笛卡儿积的结果少了 10 条。这是不相同的较多，也许会有人问，那么仅有少数呢？效果一样的。

以上可以说明有且仅有不等于的查询结果类似于笛卡儿积。一旦数据量增加，或是关联表数量增加会导致查询结果的大量增加，这样的关联可以理解为是假关联，不同于前面一个无关联的查询，这样的假关联能够在生产中见到，部分开发同学也容易忽视这类查询的问题。比如业务提一个需求，比对两个表中不一样的记录。如果照单全收的执行，那么带来的就是笛卡儿积。两个 10 条数据，只有一条不一样。现在带上关联条件，做一次正常的查询，如图 8-26 所示。

可以看到，这次查询返回了正常条数。需要后台慢日志图。请注意，这样的操作避免了笛卡儿积，不过还是全表操作。许多开发人员认为有 where 条件或者 on 条件就不是全表扫描，这种观点是不对的。只要没有起到过滤作用，扫描全表、返回全表就是全表扫描，where 后的关联仅仅是起到关联作用，避免了笛卡儿积而已。

```
mysql> select name,city from name join city on name.id=city.id;
+--------+-------+
| name   | city  |
+--------+-------+
| J3EPDo | ZEFbW |
| cwgTCW | 0NgAe |
| vkJWfV | 2jv8L |
| WlNBzR | VBDFG |
| OoJuyz | VPC2p |
| spA0jL | aZv2V |
| SZwEK7 | epEK9 |
| lMVJE7 | KIaxI |
| eaBRL1 | b03BR |
| 4MACDq | JfwL9 |
+--------+-------+
10 rows in set (0.01 sec)
```

图 8-26 两表全表关联

做一次真正的关联查询，需要对其中的一条数据做精确查询。查询语句如图 8-27 所示，name 和 city 两个表使用各自的 id 作为关联。通过 name 表的 name 列的精确等值'J3EPDo'查询，找到 name 表的 id 值为 0。由于 name 表的 id 和 city 的 id 是相等的，等于做了传递。也找到了 city 的 id 值为 0，最终通过 city 的 id 值为 0，返回 city 的 id 值为 0 的相关列。

```
mysql> select name.id nameid,name,city.id cityid,city from name join city on name.id=city.id and name.name='J3EPDo';
+--------+--------+--------+-------+
| nameid | name   | cityid | city  |
+--------+--------+--------+-------+
|      0 | J3EPDo |      0 | ZEFbW |
+--------+--------+--------+-------+
1 row in set (0.01 sec)
```

图 8-27 两表关联且有有效过滤条件

检查后台日志如图 8-28 所示。

```
# Time: 2023-05-09T02:15:18.407115Z
# User@Host: root[root] @ localhost []  Id:    38
# Query_time: 0.001373  Lock_time: 0.000012 Rows_sent: 1  Rows_examined: 20
SET timestamp=1683598518;
select name.id nameid,name,city.id cityid,city from name join city on name.id=city.id and name.name='J3EPDo';
```

图 8-28 两表关联且有有效过滤条件慢日志

从后台日志可以看到，扫描了 20 行（即 name 和 city 两个表的全表之和）才返回一行。预想检索一条数据，而实际扫描两张全表，这样的逻辑放在生产环境中是会出问题的。这还仅仅是两表关联，如果是多表关联，而且是一对多的话，关联几张表查询数量就是几张表的全量，那么这个查询就会被几十倍地放大。可以说是效率最差的查询方式。

做优化最简单也最直接的方式便是给要精确查询的列加上索引，如图 8-29 所示。

```
mysql> alter table name add index IDX_NAME_NAME (name);
Query OK, 0 rows affected (0.02 sec)
Records: 0  Duplicates: 0  Warnings: 0
```

图 8-29 谓词列增加索引

创建索引后，再次查询后对比后台日志，如图 8-30 所示。

```
# Time: 2023-05-08T08:20:59.149665Z
# User@Host: root[root] @ localhost []  Id:    29
# Query_time: 0.002183  Lock_time: 0.000013 Rows_sent: 1  Rows_examined: 11
SET timestamp=1683534059;
select name,city from name join city on name.id=city.id and name.name='J3EPDo';
```

图 8-30 谓词列增加索引后的慢日志

看到查询行数确实减少到了 11 条，也就是减少了一个表的全表（效率有所提升，真实环境如果一个表有 100 万条数据，那么这个收益就显得很大了）。但还是有一张表做了一次全表查询，那么如何才能做到效率最优的查询呢？那就是给两张表的关联列都加上索引，如图 8-31 所示。

```
mysql> alter table name add index IDX_NAME_ID (id);
Query OK, 0 rows affected (0.03 sec)
Records: 0  Duplicates: 0  Warnings: 0

mysql> alter table city add index IDX_CITY_ID (id);
Query OK, 0 rows affected (0.02 sec)
Records: 0  Duplicates: 0  Warnings: 0
```

图 8-31 关联列全部建立索引

至此，谓词建立了索引，并且两个表的关联列也分别创建了索引。创建索引后，再次查询后对比后台日志，如图 8-32 所示。

```
# Time: 2023-05-08T08:26:45.905564Z
# User@Host: root[root] @ localhost []  Id:    29
# Query_time: 0.001787  Lock_time: 0.000027 Rows_sent: 1  Rows_examined: 2
SET timestamp=1683534405;
select name,city from name join city on name.id=city.id and name.name='J3EPDo';
```

图 8-32 索引完善的慢日志

这次仅仅查询了两行，也就是两张表中的精确行。如此便达到了关联查询的最优效率。

然而在生产中仍然会碰到一种情况。即在关联列上加上了索引，做关联查询时，MySQL 却并未使用索引。实验举例再建立一张 company 表，拥有类似的表结构，如图 8-33 所示。

```
mysql> select * from company;
+----+-------------+
| id | company     |
+----+-------------+
| 0  | R6N8wsANbu  |
| 1  | SkrZGbPIRv  |
| 2  | HKjjGjnRk7  |
| 3  | YhlVFRhjZ7  |
| 4  | zZi9kJa2og  |
| 5  | z73IZ9pKm1  |
| 6  | Uk12XPPa67  |
| 7  | HiEJU9cLmr  |
| 8  | hu35ogVRj0  |
| 9  | j5nZkvOzzS  |
+----+-------------+
10 rows in set (0.01 sec)
```

图 8-33 建立 company 表

第 8 章 认识数据库中的数学（逻辑与算法）

为 company 表的关联列建立索引，如图 8-34 所示。

```
mysql> alter table company add index IDX_COMPANY_ID (id);
Query OK, 0 rows affected (0.06 sec)
Records: 0  Duplicates: 0  Warnings: 0
```

图 8-34　company 表的 id 列建立索引

将新的 company 表与之前的 name 表做一次精确关联查询。然后检查执行的慢日志，如图 8-35 所示。查询结果正如预期出现了一条，但是后台日志却显示扫描了 11 行，说明有一张表执行了全表扫描，这是为什么？

```
mysql> select name,company from name join company on name.id=company.id and name.name='J3EPDo';
+--------+-----------+
| name   | company   |
+--------+-----------+
| J3EPDo | R6N8wsANbu |
+--------+-----------+
1 row in set (0.00 sec)

# Time: 2023-05-08T08:43:51.972118Z
# User@Host: root[root] @ localhost []  Id:    29
# Query_time: 0.002307  Lock_time: 0.000017 Rows_sent: 1  Rows_examined: 11
SET timestamp=1683535431;
select name,company from name join company on name.id=company.id and name.name='J3EPDo';
```

图 8-35　company 表与 name 表关联及其慢日志

其原因就在两张表的表结构上，如图 8-36 所示。

```
mysql> desc name;
+-------+-------------+------+-----+---------+-------+
| Field | Type        | Null | Key | Default | Extra |
+-------+-------------+------+-----+---------+-------+
| id    | int         | YES  | MUL | NULL    |       |
| name  | varchar(30) | YES  | MUL | NULL    |       |
+-------+-------------+------+-----+---------+-------+
2 rows in set (0.00 sec)

mysql> desc company;
+---------+-------------+------+-----+---------+-------+
| Field   | Type        | Null | Key | Default | Extra |
+---------+-------------+------+-----+---------+-------+
| id      | varchar(10) | YES  | MUL | NULL    |       |
| company | varchar(30) | YES  |     | NULL    |       |
+---------+-------------+------+-----+---------+-------+
2 rows in set (0.01 sec)
```

图 8-36　检查关联表的表结构

从上图可以看出，两个关联列上确实都有索引。但是由于两张表关联列的字段类型不一致，导致了全表扫描的发生，这是生产中比较容易出现的一种关联查询效率低下的场景。

对此有没有对应的解决方案？也是有的。比较通俗直接的解决方案便是改变其中一列的字段类型，比如上面这个案例中 company 表的 id 列使用 varchar 类型显然是不合理的，可以将它改为 int 类型解决。

但这往往代价较大，大量数据的情况下往往要停机执行。一般来说，作为 DBA 可以提出一个更好的解决方式，那便是建立一个函数索引。函数索引的知识在 2.1.2 节已经讲过，其实现方式如图 8-37 所示（索引名字用 ID_INT 表示把数据类型转换成了整型）。

```
mysql> alter table company add index IDX_COMPANY_ID_INT ((cast(id as signed)));
Query OK, 0 rows affected (0.04 sec)
Records: 0  Duplicates: 0  Warnings: 0
```

图 8-37　建立关联列的函数索引

索引完成后再重复一次查询（查询中要带上转换函数，MySQL 中精确查询条件也需要是这个关联列，Oracle 中使用其他列精确查询也能用上关联列的索引）。执行 SQL 后，检查后台日志，如图 8-38 所示。

```
# Time: 2023-05-08T09:17:50.233336Z
# User@Host: root[root] @ localhost []  Id:     29
# Query_time: 0.001733  Lock_time: 0.000014 Rows_sent: 1  Rows_examined: 2
SET timestamp=1683537470;
select name,company from name join company on name.id=cast(company.id as signed) and name.id=1;
```

图 8-38　使用函数索引检索日志

从日志中看到只扫描了两条数据，和预期的一致。再检查一下该 SQL 的执行计划，如图 8-39 所示。

```
mysql> explain select name,company from name join company on name.id=cast(company.id as signed) and name.id=1;
+----+-------------+---------+------------+------+--------------------+--------------------+---------+-------+------+----------+-------+
| id | select_type | table   | partitions | type | possible_keys      | key                | key_len | ref   | rows | filtered | Extra |
+----+-------------+---------+------------+------+--------------------+--------------------+---------+-------+------+----------+-------+
|  1 | SIMPLE      | name    | NULL       | ref  | IDX_NAME_ID        | IDX_NAME_ID        | 5       | const |    1 |   100.00 | NULL  |
|  1 | SIMPLE      | company | NULL       | ref  | IDX_COMPANY_ID_INT | IDX_COMPANY_ID_INT | 9       | const |    1 |   100.00 | NULL  |
+----+-------------+---------+------------+------+--------------------+--------------------+---------+-------+------+----------+-------+
2 rows in set, 1 warning (0.01 sec)
```

图 8-39　函数索引关联执行计划

果然在 company 表上使用到了 IDX_COMPANY_ID_INT 的函数索引。

在 Oracle 中复刻一下相同的关联列函数转换实验。先看下两张表的结构，关联字段 ID 的类型是不同的，如图 8-40 所示。

```
SQL> desc name;
 名称                                       是否为空？ 类型
 ----------------------------------------- -------- ----------------------------
 ID                                                 NUMBER(10)
 NAME                                               VARCHAR2(30)

SQL> desc company;
 名称                                       是否为空？ 类型
 ----------------------------------------- -------- ----------------------------
 ID                                                 VARCHAR2(10)
 COMPANY                                            VARCHAR2(30)
```

图 8-40　Oracle 关联表表结构

对 company 表字符类型的 ID 列建立了函数索引后做关联查询，如图 8-41 所示。

从图中可以看到也用到了特别设计的 IDX_COMPANY_ID_NUM 函数索引。

```
SQL> explain plan for select * from name join company on name.id=cast(company.id as number) where name.name='I1MMWB';
已解释。

SQL> SELECT * FROM TABLE(DBMS_XPLAN.DISPLAY);
PLAN_TABLE_OUTPUT
---------------------------------------------------------------------------------------------------
Plan hash value: 2848190367

---------------------------------------------------------------------------------------------------
| Id  | Operation                            | Name              | Rows | Bytes | Cost (%CPU)| Time     |
---------------------------------------------------------------------------------------------------
|   0 | SELECT STATEMENT                     |                   |    1 |    67 |    4   (0)| 00:00:01 |
|   1 |  NESTED LOOPS                        |                   |    1 |    67 |    4   (0)| 00:00:01 |
|*  2 |   TABLE ACCESS FULL                  | NAME              |    1 |    30 |    3   (0)| 00:00:01 |
|   3 |   TABLE ACCESS BY INDEX ROWID BATCHED| COMPANY           |    1 |    37 |    1   (0)| 00:00:01 |
|*  4 |    INDEX RANGE SCAN                  | IDX_COMPANY_ID_NUM|    1 |       |    0   (0)| 00:00:01 |
---------------------------------------------------------------------------------------------------
```

图 8-41　Oracle 函数索引执行计划

本节主要是讲从谓词条件和索引上来避免笛卡儿积。如果在实际工作中避无可避要全表遍历，要怎么做？这里有几种方法供开发人员参考。

方法 1：SQL 改写，如图 8-42 所示（同样也适用于 Oracle 和 PostgreSQL）。

```
mysql> select * from d1 where d1.id not in (select d2.id from d2);
+----+------+------+
| id | a    | b    |
+----+------+------+
|  5 |    5 |    5 |
+----+------+------+
1 row in set (0.00 sec)
```

图 8-42　MySQL 的 SQL 改写

其原始数据如图 8-43 所示。

```
mysql> select * from d1;
+----+------+------+
| id | a    | b    |
+----+------+------+
|  1 |    1 |    1 |
|  2 |    2 |    2 |
|  3 |    3 |    3 |
|  4 |    4 |    4 |
|  5 |    5 |    5 |
+----+------+------+
5 rows in set (0.00 sec)

mysql> select * from d2;
+----+------+------+
| id | a    | b    |
+----+------+------+
|  1 |    1 |    1 |
|  2 |    2 |    2 |
|  3 |    3 |    3 |
|  4 |    4 |    4 |
+----+------+------+
4 rows in set (0.00 sec)
```

图 8-43　MySQL 两个表做差集的样本数据

方法 2：使用差值函数 minus。这个函数在 MySQL 数据库上目前不支持，如图 8-44 所示。

```
SQL> select * from d1;

        ID          A          B
---------- ---------- ----------
         1          1          1
         2          2          2
         3          3          3
         4          4          4
         5          5          5
         6          6          6

已选择 6 行。

SQL> select * from d2;

        ID          A          B
---------- ---------- ----------
         1          1          1
         2          2          2
         3          3          3
         4          4          4
         5          5          5

已选择 5 行。
```

图 8-44　Oracle 两个表做差集的样本数据

使用 minus 函数，对两个表进行关联排除，如图 8-45 所示（不是每种数据库都支持这种写法）。

```
SQL> select * from d1  minus select * from d2;

        ID          A          B
---------- ---------- ----------
         6          6          6
```

图 8-45　Oracle 函数改写

总结：

1）N 表关联一定要有关联列，关联列需要 N-1 个，这个容易疏忽。
2）关联列一定要有索引和相同的数据类型。否则会造成关联列无索引或者索引失效。
3）在满足上述条件的情况下，仅仅有一个不等于的关联列也是不行的。可以增加时间范围等其他有效过滤条件，依靠其他条件缩小结果集范围，再进行不等于的筛选。
4）SQL 可以改写但是改写的 SQL 依然是全表扫描。

8.1.5　函数算法——对 SQL 的影响

在使用 SQL 语句进行查询时，函数算法对查询的效率有着重要的影响。下面可以通过实验，探究使用函数算法对查询效率带来的影响。

对于有经验的开发人员来说，一眼就能看出有问题的函数是 SQL 谓词过滤中等号左边进行函数、算术运算或其他表达式运算。这个知识点在各种数据库上都一样，本案例以 MySQL 为实验环境。

在 MySQL 上建立实验环境，实验表结构如图 8-46 所示。

```
mysql> desc ten;
+--------+--------------+------+-----+-------------------+-------------------+
| Field  | Type         | Null | Key | Default           | Extra             |
+--------+--------------+------+-----+-------------------+-------------------+
| id     | int          | NO   | PRI | NULL              | auto_increment    |
| name   | varchar(30)  | YES  |     | NULL              |                   |
| age    | int          | YES  |     | NULL              |                   |
| weight | int          | YES  |     | NULL              |                   |
| height | int          | YES  |     | NULL              |                   |
| time   | timestamp(6) | NO   |     | CURRENT_TIMESTAMP(6) | DEFAULT_GENERATED |
+--------+--------------+------+-----+-------------------+-------------------+
6 rows in set (0.01 sec)
```

图 8-46　函数转换实验表

针对带有索引的主键列 id 进行等号左边带有函数进行查询的情况，执行计划如图 8-47 所示，执行计划中 type 是 ALL。该 SQL 最终会是一个全表扫描的过程。

```
mysql> explain select * from ten where cast(id as char)='1';
+----+-------------+-------+------------+------+---------------+------+---------+------+--------+----------+-------------+
| id | select_type | table | partitions | type | possible_keys | key  | key_len | ref  | rows   | filtered | Extra       |
+----+-------------+-------+------------+------+---------------+------+---------+------+--------+----------+-------------+
|  1 | SIMPLE      | ten   | NULL       | ALL  | NULL          | NULL | NULL    | NULL | 206406 |   100.00 | Using where |
+----+-------------+-------+------------+------+---------------+------+---------+------+--------+----------+-------------+
1 row in set, 1 warning (0.00 sec)
```

图 8-47　谓词列上做类型转换

而对于等号左边带有运算的查询，执行计划如图 8-48 所示，执行计划中 type 是 ALL。该 SQL 最终会是一个全表扫描的过程。

```
mysql> explain select * from ten where id/2=1;
+----+-------------+-------+------------+------+---------------+------+---------+------+--------+----------+-------------+
| id | select_type | table | partitions | type | possible_keys | key  | key_len | ref  | rows   | filtered | Extra       |
+----+-------------+-------+------------+------+---------------+------+---------+------+--------+----------+-------------+
|  1 | SIMPLE      | ten   | NULL       | ALL  | NULL          | NULL | NULL    | NULL | 206406 |   100.00 | Using where |
+----+-------------+-------+------------+------+---------------+------+---------+------+--------+----------+-------------+
1 row in set, 1 warning (0.00 sec)
```

图 8-48　谓词列上做除法

可以看到上面两种查询都是没有用到索引的，在大量数据查询的场景下进行全表查询会导致查询效率低下，即使是主键索引也无法避免。

这个道理其实也是数学原理。在索引章节已经讲过索引是顺序的，那么索引存储的数据必然是单调递增的（无论存储的是数值、字符串还是时间），这也是一般的 B*Tree 索引的默认排序方式（即升序索引），如图 8-49 所示。

数值和时间的顺序比较容易理解，字符串的顺序是根据 ASCII 码的顺序进行排列的。或者说索引是按照单调递减的顺序进行存储的,如图 8-50 所示。

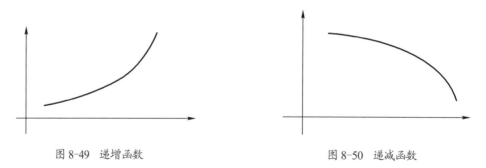

图 8-49 递增函数　　　　　　　　　图 8-50 递减函数

索引保存的是索引字段的原始值,而不是经过函数计算后的值。如果类似在图 8-48 谓词列上做除法的做法就是等于把谓词原始逻辑地址变了。等于图 8-49 递增和图 8-50 递减的原始逻辑地址变了。

而如果在谓词的等号另一端做运算,则原始逻辑地址没有改变,是进行了偏移量处理。

针对日期或者时间的函数,在查询前对日期建立了对应索引,希望针对日期查询能使用该索引,如图 8-51 所示。

```
mysql> alter table ten add index IDX_TEN_TIME (time);
Query OK, 0 rows affected (1.79 sec)
Records: 0  Duplicates: 0  Warnings: 0
```

图 8-51 为时间列建立索引

当使用函数进行查询时,想要看看 2023 年 4 月 1 日的数据,使用函数查询的执行计划如图 8-52 所示。执行计划的 type 列是 ALL,则该 SQL 将会是全表扫描。

```
mysql> explain select * from ten where year(time)=2023 and month(time)=04 and day(time)=01;
+----+-------------+-------+------------+------+---------------+------+---------+------+--------+----------+-------------+
| id | select_type | table | partitions | type | possible_keys | key  | key_len | ref  | rows   | filtered | Extra       |
+----+-------------+-------+------------+------+---------------+------+---------+------+--------+----------+-------------+
|  1 | SIMPLE      | ten   | NULL       | ALL  | NULL          | NULL | NULL    | NULL | 206406 |   100.00 | Using where |
+----+-------------+-------+------------+------+---------------+------+---------+------+--------+----------+-------------+
1 row in set, 1 warning (0.00 sec)
```

图 8-52 时间列使用函数转换

从上可以看到即使建立了索引,在谓词列使用函数的情况下,仍然没有用到这个索引。那应该怎么做?正确的时间范围查询写法如图 8-53 所示。

```
mysql> explain select * from ten where time>='2023-04-01 00:00:00' and time <'2023-04-02 00:00:00';
+----+-------------+-------+------------+-------+---------------+-------------+---------+------+------+----------+-----------------------+
| id | select_type | table | partitions | type  | possible_keys | key         | key_len | ref  | rows | filtered | Extra                 |
+----+-------------+-------+------------+-------+---------------+-------------+---------+------+------+----------+-----------------------+
|  1 | SIMPLE      | ten   | NULL       | range | IDX_TEN_TIME  | IDX_TEN_TIME| 7       | NULL |    1 |   100.00 | Using index condition |
+----+-------------+-------+------------+-------+---------------+-------------+---------+------+------+----------+-----------------------+
1 row in set, 1 warning (0.01 sec)
```

图 8-53 正确的时间范围查询写法

查询语句朴素但有效,看起来很直观。用范围限制查询时间,也如预期般使用了索引。

同样在生产中常常会看到对当前日期计算后进行比较的场景。举例：查询一个月前的数据。常常能见到两种写法，分别看看两种写法的执行计划，如图 8-54 所示。

```
mysql> explain select * from ten where DATE_ADD(time, INTERVAL 1 MONTH) = now();
+----+-------------+-------+------------+------+---------------+------+---------+------+--------+----------+-------------+
| id | select_type | table | partitions | type | possible_keys | key  | key_len | ref  | rows   | filtered | Extra       |
+----+-------------+-------+------------+------+---------------+------+---------+------+--------+----------+-------------+
|  1 | SIMPLE      | ten   | NULL       | ALL  | NULL          | NULL | NULL    | NULL | 206406 |   100.00 | Using where |
+----+-------------+-------+------------+------+---------------+------+---------+------+--------+----------+-------------+
1 row in set, 1 warning (0.00 sec)

mysql> explain select * from ten where time = DATE_SUB(NOW(),INTERVAL 1 MONTH );
+----+-------------+-------+------------+------+---------------+-------------+---------+-------+------+----------+-------+
| id | select_type | table | partitions | type | possible_keys | key         | key_len | ref   | rows | filtered | Extra |
+----+-------------+-------+------------+------+---------------+-------------+---------+-------+------+----------+-------+
|  1 | SIMPLE      | ten   | NULL       | ref  | IDX_TEN_TIME  | IDX_TEN_TIME| 7       | const |    1 |   100.00 | NULL  |
+----+-------------+-------+------------+------+---------------+-------------+---------+-------+------+----------+-------+
1 row in set, 1 warning (0.00 sec)
```

图 8-54　两种时间比较写法的执行计划对比

这两种写法能够得到相同的查询结果，但是执行效率完全不同。函数在谓词等号左边的查询，和预期一样没有使用到该索引。如今依然有很多开发人员并不知道这样做的后果是什么，所以许多公司都存在这种问题。

总结：数据库是基于物理学和数学建立起来的基础学科，而数学也是需要一定的逻辑的。所以如果要做数据库的设计和优化，必须有较好的逻辑和数学基础。如果欠缺的话，就要把这些补起来。

逻辑和算法在数据库的计算和处理中起着非常重要的作用。设计出巧妙合理的方案来操作数据库可以极大地提高应用程序的效率。

8.2　动态规划法在数据库中的应用

动态规划（Dynamic Programming）是一种基于数学原理的问题解决方法，可以用于解决各种数据库的优化问题。动态规划的核心思想是将问题分解为子问题，并利用最优子结构的特性来找到整体问题的最佳解决方案，为数据库提供了理论依据和解决方法。

动态规划不仅在数据库领域有用，也广泛应用于其他领域。我们可以用一些简单的例子来说明它的工作原理，比如爬楼梯问题和背包问题。这些问题看似不相关，但它们都可以用动态规划来解决。在这一节中，我们将详细探讨这两个经典案例，并说明它们在数据库中的应用。

8.2.1　动态规划原理 1：爬楼梯

爬楼梯问题是一个非常简单且经典的动态规划问题。最简单的爬楼梯问题描述如下：小坤正在爬楼梯，他可以选择每次爬 1 层或者 2 层台阶。如果楼梯有 5 层台阶，那么小坤爬到楼顶一共有多少种不同的方法？小坤可以选择的爬楼梯的方式如图 8-55 所示。

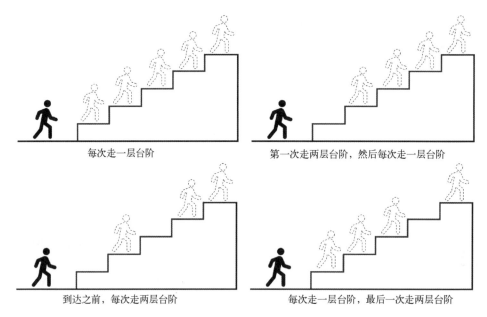

图 8-55 小坤可以选择的爬楼梯的方式

这个问题不难求解,我们使用穷举即可。一步一步往上爬,一共有 8 种不同的方式:

① 1 + 1 + 1 + 1 + 1
② 1 + 1 + 1 + 2
③ 1 + 1 + 2 + 1
④ 1 + 2 + 1 + 1
⑤ 2 + 1 + 1 + 1
⑥ 1 + 2 + 2
⑦ 2 + 1 + 2
⑧ 2 + 2 + 1

然而,当问题进一步复杂化,如果小坤是在爬泰山呢?每次爬 1~2 层台阶,有多少种方式到达山顶?要知道泰山共有 6400 层台阶,这样的穷举方式显然远远不能支持我们的要求了。

在此,需要将该问题转化为数学问题:假设总共有 n 层台阶,小坤每次可以爬 1~2 层台阶,求解小坤爬到楼顶总共有多少种方法。

下面将分别介绍递归、迭代和动态规划三种解法。

(1) 递归法

递归法是最符合人直观理解的方法,它通过将问题划分为子问题来解决。由于小坤每次只能爬 1~2 层台阶,那么到第 n 层台阶,必须要经过第 n-1 或 n-2 层台阶。因此第 n 层台阶的方法数等于爬到第 n-1 层台阶的方法数加上爬到第 n-2 层台阶的方法数。这样就可以使用如图 8-56 所示的递归法深搜暴力代码求解了。

```python
def climbStairsRecursive(n):
    if n == 1:
        return 1
    if n == 2:
        return 2
    return climbStairsRecursive(n-1) + climbStairsRecursive(n-2)

n = 5
num = climbStairsRecursive(n)
print('小坤爬{}级楼梯共有{}种方法'.format(n, num))
```

图 8-56　递归法深搜暴力代码

递归法需要自上而下调用函数本身，速度较慢。它的时间复杂度是指数级的 $O(2^n)$，因为在每个递归步骤中，需要解决两个子问题。而空间复杂度则取决于递归调用的深度，为 $O(n)$。

（2）迭代法

迭代法的核心思想与递归法是一致的，皆是类似于 Fibonacci 数列，使用递归函数 $f(n)=f(n-1)+f(n-2)$，$n>2$。利用原值去求解下一个值，然后由第一个数和第二个数去求解第三个数，以此类推。二者的区别在于执行的顺序，迭代法是从底层向上推导，避免了递归中的重复计算。

- 迭代法：使用循环结构，像做重复的工作一样，逐步接近问题的答案。从问题的起始点出发，每一步都朝着解决问题的方向前进，直到问题解决为止。

如本章的例子，已知 $f(1)=1, f(2)=2$：

$f(3)=f(1)+f(2)=1+2=3$；
$f(4)=f(2)+f(3)=2+3=5$；
$f(5)=f(3)+f(4)=3+5=8$；

- 递归法：使用函数自身的方式，将一个大问题拆分成一个或多个与原问题类似但规模较小的子问题。直到达到基本情况或终止条件，然后一层一层返回结果来解决整个问题。

如本章的例子，已知 $f(1)=1, f(2)=2$：

$$\begin{aligned}f(5)&=f(3)+f(4)\\&=f(1)+f(2)+f(2)+f(3)\\&=f(1)+f(2)+f(2)+f(1)+f(2)\\&=1+2+2+1+2\\&=8\end{aligned}$$

简而言之，迭代法是通过循环来解决问题，递归法是通过函数的嵌套调用来解决问题，而且通常在递归的情况下需要设定一个基本结束条件，以防止无限递归。不同的问题可能更适合使用其中的一种方法。

迭代法的时间复杂度为 O(n)，因为需要计算 n 层台阶的方法数。空间复杂度为 O(1)，因为只需要常数级别的额外空间来存储前面两个数。

（3）动态规划

动态规划的解法是对迭代法的优化。迭代法通常不需要存储中间结果，每次迭代都使用之前的结果来更新当前状态。动态规划使用表格或数组来存储中间结果，以避免重复计算。这种存储中间结果的特性称为"记忆化"。通常用于解决优化问题，如最短路径、最大值、最小值等。它的优势在于处理问题的重叠子问题性质，因此能够在某些情况下显著提高效率。爬楼梯问题动态规划代码如图 8-57 所示。

```python
def climbStairsDP(n):
    if n == 1:
        return 1
    if n == 2:
        return 2
    dp = [0] * (n + 1) # 创建一个数组来保存中间结果，初始值都为0
    dp[1] = 1
    dp[2] = 2
    # 从第3级楼梯开始，通过循环计算每一级楼梯的方法数
    for i in range(3, n + 1):
        dp[i] = dp[i-1] + dp[i-2]
    return dp[n] # 返回爬到第n级楼梯的方法数

n = 5
num = climbStairsDP(n)
print('小坤爬{}级楼梯共有{}种方法'.format(n, num))
```

图 8-57 爬楼梯问题动态规划代码

基于以上的描述，接下来思考动态规划在数据库的应用。在数据库的底层架构中，爬楼梯的动态规划思想可以应用于优化查询操作的性能。一种常见的例子是使用动态规划来解决关系数据库中的自连接问题，其中需要计算一个表中每一行与其他行之间的关联关系。

案例：假设有一个名为"Employees"的表，包含员工的信息，其中每个员工都有一个唯一的标识符（ID）和上级员工的标识符（ManagerID），用于建立员工之间的层次关系。现在希望查询每个员工及其直接和间接下属的数量。

传统的递归查询方法会导致性能问题，因为每次查询都需要执行多次子查询，而且重复计算了相同的子树。这就是前面递归法中出现的重叠子问题。

使用动态规划思想，可以用迭代的方式构建一个辅助表，其中每个条目表示员工及其直接和间接下属的数量。从最底层的员工开始，逐级计算每个员工的下属数量，然后根据其上级员工的下属数量来更新辅助表中的条目。

下面是一个简化的示例：

创建一个辅助表格名为"EmployeeCount"，包含两列：EmployeeID 和 SubordinateCount。EmployeeID 用于存储员工的标识符，SubordinateCount 用于存储该员工的直接和间接下属数量。接下来，可以执行以下步骤：

- 初始化辅助表格：将所有员工的 SubordinateCount 设为 0；对于没有下属的员工，将其 SubordinateCount 设为 1。
- 从最底层的员工开始，向上级递归计算下属数量：对于每个员工 A，获取其直接下属 B 的标识符。
- 在辅助表格中查找员工 B 的条目，将其 SubordinateCount 加到员工 A 的 SubordinateCount 中。
- 继续向上递归，重复上述步骤直到达到最顶层的员工。

通过这种方式进行数据库设计，避免了重复计算同一员工的下属数量，从而提高了查询操作的性能。在最后的辅助表格中，每个员工的 SubordinateCount 列就存储了其直接和间接下属的数量。

这个例子展示了动态规划思想在数据库底层架构中的应用，通过保存子问题的解，避免了重复计算，提高了查询操作的效率。

8.2.2 动态规划原理 2：背包问题

背包问题是一个经典的组合优化问题，在计算机科学和运筹学中经常被讨论和研究。该问题可以描述为：给定一个固定容量的背包和一组具有不同重量和价值的物品，选择将哪些物品放入背包中，以使得背包中物品的总价值最大化。

背包问题有三种主要的变体，分别是：

① 0-1 背包问题（0-1 Knapsack Problem）：在这个变体中，每个物品要么完整地放入背包中，要么不放入，不能进行分割。也就是说，每个物品只有一个可用的实例。目标是找到一个组合，使得物品的总重量不超过背包容量，并且总价值最大化。

② 无界背包问题（Unbounded Knapsack Problem）：又叫完全背包问题，在这个变体中，每个物品可以无限次地放入背包中。也就是说，每个物品有无限个可用的实例。目标是找到一个组合，使得物品的总重量不超过背包容量，并且总价值最大化。

③ 多重背包问题（Multiple Knapsack Problem）：在这个变体中，每个物品有多个可用的实例，每个实例都有对应的重量和价值。相比于无界背包问题，多重背包问题对每个物品的可用实例数量进行了限制。目标是找到一个组合，使得物品的总重量不超过背包容量，并且总价值最大化。

这些变体都属于 NP 完全问题，意味着在一般情况下，没有已知的有效算法可以在多项式时间内解决它们。因此，常用的解决方法是使用动态规划、贪心算法和启发式算法来近似求解或找到近似最优解。

在本节中，以 0-1 背包为例，展开讲述一下动态规划是如何实现的，以及该思想在数据库中的应用。

1. 0-1 背包问题描述

小坤是个登山爱好者，正在计划一个登山旅行。他有一个背包，这个背包空间容量比较大，但是小坤只能承载 10 公斤的重量。小坤计划带一些物品，每个物品都有自己的重量和对小坤来说的重要性。

　　物品 0：登山靴，重量 2 公斤，价值 4。

　　物品 1：登山杖，重量 3 公斤，价值 5。

　　物品 2：帐篷，重量 4 公斤，价值 8。

　　物品 3：睡袋，重量 5 公斤，价值 9。

　　小坤的目标是在保持背包重量不超过限制的情况下，选择一些物品放入背包，使得登山体验最好。这就是 0-1 背包问题，在制订登山计划时智慧地选择装备，以在有限的背包容量下找到最佳组合，以最大化小坤的满意度。

　　小坤思虑再三，始终犹豫不决。

　　① 小坤决定不带任何装备。这样他的背包是空的，重量是 0 公斤，重要性也是 0。这个选择不错，因为很轻松，但登山可能会有些困难。

　　② 小坤决定只带登山靴。这个选择会增加他的重量，背包变成 2 公斤，但重要性是 4。这是一个很好的开始，因为他现在有了一些必要的装备。

　　③ 小坤决定带登山靴和登山杖。这会让他的背包更重，达到 5 公斤，但重要性增加到 9。这个选择让他更有信心，因为有了登山杖，他可以更好地行走。

　　④ 小坤决定带登山靴、登山杖和帐篷。这使得背包更重，达到 9 公斤，但重要性为 17。现在，他有了一个安全的住所，可以在山上露营。

　　⑤ 小坤决定带所有装备，包括登山靴、登山杖、帐篷和睡袋。这将使背包重达 14 公斤，但总重要性为 26。这是最完备的选择，但这个重量已经超过了小坤承受的限制，不可取。

　　⑥ 小坤继续思考其他的组合。

　　小坤可以选择上述任何一种方式，也可以在其中做一些组合。他的目标是在保持背包不超重的情况下，选择装备，使登山经验变得最好。这就是 0-1 背包问题的穷举方式。最终，通过尝试不同的装备组合，小坤发现，带上登山靴、登山杖和睡袋，这将使背包刚好为 10 公斤，总重要性为 18。这样他将会获得最好的登山体验。

　　穷举法在处理小规模的问题时是可行的，但随着物品数量的增加，组合的数量将成指数级增长，导致计算复杂度急剧增加。在实际问题中，背包问题的物品数量通常是非常大的，穷举法很难在合理的时间内得到结果，因此需要使用更高效的方法（如动态规划）来求解 0-1 背包问题。动态规划利用了问题的重叠子问题和最优子结构性质，避免了重复计算，可以在多项式时间内高效地得到最优解。

2．动态规划

　　假如使用动态规划来解决这个 0-1 背包问题。需要构建一个二维数组 dp，可以认为这是一个特殊的表格，这个表格有两个维度，每个方格都有两个坐标，横向的坐标是 i，纵向的是 j。现在，i 表示物品的编号，比如第 1 个物品、第 2 个物品，以此类推。j 表示容量，也就是背包能承受的最大重量，比如 4 公斤、5 公斤等。

　　接下来，需要填充这个二维数组 dp。如表 8-1 所示的背包问题的 dp 空白表格见表 8-1。

表 8-1 背包问题的 dp 空白表格

	重 0	重 1	重 2	重 3	重 4	重 5	重 6	重 7	重 8	重 9	重 10
物品 0											
物品 1											
物品 2											
物品 3											

从左上角开始，填充表格。每个单元格的值表示在当前背包容量下能够获得的最大总价值。小坤只能选择每个物品一次（0 或 1），所以只有两种选择。从上到下、从左到右填充表格，使用以下规则：

- 如果当前物品的重量大于当前背包容量，那么无法放入，直接继承上一行的值。
- 否则，选择放入当前物品（价值加上上一行同列的值）或者不放入当前物品（继承上一行的值），选择较大的那个。
- 最后一个单元格（右下角）中的值就是最大的总价值。

从第一个物品开始考虑（本例中有 4 个物品，纵向编号从 0 到 3 表示这 4 个物品。而横向编号 0~10 表示具体的重量值），对于第一个物品（纵向编号是 0 的这一行），只要背包容量大于或等于它的重量 2（公斤），就可以将它放入背包中，并且在 dp[0][j]中记录下它的价值。

dp 后面第一个代表物品编号，第二个代表上限。如 dp[1][2]的含义为背包上限是 2，只能从物品 0 和物品 1 中选择。所以 dp[0][2]=4 即白底表格中第一行第三列的 4，这里的 4 是价值，不是重量单位（因为只要第一个物品放进去了，价值就是 4。而 dp[0][0]和 dp[0][1]根本不足以放任何一个物品，所以价值是 0。而无论上限是多少，如果只放物品 0 进去，那么价值只有 4，所以这一行后面的价值都是 4），这样第一排就填好了。单一物品背包 dp 表格见表 8-2。

表 8-2 单一物品背包 dp 表格

	重 0	重 1	重 2	重 3	重 4	重 5	重 6	重 7	重 8	重 9	重 10
物品 0	0	0	4	4	4	4	4	4	4	4	4
物品 1											
物品 2											
物品 3											

所以，对于物品 1 来说，在上限是 0 和 1 的时候都放不进去，所以 dp[1][0]和 dp[1][1]都是 0。其实 dp[1][2]也是放不进去的，因为物品 1 重量是 2，但是此时物品 0 是可以放进去的，所以 dp[1][2]价值还是 4，这个价值是物品 0 贡献的。dp[1][3]说明上限是 3，可以在物品 0 和 1 中选 1 个或者 2 个。如果放进去的是物品 0，价值是 4，放进去物品 1 价值是 5。选择价值高的，所以 dp[1][3]=5，背包全部数据表见表 8-3。以上都是基于上限选择了部分放入。

然后继续考虑后面的物品。对于每个物品 i，需要考虑两种情况：放入背包和不放入背包。

表 8-3 背包全部数据表

	重 0	重 1	重 2	重 3	重 4	重 5	重 6	重 7	重 8	重 9	重 10
物品 0	0	0	4	4	4	4	4	4	4	4	4
物品 1	0	0	4	5	5	9	9	9	9	9	9
物品 2	0	0	4	5	8	9	12	13	13	17	17
物品 3	0	0	4	5	8	9	12	13	14	17	18

举例，背包容量为 6（公斤）的时候，且已经放进去了两个物品，遇到第 3 个物品（物品 2）的重量为 4（公斤）的选择，要不要把它放进背包呢？需要做比较权衡。

- 如果放入背包，则 6 公斤-4 公斤，只有 2 公斤的空余，只能选择物品 0 了。物品 0 和物品 2 的价值和是 12。
- 如果不放入背包，则 6 公斤的背包，装了物品 0 和物品 1，共计 5 公斤，空余 1 公斤。物品 0 和物品 1 的价值和是 9。

因此，需要更新表格 dp[2][6]=12。后面的依次类推。

最后，当遍历完所有的物品并且考虑了所有的背包容量情况后，dp[n-1][capacity]中记录的就是在考虑所有物品并且背包容量为 capacity 的情况下，能够获得的最大总价值。背包问题动态规划代码如图 8-58 所示。

```python
def knapsack_01(weight, value, capacity):

    # 二维数组
    dp = [[0] * (capacity + 1) for _ in range(len(weight))]

    # 初始化，只要背包容量超过第一个物品的重量，就可以放入第一个物品
    for j in range(weight[0], capacity + 1):
        dp[0][j] = value[0]

    # weight数组的长度是物品个数
    for i in range(1, len(weight)):  # 遍历物品
        for j in range(capacity + 1):  # 遍历背包容量
            if j < weight[i]:
                dp[i][j] = dp[i - 1][j]
            else:
                dp[i][j] = max(dp[i - 1][j], dp[i - 1][j - weight[i]] + value[i])

    return dp[len(weight) - 1][capacity]
```

图 8-58 背包问题动态规划代码

与穷举法对比：

1）时间复杂度。穷举法的时间复杂度为 $O(2^n)$，随着物品数量的增加，计算复杂度成指数级增长。而动态规划的时间复杂度为 $O(nC)$，其中 n 为物品数量，C 为背包容量。动态规划的时间复杂度较低，可以在多项式时间内解决问题。

2）重复计算。穷举法没有充分利用重叠子问题的性质，会进行大量的重复计算。而动态规划通过存储中间结果，避免了重复计算，提高了计算效率。

3）解空间搜索。穷举法需要搜索所有可能的组合，而动态规划通过状态转移方程，将问题拆解为多个子问题，并逐步填充动态规划表，不需要搜索整个解空间，只需计算特定的子问题。

3．动态规划的应用

在数据库中，类似于 0-1 背包问题的动态规划思想有一些重要的应用，这些应用通常涉及在数据库查询、处理和优化过程中做出最优选择。

（1）查询优化

在数据库查询优化中，经常面临在给定资源限制（如内存、CPU 等）下，如何选择最佳的执行计划。这类问题可以抽象成 0-1 背包问题。每个查询计划可以看作一个物品，它有一个对应的执行代价（如 IO 消耗、CPU 消耗等）和价值（如查询性能）。我们的目标是在资源限制下选择一组查询计划，使得总执行性能最大化。动态规划可以帮助我们找到最优的查询计划组合，以获得最佳性能。

（2）缓存替换策略

在数据库中，常常使用缓存来加速查询和数据访问。然而，缓存的大小是有限的，因此需要选择合适的缓存替换策略。这类问题可以转化为 0-1 背包问题，其中每个缓存页可以视为一个物品，它有一个对应的大小（内存占用）和价值（命中频率或访问频率）。我们的目标是在有限的缓存空间下，选择一组缓存页使得缓存命中率最大化，从而提高查询性能。

（3）资源分配问题

在数据库系统中，有时需要对资源进行合理的分配，如磁盘空间、内存、连接数等。这类问题可以看作 0-1 背包问题，其中每个资源可以看作一个物品，它有一个对应的限制量和价值（资源的重要性或需求程度）。目标是在有限的资源约束下，选择合适的资源分配方案，使得整个系统的性能或利用率最优。

（4）优先级调度问题

在数据库中，一些任务可能有不同的优先级，需要进行合理的调度。这类问题也可以转化为 0-1 背包问题，其中每个任务可以看作一个物品，它有一个对应的优先级（权重）和执行时间。我们的目标是在有限的时间（资源）内，选择合适的任务集合执行，使得任务的权重和最大化。

案例 1：钢材贸易商采购问题

假设有一个钢铁交易平台，其中存储了不同品种的钢材，来自不同供应商发布的资源。每种资源都有不同的品种、规格、采购价、预估销售价等信息。现在有一个钢材贸易商，在搜索栏中输入了一组关键词条件：

钢材品种为"冷轧卷板"；

钢材厚度为"2.5～3.5mm"；

宽度为"800～1200mm"；

价格范围为"3000～5000元"。

该贸易商希望以100万的预算，采购满足上述条件的一批钢材，使得这组钢材总的预估销售价格最大。

该电商平台的数据库中存储了成千上万个供应商发布的钢材信息，每个供应商可能有多个品种和规格的钢材，钢材数量更是非常庞大，直接进行全表扫描来找到满足用户关键词条件的钢材组合是非常耗时的。

为了优化查询性能，我们可以采用动态规划的0-1背包算法来解决这个问题。我们可以将每个符合用户输入的关键词的钢材视为一个物品，背包的容量是100万，它对应的预估销售价作为物品的价值。步骤如下：

1）根据用户输入的关键词条件，在数据库中进行索引查找，筛选出满足关键词条件的钢材信息。这一步可以利用数据库的索引来加速查询，减少不必要的数据读取。

2）构建动态规划表：创建一个二维数组dp，其中dp[i][j]表示在前i个钢材中，选择总采购价不超过j的情况下，钢材的销售价格最大值。

3）遍历满足关键词条件的钢材信息：根据钢材的采购价和预估销售价更新dp数组的值。具体地，对于第i个钢材，如果当前背包容量j大于或等于钢材的采购价，我们可以选择将该钢材放入背包，然后更新dp[i][j]为前一个状态dp[i-1][j-钢材采购价]+当前钢材的销售价，与前一个状态dp[i-1][j]取最大值；如果当前背包容量j小于钢材的采购价，则dp[i][j]保持不变，即dp[i][j]=dp[i-1][j]。

4）找到最大销售价格：在dp数组中找到最大的值，即为所需的结果。

通过动态规划的0-1背包算法，结合数据库索引和数据筛选，可以在较短的时间内找到满足用户关键词条件的一组钢材，使得这组钢材的销售价格最大，同时不超过钢材贸易商的采购预算。这样可以大大提高查询性能，节省数据库查询时间，同时优化钢材贸易的囤货决策，满足客户需求，最大化销售收益。

案例2：电商广告投放问题

假设某电商网站有一个广告投放系统，广告商可以选择在网站的不同位置投放广告。每个广告位都有不同的曝光量和点击率，不同广告的投放会消耗一定的广告预算。广告商希望在有限的广告预算内，获得最大化的广告点击量，从而获得更好的广告效果。此时，查询优化的目标是帮助广告商选择合适的广告投放方案，以获得最大化的广告点击量，并在预算范围内实现最佳的广告投放策略。

我们可以采用动态规划的思想来解决这个问题。将每个广告位视为一个物品，它有一个对应的花费（广告预算）和收益（广告点击量）。为了应用动态规划解决这个问题，我们可以按照以下步骤进行。

1）构建动态规划表：创建一个二维数组dp[i][j]，其中i表示前i个广告位，j表示当前广告预算限制。dp[i][j]表示在前i个广告位中，在广告预算限制为j的情况下，能够获得的最大广告点击量。

2）状态转移方程：根据当前广告位的花费和收益，可以得到状态转移方程为
dp[i][j]=max(dp[i-1][j],dp[i-1][j-cost[i]]+clicks[i])

其中，cost[i]表示第 i 个广告位的花费（广告预算），clicks[i]表示第 i 个广告位的收益（广告点击量）。

3）填充动态规划表：从第一个广告位开始，逐步填充动态规划表。遍历每个广告位和每个广告预算限制，根据状态转移方程更新 dp[i][j]的值。

4）找到最优解：在填充完动态规划表后，最终的最优解将位于 dp[n][budget]，其中 n 表示广告位的数量，budget 表示广告商的预算限制。这个值即为最大广告点击量。

通过动态规划的优化，广告商可以得到最佳的广告投放策略，在预算范围内实现广告点击量最大化，从而提高广告效果和回报率。这个案例充分体现了动态规划思想，通过子问题的最优解推导出整体的最优解，解决了查询优化问题。

8.2.3 动态规划的五部曲

动态规划的五部曲是指解决动态规划问题时常用的五个步骤。这些步骤有助于我们在理解问题、设计算法以及编写代码时更加清晰地思考。下面是动态规划的五部曲。

1）定义状态：首先要明确问题的状态，即用一些变量来表示问题的不同维度或特征。状态的定义需要能够简洁地描述问题的子问题，以便在后续步骤中进行处理。

2）确定状态转移方程：状态转移方程是动态规划问题的核心，它描述了问题状态之间的转移关系。通过状态转移方程，我们可以从子问题的解推导出更大规模问题的解。状态转移方程通常基于问题的递推关系来定义。

3）初始化：初始化是为了处理边界情况，确保状态转移方程在处理第一个状态或某些特殊状态时能够正确运行。通过初始化，我们可以将问题的边界情况转化为基本的子问题，为后续计算打下基础。

4）递推或填表：根据已知的初始状态和状态转移方程，我们可以使用递推或填表的方式来计算所有可能状态的值。这是动态规划的重要步骤，通过计算得到的值将用于解决最终问题。

5）求解最优解：在计算得到所有状态的值后，我们可以根据问题的要求，从中找出最优解。最优解可能是最大值、最小值或其他特定的目标函数值，取决于问题的具体要求。

接下来采用动态规划的五部曲来解决以上述贸易商采购问题。

（1）定义状态

dp[i][j]表示在前 i 个钢材中选择总采购价不超过 j 的情况下，钢材的预估销售价格最大值。其中，i 表示钢材的数量，j 表示当前背包的容量（采购价预算）。

（2）确定状态转移方程

对于每个钢材 i，它有两种选择：选或者不选。

- 如果选择第 i 个钢材（前提是，当 j≥钢材 i 的采购价，否则背包容量不足，不能选该钢材），则

dp[i][j]=dp[i-1][j-钢材 i 的采购价]+钢材 i 的预估销售价

- 如果不选择第 i 个钢材，则

dp[i][j]=dp[i-1][j]

综合以上两种情况，状态转移方程可以表示为

dp[i][j]=max(dp[i-1][j],dp[i-1][j-钢材 i 的采购价]+钢材 i 的预估销售价)

（3）初始化

将 dp 数组初始化为 0，表示在没有任何钢材可选的情况下，预估销售价格为 0，即 dp[i][0]=dp[0][j]=0

（4）递推或填表

根据状态转移方程，从前往后依次计算 dp 数组中的每个值，填充表格。依次遍历所有钢材和采购价预算的组合，计算出在满足用户关键词条件下，选择一组钢材时的预估销售价格。

（5）求解最优解

在 dp 数组填表完成后，可以从 dp[n][100 万]（n 为钢材数量）处找到预估销售价格的最大值，即为所需的结果。

通过动态规划的五部曲，我们可以在给定的关键词条件下，智能地选择钢材供应，以最大化钢材的预估销售价格，提供更优质的服务和产品，同时优化成本和利润，满足采购预算。这对于钢材贸易商来说是一个关键的优化问题，动态规划为其提供了一种高级、有效的优化方法。

8.3 数据库开发中的逻辑思维

数据库开发的逻辑是指，在设计和实现 SQL 的编写上采用最优解，具体逻辑思维如下。

8.3.1 元数据的空与非空

首先一个问题 NULL 是什么？可能大家会说就是空。如果这样简单可能这里就不用提及了，因为答案是 NULL，而不是空。由于不同数据库在优化器上处理的不一样，可能要分开说。但是最起码的逻辑还是都一致的，那就是 NULL 不是空。

首先看一下 Oracle 对于 NULL 值存储的处理方式，给表中插入空值数据如图 8-59 所示。

```
SQL> select * from pp;

        ID          A
---------- ----------
         1          0
         3          1

SQL> insert into pp values (2,null);

已创建 1 行。
```

图 8-59　给表中插入空值数据

下面要做一个说明，以此来说明 NULL 和""的关系。这里的""是空字符串的意思。给 pp

这个表继续追加一条记录，如图 8-60 所示。

```
SQL> insert into pp values (4,'');

已创建 1 行。

SQL> commit;

提交完成。

SQL> select * from pp;

        ID          A
---------- ----------
         1          0
         2
         4
         3          1
```

图 8-60　给表中写入空字符串

最终的实验数据就是 4 条。查询 A 不等于 0 的数据，如图 8-61 所示。

```
SQL> select * from pp where a!=0;

        ID          A
---------- ----------
         3          1
```

图 8-61　查询 A 不等于 0 的数据

从理论上 A 不等于 0 的数据应该是 ID 为 2、3、4。但是实际情况大相径庭。继续实验，对 A 列求和以及对 A 列求平均，如图 8-62 所示，A 列的和是 0+1=1，平均是 (1+0)/2。

```
SQL> select sum(a) from pp;

    SUM(A)
----------
         1

SQL> select avg(a) from pp;

    AVG(A)
----------
        .5
```

图 8-62　对 A 列求和以及对 A 列求平均

对于求和是 1，没有争议。但是对于平均似乎又觉得为什么不是除以 4？其实很好理解，求和都没有带 NULL，求平均值也不会带 NULL 了。那么对 A 列进行运算呢？对 A 列进行加运算再求平均如图 8-63 所示。

通过实验可见，NULL 和空字符串都没有参与到运算中来。而且空字符串在 Oracle 中认为就是 NULL。那么为什么所谓的空（也就是 NULL）没有参与运算？这就是回答最开始的问题了，因为在数据库中 NULL 不是空，而是不确定。

```
SQL> update pp set a=a+10;

已更新 4 行。

SQL> commit;

提交完成。

SQL> select * from pp;

        ID          A
---------- ----------
         1         10
         2
         4
         3         11

SQL> select avg(a) from pp;

    AVG(A)
----------
      10.5
```

图 8-63　对 A 列进行加运算再求平均

明白了这个道理以后便可以向广大开发人员以及很多领导解释一下为什么在 Oracle 11g 以前以及 MySQL8 以前加字段是非常重量级的操作，而这两个版本以后的数据库再也不怕加字段了。如图 8-64 所示，在 Excel 中增加一列数据很容易。因为 Excel 本身是行列规矩的表格，又是在内存操作，数据量也不大。

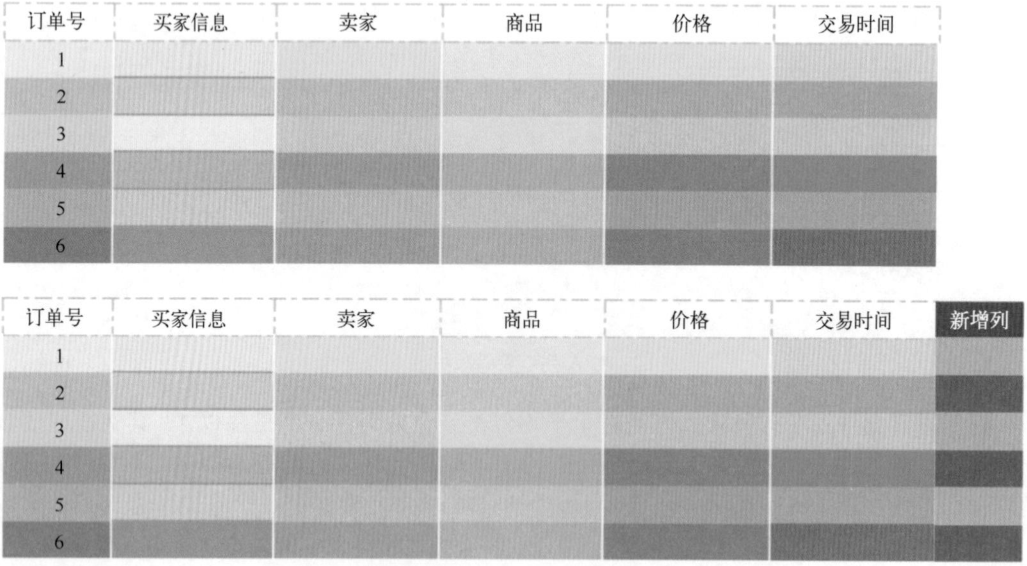

图 8-64　在 Excel 中增加一列数据

而在关系数据库中数据库存储方式是行首尾相连的，如图 8-65 所示。

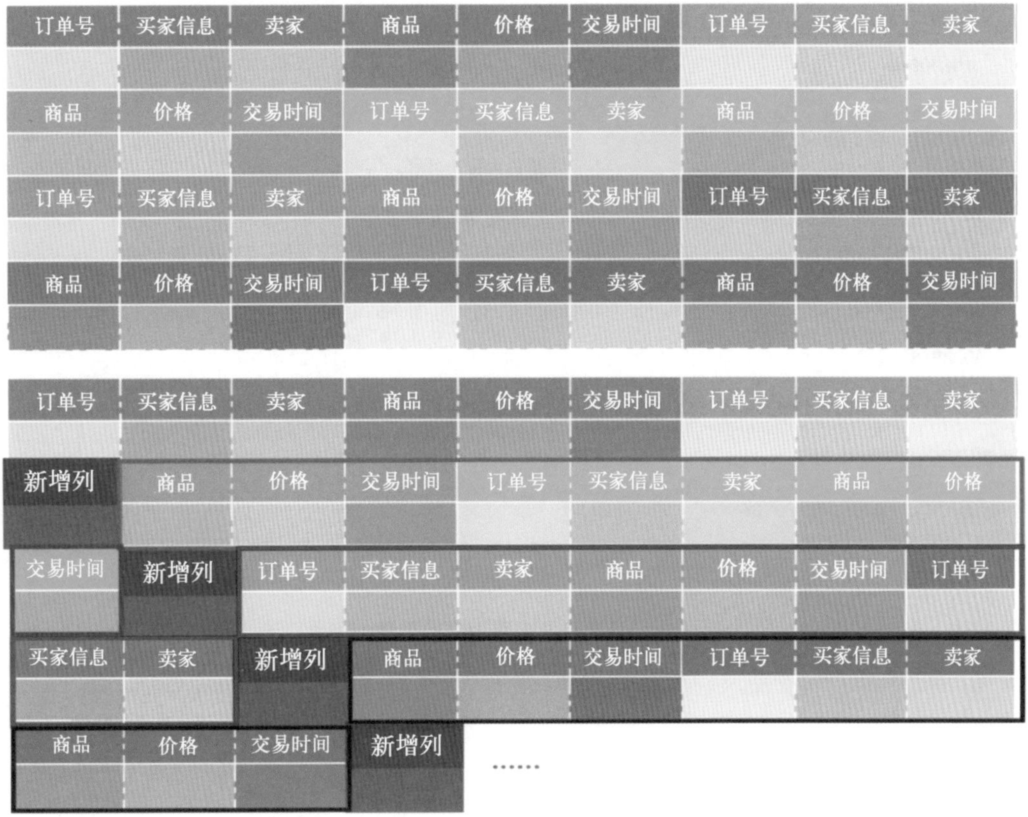

图 8-65　数据库存储方式

如果从前的数据库增加一列，那么黄色的那条记录增加一列，黄色整行向后移动一格。绿色的那条记录自己增加一列，整行向后移动两格。蓝色的那条记录自己增加一列，整行向后移动三格。以此类推，可以说几乎是把整个表做了一次移动，代价和重建一次表差不多。

为了解决这个问题，数据库采用了元数据的处理方法。即加字段的时候，没有给表中的数据进行移动，可以理解只加了表结构，只是每行上的数据其实没有加上去。千万级别的表进行加字段如图 8-66 所示，用 60ms 完成了变更。

最终的效果是，表结构变了，但是数据没加上去，而查询的那一刻才把字段的内容放进去。读取数据为实体数据+元数据如图 8-67 所示，这里的 TEL 列是新加的，这里特意用了"空"，而不是 NULL，因为空是业务逻辑上的空，而不是数据库上的不确定。

```
SQL> select count(*) from thousand;

  COUNT(*)
----------
  10000000

已用时间: 00: 00: 00.97
SQL> desc thousand;
 名称                                      是否为空？ 类型
 ----------------------------------------- -------- ----------------------------
 ID                                        NOT NULL NUMBER(20)
 NAME                                               VARCHAR2(30)
 AGE                                                NUMBER(3)
 HEIGHT                                             NUMBER(4)
 WEIGHT                                             NUMBER(4)
 TIME                                               TIMESTAMP(6)
 ADDRESS                                   NOT NULL VARCHAR2(20)

SQL> alter table thousand add tel varchar2(20) default '空' not null;

表已更改。

已用时间: 00: 00: 00.06
```

图 8-66　千万级别的表进行加字段

```
SQL> desc thousand;
 名称                                                是否为空？ 类型
 --------------------------------------------------- -------- ----------------------------
 ID                                                  NOT NULL NUMBER(20)
 NAME                                                         VARCHAR2(30)
 AGE                                                          NUMBER(3)
 HEIGHT                                                       NUMBER(4)
 WEIGHT                                                       NUMBER(4)
 TIME                                                         TIMESTAMP(6)
 ADDRESS                                             NOT NULL VARCHAR2(20)
 TEL                                                 NOT NULL VARCHAR2(20)


SQL> select * from thousand FETCH FIRST 5 ROWS ONLY;

  ID NAME      AGE HEIGHT WEIGHT TIME                              ADDRESS    TEL
----- -------- ---- ------ ------ -------------------------------- ---------- -----
  746 hO0Bq1    18    150    180 26-4月 -23 04.40.32.853752 下午    无          空
  747 Hx6ZIo   18    150    180 26-4月 -23 04.40.32.853752 下午    无          空
  748 pXFrIh   18    150    180 26-4月 -23 04.40.32.853752 下午    无          空
  749 7ig3g0   18    150    180 26-4月 -23 04.40.32.853752 下午    无          空
  750 QiAu6G   18    150    180 26-4月 -23 04.40.32.853752 下午    无          空
```

图 8-67　读取数据为实体数据+元数据

当年 Oracle 要有默认值且字段不为 NULL 的原因就是，首先排除了不确定。其次既然不是 NULL，而且又有默认值，那么查询的时候就显示默认值。

如果永远不查询,那么这个字段不会被观测,也就不需要赋值。这个是在观测时才进行赋值。讲到这里,开发人员说难道是这是量子力学吗?其实这就是量子力学的理念。在 Oracle 11g 的时候做到了,而后来 MySQL8 也借鉴了这个思路。

现如今的数据库增加字段已经不需要 NOT NULL 和 DEFAULT 的加持也能实现快速加字段了。主要是数据库用逻辑解决了很多从前比较棘手或难度较大的问题。一般来说,Oracle 数据库中 where 条件的谓词对于 IS NULL 是无法用到索引的,因为 NULL 根本不存在于索引之中。如图 8-68 所示,Oracle 中谓词为空的判断导致全表扫描。

```
SQL> desc car;
名称                                      是否为空?  类型
--------------------------------------- -------- ----------------
ID                                                NUMBER(38)
CARNO                                             VARCHAR2(10)
T                                                 DATE

SQL> explain plan for select * from car where carno is null;

已解释。

SQL> SELECT * FROM TABLE(DBMS_XPLAN.DISPLAY);

PLAN_TABLE_OUTPUT
--------------------------------------------------------------------------------
Plan hash value: 704907121

--------------------------------------------------------------------------------
| Id  | Operation          | Name | Rows  | Bytes | Cost (%CPU)| Time     |
--------------------------------------------------------------------------------
|   0 | SELECT STATEMENT   |      |     1 |    22 |     0   (0)|          |
|*  1 |  FILTER            |      |       |       |            |          |
|   2 |   TABLE ACCESS FULL| CAR  |  110K |  2363K|   137   (1)| 00:00:01 |
--------------------------------------------------------------------------------
```

图 8-68 Oracle 中谓词为空的判断导致全表扫描

此时将谓词的字段设置为非空属性,再次检查 SQL 执行计划,Oracle 谓词属性非空的执行计划如图 8-69 所示。

```
SQL> alter table car modify carno varchar2(10) not null;


SQL> explain plan for select * from car where carno is null;

已解释。

SQL> SELECT * FROM TABLE(DBMS_XPLAN.DISPLAY);

PLAN_TABLE_OUTPUT
--------------------------------------------------------------------------------
Plan hash value: 704907121

--------------------------------------------------------------------------------
| Id  | Operation          | Name | Rows  | Bytes | Cost (%CPU)| Time     |
--------------------------------------------------------------------------------
|   0 | SELECT STATEMENT   |      |     1 |    22 |     0   (0)|          |
|*  1 |  FILTER            |      |       |       |            |          |
|   2 |   TABLE ACCESS FULL| CAR  |  110K |  2363K|   137   (1)| 00:00:01 |
--------------------------------------------------------------------------------
```

图 8-69 Oracle 谓词属性非空的执行计划

对比起来没什么不同。真实执行非空约束的空值判断，如图 8-70 所示，虽然执行计划显示全表。但是实际执行却没有因为 is null 而触发全表扫描，反而是几乎没有任何资源消耗。物理读和逻辑读都是 0。

```
SQL> desc car;
 名称                                      是否为空? 类型
 ----------------------------------------- -------- ----------------------------
 ID                                                 NUMBER(38)
 CARNO                                     NOT NULL VARCHAR2(10)
 T                                                  DATE

SQL> set timing on;
SQL> select * from car where carno is null;

未选定行

已用时间:  00: 00: 00.01

执行计划
----------------------------------------------------------
Plan hash value: 704907121

---------------------------------------------------------------------------
| Id  | Operation          | Name | Rows  | Bytes | Cost (%CPU)| Time     |
---------------------------------------------------------------------------
|   0 | SELECT STATEMENT   |      |     1 |    22 |     0   (0)|          |
|*  1 |  FILTER            |      |       |       |            |          |
|   2 |   TABLE ACCESS FULL| CAR  |  110K |  2363K|   137   (1)| 00:00:01 |
---------------------------------------------------------------------------

Predicate Information (identified by operation id):
---------------------------------------------------

   1 - filter(NULL IS NOT NULL)


统计信息
----------------------------------------------------------
          0  recursive calls
          0  db block gets
          0  consistent gets
          0  physical reads
          0  redo size
        500  bytes sent via SQL*Net to client
        586  bytes received via SQL*Net from client
          1  SQL*Net roundtrips to/from client
          0  sorts (memory)
          0  sorts (disk)
          0  rows processed
```

图 8-70 真实执行非空约束的空值判断

之所以这样，全是因为中间的 filter（NULL IS NOT NULL），从逻辑上直接否定了 SQL 执行的必要性，从而从根本上避免了无用功。

在日常工作中大家也需要有类似的想法和思路。真正的优化不是简单的加索引，而是从逻辑上进行优化。

8.3.2 优化器处理极值和极限：加速查询速度

这里的极值和之前说的 max、min 有些类似，不同的是完全和业务贴近。多年前笔者在公安行业时，在一个省公安厅项目中有这样一个需求。

其背景是这样的：全省有 5000 个以上的车辆道路监控设备，俗称电子警察。这些电子警察或者在路口（监控东、南、西、北四个方向的 N 个车道的每一辆通过的车辆），或者在路段（监控每个车道每一辆通过的车辆），每个电子警察负责 4～8 个车道。由我国当今的交通情况可以得知，一个设备高峰时平均每秒产生 4～5 条数据。而在夜间，一分钟一条数据也是正常的。省公安厅为了监控每个设备的正常运行状况，希望通过产生的数据来判断每个采集设备是不是正常工作。其判断逻辑是，30min 内如果一条数据也没有（几乎不可能），那么就将这些无数据的设备记录为疑似故障设备。每天会检查有哪些设备故障，安排人员检修。

从需求上讲这是合理的，给出的设计逻辑也没有太大问题。

开发人员拿到这个需求立即就开始做了，做好了以后就开始通过查找数据库中的实时车辆数据，去巡检这些设备。当开发人员的这个功能开启的时候，笔者就发现整个数据库运行的负荷上升了。不过好在 Oracle 强大的处理能力并没有让数据库瘫痪。过了一段时间，开发找到笔者说，刚做的功能无法达到预期效果，需要优化。

开发人员描述，这个功能计划每半个小时执行一次，而实际情况是执行一次不止半个小时。也就是说，按照他现在的做法，在规定时间内根本无法完成。然后下一次定时巡检就会再次发起，导致永远无法完成。而且后面发起的巡检可能对数据库产生叠加的冲击。

开发人员具体的 SQL 逻辑是 SELECT COUNT(*) FROM TXCL WHERE TGSJ>SYSDATE-30/1440 AND TGSJ<SYSDATE AND SBBH IN (SELECT SBBH FROM SSSB);

对以上 SQL 语句进行解释。

- TXCL：通行车辆表。每过一辆车记录一条数据。一般来说这个表是几十亿级别的。
- TGSJ：通过时间。代表每条记录产生的时间（也就是抓拍到车辆经过采集点的时间）。
- SYSDATE -30/1440：从现在往前的 30min（一天是 1440min）。即统计任务开始前半小时。
- SYSDATE：现在时刻。即统计任务开始的时间。
- SBBH：设备编号。每条记录都会有一个字段记录其来自哪个采集设备。
- 子查询中的 SSSB：设施设备。全省所有采集设备的目录。

以上的语句翻译一下就是在有几十亿数据的通行车辆表中，用设备表中 5000 个设备依次（等于循环 5000 次）去查一下半个小时内有没有数据。半小时的数据量大约为 200 万～400 万。

通过开发人员的描述，相信读者已经大致知道了，这样的运算量确实有点大。而且这个数据库一边要处理每秒海量并发的写入，一边要处理这样的一个统计分析任务。可能有人会说，这种应该放在大数据中去做，估计不少人会这样做。事实上，在十几年前就在这个单机的环境下就可以轻松处理这个需求。

首先对需求进行分析，需求明确，而且已经给出了较好的设计。采集设备 30min 内没有数据没有就是设备可能损坏，那么这句话的等价逻辑就是：只要有 1 条数据，设备就没有出问题。这是解题的关键，很遗憾开发人员没有领会到，许多有要使用大数据想法的人可能也没有想到。那么处理 1 条数据很难吗？一点都不难，解题方案有很多。

方案 1：分钟级别统计。30min 有 200 万～400 万条数据，那么 1min 的数据大致是 10 万条。那么可以设计成只查最后 1min 的数据。如果最后 1min 有数据，哪怕只有 1 条数据，那么这个设备就判定是好的。至于前面 29min 是不是有数据都不重要。数据量缩小为原来的 1/30，查询速度自然提升。最后 1min 可以将这 5000 个设备的 95%甚至 100%判定完毕。

如果出现最后 1min 少数设备不在范围内的情况，则还应该做第二步，将这 1min 判定完毕的设备（近似 5000 个）与设施设备（5000 个）进行差集比较。得到的就是最后 1min 还没有数据的设备，这个运算量相比较几百万来说，可以忽略不计了。并将这些设备编号的设备记录在临时表中。之后用这些少量设备去一分钟一分钟地往前查，依次迭代逐步缩小疑似故障设备的数量，直到排除所有设备。当然最不幸的就是循环了 30 次（第 2 次到第 30 次运算量远比第 1 次小）才判断完成。但是这个比起循环 5000 次的实现方式来说，效率提升太多了。而且即使将来 5000 个设备变成了 10000 个设备，方案 1 还是循环 30 次，运算量没有大的变化。而采用开发人员的原始方案就要循环 10000 次，可能几个小时也未必有结果。

方案 1 的实际耗时为 6～11s。因为有的时候循环是多了几次，有的时候可能 1～2 次就结束了，所以是一个区间。粗略估计采用巧妙设计使得效率至少提升了 300 倍。方案 1 是用存储过程实现的，但是许多人对存储过程有点偏见。

偏见 1：不好调试。这是错误的，可以调试。

偏见 2：不好维护。这是错误的，10 行的存储过程比起 Python 都简单，更加别说是 Java 了。如果写几千行的代码，必然不好维护。但是一旦存储过程超过 50 行了，那么必然是设计错了。存储过程一般不要超过 20 行。以上的方案 1 笔者用了十几行就完成了。

偏见 3：不方便管理。这是错误的，按照版本管理，是很好管理的。

整个过程完成后给开发人员进行了讲解和演示，请他按照这个思路进行重构。开发人员表示直接调用我的存储过程。

方案 2：布尔统计。通过需求是通过 1 条数据就可以判定，那么一个设备中半个小时的数据量是 10000 条和 1 条，对于判定的结果来说没有区别。因为需求不是要求精确的数值，而是要求的布尔的逻辑，0 或者 1。换句话说，通过 COUNT 计算设备半小时总数的实现方式和需求不等价，大大扩大了需求范围。这也就是为什么需求是合理的，但是结果存在问题。所以只需要每个设备统计时最后加上 ROWNUM<2。这样的话 5000 个设备只要计算 5000 次就可以得到 5000 个 0 或者 1。

方案 3：极值统计。通过索引的极值可以非常快速地得到数据。即直接通过 SBBH+TGSJ 得到半小时内每个设备的最早时间和最新时间，需要 5000 次的运算。

综上所述，没有增加任何资源，没有使用到任何大数据技术，没有把架构搞得复杂，整个效率就有几百到上千倍的提升。这个案例的经验就是，DBA 一定要去把控需求和设计，这样才能做得好。SQL 越高效，数据库越稳定。

8.3.3 并发处理热点的逻辑：避免锁

在 Oracle 23c 中有一个锁，它借鉴了阿里双十一的经验。其实锁本身就是防止并发出错的。相信大家非常熟悉微信红包的场景，群里的一个红包可能几秒就被几十个人抢完了。对于腾讯是怎么具体实现的不得而知，这里仅有一些数据：

2016 年微信红包峰值为每秒 50 万次。

2017 年微信红包峰值为每秒 98 万次。

在此之后没有公布过峰值数据，后台多少台机器、怎么运作的也是秘密。微信红包的架构和设计不敢妄测。

那么如果所在的企业和公司也想设计一个活动页面或者在企业的 App 中设计一次促销活动，请用户在某一时刻单击按钮获得类似红包的并发场景，该如何去做？是不是又是需要海量的机器，以及复杂的架构把消息队列、缓存等都用上呢？完全不必要，任何一个数据库都可以做到。本案例仅仅展示一种思路，此思路与微信红包无关。

首先建立一个派发红包的表，如图-红包表所示。

在活动开始前执行一万个红包的初始化，如图 8-71 所示。创建一个存储过程，该存储过程产生 1 万个随机金额的红包（当然这里的 1 万个红包的总金额是不能超过活动的总预算的）。这就像在群里发随机红包时输入人数和输入金额的动作。即在活动开始以前几分钟甚至几个小时已经把红包进行了预分配。这是核心设计思路，不是说活动开始并发地进行确定红包总数的操作，这一切在开始之前已经确定了。否则一定会产生并发的锁。

```
create or replace procedure pred
as
m int;
begin
  for i in 1..10000 loop
    SELECT TRUNC (DBMS_RANDOM.VALUE (0, 100)) into m FROM DUAL;
    insert into red values (i,m,trunc(sysdate));
  end loop;
  commit;
end;

SQL> set timing on;
SQL> exec pred;

PL/SQL procedure successfully completed

Executed in 2.215 seconds
```

图 8-71 初始化红包

经过 2s 左右 1 万个红包就已经生成了。初始化的结果如图 8-72 所示，已分配待领取红包金额的前 10 个分别是 23、40、46、80、52、31、1、67 和 89。也就是说，第一个点击的用户红包是 23 元，第二个点击的用户红包是 40 元，以此类推。

```
SQL> select id,money,to_char(time,'YYYY-MM-DD HH24:MI:SS') from red where id<10;

        ID      MONEY TO_CHAR(TIME,'YYYY-
---------- ---------- -------------------
         1         23 2023-06-24 00:00:00
         2         40 2023-06-24 00:00:00
         3         46 2023-06-24 00:00:00
         4         80 2023-06-24 00:00:00
         5         52 2023-06-24 00:00:00
         6         31 2023-06-24 00:00:00
         7          1 2023-06-24 00:00:00
         8         67 2023-06-24 00:00:00
         9         89 2023-06-24 00:00:00

已选择 9 行。

已用时间:  00: 00: 00.00
```

图 8-72 已分配待领取红包金额

剩下的就等活动开始时用户发起点击的动作。模拟抢红包动作如图 8-73 所示，同样，模拟 1 万个用户（以手机号为用户标识），发起点击动作。

```
create or replace procedure red_get
as
m int;
begin
  for i in 1..10000 loop
    SELECT TRUNC (DBMS_RANDOM.VALUE (13000000000,13900000000 )) into m FROM DUAL;
    insert into red values (i,m,trunc(sysdate));
  end loop;
  commit;
end;
/


SQL> exec red_get;

PL/SQL procedure successfully completed


Executed in 3.555 seconds
```

图 8-73 模拟抢红包动作

读者可以看到这是 1 万个 INSERT 的动作，并没有并发的扣红包总金额的 UPDATE 的动作，所以不会产生任何的锁。最终点击用户行为记录如图 8-74 所示，尾号 7736 的用户是第一个，尾号 4014 的用户是第二个，以此类推。

最后将红包生成的顺序和点击红包的顺序关联起来，关联得出最终用户红包金额如图 8-75 所示，用户按照系统预先规划好的生成顺序对应自己在系统中的点击顺序得出最终结果。

```
SQL> select id,tel,to_char(time,'YYYY-MM-DD HH24:MI:SS') from red_person where id<10;

        ID TEL        TO_CHAR(TIME,'YYYY-
---------- ---------- -------------------
         1 137   7736 2023-06-24 00:00:00
         2 136   4014 2023-06-24 00:00:00
         3 132   9887 2023-06-24 00:00:00
         4 133   9895 2023-06-24 00:00:00
         5 136   8643 2023-06-24 00:00:00
         6 132   9473 2023-06-24 00:00:00
         7 134   4423 2023-06-24 00:00:00
         8 138   8280 2023-06-24 00:00:00
         9 136   3633 2023-06-24 00:00:00

已选择 9 行。

已用时间:  00: 00: 00.01
```

图 8-74　最终点击用户行为记录

```
SQL> select a.id,money,b.tel from red a,red_person b where a.id=b.id and b.id<10;

        ID      MONEY TEL
---------- ---------- -----------
         1         23 137   7736
         2         40 136   4014
         3         46 132   9887
         4         80 133   9895
         5         52 136   8643
         6         31 132   9473
         7          1 134   4423
         8         67 138   8280
         9         89 136   3633

已选择 9 行。

已用时间:  00: 00: 00.05
```

图 8-75　关联得出最终用户红包金额

整个过程关键的设计就是预先分配并且固化了生成红包的顺序，而不是点击抢红包的时间才去扣减红包余额，消除了万人并发的锁。该设计思路在 2～3s 内完成上万人的并发场景。后续可以根据最终用户红包金额关联到用户账户的余额进行更新，一般的数据库 1～2s 即可更新完毕。

这个案例的重点是设计实现逻辑，不能采用传统的在一个事务中 UPDATE 并发更新热点的红包总额表，从而造成严重的锁等待，而是应该采用无锁的设计来解决上万并发的问题。

8.3.4　减库存的逻辑：从设计出发防止超卖

有一次开发人员小 G 找到笔者，咨询关于锁的问题。他的需求很简单，减库存时怎么能避免锁提高并发。小 G 想到的是，每次去减库存时先看看还有没有库存，以及还够不够扣除这次的量。比如某商品库存为 100 件，经过一系列的销售，目前只剩下 10 件。小 G 说如果此时一个用户 A 要下单买 12 件,他会先到数据库中查一下商品库存,12>10,那么无法下单。如果用户 A 要下单买 8 件,他也会到数据库中查一下商品库存,8<10,可以下单。但是存在如下问题:

- 每次更新之前都查一次再更新,比直接更新多了一步操作,效率上降低了。但是如果不检查一次,担心最后库存变成负数导致超卖。
- 即使每次查询到有库存,但是不能保证再去更新的时候数据不变。这类似于看到有火车票,等再去支付的时候发现没票了。

听了他的描述,笔者给了小 G 的一个建议。将商品库存的字段属性设置为非负数,类似 MySQL 中的无符号型整数,以及 Oracle 中加上 check>0 的检查。

这样在数据库底层可以保证这个数据不可能出现负数。该设计的好处是可以省去更新前的检查步骤而且也未必能保证检查得准。

可以大胆地去更新应用程序,因为数据库的写入是串行的,即使几个更新并发,最终也是依次减库存,直到某一个更新因为数据不满足字段范围导致失败。应用程序只要捕获异常,将这个错误通知用户,没有库存即可。

小 G 听后觉得有道理,就这样去做了,解决了他的难处。不仅效率提升了,而且开发的逻辑也简化了。最重要的是不会出现问题。

过了一段时间,小 G 愁眉苦脸地再次找到我说,现在店铺允许超卖。我听后觉得那么把字段属性的约束放开就好了。小 G 说:"没那么简单,允许是允许,但是只允许 5 个,多了还是不行。有没有设置大于或等于-5 的方法?"数据库是没有这种方式的,而且这个需求有点不科学了,不过还是有方法的。笔者问:"那么是不是多卖 5 个就可以?"小 G 说:"是的,有办法解决吗?"笔者说:"有!比如你现在库存是 100,那么就把需要超卖 5 个的商品,库存直接定在 105 而不是 100。"小 G 一听,恍然大悟。

至此,这个案例就结束了,在这个问题上小 G 再也没有找过笔者了。从本案例可以得出,最好的优化方式就是设计,一个好的数据库模型设计让开发人员工作量减少很多,而且能保证不出错。而面对有些需求时,进行反向思维。从应对-5 的问题,改成整体平移+5,应用这种数学思维就让问题重新回到了原点。

第 9 章

DBA 的日常：数据库管理及开发的最佳实践

数据库管理及应用开发中有很多最佳实践，示范性和启发性很强，通过这些实践可以解决大家在数据库开发、运维、构建上的许多问题。本章就将集中分享笔者工作中收集和使用的一些实践，供广大读者参考。

本章内容
七个针对数据库特性的最佳实践
面向执行器和优化器的最佳实践
数据库的复制（克隆）与高可用受控切换的最佳实践
三个 SQL 编写的最佳实践
时序数据库使用的最佳实践

9.1 七个针对数据库特性的最佳实践

9.1.1 Oracle 的 DML 重定向

在正式环境中,通常数据库不是单点的,会有高可用的架构来保障。Oracle 的高可用分为双活高可用 RAC 和容灾高可用 ADG。当然这两个可以配合起来使用。本节是针对容灾高可用 ADG 来讲的。

一般来说,我们是主库写入数据,然后传给备库。这样备库也有了和主库一样的变化。这个过程是单向的,因为从库(也就是备库)是只读的,它只能接受主库送过来的数据,不能接受其他的变更。很简单,比如主库现在有 3 条数据:

1A
2B
3C

假设把主库 1 的 A 改成 AA。那么从库也是 1 的 A 改成 AA。

那么试想一下,如果把从库的 1 改成 AAA,那么主库会变吗?首先无法修改,其次如果可以改,那么主库 1A 和从库 1AAA 会出现不一致。在 MySQL 的主从情况下,经常有从库被写入然后数据不一致,在某一时刻导致主从中断的情况。所以 Oracle 和 PostgreSQL 两个数据库的从库(备库)是不可写的。

当然以上都是过去式了,Oracle 从 19c 开始可以修改了。刚才说的推翻了吗?这不是不一致了吗?其实没有推翻,这都是源于一个好的特性设计,就是从库(备库)依然不能被直接写入。从库(备库)在接收到写入命令时,把自己接收到的 SQL,先发给主库,主库执行完毕,再提交给从库(备库)。这样的设计让使用者没有感到异样,又保证了数据的一致性,两全其美。

这种特性设计就叫作 ADG 的重定向或者说 DML 重定向,本节就来实践一下这个特性,并且给出 DML 重定向的最佳实践,简单分析该特性适用于什么场景以及不适合什么场景。

DML 重定向特性由一个参考控制。打开这个 DML 重定向参数,Oracle 的 DML 重定向参数打开如图 9-1 所示。对应的命令如下。

```
SQL> show parameter adg_redirect_dml;

NAME                                 TYPE        VALUE
------------------------------------ ----------- ------------------------------
adg_redirect_dml                     boolean     FALSE

SQL> alter system set adg_redirect_dml = true scope = both;

系统已更改。

SQL> show parameter adg_redirect_dml;

NAME                                 TYPE        VALUE
------------------------------------ ----------- ------------------------------
adg_redirect_dml                     boolean     TRUE
```

图 9-1 打开 Oracle 的 DML 重定向参数

show parameter adg_redirect_dm;——检查参数是否打开

alter system set adg_redirect_dml = true scope = both;——开启参数即时生效,无须重启数据库实例。

在会话级别开启 ADG 的 DML 重定向功能。对备库进行 DML 设置,如图 9-2 所示,在备库的角色为 PHYSICAL STANDBY,并且也可以设置 ADG_REDIRECT_DML,可以看到改动生效了,说明这个参数的开启不用重启数据库实例。

```
SQL> alter session enable ADG_REDIRECT_DML;

会话已更改。

SQL> select open_mode,database_role from v$database;

OPEN_MODE            DATABASE_ROLE
-------------------- ----------------
READ ONLY WITH APPLY PHYSICAL STANDBY
```

图 9-2 备库进行 DML 设置

在从库(备库)上登录一个 PDB,启动 DML 重定向。顺便检查一下现在数据库的角色,说明是在从库上。如果是在主库上,那么角色应该是如图 9-3 所示,展示主库与从库的角色区别。

```
SQL> select open_mode,database_role from v$database;

OPEN_MODE            DATABASE_ROLE
-------------------- ----------------
READ WRITE           PRIMARY
```

图 9-3 展示主库与从库的角色区别

再分别检查一下数据在主库和从库(备库)的状态,主库状态和主库实验数据如图 9-4 所示,从库状态和从库实验数据如图 9-5 所示。此数据为之前自带数据。

```
SQL> select open_mode,database_role from v$database;

OPEN_MODE            DATABASE_ROLE
-------------------- ----------------
READ WRITE           PRIMARY


SQL> select * from t;

        ID A          T
---------- ---------- ------------------
         2 2          2022:08:1615:55:33
         3 3          2022:08:1615:57:12
         1 1          2022:08:1614:16:42
```

图 9-4 主库状态和主库实验数据

```
SQL> select open_mode,database_role from v$database;

OPEN_MODE              DATABASE_ROLE
--------------------   ----------------
READ ONLY WITH APPLY   PHYSICAL STANDBY

SQL> select * from t;

       ID A            T
---------- ----------  ------------------
         2 2           2022:08:1615:55:33
         3 3           2022:08:1615:57:12
         1 1           2022:08:1614:16:42
```

图 9-5 从库状态和从库实验数据

实验开始,在从库上写入数据,如图 9-6 所示。

```
SQL> select open_mode,database_role from v$database;

OPEN_MODE              DATABASE_ROLE
--------------------   ----------------
READ ONLY WITH APPLY   PHYSICAL STANDBY

SQL> insert into t values (4,'4',sysdate);

已创建 1 行。

SQL> commit;

提交完成。

SQL> select * from t;

       ID A            T
---------- ----------  ------------------
         2 2           2022:08:1615:55:33
         3 3           2022:08:1615:57:12
         4 4           2023:05:0913:35:11
         1 1           2022:08:1614:16:42
```

图 9-6 从库上执行写入数据的命令

在从库上的 SQL 执行后,在主库上进行查询,如图 9-7 所示。

```
SQL> select open_mode,database_role from v$database;

OPEN_MODE              DATABASE_ROLE
--------------------   ----------------
READ WRITE             PRIMARY

SQL> select * from t;

       ID A            T
---------- ----------  ------------------
         2 2           2022:08:1615:55:33
         3 3           2022:08:1615:57:12
         4 4           2023:05:0913:35:11
         1 1           2022:08:1614:16:42
```

图 9-7 从库写入数据后在主库上进行查询

实验完毕，这个很简单。不过要说明的是，这个场景一定是使用在偶尔写入的场景中。因为它比起直接写入多了一个来回，有一定的延迟，虽然是毫秒级别的。若直接写主库每秒能达到 1 万次，这样通过 DML 重定向可能只能写 1 千次，当然这和机器配置网络条件都有关系，在不同硬件条件下的数据不一样。可能衰减为不到 1/10，也可能大约为 1/10。一切都要以具体环境而论，使用场景上，一定是偶尔写入，而不是当作正常的主力写入。

在本地模拟应用连接数据库服务器做批量写入实验,实验对象测试表结构如图 9-8 所示，其中 system_date 字段使用默认值为系统的当前时间。

```
SQL> create table temp(id number(10), system_date date default sysdate);
表已创建。
```

图 9-8 DML 循环写入实验表

本地编写 Java 代码使用 JDBC 连接，Java 代码如图 9-9 所示。主库的 IP 地址最后一位是 103，模拟 100 条数据写入，写入方式为单条提交。100 条写入耗时约 3s。

图 9-9 使用 Java 循环写入主库

通过改变连接目标的 IP，向从库写入一百条数据。使用 Java 代码循环写入从库，如图 9-10 所示，从库的 IP 地址最后一位是 105，写入 100 条耗时约 210s。

```java
package db;

import java.sql.*;
import java.util.Date;

public class oracle_insert_single_test {
    public static void main(String[] args) {
        Connection conn = null;
        try{
            Class.forName("oracle.jdbc.driver.OracleDriver");
            conn = DriverManager.getConnection( url: "jdbc:oracle:thin:@10.60.143.105:1521/pdb1", user: "xxg", password: "xxg");
            Statement sql = conn.createStatement();
            Date startTime = new Date();
            for (int i=100; i < 200; i++){
                sql.executeUpdate( sql: "INSERT INTO temp (id,insert_date) " +
                        "VALUES ("+ i +")");
            }
            Date endTime = new Date();
            System.out.println("写入总耗时: " + (endTime.getTime()-startTime.getTime()) + "ms");

        }catch (Exception e){
            e.printStackTrace();
        }finally {
            try {
                if (conn != null) conn.close();
            }catch (Exception e){
                e.printStackTrace();
            }
        }
    }
}

Run: oracle_insert_single_test ×
/Library/Java/JavaVirtualMachines/zulu-8.jdk/Contents/Home/bin/java ...
写入总耗时: 210232ms

Process finished with exit code 0
```

图 9-10　使用 Java 循环写入从库

最终写入的数据过程是主库直接写入 ID 从 0~99，从库上直接写入 ID 从 100~199。在从数据库服务器上进行查询，可以看到时间数据中主库从库写入对比，如图 9-11 所示。从图中可以看到 ID 为 0~99 的数据从主库写入，一百条数据单条写入总耗时约为 3s，而 ID 从 100~199 的数据由从库连接写入，一百条数据的写入时间约为 207s，在本次的实验环境中有 70 倍左右的差距，可见即使有 DML 这样的技术，从库也不宜作为生产环境中的主力写入渠道。

```
SQL> select * from temp where id in(0,99,100,199);

        ID SYSTEM_DATE
---------- --------------------
         0 2023:05:0914:09:54
        99 2023:05:0914:09:57
       100 2023:05:0914:12:01
       199 2023:05:0914:15:28
```

图 9-11　主库从库批量写入对比

所以这个功能的最佳实践是：从库少量写入的场景，可以用 DML 重定向实现双活架构。这个少量的定义是频率在两秒一条的场景下。本节是一对普通虚拟机的实验环境，读者的真实环境应该比笔者的要好。两秒一条的速率是比较保守的估计，实际环境会比这乐观，在上线前请结合实际环境进行一下测试得出真实环境的性能。

这个 DML 重定向的特性从侧面说明了一个分库分表和分布式的问题。为什么在从库写入仅仅多了半个来回，总耗时就为原来的几十倍？因为多了一层网络，分库分表和分布式在网络上的消耗很多。甚至在有些场景下，分布式数据库的性能还不如单机的 MySQL、PostgreSQL 的性能。就是因为网络因素有较大的制约因素。

9.1.2　Oracle 的资源隔离

数据库的资源隔离是指在多租户的应用程序使用各自的数据库时，通过控制和管理数据库的资源分配，确保各个租户数据之间的数据库互不干扰，从而保证数据库的稳定性、可靠性和性能。这个功能，各大公有云上的数据库做得很好。本节主要谈一下私有云本地化部署的资源隔离。

由于原生的 MySQL 和 PostgreSQL 没有租户概念，所以仅谈及有租户概念的 Oracle 数据库的资源隔离。

Oracle 的资源隔离分为在 Exadata 下的资源隔离和非 Exadata 的资源隔离。首先申明一点，这个非 Exadata 是指数据库安装在物理机上，而不是虚拟机。在虚拟设备上运行生产环境，不是 Oracle 的最佳实践，Oracle 也不提供官方支持。如果是虚拟机说到这里想必大家知道了，在生产环境不要使用 VMware 来部署，至少是一个物理服务器，否则如果出了问题，即使购买了官方服务，官方也不会提供直接支持。Oracle 可以提供技术支持，如果最终定位到不是 Oracle 的问题，那么会转到 VMware 的官方去处理。

在 Exadata 下，Oracle 有自己的硬件和算法来控制资源的隔离，这种属于软硬结合的控制，效果比较好。但是一般物理机下，是通过参数来进行控制的。这属于纯软控制，比起一体机的控制效果要差一些，但是比没有控制和隔离要好很多。

官方文档中对这些设置的说明并不是很清楚，而本节通过实验进行整理，比较适合读者学习。还有一点要明确，就是在虚拟机下最好不要做资源隔离，因为通常虚拟机的资源不够大，甚至不足。在一个非常有限的资源情况下，可能全局资源都不足，在捉襟见肘的情况下是无法调配资源的。资源隔离通常是指资源充沛情况下进行隔离和熔断，而不是全局资源枯竭还限制隔离。

资源隔离一定要在 CDB 下实施，建立资源控制计划。它可以给不同的 PDB 设置不同的资源隔离策略，这些策略涉及 CPU、IO。即刻生效无须重启个体 PDB 数据库，更加无须重启整个 CDB 数据库。这与虚拟机增加或减少资源时重启有较大不同。而内存的隔离限制就是采用最传统的 SGA 和 PGA 的设置来完成，同样即刻生效无须重启个体 PDB 数据库，更加无须重启整个 CDB 数据库。

通过 SQLPLUS 用命令行登录 CDB，如图 9-12 所示。打出 show pdbs 的命令，说明在 CDB 下，而且有管理员权限。

```
SQL> show pdbs;

    CON_ID CON_NAME                       OPEN MODE  RESTRICTED
    ------ ------------------------------ ---------- ----------
         2 PDB$SEED                       READ ONLY  NO
         3 XXG                            READ WRITE NO
         4 XXG2                           READ WRITE NO
         5 TU                             READ WRITE NO
```

图 9-12　用命令行登录 CDB

在 SQL>的提示符下继续执行如下命令：

exec DBMS_RESOURCE_MANAGER.CREATE_PENDING_AREA();

该命令的作用是创建一个资源管理的区域，使用 CREATE_PENDING_AREA 步骤创建挂起区域。

在 SQL>的提示符下继续执行如下命令：

exec DBMS_RESOURCE_MANAGER.VALIDATE_PENDING_AREA();

该命令的作用是使刚才创建的区域生效。两个命令的执行效果如图 9-13 所示。

```
SQL> exec DBMS_RESOURCE_MANAGER.CREATE_PENDING_AREA();

PL/SQL 过程已成功完成。

SQL> exec DBMS_RESOURCE_MANAGER.VALIDATE_PENDING_AREA();

PL/SQL 过程已成功完成。
```

图 9-13　创建并且生效资源隔离区域

接下来根据资源隔离的步骤，设定一个资源隔离计划，后续资源隔离的策略都是基于这个计划的。在 SQL>的提示符下继续执行如下命令：

exec DBMS_RESOURCE_MANAGER.CREATE_CDB_PLAN(plan => 'test',comment => 'temp cdb plan for resource isolation');

这句命令的含义是创建一个隔离计划。该隔离计划的名字是 test，注释是自定义，这里注释的意思是临时 CDB 资源隔离。

创建资源隔离计划，命令执行效果如图 9-14 所示。

```
SQL> begin
  2  DBMS_RESOURCE_MANAGER.CREATE_CDB_PLAN
  3  (plan    => 'test',comment => 'test');
  4  end;
  5  /
```

图 9-14　创建资源隔离计划命令执行效果

在以上的资源隔离计划建立完毕以后，开始正式配置资源隔离。

CPU 是计算机中的重要资源，所以需要为不同重要级别的数据库设定不同的待遇。重要的数据库多分配一些资源，非核心的数据库少分配一些资源，按照等级合理划分。建立了三个等级的资源隔离计划如图 9-15 所示。三个隔离计划都是基于 test 的隔离计划，定义级别分别是 high（高）、middle（中）、low（低）。

```
SQL> begin
  2  DBMS_RESOURCE_MANAGER.CREATE_CDB_PROFILE_DIRECTIVE
  3  (plan => 'test',profile => 'high',shares => 3,utilization_limit => 40);
  4  end;
  5  /

SQL> begin
  2  DBMS_RESOURCE_MANAGER.CREATE_CDB_PROFILE_DIRECTIVE
  3  (plan => 'test',profile => 'middle',shares => 2,utilization_limit => 30);
  4  end;
  5  /

SQL> begin
  2  DBMS_RESOURCE_MANAGER.CREATE_CDB_PROFILE_DIRECTIVE
  3 (plan => 'test',profile => 'low',shares => 1,utilization_limit => 10);
  4  end;
  5  /
```

图 9-15　建立了三个等级的资源隔离计划

除了图 9-15 中的 shares 这个参数，其他都不难理解。这里要注意的是整个命令其实是为了组设立的。如果这里只有 3 组数据库，分别指定了 P1 组是 3，P2 组是 2，P3 组是 1，那么 P1 组被分得 3/(1+2+3)，即 50%；P2 组被分得 2/(1+2+3)，即 33%；P3 组被分得 1/(1+2+3)，即 17%。这样理解的确不好记忆，需要读者多多回味。可以这样理解：参数 shares 就是一个占比相对值。起到的作用是限制最高使用的比例。当使用到的资源达到这个最高值就触发资源隔离的条件，禁止超过资源上限，防止影响到其他数据库实例的使用。

数据库资源隔离设置了最高上限确保不会侵害到其他实例，那么有没有最低的资源保障呢？数据库的资源隔离就是从防止"贫富差距"过大而出发的。

图 9-15 所示的 utilization_limit 参数是最低配额，这里的配额是百分比，而参数 shares 就是一个占比相对值。

所以以上代码就可以解读为：

high 组的策略是这个组内总共的 CPU 使用率不能超过 50%，但是也要保证不低于 40% 的供给预留，在极端情况下有 40% 的 CPU 资源只能给 high 组用，不能被抢占。

middle 组的策略是这个组内总共的 CPU 使用率不能超过 33%，但是至少也要保证不低于 30% 的供给预留，在极端情况下有 30% 的 CPU 资源只能给 middle 组用，不能被抢占。

low 组的策略是这个组内总共的 CPU 使用率不能超过 16%，但是至少也要保证不低于

10%的供给预留,在极端情况下有 20%的 CPU 资源只能给 low 组用,不能被抢占。

如果只有 3 个数据库,那么 PDB1、PDB2 和 PDB3 这 3 个 PDB 就是自己的组。当然如果没有组的概念,也可以用下面的命令来修改自己的 CPU 隔离限制值。修改最小的 CPU 使用个数,如图 9-16 所示。

```
SQL> show parameter cpu

NAME                                 TYPE        VALUE
------------------------------------ ----------- ------------------------------
cpu_count                            integer     16
cpu_min_count                        string      16
parallel_threads_per_cpu             integer     1
resource_manager_cpu_allocation      integer     16
SQL> alter system set cpu_min_count=8 scope=both;

系统已更改。

SQL> show parameter cpu

NAME                                 TYPE        VALUE
------------------------------------ ----------- ------------------------------
cpu_count                            integer     16
cpu_min_count                        string      8
parallel_threads_per_cpu             integer     1
resource_manager_cpu_allocation      integer     16
```

图 9-16 修改最小的 CPU 使用个数

设置 CPU 后,需要设置内存的资源隔离。内存没有组的概念,只有单独实例设置。Oracle 数据库设有系统全局区(System Global Area,SGA)和程序全局区(Program Global Area,PGA),系统全局区由所有服务进程和后台进程共享;程序全局区由每个服务进程、后台进程专有,每个进程都有一个 PGA。设置 SGA,如图 9-17 所示。

```
SQL> show pdbs;

    CON_ID CON_NAME                       OPEN MODE  RESTRICTED
---------- ------------------------------ ---------- ----------
         2 PDB$SEED                       READ ONLY  NO
         3 XXG                            READ WRITE NO
         4 XXG2                           READ WRITE NO
         5 TU                             READ WRITE NO
SQL> show parameter sga

NAME                                 TYPE        VALUE
------------------------------------ ----------- ------------------------------
allow_group_access_to_sga            boolean     FALSE
lock_sga                             boolean     FALSE
pre_page_sga                         boolean     TRUE
sga_max_size                         big integer 9648M
sga_min_size                         big integer 0
sga_target                           big integer 9648M
unified_audit_sga_queue_size         integer     1048576
```

图 9-17 设置 SGA

登录 CDB 通过 show pdbs 看到了所有 PDB,也就是说,此时命令行停留在 CDB 实例下。CDB 的 SGA 大约 9.6GB,这是由 sga_max_size 参数和 sga_target 所决定的。sga_target 的值

不能超过 sga_max_size 值的大小，sga_max_size 是启动时候设置好的，值目前不能动态调整。但是 sga_target 只要不超过 sga_max_size 都是可以调整的。

在 PDB 中的 SGA 参数修改如图 9-18 所示。

```
SQL> show parameter sga

NAME                                 TYPE         VALUE
------------------------------------ ------------ ------------
allow_group_access_to_sga            boolean      FALSE
lock_sga                             boolean      FALSE
pre_page_sga                         boolean      TRUE
sga_max_size                         big integer  9648M
sga_min_size                         big integer  0
sga_target                           big integer  0
unified_audit_sga_queue_size         integer      1048576
SQL> alter system set sga_target=4000M scope=both;

系统已更改。

SQL> show parameter sga

NAME                                 TYPE         VALUE
------------------------------------ ------------ ------------
allow_group_access_to_sga            boolean      FALSE
lock_sga                             boolean      FALSE
pre_page_sga                         boolean      TRUE
sga_max_size                         big integer  9648M
sga_min_size                         big integer  0
sga_target                           big integer  4000M
unified_audit_sga_queue_size         integer      1048576
```

图 9-18　在 PDB 中的 SGA 参数修改

上图中通过命令 alter system set sga_target=4000M scope=both;将 PDB 的内存大小限制在了 4000MB。设置参数以后重新查询，可以看出即时生效。也就是说，只要在初始范围内的加降内存资源都可以通过一个命令动态完成。

同样，也可以执行修改 PGA 的命令：alter system set pga_aggregate_target=3G scope=both;来修改 PGA 的大小。在 PDB 中的 PGA 参数修改如图 9-19 所示。

```
SQL> show parameter pga

NAME                                 TYPE         VALUE
------------------------------------ ------------ ------------
pga_aggregate_limit                  big integer  10G
pga_aggregate_target                 big integer  4G
SQL> alter system set pga_aggregate_target=3G scope=both;

系统已更改。

SQL> show parameter pga

NAME                                 TYPE         VALUE
------------------------------------ ------------ ------------
pga_aggregate_limit                  big integer  10G
pga_aggregate_target                 big integer  3G
```

图 9-19　在 PDB 中的 PGA 参数修改

最后，讲述一下 IO 的资源隔离。IO 是数据库中最容易出现瓶颈的资源，所以一般情况下，控制好 IO 就能控制住负荷。但是在实际工作中发现只控制 IO 不行，极端情况下会引起 CPU 内核态的问题。

所以要一起控制 IO 和 CPU。结合不同数据库调用 IO 的规模，为不同重要级别的数据库设定不同的级别。IO 也没有组的概念，只有单独实例设置。

可以在 CDB 或 PDB 级别去设置下列参数来控制 PDB 级别时 IO 阈值。

- MAX IOPS：PDB 中每秒最大的 IO 操作次数。默认值为 0，不建议设置小于 100 的 IOPS。
- MAX MBPS：PDB 中每秒最大的 IO 带宽（MB）。默认值为 0，不建议设置小于 25MB 的 MBPS。

关于这两个参数的使用，需要考虑如下几个要点。

- 这两个参数是独立的，可以设置一个、两个，或者一个都不设置。
- 当这两个参数在 CDB 根中设置后，它们将变成所有 PDB 的默认值。
- 当在 PDB 级别设置了这两个参数后，PDB 中的值可以覆盖默认值。
- 如果在 CDB 和 PDB 中这两个参数值均为 0，则没有 IO 阈值。
- 某些常规函数的关键 IO 不会受到限制，但依然会计算到总 IO 中，所以，实际 IO 是有可能瞬时超过设定的阈值的。
- 该参数只有多租户环境下可以使用。
- Exadata 无法使用该特性（一体机的 IO 隔离是一体机的存储来完成的）。
- 达到了设定阈值会造成一个等待事件，叫作 resmgr:I/O rate limit。

修改具体 PDB 的 IO 方式如下：从 CDB 切换到相应的 PDB 中。通过参数查看到现在两个参数都是 0，即不限制。未设定 IO 限制的状态如图 9-20 所示。

```
SQL> show pdbs;

    CON_ID CON_NAME                       OPEN MODE  RESTRICTED
---------- ------------------------------ ---------- ----------
         2 PDB$SEED                       READ ONLY  NO
         3 XXG                            READ WRITE NO
         4 XXG2                           READ WRITE NO
         5 TU                             READ WRITE NO
SQL> alter session set container=tu;

会话已更改。

SQL> show parameter max_mbps

NAME                                 TYPE        VALUE
------------------------------------ ----------- ------------------------------
max_mbps                             integer     0
SQL> show parameter max_iops

NAME                                 TYPE        VALUE
------------------------------------ ----------- ------------------------------
max_iops                             integer     0
```

图 9-20　未设定 IO 限制的状态

下面执行两个命令（这两个参数没有显示指定单位）：

```
alter system set max_iops=5000 scope=both;
alter system set max_mbps=100 scope=both;
```

该命令的含义是每秒 5000 次物理 IO 和每秒 100MB 的 IO 吞吐。先达到哪个阈值，哪个就先起作用。执行 IO 限制设置命令，检查设置效果，如图 9-21 所示。

```
SQL> alter system set max_mbps=100 scope=both;

系统已更改。

SQL> alter system set max_iops=5000 scope=both;

系统已更改。

SQL> show parameter max_mbps

NAME                                 TYPE        VALUE
------------------------------------ ----------- ----------
max_mbps                             integer     100
SQL> show parameter max_iops

NAME                                 TYPE        VALUE
------------------------------------ ----------- ----------
max_iops                             integer     5000
```

图 9-21　执行 IO 限制设置命令

执行完毕检查，可以看到参数生效。如果单一设置 IO 限制，而不对 CPU 进行资源限制，也可能造成问题。笔者经历过一个案例是，突然发生对一个表的全表高并发出报表，每次将近 300 万次的 IO 物理读，这样大量的并发访问某些资源，产生大量 IO 等待，造成系统 CPU 的异常升高，从用户态转移到内核态，从而影响了全局。

所以仅仅限制 IO 不能解决所有问题，IO 和 CPU 要一起设置。

同时在生产环境等需要进行 IO 资源隔离的场景中，可以查看数据库已有的 IOPS 和 MBPS 的运行数据。结合实际数据设置这两个参数，SQL 中时间是结合实际业务高峰时的情况进行采集的。

```
SELECT r.snap_id,
r.con_id,
p.pdb_name,
to_char(r.begin_time,'yyyy-mm-dd hh24:mi:ss') begin_time,
to_char(r.end_time,'yyyy-mm-dd hh24:mi:ss') end_time,
round(r.iops,1) iops,
round(r.iombps,1) iombps,
round(r.iops_throttle_exempt,1) iops_throttle_exempt,
round(r.iombps_throttle_exempt,1) iombps_throttle_exempt,
```

```
round(r.avg_io_throttle,1) avg_io_throttle
FROM dba_hist_rsrc_pdb_metric r,
cdb_pdbs p
WHERE r.con_id = p.con_id and begin_time>to_date('2022-06-27 10:00:00', 'yyyy-mm-dd hh24:mi:ss')
ORDER BY p.pdb_name,to_char(r.begin_time,'yyyy-mm-dd hh24:mi:ss');
```

执行该 SQL 后可以看到 PDB 每个时间段的数据统计，知道了数据库运行一般需要多少 IO，限制在日常的基础上多一点，既能满足业务，又能防止过大使用 IO。

在 CPU、内存和 IO 设置完毕以后就完成了官方给出的资源隔离方案。但是从实际角度来说，个人觉得这并不是万无一失的。

就最佳实践而言，在虚拟机上做以上提及的资源限制，效果并不理想。其原因并不是数据库本身的问题，而是虚拟机通常资源并不充裕，所以设置资源的限制是很难把握的，在紧张的资源环境中遇到资源竞争时可能会捉襟见肘。

通常该特性至少要在性能较好的物理机上实施，以物理机的充足资源可以动态地调配。当然最好是在 Oracle 的 Exadata 上实施。那么在没有 Exadata 的情况下有没有低成本的实现方式？传统做法是使用概要文件，设置和修改概要文件，如图 9-22 所示。创建一个概要文件限制一次 SQL 最大可以使用 1000 个 IO，根据文档显示这 1000 个是逻辑 IO 和物理 IO 的组合。也可以随时修改这个限制，并且通过 alter user xxx profile xxgtest;来将这个限制资源的概要文件和用户 xxx 关联起来。

```
SQL> create profile xxgtest limit logical_reads_per_session 1000 ;
```
配置文件已创建。
```
SQL> alter profile xxgtest limit logical_reads_per_session 2000 ;
```
配置文件已更改。
```
SQL> create user xxx identified by xxx;
```
用户已创建。
```
SQL> grant dba to xxx;
```
授权成功。
```
SQL> alter user xxx profile xxgtest;
```
用户已更改。

图 9-22 设置和修改概要文件

当用户 xxx 登录后，限制生效。触发限制被阻止，如图 9-23 所示。当一个笛卡儿积的 SQL 调用大量 IO 资源时，数据库将其阻断。

```
[oracle@dev-dbteam-143101 ~]$ sqlplus xxx/xxx@xxg

连接到:
Oracle Database 19c Enterprise Edition Release 19.0.0.0.0 - Production
Version 19.3.0.0.0


SQL> select count(*) from xxg.d;

  COUNT(*)
----------
    300000


SQL> select *  from xxg.c ,xxg.d;
select *  from xxg.c ,xxg.d
                     *
第 1 行出现错误:
ORA-02394: 超出 IO 使用的会话限制,您将被注销
```

图 9-23 触发限制被阻止

在没有充足硬件资源下,可以用概要文件缓解不良 SQL 对数据库的冲击。而当有充足资源时,可以考虑资源隔离。

9.1.3 MySQL 的延迟复制——挽救误删除

MySQL 建立主从高可用是用来作为容灾的解决方案。不过如果主库上执行了误删除操作,那么从库上也会同步支持误删除。如何解决这种问题?本节将介绍使用延迟复制的方法,挽救误删除数据,从而保证数据的安全性。

下面用实验环境演示主从延迟复制的搭建,以及当误删除发生时,利用延迟数据库上延迟复制的特性,使得从库不会及时得到主库误删除的指令,运维人员有时间、有机会来修复误删数据,而从将数据保留了下来。生产环境推荐一主一从一延迟,本实验省略从库(在本实验中从库没有演示意义),为主库+延迟从库。

此次实验使用 GTID 的方式搭建,主库 IP 地址尾号为 92,延迟从库 IP 地址尾号为 32。在主库 92 上先创建用于复制的账户,如图 9-24 所示。

```
mysql> create user 'repl'@'%' identified by '1';
Query OK, 0 rows affected (0.01 sec)

mysql> grant replication slave on *.* to 'repl'@'%';
Query OK, 0 rows affected (0.00 sec)
```

图 9-24 在主库建立复制账户

在主库 92 上导出全库的数据并传给延迟从库,并修改配置文件 my.cnf 加上 GTID 所需的参数。设置 GTID 重启数据库,如图 9-25 所示。

```
[root@test ~]# mysqldump –u root –p  -A > 92.sql
Enter password:
[root@test ~]# scp 92.sql root@10.60.143.32:/root/
[root@test ~]# vim /etc/my.cnf
.
.
.
gtid_mode=ON
log-slave-updates=ON
enforce-gtid-consistency=ON
.
.
.
[root@test ~]# service mysqld restart
Stopping mysqld:                                        [  OK  ]
Starting mysqld:                                        [  OK  ]
```

图 9-25　设置 GTID 重启数据库

同样在延迟从库上修改配置文件，导入主库数据，重启数据库。延迟数据库配置并重启，如图 9-26 所示。

```
[root@MySQL8M ~]# mysql –uroot –p < /root/92.sql
[root@test ~]# vim /etc/my.cnf
.
.
.
gtid_mode=ON
log-slave-updates=ON
enforce-gtid-consistency=ON
.
.
.
[root@test ~]# service mysqld restart
Stopping mysqld:                                        [  OK  ]
Starting mysqld:                                        [  OK  ]
```

图 9-26　延迟数据库配置并重启

在延迟数据库上设置延迟并且开启复制线程，设置延迟时长，如图 9-27 所示。

```
mysql> stop slave;
Query OK, 0 rows affected, 2 warnings (0.00 sec)

mysql> CHANGE REPLICATION SOURCE to
SOURCE_HOST='10.60.143.92',SOURCE_USER='repl',SOURCE_PASSWORD='1',SOURCE_AUTO_POSITION=1,MASTER_DELAY = 600;
Query OK, 0 rows affected, 3 warnings (0.06 sec)

mysql> start slave;
Query OK, 0 rows affected, 1 warning (0.02 sec)
```

图 9-27　设置延迟时长

检查主从复制状态和延迟时间，如图 9-28 所示。

```
mysql> show slave status\G;
*************************** 1. row ***************************
               Slave_IO_State: Waiting for source to send event
                  Master_Host: 10.60.143.92
                  Master_User: repl
                  Master_Port: 3306
                Connect_Retry: 60
              Master_Log_File: mysql-bin.000005
          Read_Master_Log_Pos: 156
               Relay_Log_File: MySQL8M-relay-bin.000002
                Relay_Log_Pos: 372
        Relay_Master_Log_File: mysql-bin.000005
             Slave_IO_Running: Yes
            Slave_SQL_Running: Yes
                      .
                      .
                      .
              Replicate_Do_DB:
          Replicate_Ignore_DB:
           Replicate_Do_Table:
       Replicate_Ignore_Table:
      Replicate_Wild_Do_Table:
  Replicate_Wild_Ignore_Table:
                   Last_Errno: 0
                   Last_Error:
                 Skip_Counter: 0
          Exec_Master_Log_Pos: 156
              Relay_Log_Space: 584
              Until_Condition: None
               Until_Log_File:
                Until_Log_Pos: 0
           Master_SSL_Allowed: No
           Master_SSL_CA_File:
           Master_SSL_CA_Path:
              Master_SSL_Cert:
            Master_SSL_Cipher:
               Master_SSL_Key:
        Seconds_Behind_Master: 0
Master_SSL_Verify_Server_Cert: No
                Last_IO_Errno: 0
                Last_IO_Error:
               Last_SQL_Errno: 0
               Last_SQL_Error:
  Replicate_Ignore_Server_Ids:
             Master_Server_Id: 101
                  Master_UUID: 09fd4721-3747-11ed-973b-00505696ce1c
             Master_Info_File: mysql.slave_master_info
                    SQL_Delay: 600
          SQL_Remaining_Delay: NULL
      Slave_SQL_Running_State: Replica has read all relay log; waiting for more updates
           Master_Retry_Count: 86400
                  Master_Bind:
      Last_IO_Error_Timestamp:
     Last_SQL_Error_Timestamp:
               Master_SSL_Crl:
           Master_SSL_Crlpath:
           Retrieved_Gtid_Set:
            Executed_Gtid_Set: 790b8b3a-8b09-11ed-94e4-0050569684a9:1-25278
                Auto_Position: 1
         Replicate_Rewrite_DB:
                 Channel_Name:
           Master_TLS_Version:
       Master_public_key_path:
        Get_master_public_key: 0
            Network_Namespace:
1 row in set, 1 warning (0.01 sec)
```

图 9-28 检查主从复制状态和延迟时间

图 9-28 中各项状态都正常，SQL_Delay 显示 600 表示延迟复制已经建立成功了，并且有 600s 的延迟，表示从库的所有行为会落后主库 10min。即在主库上发生误删操作的 10min 内（生产环境 10min 可能来不及，可以推荐 3~4h），从库上的数据是尚未删除的。

模拟一个典型场景，演示一下误删数据的恢复。在主库上，假设有一个很久之前建立的待被误删除表如图 9-29 所示。该表在从库中也有一样的数据。

```
mysql> select * from t;
+------+----------+
| id   | time     |
+------+----------+
|    0 | 14:39:45 |
|    1 | 14:39:45 |
|    2 | 14:39:45 |
+------+----------+
3 rows in set (0.00 sec)
```

图 9-29　待被误删除表

手动刷新一下 binlog 日志，新生成一个全新的 binlog 日志。并且删除该日志之前的其他 binlog 日志，清理所有 binlog，如图 9-30 所示。演示极端情况下生产中定期自动删除历史日志的动作。

```
mysql> flush logs;
Query OK, 0 rows affected (0.07 sec)

mysql> PURGE MASTER LOGS BEFORE '2023-06-27 14:50:00';
Query OK, 0 rows affected (0.00 sec)
```

图 9-30　清理所有 binlog

在清理历史 binlog 日志后，发生了两笔业务。增量数据写入即将被误删除表如图 9-31 所示，在数据库中表现为写入了两条数据。

```
mysql> insert into t values(3,sysdate()),(4,sysdate());
Query OK, 2 rows affected (0.00 sec)
Records: 2  Duplicates: 0  Warnings: 0

mysql> select * from t;
+------+----------+
| id   | time     |
+------+----------+
|    0 | 14:39:45 |
|    1 | 14:39:45 |
|    2 | 14:39:45 |
|    3 | 14:51:20 |
|    4 | 14:51:20 |
+------+----------+
5 rows in set (0.00 sec)
```

图 9-31　增量数据写入即将被误删除表

第 9 章 DBA 的日常：数据库管理及开发的最佳实践

随即发生了误操作，最严重的表级误删除，如图 9-32 所示，直接进行了 drop 表的误操作。

```
mysql> drop table t;
Query OK, 0 rows affected (0.01 sec)

mysql> select * from t;
ERROR 1146 (42S02): Table 'c.t' doesn't
```

图 9-32 最严重的表级误删除

得知误删除，第一反应去 binlog 中翻找日志，但是 binlog 的历史日志已经被清理，MySQL 的 binlog 解析如图 9-33 所示，只能找到最新的两条数据的写入以及一次误删除动作。而且此时实时从库（如果有的话）也执行了同样的 drop 表操作。

```
[root@test mysql]# mysqlbinlog mysql-bin.000011 --base64-output=decode-rows -v
.
.
.
.
.
BEGIN
/*!*/;
# at 355
#230627 14:51:20 server id 101  end_log_pos 401 CRC32 0xcc8c7aff    Table_map: `c`.`t` mapped to number 106
# at 401
#230627 14:51:20 server id 101  end_log_pos 452 CRC32 0xa43d8f8c    Write_rows: table id 106 flags:
STMT_END_F
### INSERT INTO `c`.`t`
### SET
###   @1=3
###   @2='14:51:20'
### INSERT INTO `c`.`t`
### SET
###   @1=4
###   @2='14:51:20'
# at 452
#230627 14:51:20 server id 101  end_log_pos 483 CRC32 0x4c07b5db    Xid = 240
COMMIT/*!*/;
# at 483
#230627 14:52:36 server id 101  end_log_pos 560 CRC32 0xbaf964ab    GTID     last_committed=1
sequence_number=2      rbr_only=no    original_committed_timestamp=1687848756187879
immediate_commit_timestamp=1687848756187879     transaction_length=197
# original_commit_timestamp=1687848756187879 (2023-06-27 14:52:36.187879 CST)
# immediate_commit_timestamp=1687848756187879 (2023-06-27 14:52:36.187879 CST)
/*!80001 SET @@session.original_commit_timestamp=1687848756187879*//*!*/;
/*!80014 SET @@session.original_server_version=80026*//*!*/;
/*!80014 SET @@session.immediate_server_version=80026*//*!*/;
SET @@SESSION.GTID_NEXT= '09fd4721-3747-11ed-973b-00505696ce1c:29'/*!*/;
# at 560
#230627 14:52:36 server id 101  end_log_pos 680 CRC32 0xde5b7805    Query    thread_id=12
exec_time=0    error_code=0    Xid = 242
use `c`/*!*/;
SET TIMESTAMP=1687848756/*!*/;
DROP TABLE `t` /* generated by server */
```

图 9-33 MySQL 的 binlog 解析

此时由于历史日志的缺失，想要通过日志恢复此表的表结构以及历史数据已经不可能了。好在从架构设计上考虑到了事先有带有延迟的从库，在延迟从库上检查，发现延迟从库有误删除之前的全表记录，如图 9-34 所示。

```
mysql> select * from t;
+----+----------+
| id | time     |
+----+----------+
|  0 | 14:39:45 |
|  1 | 14:39:45 |
|  2 | 14:39:45 |
+----+----------+
3 rows in set (0.00 sec)
```

图 9-34　延迟从库有误删除之前的全表记录

先暂停延迟从库的同步，防止删除语句同步过来，如图 9-35 所示。

```
mysql> stop slave;
Query OK, 0 rows affected, 1 warning (0.01 sec)
```

图 9-35　及时止损停止同步

在延迟从库上将还未被误删除的表全表导出，并将数据文件回传给主库，如图 9-36 所示。

```
[root@MySQL8M ~]# mysqldump -uroot -p1 --set-gtid-purged=off c t  > ct.sql
mysqldump: [Warning] Using a password on the command line interface can be insecure.

[root@MySQL8M ~]# scp ct.sql root@10.60.143.92:/root/
root@10.60.143.92's password:
ct.sql                                        100% 1820     2.2MB/s   00:00
```

图 9-36　回传主库数据

在主库上，通过导入数据恢复该表的表结构以及历史数据，如图 9-37 所示。

```
[root@test mysql]# mysql -u root -p c  < /root/ct.sql
Enter password:
```

图 9-37　导入被误删的整表数据

导入完成检查数据，表结构以及历史数据已经恢复到实验开始环境，如图 9-38 所示。

```
mysql> select * from t;
+----+----------+
| id | time     |
+----+----------+
|  0 | 14:39:45 |
|  1 | 14:39:45 |
|  2 | 14:39:45 |
+----+----------+
3 rows in set (0.00 sec)
```

图 9-38　表恢复到实验开始环境

再通过主库上最新的 binlog 日志，恢复新发生的交易。这里要记得通过设置恢复时间规避误删的动作。主库解析 binlog 回写增量数据如图 9-39 所示。由于是 GTID 模式的，所以要增加跳过 GTID 的选项，否则数据无法写入。

```
[root@test mysql]# mysqlbinlog --skip-gtids mysql-bin.000011 --stop-datetime='2023-06-27 14:52:00' | mysql
```

图 9-39　主库解析 binlog 回写增量数据

恢复完成，数据已经完整恢复至误删前的状态，如图 9-40 所示。

```
mysql> select * from t;
+------+----------+
| id   | time     |
+------+----------+
|    0 | 14:39:45 |
|    1 | 14:39:45 |
|    2 | 14:39:45 |
|    3 | 14:51:20 |
|    4 | 14:51:20 |
+------+----------+
5 rows in set (0.00 sec)
```

图 9-40　回到误删除之前时间点数据无丢失

此时延迟复制还是关闭的，可以到延迟从库上开启延迟复制，延迟从库会重放刚才的补救动作，再次承担起保护数据的工作，如图 9-41 所示。

```
mysql> start slave;
Query OK, 0 rows affected, 1 warning (0.02 sec)
```

图 9-41　恢复延迟复制线程

在使用延迟复制功能的情况下，无须备份还原误删除的数据的工作。

9.1.4　PostgreSQL 的延迟复制

PostgreSQL 也同样有延迟复制的功能。但是在实际使用上和 MySQL 有较大的区别。本节介绍一下其延迟复制的功能和适用场景。

PostgreSQL 的延迟复制功能通过设置从库延迟参数 recovery_min_apply_delay 来完成，如图 9-42 所示。在从库上编辑 postgresql.conf 文件，向文件中设置延迟的时长。本案例设置为 10min 的延迟，即主从数据库的同步都滞后 10min。

```
[pg15@dev-dbteam-148181 pg_root]$ vim postgresql.conf

[pg15@dev-dbteam-148181 pg_root]$ cat postgresql.conf |grep "apply_delay"
#recovery_min_apply_delay = 0           # minimum delay for applying changes during recovery
recovery_min_apply_delay = 1min


[pg15@dev-dbteam-148181 pg_root]$ pg_ctl reload
server signaled

[pg15@dev-dbteam-148181 pg_root]$ psql
psql (15.3)
Type "help" for help.

postgres=# \c xxg
You are now connected to database "xxg" as user "postgres".
xxg=# show recovery_min_apply_delay;
 recovery_min_apply_delay
--------------------------
 1min
(1 row)
```

图 9-42　设置 PostgreSQL 从库延迟参数

设置完成后无须重启数据库，执行 pg_ctl reload 重新加载参数即可生效。检查延迟参数，如图 9-43 所示。通过执行 psql 命令检查参数。主库上该参数是 0，从库是 10min。-h 后面是连接的数据库主机 IP，-c 代表执行，引号内为需要执行的命令"show recovery_min_apply_delay"。

```
[pg15@dev-dbteam-148180 ~]$ psql -h 10.60.148.180  -d xxg  -c "show recovery_min_apply_delay"
 recovery_min_apply_delay
--------------------------
 0
(1 row)

[pg15@dev-dbteam-148180 ~]$ psql -h 10.60.148.181  -d xxg  -c "show recovery_min_apply_delay"
 recovery_min_apply_delay
--------------------------
 10min
(1 row)

[pg15@dev-dbteam-148180 ~]$
```

图 9-43　检查延迟参数

至此完成了 PostgreSQL 数据库的主从延迟，正式环境推荐一主一从一延迟，实验时一主一延迟就可以达成演示的目的。

在主库进行数据写入，主库初始数据如图 9-44 所示，建立实验表，写入带有时间戳的数据，并且查询当前时间。

```
xxg=# create table x.xue (id int,a varchar(5),t  timestamp);
CREATE TABLE
xxg=# insert into x.xue values (1,'a',now());
INSERT 0 1
xxg=# insert into x.xue values (2,'b',now());
INSERT 0 1
xxg=# insert into x.xue values (3,'c',now());
INSERT 0 1
xxg=# select * from x.xue;
 id | a |             t
----+---+----------------------------
  1 | a | 2023-07-06 16:23:00.577097
  2 | b | 2023-07-06 16:23:00.582098
  3 | c | 2023-07-06 16:23:00.584826
(3 rows)

xxg=# select now();
              now
-------------------------------
 2023-07-06 16:23:18.430008+08
(1 row)

xxg=# exit
```

图 9-44　主库初始数据

在操作系统层面同时查询主库和从库数据。主库和从库数据对比如图 9-45 所示，从图中看到当前操作系统时间（date 显示 16:25:49）由于还未到 10min，所以数据只在 180 的主库上，还没有在 181 的从库上。

```
[pg15@dev-dbteam-148180 pg_root]$ date
Thu Jul  6 16:25:49 CST 2023

[pg15@dev-dbteam-148180 pg_root]$ psql -h 10.60.148.180  -d xxg  -c "select * from x.xue"
 id | a |             t
----+---+----------------------------
  1 | a | 2023-07-06 16:23:00.577097
  2 | b | 2023-07-06 16:23:00.582098
  3 | c | 2023-07-06 16:23:00.584826
(3 rows)

[pg15@dev-dbteam-148180 pg_root]$ psql -h 10.60.148.181  -d xxg  -c "select * from x.xue"
ERROR:  relation "x.xue" does not exist
LINE 1: select * from x.xue
                      ^
```

图 9-45 主库和从库数据对比

当达到延迟的时间后,主从延迟完成对比,如图 9-46 所示。从图中看到当前操作系统时间(date 显示 16:34:18)已经超过了写入的时间(16:23:00),所以数据在 180 的主库上和 181 的从库上都存在。

```
[pg15@dev-dbteam-148180 pg_root]$ date
Thu Jul  6 16:34:18 CST 2023
[pg15@dev-dbteam-148180 pg_root]$ psql -h 10.60.148.180  -d xxg  -c "select * from x.xue"
 id | a |             t
----+---+----------------------------
  1 | a | 2023-07-06 16:23:00.577097
  2 | b | 2023-07-06 16:23:00.582098
  3 | c | 2023-07-06 16:23:00.584826
(3 rows)

[pg15@dev-dbteam-148180 pg_root]$ psql -h 10.60.148.181  -d xxg  -c "select * from x.xue"
 id | a |             t
----+---+----------------------------
  1 | a | 2023-07-06 16:23:00.577097
  2 | b | 2023-07-06 16:23:00.582098
  3 | c | 2023-07-06 16:23:00.584826
(3 rows)
```

图 9-46 主从延迟完成对比

根据上一节 MySQL 数据延迟的实验思路,在主库上执行误删除表以后,再如法炮制地在延迟库上进行导出并恢复。主库删除表查询延迟库如图 9-47 所示,发现主库上延迟应用已经有 25s 了。查询主库数据,表已经被删除。

但是查询从库数据时长时间无响应,最终只得取消。其原因是延迟复制只是延迟了 replay,而不是延迟 wal 传输。所以从库出现了上 drop 一张表的操作和 select 同一张表的操作,产生了锁等待。

登录从库检查主从进度,延迟复制从库进度如图 9-48 所示。replay_lag 已经有将近 9min 的延迟应用。而逻辑导出也不出意外地因为无法 select,导致导出动作被阻塞。

```
xxg=# select * from pg_stat_replication ;
-[ RECORD 1 ]----+------------------------------
pid              | 52734
usesysid         | 10
usename          | postgres
application_name | walreceiver
client_addr      | 10.60.148.181
client_hostname  |
client_port      | 37812
backend_start    | 2023-07-06 15:00:06.179136+08
backend_xmin     |
state            | streaming
sent_lsn         | 0/8036A78
write_lsn        | 0/8036A78
flush_lsn        | 0/8036A78
replay_lsn       | 0/80355B8
write_lag        | 00:00:00.000919
flush_lag        | 00:00:00.001539
replay_lag       | 00:00:25.567928
sync_priority    | 0
sync_state       | async
reply_time       | 2023-07-06 16:35:37.916106+08

xxg=# exit


[pg15@dev-dbteam-148180 pg_root]$ psql -h 10.60.148.180 -d xxg  -c "select * from x.xue"
ERROR:  relation "x.xue" does not exist
LINE 1: select * from x.xue
                      ^
[pg15@dev-dbteam-148180 pg_root]$ psql -h 10.60.148.181 -d xxg  -c "select * from x.xue"
^CCancel request sent
ERROR:  canceling statement due to user request
```

图 9-47　主库删除表查询延迟库

```
postgres=# select * from pg_stat_replication ;
-[ RECORD 1 ]----+------------------------------
pid              | 52734
usesysid         | 10
usename          | postgres
application_name | walreceiver
client_addr      | 10.60.148.181
client_hostname  |
client_port      | 37812
backend_start    | 2023-07-06 15:00:06.179136+08
backend_xmin     |
state            | streaming
sent_lsn         | 0/8036A78
write_lsn        | 0/8036A78
```

图 9-48　延迟复制从库进度

```
flush_lsn        | 0/8036A78
replay_lsn       | 0/80355B8
write_lag        | 00:00:00.000919
flush_lag        | 00:00:00.001539
replay_lag       | 00:08:57.208669
sync_priority    | 0
sync_state       | async
reply_time       | 2023-07-06 16:44:09.556786+08

postgres=# exit

[pg15@dev-dbteam-148181 pg_root]$ pg_dump -d xxg  -t x.xue  >/setup/xue.txt
^Cpg_dump: terminated by user
```

图 9-48 延迟复制从库进度（续）

而当时间达到 10min 后，检查主从数据表，如图 9-49 所示，主库上发起对主库 180 的查询和从库 181 的查询，都显示表已经被删除了，而在从库 181 上的逻辑备份没有备份到被误删除的表。

```
[pg15@dev-dbteam-148180 pg_root]$ date
Thu Jul  6 17:01:18 CST 2023
[pg15@dev-dbteam-148180 pg_root]$ psql -h 10.60.148.180  -d xxg  -c "select * from x.xue"
ERROR:  relation "x.xue" does not exist
LINE 1: select * from x.xue
                      ^
[pg15@dev-dbteam-148180 pg_root]$ psql -h 10.60.148.181  -d xxg  -c "select * from x.xue"
ERROR:  relation "x.xue" does not exist
LINE 1: select * from x.xue
                      ^

[pg15@dev-dbteam-148181 pg_root]$ date
Thu Jul  6 17:01:31 CST 2023
[pg15@dev-dbteam-148181 pg_root]$ pg_dump -d xxg  -t x.xue  >/setup/xue.txt
pg_dump: error: no matching tables were found
```

图 9-49 检查主从数据表存在

现阶段在原生的 PostgreSQL 的延迟从库上未能达到和 MySQL 一样的预期效果。根据该延迟参数建立的延迟从库主要用于 DML 操作，而不能用于 DDL 操作。

当然 PostgreSQL 也有很多种解决方案，这里主要是说明该方案并不能直接达到预期，在实际工作中要对此有一个预见。

这里介绍一下另外一个数据恢复方案，该方案在主库上即可完成对少量数据误操作的恢复。安装 walminer 扩展插件，如图 9-50 所示，会报错，因为该方案需要手工安装的这个扩展功能插件，PostgreSQL 本身没自带。

```
postgres=# create extension walminer ;
ERROR:  extension "walminer" is not available
DETAIL:  Could not open extension control file "/opt/pgsql_15.3/share/extension/walminer.control
HINT:  The extension must first be installed on the system where PostgreSQL is running.
```

图 9-50　安装 walminer 扩展插件

该扩展插件的下载地址是：

https://gitee.com/movead/XLogMiner/tree/walminer_3.0_stable/#https://gitee.com/movead/XLogMiner/releases/download/walminer_dev_4.1.0_20221113/walminer_x86_64_centos_v4.1.0.tar.gz。下载完成后上传到 PostgreSQL 数据库服务器上。按照报错提示，将加压后的 walminer 目录复制到编译环境中。寻找编译环境路径，全目录复制到指定位置，如图 9-51 所示。

```
[root@MySQL-M-connect walminer]# find / -name contrib
/setup/postgresql-15.3/contrib


[root@MySQL-M-connect XLogMiner-master]# cp -r  walminer /setup/postgresql-15.3/contrib
```

图 9-51　寻找编译环境路径

确认进入 walminer 目录，执行 make && make install 的编译指令，如图 9-52 所示。整个过程无须停机。

```
[root@MySQL-M-connect walminer]# pwd
/setup/postgresql-15.3/contrib/walminer

[root@MySQL-M-connect walminer]# make && make install
```

图 9-52　确认进入 walminer 目录

编译完成后，直接可以使用。再次安装扩展，在失败的命令后再次执行安装扩展功能插件，执行成功，如图 9-53 所示。

```
postgres=# create extension walminer ;
ERROR:  extension "walminer" is not available
DETAIL:  Could not open extension control file "/opt/pgsql_15.3/share/extension/walminer.control
HINT:  The extension must first be installed on the system where PostgreSQL is running.

postgres=# create extension walminer ;
CREATE EXTENSION
postgres=#
```

a)

```
postgres=# \c xxg
You are now connected to database "xxg" as user "postgres".

xxg=# create table x.x (id int,a varchar(5),t  timestamp);
CREATE TABLE

xxg=# insert into x.x values (1,'a',now());
INSERT 0 1
xxg=# insert into x.x values (2,'b',now());
INSERT 0 1
xxg=# insert into x.x values (3,'c',now());
INSERT 0 1
```

b)

图 9-53　再次安装扩展

查询当前日志路径，如图 9-54 所示。先通过 show data_directory;命令获得主数据目录，再通过 select pg_walfile_name(pg_current_wal_lsn());获得当前日志文件名。

```
xxg=# show data_directory;
    data_directory
----------------------
 /pgdata/pg15/pg_root
(1 row)


xxg=# select pg_walfile_name(pg_current_wal_lsn());
     pg_walfile_name
--------------------------
 000000060000000000000000C
(1 row)
```

图 9-54　查询当前日志路径

执行 select walminer_wal_add 命令在 walminer 中添加分析日志，如图 9-55 所示。括号内的参数为刚刚查出来的 data_directory 的绝对路径+pg 的归档目录（pg_wal）+刚才查出来的 pg_walfile_name 文件名。这里可以加多个文件，命令分多次执行，而不是在一个命令中完成。也可以一次性添加所有文件，只需要输入到 pg_wal 目录即可。类似的命令是 select walminer_wal_add('/pgdata/pg15/pg_root/pg_wal/')。

```
xxg=#  select walminer_wal_add('/pgdata/pg15/pg_root/pg_wal/000000060000000000000000C');
    walminer_wal_add
---------------------
 1 file add success
(1 row)
```

图 9-55　在 walminer 中添加日志

添加好待分析的日志后，执行 select walminer_all();解析日志的命令，walminer 解析日志如图 9-56 所示。这个命令在不同的版本中不一样。

```
xxg=#  select walminer_all();
NOTICE:  Switch wal to 000000060000000000000000C on time 2023-07-07 11:20:40.961881+08
    walminer_all
---------------------
 pg_minerwal success
(1 row)
```

图 9-56　walminer 解析日志

最后解析 wal 日志得到了已经执行的 SQL 以及该 SQL 的逆运算。可恢复的 SQL 如图 9-57 所示。在 undo_text 下列出了与 op_text 已经执行的 SQL 对应的反向 SQL，如果 op_text 是 delete 的操作，则 up_text 就是 insert 的操作。对于误操作的 delete，就可以复制 undo_text 的 insert 进行恢复。但是当表被误删后，walminer_contents 表内无法查到过去的 SQL。

```
xxg=# select op_text,undo_text from walminer_contents;
                         op_text                          |                         undo_text
----------------------------------------------------------+----------------------------------------------------------
 INSERT INTO x.x(id ,a ,t) VALUES(1 ,'a' ,'2023-07-07 11:17:54.722237') | DELETE FROM x.x WHERE id=1 AND a='a' AND t='2023-07-07 11:17:54.722237'
 INSERT INTO x.x(id ,a ,t) VALUES(2 ,'b' ,'2023-07-07 11:17:54.726782') | DELETE FROM x.x WHERE id=2 AND a='b' AND t='2023-07-07 11:17:54.726782'
 INSERT INTO x.x(id ,a ,t) VALUES(3 ,'c' ,'2023-07-07 11:17:54.729335') | DELETE FROM x.x WHERE id=3 AND a='c' AND t='2023-07-07 11:17:54.729335'
(3 rows)
```

图 9-57　得到可恢复的 SQL

官方给出的原因是：在 PostgreSQL 数据库中，表以文件的形式采用 OID 命名规则存储于 $PGDATA/base/数据库 ID/relfilenode 目录中。当进行 DROP TABLE 操作时，会将文件整体删除。由于在操作系统中表文件已经不存在，所以只能采用恢复磁盘的方法进行数据恢复。

对于数据误删的恢复还有一个解决方案——pgtrashcan，不过在撰写本章时对其实验发现，该项目已经没有维护了。所以下载源码后在进行编译时，pgtrashcan 不支持新版本编译，在新版本中无法编译通过，如图 9-58 所示。使得对于误删的恢复只能使用传统的备份手段。

```
[root@MySQL-M-connect pgtrashcan-master]# make PG_CONFIG=/opt/pgsql_15.3/bin/pg_config
gcc -std=gnu99 -Wall -Wmissing-prototypes -Wpointer-arith -Wdeclaration-after-statement -Werror=vla -Wendif-labels -Wmissing-format-attribute
-Wformat-security -fno-strict-aliasing -fwrapv -fexcess-precision=standard -O2 -fPIC -I. -I./ -I/opt/pgsql_15.3/include/internal -D_GNU_SOURCE  -c -o pgtrashcan.o pgtrashcan.c
pgtrashcan.c: In function '_PG_init':
pgtrashcan.c:45:22: warning: assignment from incompatible pointer type [enabled by default]
  ProcessUtility_hook = pgtrashcan_ProcessUtility;
                      ^
pgtrashcan.c: In function 'pgtrashcan_ProcessUtility':
pgtrashcan.c:149:5: error: 'RangeVar' has no member named 'inhOpt'
     r->inhOpt = INH_YES;
      ^
pgtrashcan.c:149:16: error: 'INH_YES' undeclared (first use in this function)
     r->inhOpt = INH_YES;
                 ^
pgtrashcan.c:149:16: note: each undeclared identifier is reported only once for each function it appears in
pgtrashcan.c:162:2: warning: passing argument 1 of 'prev_ProcessUtility' from incompatible pointer type [enabled by default]
  (*prev_ProcessUtility) (parsetree, queryString, context, params, dest, completionTag);
  ^
pgtrashcan.c:162:2: note: expected 'struct PlannedStmt *' but argument is of type 'struct Node *'
pgtrashcan.c:162:2: error: incompatible type for argument 4 of 'prev_ProcessUtility'
pgtrashcan.c:162:2: note: expected 'ProcessUtilityContext' but argument is of type 'ParamListInfo'
pgtrashcan.c:162:2: warning: passing argument 5 of 'prev_ProcessUtility' from incompatible pointer type [enabled by default]
pgtrashcan.c:162:2: note: expected 'ParamListInfo' but argument is of type 'struct DestReceiver *'
pgtrashcan.c:162:2: warning: passing argument 6 of 'prev_ProcessUtility' from incompatible pointer type [enabled by default]
pgtrashcan.c:162:2: note: expected 'struct QueryEnvironment *' but argument is of type 'char *'
pgtrashcan.c:162:2: error: too few arguments to function 'prev_ProcessUtility'
make: *** [pgtrashcan.o] Error 1
```

图 9-58 pgtrashcan 不支持新版本编译

受限于个人目前水平，在 PostgreSQL 数据库上最佳实践是执行 DROP 时候一定要确认是不是可以 DROP。对于一般的 DELETE 误删除，还是可以通过延迟库或 walminer 来解决的。

9.1.5 数据库对于磁盘 IO 吞吐的要求

数据库是一个 IO 密集型产品，对磁盘 I/O（输入/输出）有着特定的要求。这是因为数据库需要频繁地读取和写入数据到磁盘。以下是数据库对磁盘 IO 的一些要求和注意事项。

1）高吞吐量：数据库通常需要处理大量的读写操作，因此对磁盘的吞吐量要求较高。磁盘的读写速度应该足够快，以便能够快速处理数据库的请求。

2）低延迟：数据库对于磁盘 IO 的延迟要求也很高。延迟是指从发送请求到接收到响应之间的时间间隔。较低的延迟可以提高数据库的响应速度和性能。

3）随机访问性能：数据库通常需要随机地访问磁盘上的数据，而不仅仅是顺序读写。因此，磁盘的随机访问性能也是数据库关注的重点之一。

基于以上几点，数据库一般来说禁止使用 NAS（网络附加存储）这种存储介质，因为 NAS 对于大文件的读写较为友好，适用于网盘等场景。而数据库的 IO 密集体现在对数据块的读写上，若 NAS 存储用在数据库上，会使得数据库性能很差。此外数据库对 IO 还有其他几个要求。

1）缓存管理：数据库通常会使用缓存来减少对磁盘的访问次数，提高性能。数据库会将经常访问的数据缓存在内存中，以便更快地响应读取请求。因此，对于磁盘 IO 来说，缓存管理也是重要的考虑因素。

2）可靠性和持久性：数据库对于数据的可靠性和持久性要求很高。磁盘 IO 能够最大限度确保数据在写入时不会丢失，并且在发生故障或断电时能够进行恢复。

大部分数据库的体系和优化技巧都是围绕着如何减少 IO 而制定的，所以 IO 矛盾是数据库的首要矛盾。这个矛盾在 SQL 高质量的环境中并不明显，甚至可能极少出现。但是如果 SQL 普遍质量不高的话，那么这个矛盾就非常突出了。如果开发团队缺少设计和优化经验，或者即使有解决方案但是无法去整改的话，就只能通过提升 IO 能力来缓解问题。在应用程序没有办法优化或整改的情况下，可以替换高性能的硬件，使得诸如并发读取、大量写入甚至是互斥锁等问题都得到了解决。

提升 IO 能力的途径有以下几种。

1）使用高性能的磁盘驱动器或存储设备，如固态硬盘（SSD）或 RAID 阵列，以提供更快的读写速度和更好的可靠性。在硬件日新月异的今天，IO 能力已经比几十年前提升了很多。每秒几十 GB 到每秒几百 GB 的硬件已经比较成熟，成为主流配置。

2）配置适当的磁盘阵列和缓存设置，以平衡吞吐量、延迟和可靠性，如 Oracle 的 Exadata 等是可以依靠增加阵列的硬盘来达到增加吞吐量的。

3）定期进行磁盘维护和优化，如碎片整理和磁盘清理，以提高性能。一般来说，在关系数据库中 DELETE 的操作是不释放空间的，久而久之会造成碎片，所以在设计应用程序时应该避免使用 DELETE 的操作。

4）根据数据库的工作负载和访问模式进行调优，如调整缓存大小、调整日志写入策略等。众所周知，内存越大，缓存的数据也就越多，可以支持更多的数据加载到内存中，从而使用内存读来代替磁盘读。这样即使硬盘的 IO 能力不高，数据库整体能力也会比较高。在应用程序没有办法优化或暂时无法整改而基础 IO 能力又不足的双重制约下，增加内存并且调大相关参数，使得大量数据在内存中完成检索和排序，也能解决许多问题。

一些疑难杂症有的时候就是因为换了高性能的磁盘或者是增加了内存，都得到了缓解甚至是解决。可见 IO 能力对数据库的稳定性是多么至关重要。在信通院发布的数据库趋势中，利用新兴硬件就是趋势之一。

9.2 面向执行器和优化器的最佳实践

数据库的优化器和执行器是核心组件。不同的数据库的组件功能和特性是不一样的，所以最佳实践也不一样。这部分也是与应用开发人员交互最多的，希望可以对广大应用开发人员起到帮助。

9.2.1 主流数据库的事务异常处理机制对比（MySQL、PostgreSQL、Oracle）

不同数据库对事务异常的处理存在区别，下面对比一下这几个主流数据库的差异。首先在 MySQL 中，建立两张表结构一模一样的事务实验基础表，如图 9-59 所示。

```
mysql> create table xxgt (id int ,a int ,b int, primary key (id));
Query OK, 0 rows affected (0.03 sec)

mysql> create table tp (id int ,a int ,b int, primary key (id));
Query OK, 0 rows affected (0.03 sec)
```

图 9-59 MySQL 建立两张事务实验基础表

分别给两张带主键的表写入初始化数据，如图 9-60 所示。

```
mysql> insert into xxgt values (1,1,1),(2,2,2),(3,3,3);
Query OK, 3 rows affected (0.01 sec)
Records: 3  Duplicates: 0  Warnings: 0

mysql> insert into tp values (1,1,1);
Query OK, 1 row affected (0.01 sec)

mysql> select * from xxgt;
+----+------+------+
| id | a    | b    |
+----+------+------+
|  1 |    1 |    1 |
|  2 |    2 |    2 |
|  3 |    3 |    3 |
+----+------+------+
3 rows in set (0.00 sec)

mysql> select * from tp;
+----+------+------+
| id | a    | b    |
+----+------+------+
|  1 |    1 |    1 |
+----+------+------+
1 row in set (0.00 sec)
```

图 9-60 MySQL 两张事务表初始化数据

分别模拟对两个表写入数据。其中一个表写入事务正常，写入成功；而另一个表在写入时制造写入异常的场景，使得写入失败。事务部分成功，如图 9-61 所示。

```
mysql> begin;
Query OK, 0 rows affected (0.00 sec)

mysql> insert into xxgt values (4,4,4);
Query OK, 1 row affected (0.00 sec)

mysql> insert into tp values (1,1,1);
ERROR 1062 (23000): Duplicate entry '1' for key 'tp.PRIMARY'
mysql> commit;
Query OK, 0 rows affected (0.00 sec)
```

图 9-61 事务部分成功

最终检查两个表的数据是部分成功。因为在事务提交后发现 xxgt 表预计写入 id 为 4 的数据成功，而 tp 表写入 id 为 1 的数据失败。而这两条数据是在一个事务中完成的。事务结束后查询如图 9-62 所示。那这个是不是说明关系数据库的 ACID 被打破了？或者说是不严谨呢？

```
mysql> select * from xxgt;
+----+------+------+
| id | a    | b    |
+----+------+------+
|  1 |    1 |    1 |
|  2 |    2 |    2 |
|  3 |    3 |    3 |
|  4 |    4 |    4 |
+----+------+------+
4 rows in set (0.00 sec)

mysql> select * from tp;
+----+------+------+
| id | a    | b    |
+----+------+------+
|  1 |    1 |    1 |
+----+------+------+
1 row in set (0.00 sec)
```

图 9-62 事务结束后查询

其实不能这样说，这里要注意的是数据库（不仅仅是 MySQL）的处理异常机制。当事务遇到失败（异常）时，正确的数据库操作应该执行的是回退，而不是不管过程强制执行提交。

这里用同样的方法检查 PostgreSQL 数据库的处理过程。与 MySQL 环境一样建立同名的两个表以及初始化一模一样的数据。PostgreSQL 实验表准备如图 9-63 所示。

```
postgres=# \c x
You are now connected to database "x" as user "postgres".

x=# create table xxgt (id int ,a int ,b int, primary key (id));
CREATE TABLE

x=# create table tp (id int ,a int ,b int, primary key (id));
CREATE TABLE

x=# insert into xxgt values (1,1,1),(2,2,2),(3,3,3);
INSERT 0 3

x=# insert into tp values (1,1,1);
INSERT 0 1

x=# select * from xxgt;
 id | a | b
----+---+---
  1 | 1 | 1
  2 | 2 | 2
  3 | 3 | 3
(3 rows)

x=# select * from tp;
 id | a | b
----+---+---
  1 | 1 | 1
(1 row)
```

图 9-63 PostgreSQL 实验表准备

开启事务后，同样采用部分成功的方式模拟操作数据。PostgreSQL 事务处理如图 9-64 所示，PostgreSQL 数据库的执行器在执行提交的时候，发现了事务中存在异常。最终 PostgreSQL 数据库真实执行的其实是回滚操作（在 commit 之后隐式执行了 ROLLBACK）。也就是说，PostgreSQL 数据库其实没有执行 commit。

```
x=# begin;
BEGIN
x=*# insert into xxgt values (4,4,4);
INSERT 0 1
x=*# insert into tp values (1,1,1);
ERROR:  duplicate key value violates unique constraint "tp_pkey"
DETAIL:  Key (id)=(1) already exists.
x=!# commit;
ROLLBACK
```

图 9-64 PostgreSQL 事务处理

Oracle 对于以上事务的处理方式和 MySQL 是一样的。即事务出现异常应该由用户或应用程序主动全部回滚，而不是执行提交。这里不再占用篇幅说明。所以并非是在基础的 ACID 原理上发生了问题，而是数据库显示操作或隐式操作的区别。而事务遇到异常，无论应用程序还是人工执行，都应该执行回退，这个习惯要养成。

9.2.2 MySQL 和达梦两个数据库优化器的对比

有一次在一个群中有人发出求助。求助的朋友说他们的系统从 MySQL 数据库切换到达梦数据库后，反馈下面有个 SQL 原来在 MySQL 数据库中执行很快，在达梦数据库中执行就卡住了。他的反馈是执行很长时间没有结果或者无响应。

这个 SQL 看上去很简单：SELECT * FROM T ORDER BY ID LIMIT 100。这里的 ID 是该表的自增主键，因为是从 MySQL 数据库切换到达梦数据库，所以应该保留。在达梦数据库中的表结构如图 9-65 所示，与 MySQL 中分区的表结构非常相似。

```
"id" INT NOT NULL,
"is_accompanyvisit" INT NOT NULL,
"is_consistent" INT NOT NULL,
"type" INT,
"service_result" INT,
"bjkhsj" TIMESTAMP,
"health_test_state" VARCHAR(30),
"next_reservation_date" TIMESTAMP,
"health_test_date" TIMESTAMP,
"health_test_num" VARCHAR(15),
"microsand_column1" VARCHAR(60),
"microsand_column2" VARCHAR(60),
"is_activation" VARCHAR(15),
"mb_znmf_meeting_id" BIGINT,
CONSTRAINT CONS134219181 PRIMARY KEY ("id","scan_date")
);
CREATE UNIQUE INDEX INDEX33559251 ON RPT."t_activity_statistics" ("id", "scan_date");
```

图 9-65 达梦数据库的表结构

从 SQL 上看是要查这个表最早的 100 条记录。在 MySQL 数据库中优化器的逻辑就是使用主键，找到 1 然后提取 1～100 的这些数据，而不是全表扫描，也不涉及排序，所以很快。

而在达梦数据库中，执行计划如图 9-66 所示，显示是一个全表扫描后全表排序。在达梦的优化器给出的执行计划中是查了（图中结果集）1.6 亿行数据。如果基于这个执行计划执行那么一定不快。这也就是有人反馈无响应以至于到群里求助的原因。

名称	附加信息	代价	结果集	行数据处理长度	描述
NSET2		8672621	100	1624	RESULT SET
PRJT2	exp_num(53), is_atom(FALSE)	8672621	100	1624	PROJECT OPERATION
SORT3	key_num(2), is_distinct(FALSE), top_flag(1), is_adaptive(0)	8672621	100	1624	SORT
PARALLEL	scan_type(FULL), key_num(0, 0, 0), simple(0)	83587	161910204	1624	PARALLEL
CSCN2	INDEX33569709(t_activity_statistics as t_activity_statistics)	83587	161910204		

图 9-66 达梦执行计划

在参考大家给出的各种建议后虽然性能有些提升，但是提升的效率有限，根本达不到用户的要求。笔者给出了如下的建议，将 SQL 进行改写。改写的 SQL 为：SELECT * FROM T WHERE ID<1000 ORDER BY ID LIMIT 100，改写前后仅仅差了一个 WHERE ID<1000。之所以这样改写，是因为 SQL 的语义是要在 1.6 亿的数据中找到最早的 100 条，既然数据库的优化器识别不出来，那么就人为地告知优化器，在最早的 1000 条中找出来。因为 ID 有主键索引，所以 ID<1000，一定是可以使用到主键索引的。使用到索引就避免了全表，从而性能大大提升。

听从笔者建议改写后，反馈效果和在 MySQL 中一模一样了。其结论是如果使用达梦数据库，要达到原来使用 MySQL 一样的效果，则要对 SQL 进行改写。以上还是一个小的场景，甚至可以说一个单表的简单场景。

这里并非说哪个数据库更加好，而是看哪个数据库的优化器更加强大。越强大的优化器越是能适应各种复杂的场景。这里延伸一下，任意一种数据库的切换，带来的都是 SQL 重新进行适配的改写。有的是少量改写，有的可能是完全重构。

9.2.3 SQL 语句的解析过程分析

一般的关系数据库对于 SQL 都有一个解析过程。Oracle 有硬解析、软解析和软软解析三种。其他数据库有的没有这么详细的分类，但是无论如何解析都是要进行的。如果一个 SQL 被执行过了，说明它的语法是没有问题的，那么它就可以跳过解析阶段去执行了。否则就要先去解析语法，看看语法有没有问题，而解析语法是要将整个 SQL 全部执行一遍。接下来写一句有语法问题的 SQL：selec t id1 ,name f rom ten2 whe re   i d=1，故意写错看看数据库如何处理。Oracle 语法解析 1 如图 9-67 所示，因为拼写错误所以报错。

```
连接到:
Oracle Database 19c Enterprise Edition Release 19.0.0.0.0 - Production
Version 19.3.0.0.0

SQL> desc ten;
 名称                                      是否为空? 类型
 ----------------------------------------- -------- ----------------------------
 ID                                                 NUMBER(20)
 NAME                                               VARCHAR2(30)
 AGE                                                NUMBER(3)
 HEIGHT                                             NUMBER(4)
 WEIGHT                                             NUMBER(4)
 TIME                                               TIMESTAMP(6)

SQL> selec t id1 ,name f rom ten2 whe re    i d=1;
SP2-0734: 未知的命令开头 "selec t id..." - 忽略了剩余的行。
```

图 9-67　Oracle 语法解析 1

Oracle 需要知道第一个关键字是什么，是 SELECT 这样的 DML 语句，还是 ALTER 的 DDL 语句。随后修正了 SELECT 以后，Oracle 语法解析 2 如图 9-68 所示，根据解析优先级依次验证 FROM、WHERE 和等号（由于"="之前有个空格，执行器不知道怎么处理了）。

```
SQL> select  id1 ,name f rom ten2 whe re    i d=1;
select       id1 ,name f rom ten2 whe re    i d=1
                           *
第 1 行出现错误:
ORA-00923: 未找到要求的 FROM 关键字


SQL>
SQL> select  id1 ,name from ten2 whe re    i d=1;
select       id1 ,name from ten2 whe re    i d=1
                                  *
第 1 行出现错误:
ORA-00933: SQL 命令未正确结束


SQL> select  id1 ,name from ten2 where   i d=1;
select       id1 ,name from ten2 where   i d=1
                                    *
第 1 行出现错误:
ORA-00920: 无效的关系运算符
```

图 9-68　Oracle 语法解析 2

语法问题解决后，随后开始了语义和权限的解析。Oracle 语义解析如图 9-69 所示，根据解析优先级依次修正了表名、列名后终于执行出正确的结果。

```
SQL> select   id1 ,name from ten2 where  id=1;
select        id1 ,name from ten2 where  id=1
                                *
第 1 行出现错误：
ORA-00942: 表或视图不存在


SQL> select   id1 ,name from ten where  id=1;
select        id1 ,name from ten where  id=1
              *
第 1 行出现错误：
ORA-00904: "ID1": 标识符无效


SQL> select   id ,name from ten where  id=1;

       ID NAME
---------- ------------------------------
        1 test_batch_10000
        1 test_single_10000
        1 Y8WCUt
```

图 9-69 Oracle 语义解析

因为 SQL 是用户与数据库打交道的唯一语言，数据库其实并不理解输入 SELECT * FROM XXXX 是什么意思。其实有时候看别人写的 SQL 时我们自己也不理解，比如上面的 XXXX，问了开发才知道是详细信息的汉语拼音缩写。

之所以要跳过解析阶段，是为了避免解析，其道理类似于开发中的复用原则，代价最小。在 MySQL 中也是类似的解析过程，MySQL 解析过程如图 9-70 所示。所不同的是 Oracle 中每个错误提示到位；而 MySQL 采用了语法提示，每个问题前面都有 near，用户可以根据提示定位当前问题点。

```
mysql> desc ten;
+--------+--------------+------+-----+-------------------+-------------------+
| Field  | Type         | Null | Key | Default           | Extra             |
+--------+--------------+------+-----+-------------------+-------------------+
| id     | int          | NO   | PRI | NULL              | auto_increment    |
| name   | varchar(30)  | YES  |     | NULL              |                   |
| age    | int          | YES  |     | NULL              |                   |
| weight | int          | YES  |     | NULL              |                   |
| height | int          | YES  |     | NULL              |                   |
| time   | timestamp(6) | NO   |     | CURRENT_TIMESTAMP(6) | DEFAULT_GENERATED |
+--------+--------------+------+-----+-------------------+-------------------+
6 rows in set (0.01 sec)

mysql> selec t id1 ,name f rom ten2 whe re   i d=1;
ERROR 1064 (42000): You have an error in your SQL syntax; check the manual that corresponds to
 your MySQL server version for the right syntax to use near 'selec t id1 ,name f rom ten2 whe re   i d=1' at line 1

mysql> select id1 ,name f rom ten2 whe re   i d=1;
ERROR 1064 (42000): You have an error in your SQL syntax; check the manual that corresponds to
 your MySQL server version for the right syntax to use near 'rom ten2 whe re   i d=1' at line 1

mysql> select id1 ,name from ten2 whe re   i d=1;
ERROR 1064 (42000): You have an error in your SQL syntax; check the manual that corresponds to
 your MySQL server version for the right syntax to use near 're   i d=1' at line 1
```

图 9-70 MySQL 解析过程

最终也是经过表和列的校验修正，MySQL 解析正确，如图 9-71 所示。

```
mysql> select id1 ,name from ten2 where  id=1;
ERROR 1146 (42S02): Table 'x.ten2' doesn't exist

mysql> select id1 ,name from ten where  id=1;
ERROR 1054 (42S22): Unknown column 'id1' in 'field list'

mysql> select id ,name from ten where  id=1;
+----+------+
| id | name |
+----+------+
|  1 | xxg  |
+----+------+
1 row in set (0.01 sec)
```

图 9-71　MySQL 解析正确

在 PostgreSQL 数据库中解析时，PostgreSQL 解析过程如图 9-72 所示，前半部分和 MySQL 的经过以及提示一模一样。

```
xxg=# \d ten
                        Table "public.ten"
 Column |            Type             | Collation | Nullable |      Default
--------+-----------------------------+-----------+----------+-------------------
 id     | integer                     |           | not null |
 name   | character varying(20)       |           |          |
 age    | integer                     |           |          |
 height | integer                     |           |          |
 weight | integer                     |           |          |
 time   | timestamp without time zone |           |          | CURRENT_TIMESTAMP
Indexes:
    "ten_pkey" PRIMARY KEY, btree (id)

xxg=# selec t id1 ,name f rom ten2 whe re  i d=1;
ERROR:  syntax error at or near "selec"
LINE 1: selec t id1 ,name f rom ten2 whe re  i d=1;
        ^
xxg=#
xxg=# select id1 ,name f rom ten2 whe re  i d=1;
ERROR:  syntax error at or near "rom"
LINE 1: select id1 ,name f rom ten2 whe re  i d=1;
                          ^
xxg=#
xxg=# select id1 ,name from ten2 whe re  i d=1;
ERROR:  syntax error at or near "re"
LINE 1: select id1 ,name from ten2 whe re  i d=1;
                                       ^
```

图 9-72　PostgreSQL 解析过程

而后半部分在列名写得和真实列名有点偏差时，PostgreSQL 解析提示中看到了类似大数据猜你喜欢的这种提示（HINT:　Perhaps you meant to reference the column "ten.id".），如图 9-73 所示。而如果是根本不存在的列，则会直接报出列不存在的错误。

由上述可见，无论什么数据库都是要对 SQL 进行解析，而且每个关键字和字段都要去验证。那么可以得出一个结论，SQL 写得越短越好，越简单越好。虽然 CPU 计算不是问题，但是如果能把 SQL 优化好，那么对数据库来说是锦上添花的。

```
xxg=# select id1 ,name from ten where  id=1;
ERROR:  column "id1" does not exist
LINE 1: select id1 ,name from ten where  id=1;
               ^
HINT:  Perhaps you meant to reference the column "ten.id".
xxg=#
xxg=#
xxg=# select a ,name from ten where  id=1;
ERROR:  column "a" does not exist
LINE 1: select a ,name from ten where  id=1;
               ^
xxg=#
xxg=#
xxg=# select id ,name from ten where  id=1;
 id |  name
----+--------
  1 | XO6T5Z
(1 row)
```

图 9-73　PostgreSQL 解析提示

9.2.4　优化器的多表关联

来看两个 SQL 对比，这里的表名用字母代替。A 和 B 都不知道代表什么，这才是最符合数据库角度去看问题。因为一般开发人员知道自己所写的订单表关联用户表，知道哪个表数据多，哪个表数据少，从表名就可以判断是 USER 表或 ORDER 表，但是数据库不知道。对数据库来说就是两个 TABLE 的对象。就像下面两句，完全不知道 A 和 B 的业务关系。

SELECT * FROM A,B WHERE A.ID=B.ID AND A.ID=1;--第一句
SELECT * FROM A,B WHERE A.ID=B.ID AND A.ID=1 AND B.ID=1;--第二句

请问第一句和第二句是不是等价？

第一句是说 A 和 B 两个表关联，关联列是 ID，要找出 A 表 ID=1 的数据，由于 A 表 ID 要和 B 表的 ID 相等，那么 B 表的 ID=1 的数据也被找出来，一并返回给客户端。这么说来，第一句和第二句就是等价的了。因为 A 表 ID=1 的同时，B 表 ID=1 也成立。其实这就是数据库优化器的等价改写。至于先找 A 表 ID=1 还是先找 B 表 ID=1，仅凭借现在的条件是无法判断的。但是有一点可以肯定，就是只有先 A 表 ID=1 或者先 B 表 ID=1 两种可能。

那么继续做一个推论，如果像下面这两句一样，又增加了一个 C 表。请问第三句和第四句是不是等价？

SELECT * FROM A,B,C WHERE A.ID=B.ID AND A.ID=C.ID AND C.ID=1;--第三句
SELECT * FROM A,B,C WHERE A.ID=B.ID AND A.ID=C.ID AND C.ID=1 AND B.ID=1 AND A.ID=1;
--第四句

通过刚才第一句和第二句的论证，可以推论第三句和第四句也同样是等价的。所不同的是有以下几种可能性：

- 先 A 表 ID=1 再 B 表 ID=1 再 C 表 ID=1。
- 先 A 表 ID=1 再 C 表 ID=1 再 B 表 ID=1。
- 先 B 表 ID=1 再 A 表 ID=1 再 C 表 ID=1。

- 先 B 表 ID=1 再 C 表 ID=1 再 A 表 ID=1。
- 先 C 表 ID=1 再 A 表 ID=1 再 B 表 ID=1。
- 先 C 表 ID=1 再 B 表 ID=1 再 A 表 ID=1。

共计 6 种可能。也就是说，两个表关联是 2 种可能，三个表关联是 6 种可能，这个关系不是一维的相加，而是阶乘。

2！=1×2

3！=1×2×3

那么 N 表的关联就是 N 的阶乘，笔者在工作中见过 A 表关联 B 表再关联 C 表依次按照 26 个英文字母排列，最后一个表是 P。可想而知，这样的 SQL 复杂度非常高，而这在实际工作中表并不是最多的，30 多个表关联的情况也是存在的。

那么数据库遇到这么多种可能性到底如何选择？这就要靠数据库的大脑——优化器了。每个数据库产品的优化器不一样，目前做得最好的是 Oracle。优化器算法结合准确的统计信息构建了强大的数据库 SQL 解析能力。这里最难的还是优化器的算法，其底层都是基础数据。

由于每个数据库优化器中算法的支持不一，所以优化器有强弱之分。在开源数据库产品上没有商用数据库强，那么是不是使用者需要在使用过程中考量一下？在设计的时候尽可能减少表的定义以及表的关联，在高并发场景下使用反范式的设计。

越是核心的请求最终对应的 SQL 一定是极致简洁的。因为越是核心的 SQL 越是高频，越是简洁的 SQL 越能承受高并发。

商用数据库也是同样的道理，能减少一层关联就减少一层，很多涉及 SQL 改写的逻辑也是从这个角度出发的。

30 多个表乃至以上关联的都是违反开发规范的，即使在 Impala 中也不乏遇到过 30 多个表关联导致崩溃的。

9.2.5 SQL 使用 in 进行子查询

从一个故障谈起，笔者多年前遇到的一个故障，当时数据库 CPU 占用率达到 100%。当时一个下午，整个系统均处于几乎不可用状态。在遇到问题后就重启应用，但是发现重启应用依然毫无作用。

出现问题的是会员模块，大致脱敏 SQL（为了可读性，将该脱敏 SQL 用工具美化后）如下。

```
SELECT
    *
FROM
    USERPROCESS
WHERE
    PROCESSID IN (SELECT
            MAX(PROCESSID)
        FROM
            USERPROCESS
```

```
WHERE
            USERID = ?)
```

数据库是 MySQL 的 5.7。以上 SQL 效果是全表扫描，而这个是会员登录的必备一步，所以可想而知为什么系统几乎不可用，也知道为什么重启应用不起作用了。SQL 的意思是查到用户登录日志中最新一条日志 ID，再根据这个 ID 找到这一行数据。其中 USERID 列和 PROCESSID 均建立了索引，它的问题在于最外层的 IN。

分析问题以后发现是 SQL 写法的不同导致优化器执行得天差地别。而每种数据库的优化器又都不一样。本节是从 MySQL 优化器出发，分析问题，同时对比 MySQL 的兼容产品 TiDB 的同场景表现情况。因为 TiDB 是兼容 MySQL 的，检查一下在优化器方面是借鉴还是自立门户。

模拟故障场景，建立 MySQL 实验环境，如图 9-74 所示。MySQL 的 xxg1 表中，id 是主键，a 列是索引，写入了 5 条数据。

```
mysql> CREATE TABLE xxg1 (
    ->   id INT NOT NULL,
    ->   a INT,
    ->   b VARCHAR(10),
    ->   c DATETIME,
    ->   PRIMARY KEY (id),
    ->   INDEX w1 (a)
    -> );
Query OK, 0 rows affected (0.03 sec)

mysql> INSERT INTO xxg1 (id, a, b, c) VALUES (1, 1, '1', NOW());
Query OK, 1 row affected (0.01 sec)

mysql> INSERT INTO xxg1 (id, a, b, c) VALUES (2, 2, '2', NOW());
Query OK, 1 row affected (0.01 sec)

mysql> INSERT INTO xxg1 (id, a, b, c) VALUES (3, 3, '3', NOW());
Query OK, 1 row affected (0.00 sec)

mysql> INSERT INTO xxg1 (id, a, b, c) VALUES (4, 4, '4', NOW());
Query OK, 1 row affected (0.00 sec)

mysql> INSERT INTO xxg1 (id, a, b, c) VALUES (5, 5, '5', NOW());
Query OK, 1 row affected (0.00 sec)

mysql>
mysql> select * from xxg1;
+----+------+------+---------------------+
| id | a    | b    | c                   |
+----+------+------+---------------------+
|  1 |    1 | 1    | 2023-04-26 20:02:38 |
|  2 |    2 | 2    | 2023-04-26 20:02:38 |
|  3 |    3 | 3    | 2023-04-26 20:02:38 |
|  4 |    4 | 4    | 2023-04-26 20:02:38 |
|  5 |    5 | 5    | 2023-04-26 20:02:38 |
+----+------+------+---------------------+
5 rows in set (0.00 sec)
```

图 9-74 建立 MySQL 实验环境

由于 a 列有索引，那么使用 a 列作为查询条件，MySQL 不同等值执行计划如图 9-75 所示，都用到了索引 w1。

```
mysql> explain select * from xxg1 where a=5;
+----+-------------+-------+------------+------+---------------+------+---------+-------+------+----------+-------+
| id | select_type | table | partitions | type | possible_keys | key  | key_len | ref   | rows | filtered | Extra |
+----+-------------+-------+------------+------+---------------+------+---------+-------+------+----------+-------+
|  1 | SIMPLE      | xxg1  | NULL       | ref  | w1            | w1   | 5       | const |    1 |   100.00 | NULL  |
+----+-------------+-------+------------+------+---------------+------+---------+-------+------+----------+-------+
1 row in set, 1 warning (0.00 sec)

mysql> explain select * from xxg1 where a in (5);
+----+-------------+-------+------------+------+---------------+------+---------+-------+------+----------+-------+
| id | select_type | table | partitions | type | possible_keys | key  | key_len | ref   | rows | filtered | Extra |
+----+-------------+-------+------------+------+---------------+------+---------+-------+------+----------+-------+
|  1 | SIMPLE      | xxg1  | NULL       | ref  | w1            | w1   | 5       | const |    1 |   100.00 | NULL  |
+----+-------------+-------+------------+------+---------------+------+---------+-------+------+----------+-------+
1 row in set, 1 warning (0.00 sec)
```

图 9-75　MySQL 不同等值执行计划

由于涉及 MAX 的极值函数，所以看看 MySQL 极值函数的执行计划，如图 9-76 所示，在 type 列并没有出现 ALL，即不是全表扫描。其中 "Select tables optimized away" 表示已经是优化过的了，没有必要再优化了。从第 3 章索引的概念得知，MAX 和 MIN 这种极值就是找到索引的最左边和最右边，没有比这个更加优化的了。在 MySQL 官方站点翻到两段相关的描述，印证了上述观点，原文如下：

For explains on simple count queries (i.e. explain select count(*) from people) the extra section will read "Select tables optimized away." This is due to the fact that MySQL can read the result directly from the table internals and therefore does not need to perform the select.

```
mysql> explain select max(a) from xxg1;
+----+-------------+-------+------------+------+---------------+------+---------+------+------+----------+------------------------------+
| id | select_type | table | partitions | type | possible_keys | key  | key_len | ref  | rows | filtered | Extra                        |
+----+-------------+-------+------------+------+---------------+------+---------+------+------+----------+------------------------------+
|  1 | SIMPLE      | NULL  | NULL       | NULL | NULL          | NULL | NULL    | NULL | NULL |     NULL | Select tables optimized away |
+----+-------------+-------+------------+------+---------------+------+---------+------+------+----------+------------------------------+
1 row in set, 1 warning (0.00 sec)
```

图 9-76　MySQL 极值函数执行计划

模拟故障场景 SQL，MySQL 的 IN 子查询执行计划如图 9-77 所示，第一句使用=连接子查询的 SQL 在父查询中用到了索引，type 是 ref。而第二句使用 IN 连接子查询的 SQL 在父查询中没有用到索引，type 是 ALL，发生了全表扫描。注意第二个执行计划的 select_type 显示的是 DEPENDENT SUBQUERY，意思是父查询要依赖子查询。实际执行就会先执行外部的查询，然后再循环执行内部的查询。

```
mysql> explain select * from xxg1 where a=(select max(a) from xxg1);--使用索引
+----+-------------+-------+------------+------+---------------+------+---------+-------+------+----------+------------------------------+
| id | select_type | table | partitions | type | possible_keys | key  | key_len | ref   | rows | filtered | Extra                        |
+----+-------------+-------+------------+------+---------------+------+---------+-------+------+----------+------------------------------+
|  1 | PRIMARY     | xxg1  | NULL       | ref  | w1            | w1   | 5       | const |    1 |   100.00 | Using where                  |
|  2 | SUBQUERY    | NULL  | NULL       | NULL | NULL          | NULL | NULL    | NULL  | NULL |     NULL | Select tables optimized away |
+----+-------------+-------+------------+------+---------------+------+---------+-------+------+----------+------------------------------+
2 rows in set, 1 warning (0.00 sec)

mysql> explain select * from xxg1 where a in (select max(a) from xxg1);--全表
+----+--------------------+-------+------------+------+---------------+------+---------+------+------+----------+------------------------------+
| id | select_type        | table | partitions | type | possible_keys | key  | key_len | ref  | rows | filtered | Extra                        |
+----+--------------------+-------+------------+------+---------------+------+---------+------+------+----------+------------------------------+
|  1 | PRIMARY            | xxg1  | NULL       | ALL  | NULL          | NULL | NULL    | NULL |    5 |   100.00 | Using where                  |
|  2 | DEPENDENT SUBQUERY | NULL  | NULL       | NULL | NULL          | NULL | NULL    | NULL | NULL |     NULL | Select tables optimized away |
+----+--------------------+-------+------------+------+---------------+------+---------+------+------+----------+------------------------------+
2 rows in set, 1 warning (0.00 sec)
```

图 9-77　MySQL 的 IN 子查询执行计划

从实验数据得知子查询的 max(a)应该是 5，而在图 9-75 中可以看到 IN 5 和=5 都是可以用到索引的，但是当 IN 遇到了子查询就和=子查询表现出完全不一样的效果。

在这里看起来全表扫描就只有 5 行数据，可是在真实环境中想象一下，如果这个表有几百万到一千万行数据呢？每个用户登录都会扫描百万级别的全表，那么无怪当时系统不可用了。

之所以会发生这样的问题，主要原因可能是从 Oracle 转到 MySQL。MySQL 不是用来替代 Oracle 的选择方案，在介绍数据库选型时也讨论过这个问题。在 Oracle 场景下同样的实验，对于 IN 和等号的处理是无差别的。Oracle 极值函数执行计划如图 9-78 所示，采用了 INDEX FULL SCAN (MIN/MAX)的方式。而且本例中使用的字段为字符串，MySQL 案例中使用的是 INT。

```
SQL> explain plan for select max(name) from ten ;

Explained

SQL> SELECT * FROM TABLE(DBMS_XPLAN.DISPLAY);

PLAN_TABLE_OUTPUT
--------------------------------------------------------------------------------
Plan hash value: 1293465522
--------------------------------------------------------------------------------
| Id  | Operation                  | Name                  | Rows  | Bytes | Cost
--------------------------------------------------------------------------------
|   0 | SELECT STATEMENT           |                       |     1 |     9 |
|   1 |  SORT AGGREGATE            |                       |     1 |     9 |
|   2 |   INDEX FULL SCAN (MIN/MAX)| SYS_AI_fhcgc0pqah0zn  |     1 |     9 |
--------------------------------------------------------------------------------

9 rows selected
```

图 9-78 Oracle 极值函数执行计划

使用=关联子查询，Oracle 极值函数子查询如图 9-79 所示，子查询用到了索引，通过极值函数返回父查询，父查询用到了索引范围扫描。

```
SQL> explain plan for select * from ten  where name= (select max(name) from ten) ;

Explained

SQL> SELECT * FROM TABLE(DBMS_XPLAN.DISPLAY);

PLAN_TABLE_OUTPUT
--------------------------------------------------------------------------------
Plan hash value: 2053661348
--------------------------------------------------------------------------------
| Id  | Operation                             | Name                 | Rows  | Byt
--------------------------------------------------------------------------------
|   0 | SELECT STATEMENT                      |                      |     1 |
|   1 |  TABLE ACCESS BY INDEX ROWID BATCHED  | TEN                  |     1 |
|*  2 |   INDEX RANGE SCAN                    | SYS_AI_fhcgc0pqah0zn |     1 |
|   3 |    SORT AGGREGATE                     |                      |     1 |
|   4 |     INDEX FULL SCAN (MIN/MAX)         | SYS_AI_fhcgc0pqah0zn |     1 |
--------------------------------------------------------------------------------
Predicate Information (identified by operation id):
---------------------------------------------------
   2 - access("NAME"= (SELECT MAX("NAME") FROM "TEN" "TEN"))

16 rows selected
```

图 9-79 Oracle 极值函数子查询 1

使用 IN 关联子查询，Oracle 极值函数子查询如图 9-80 所示，子查询用到了索引，通过极值函数返回父查询，父查询用到了索引范围扫描。可见子查询的函数并没有对父查询起到干扰作用。通过同一家公司的两个产品在这个场景上的使用情况可见优化器处理的逻辑是不一样的，只能说明 Oracle 的优化器强，更加智能。

```
SQL> explain plan for select * from ten  where name in (select max(name) from ten) ;

Explained

SQL> SELECT * FROM TABLE(DBMS_XPLAN.DISPLAY);

PLAN_TABLE_OUTPUT
--------------------------------------------------------------------------------
Plan hash value: 2053661348

--------------------------------------------------------------------------------
| Id  | Operation                           | Name                  | Rows | Byt
--------------------------------------------------------------------------------
|   0 | SELECT STATEMENT                    |                       |    1 |
|   1 |  TABLE ACCESS BY INDEX ROWID BATCHED| TEN                   |    1 |
|*  2 |   INDEX RANGE SCAN                  | SYS_AI_fhcgc0pqah0zn  |    1 |
|   3 |    SORT AGGREGATE                   |                       |    1 |
|   4 |     INDEX FULL SCAN (MIN/MAX)       | SYS_AI_fhcgc0pqah0zn  |    1 |
--------------------------------------------------------------------------------

Predicate Information (identified by operation id):
--------------------------------------------------------

   2 - access("NAME"= (SELECT MAX("NAME") FROM "TEN" "TEN"))

16 rows selected
```

图 9-80　Oracle 极值函数子查询 2

那么换到兼容 MySQL 的国产数据库上呢？这是本节的重点。TiDB 实验表情况如图 9-81 所示，表中有 8 条记录。由于 TiDB 兼容 MySQL，所以命令和语法基本一致，从 SELECT 语句看不出这是在 TiDB 环境下。而从表结构的命令中看到了 /*T![clustered_index] CLUSTERED */，这是 TiDB 特有的。

```
mysql> select * from s;
+-----+------+------+
| id  | name | b    |
+-----+------+------+
|   1 | a    |   10 |
|   2 | e    |   10 |
|   3 | f    |   10 |
|   4 | g    |   10 |
|   5 | h    |   10 |
| 125 | b    |   10 |
| 126 | c    |   10 |
| 127 | d    |   10 |
+-----+------+------+
8 rows in set (0.01 sec)
```

图 9-81　TiDB 实验表情况

```
mysql> show create table s\G
*************************** 1. row ***************************
       Table: s
Create Table: CREATE TABLE `s` (
  `id` tinyint(4) NOT NULL AUTO_INCREMENT,
  `name` varchar(30) DEFAULT NULL,
  `b` int(11) DEFAULT '10',
  PRIMARY KEY (`id`) /*T![clustered_index] CLUSTERED */,
  KEY `n1` (`name`),
  KEY `b3` (`b`)
) ENGINE=InnoDB DEFAULT CHARSET=utf8mb4 COLLATE=utf8mb4_bin AUTO_INCREMENT=30001
1 row in set (0.03 sec)
```

图 9-81　TiDB 实验表情况（续）

先熟悉一下 TiDB 执行计划，TiDB 基本扫描示例如图 9-82 所示。id 是主键。单行精确检索是根据主键查询返回一行。这种在关系数据库中通常叫作点查，图 9-82 中的 Point_Get_1 的意思就是点查。如果返回值不是单条，而是一个区间，这种在关系数据库中叫作范围查询，就是图中的 TableRangeScan_5。而全表扫描就是图中的 TableFullScan_4。

```
mysql> explain select * from s where id = 1;
+-------------+---------+------+--------------+---------------+
| id          | estRows | task | access object| operator info |
+-------------+---------+------+--------------+---------------+
| Point_Get_1 | 1.00    | root | table:s      | handle:1      |
+-------------+---------+------+--------------+---------------+
1 row in set (0.00 sec)

mysql> explain select * from s where id <3;
+-------------------------+---------+-----------+---------------+----------------------------------------------+
| id                      | estRows | task      | access object | operator info                                |
+-------------------------+---------+-----------+---------------+----------------------------------------------+
| TableReader_6           | 2.67    | root      |               | data:TableRangeScan_5                        |
| └─TableRangeScan_5      | 2.67    | cop[tikv] | table:s       | range:[-inf,3), keep order:false, stats:pseudo |
+-------------------------+---------+-----------+---------------+----------------------------------------------+
2 rows in set (0.01 sec)

mysql> explain select * from s;
+------------------------+---------+-----------+---------------+--------------------------------------+
| id                     | estRows | task      | access object | operator info                        |
+------------------------+---------+-----------+---------------+--------------------------------------+
| TableReader_5          | 8.00    | root      |               | data:TableFullScan_4                 |
| └─TableFullScan_4      | 8.00    | cop[tikv] | table:s       | keep order:false, stats:pseudo       |
+------------------------+---------+-----------+---------------+--------------------------------------+
2 rows in set (0.00 sec)
```

图 9-82　TiDB 基本扫描示例

TiDB 极值函数执行计划如图 9-83 所示，存储节点 tikv 的执行计划是全表扫描。

```
mysql> explain select max(id) from s;
+-----------------------------------+---------+-----------+---------------+------------------------------------------+
| id                                | estRows | task      | access object | operator info                            |
+-----------------------------------+---------+-----------+---------------+------------------------------------------+
| StreamAgg_9                       | 1.00    | root      |               | funcs:max(test.s.id)->Column#4           |
| └─Limit_13                        | 1.00    | root      |               | offset:0, count:1                        |
|   └─TableReader_22                | 1.00    | root      |               | data:Limit_21                            |
|     └─Limit_21                    | 1.00    | cop[tikv] |               | offset:0, count:1                        |
|       └─TableFullScan_20          | 1.00    | cop[tikv] | table:s       | keep order:true, desc, stats:pseudo      |
+-----------------------------------+---------+-----------+---------------+------------------------------------------+
5 rows in set (0.01 sec)
```

图 9-83　TiDB 极值函数执行计划

TiDB IN 连接子查询的执行计划如图 9-84 所示，和 MySQL 结果类似，但是过程不一样。因为 TiDB 单独极值函数就是全表。

```
mysql> explain select * from s where id in (select max(id) from s);
+--------------------------------+---------+-----------+---------------+----------------------------------------------------------------------------------------------+
| id                             | estRows | task      | access object | operator info                                                                                |
+--------------------------------+---------+-----------+---------------+----------------------------------------------------------------------------------------------+
| IndexJoin_19                   | 1.25    | root      |               | inner join, inner:TableReader_16, outer key:Column#7, inner key:test.s.id, equal cond:eq(Column#7, test.s.id) |
| ├─HashAgg_26(Build)            | 1.00    | root      |               | group by:Column#7, funcs:firstrow(Column#7)->Column#7                                        |
| │ └─Selection_27               | 0.80    | root      |               | not(isnull(Column#7))                                                                        |
| │   └─StreamAgg_29             | 1.00    | root      |               | funcs:max(test.s.id)->Column#7                                                               |
| │     └─Limit_33               | 1.00    | root      |               | offset:0, count:1                                                                            |
| │       └─TableReader_42       | 1.00    | root      |               | data:Limit_41                                                                                |
| │         └─Limit_41           | 1.00    | cop[tikv] |               | offset:0, count:1                                                                            |
| │           └─TableFullScan_40 | 1.00    | cop[tikv] | table:s       | keep order:true, desc, stats:pseudo                                                          |
| └─TableReader_16(Probe)        | 1.00    | root      |               | data:TableRangeScan_15                                                                       |
|   └─TableRangeScan_15          | 1.00    | cop[tikv] | table:s       | range: decided by [Column#7], keep order:false, stats:pseudo                                 |
+--------------------------------+---------+-----------+---------------+----------------------------------------------------------------------------------------------+
10 rows in set (0.01 sec)
```

图 9-84　TiDB IN 连接子查询的执行计划

TiDB =连接子查询的执行计划如图 9-85 所示，和 MySQL 结果完全不一样，其效果基本和点查一样。

```
mysql> explain select * from s where id = (select max(id) from s);
+---------------+---------+------+---------------+---------------+
| id            | estRows | task | access object | operator info |
+---------------+---------+------+---------------+---------------+
| Point_Get_38  | 1.00    | root | table:s       | handle:127    |
+---------------+---------+------+---------------+---------------+
1 row in set (0.02 sec)

mysql>
```

图 9-85　TiDB =连接子查询的执行计划

而 PostgreSQL 数据库的和 Oracle 数据库在这个场景也不完全一样。PostgreSQL 极值函数执行计划如图 9-86 所示，从 Index Only Scan 可以看出，PostgreSQL 在极值时使用到了索引扫描。

```
xxg=# explain select max(id) from ten;
                                        QUERY PLAN
----------------------------------------------------------------------------------------------
 Result  (cost=0.32..0.33 rows=1 width=4)
   InitPlan 1 (returns $0)
     ->  Limit  (cost=0.29..0.32 rows=1 width=4)
           ->  Index Only Scan Backward using ten_pkey on ten  (cost=0.29..2854.29 rows=100000 width=4)
                 Index Cond: (id IS NOT NULL)
(5 rows)
```

图 9-86　PostgreSQL 极值函数执行计划

在 PostgreSQL 中使用 IN 和=连接子查询的执行计划如图 9-87 所示。既不像 MySQL 那种只有一个用到，也不像 Oracle 那样用到了索引而且一模一样，而是都使用到了索引但是处理逻辑不一样。

通过以上几个实验发现四种数据库有四种做法，这还是一个非常简单的场景，那么复杂场景下呢？一定会变得超出想象。所以在用一种数据库去替代另外一种数据库的情况下，如果应用不做改写其实很难做到一模一样。

```
xxg=# explain select * from ten where id = (select max(id) from ten);
                                    QUERY PLAN
-------------------------------------------------------------------------------------------------
 Index Scan using ten_pkey on ten  (cost=0.62..8.64 rows=1 width=31)
   Index Cond: (id = $1)
   InitPlan 2 (returns $1)
     ->  Result  (cost=0.32..0.33 rows=1 width=4)
           InitPlan 1 (returns $0)
             ->  Limit  (cost=0.29..0.32 rows=1 width=4)
                   ->  Index Only Scan Backward using ten_pkey on ten ten_1  (cost=0.29..2854.29 rows=100000 width=4)
                         Index Cond: (id IS NOT NULL)
(8 rows)

xxg=# explain select * from ten where id in (select max(id) from ten);
                                    QUERY PLAN
-------------------------------------------------------------------------------------------------
 Nested Loop  (cost=0.61..8.65 rows=1 width=31)
   ->  Result  (cost=0.32..0.33 rows=1 width=4)
         InitPlan 1 (returns $0)
           ->  Limit  (cost=0.29..0.32 rows=1 width=4)
                 ->  Index Only Scan Backward using ten_pkey on ten ten_1  (cost=0.29..2854.29 rows=100000 width=4)
                       Index Cond: (id IS NOT NULL)
   ->  Index Scan using ten_pkey on ten  (cost=0.29..8.31 rows=1 width=31)
         Index Cond: (id = ($0))
(8 rows)
```

图 9-87　PostgreSQL 两种子查询连接

9.2.6　数据归档和数据迁移

数据量多其实对于使用来说只要 SQL 正确使用索引就不会出现问题。不过对于存储的要求就不一样了，数据是线性增加，对空间的要求也在不断增加。当空间不足时，要么采用开源策略（如增加磁盘），要么采用节流方案（如压缩数据或数据归档）。其中，增加磁盘并不属于有技术难度的操作，也和本节的归档无关，这里不做介绍。这里介绍数据压缩和数据迁移归档。

注意：数据删除不属于数据归档，而且最重要的是删除数据并不能释放磁盘空间，甚至在有些数据库上对性能毫无改善，而即使今日还有很多人不知道这个事实。下面通过实验带大家看看这个情况。

首先构造一个 1000 万级别的数据表，如图 9-88 所示，其表结构除了 ID，都是随机产生。

收集该表统计信息，如图 9-89 所示，查看表的统计信息，XXGTHOUSAND 表有 1000 万行，占用 76567 个数据块。

```
create table xxgthousand as
SELECT ROWNUM AS ID, TO_CHAR(SYSDATE + ROWNUM/24/3600,'yyyy-mm-dd hh24:mi:ss') as tdate,
TRUNC(DBMS_RANDOM.VALUE(0,100)) AS RANDOM_ID,
DBMS_RANDOM.STRING('X',20) RANDOM_STRING FROM xmltable('1 to 10000000');
```

```
SQL> desc xxgthousand;
Name           Type           Nullable Default Comments
-------------- -------------- -------- ------- --------
ID             NUMBER         Y
TDATE          VARCHAR2(19)   Y
RANDOM_ID      NUMBER         Y
RANDOM_STRING  VARCHAR2(4000) Y
```

图 9-88　构造 1000 万级别的数据表

```
SQL> select table_name,num_rows,blocks from dba_tab_statistics where table_name='XXGTHOUSAND';

TABLE_NAME                                                              NUM_ROWS     BLOCKS
------------------------------------------------------------------- ----------  ---------
XXGTHOUSAND                                                          10000000      76567

Executed in 0.916 seconds


SQL> select count(*) from  xxgthousand;

  COUNT(*)
----------
  10000000

Executed in 2.974 seconds
```

图 9-89　查看表的统计信息

对 XXGTHOUSAND 的 ID 列建立索引，根据 ID 查询及其执行计划如图 9-90 所示，SQL 使用到了 ID 列的索引，执行计划的 Cost 是 3。

```
SQL> select id from xxgthousand where id=2000;

        ID
----------
      2000

已用时间：  00: 00: 00.02

执行计划
----------------------------------------------------------
Plan hash value: 3645589350

--------------------------------------------------------------------
| Id  | Operation         | Name | Rows  | Bytes | Cost (%CPU)| Time     |
--------------------------------------------------------------------
|   0 | SELECT STATEMENT  |      |     1 |     6 |     3   (0)| 00:00:01 |
|*  1 |  INDEX RANGE SCAN | A_ID |     1 |     6 |     3   (0)| 00:00:01 |
--------------------------------------------------------------------
```

图 9-90　根据 ID 查询及其执行计划

由于使用到了索引，只产生了很少的磁盘读。使用索引的实际 IO 发生情况如图 9-91 所示，发生了 1 次磁盘物理读。

```
统计信息
----------------------------------------------------------
          1  recursive calls
          0  db block gets
          4  consistent gets
          1  physical reads
          0  redo size
        552  bytes sent via SQL*Net to client
        402  bytes received via SQL*Net from client
          2  SQL*Net roundtrips to/from client
          0  sorts (memory)
          0  sorts (disk)
          1  rows processed
```

图 9-91　使用索引的实际 IO 发生情况

再进行一个无索引的查询，全表扫描 SQL 执行计划如图 9-92 所示，使用的是 TABLE ACCESS FULL 的全表扫描，并且执行计划的 Cost 达到了 20845。

```
SQL> select RANDOM_STRING from xxgthousand where RANDOM_STRING='VMXIOB5085MB427S3B8P';

RANDOM_STRING
--------------------------------------------------------------------------------
VMXIOB5085MB427S3B8P

已用时间：  00: 00: 02.68

执行计划
----------------------------------------------------------
Plan hash value: 3397830195

---------------------------------------------------------------------------------
| Id  | Operation         | Name         | Rows  | Bytes | Cost (%CPU)| Time     |
---------------------------------------------------------------------------------
|   0 | SELECT STATEMENT  |              |     1 |    21 | 20845   (1)| 00:00:01 |
|*  1 |  TABLE ACCESS FULL| XXGTHOUSAND  |     1 |    21 | 20845   (1)| 00:00:01 |
---------------------------------------------------------------------------------
```

图 9-92　全表扫描 SQL 执行计划

最终由于没有建立索引，无法使用索引，执行了全表扫描。无索引的实际 IO 发生情况如图 9-93 所示，产生了 76164 的磁盘读，76164 个数据块和统计信息 76567 个数据块相差不多。

```
Predicate Information (identified by operation id):
---------------------------------------------------

   1 - filter("RANDOM_STRING"='VMXIOB5085MB427S3B8P')


统计信息
----------------------------------------------------------
          0  recursive calls
          0  db block gets
      76169  consistent gets
      76164  physical reads
          0  redo size
        581  bytes sent via SQL*Net to client
        442  bytes received via SQL*Net from client
          2  SQL*Net roundtrips to/from client
          0  sorts (memory)
          0  sorts (disk)
          1  rows processed
```

图 9-93　无索引的实际 IO 发生情况

而如果删除 50%的数据后,使用索引的 SQL 依旧使用索引。删除数据后使用索引 SQL 的执行计划如图 9-94 所示,其执行计划和未删除数据一致,Cost 依然是 3。

```
SQL> delete from xxgthousand where id<5000000;

已删除 4999999 行。

已用时间:  00: 03: 12.65

SQL> commit;

提交完成。


SQL> select id from xxgthousand where id=6000000;

        ID
----------
   6000000

已用时间:  00: 00: 00.02

执行计划
----------------------------------------------------------
Plan hash value: 3645589350

--------------------------------------------------------------------
| Id  | Operation        | Name | Rows  | Bytes | Cost (%CPU)| Time     |
--------------------------------------------------------------------
|   0 | SELECT STATEMENT |      |     1 |     6 |     3   (0)| 00:00:01 |
|*  1 |  INDEX RANGE SCAN| A_ID |     1 |     6 |     3   (0)| 00:00:01 |
--------------------------------------------------------------------

Predicate Information (identified by operation id):
---------------------------------------------------

   1 - access("ID"=6000000)
```

图 9-94　删除数据后使用索引 SQL 的执行计划

最终使用 SQL 的执行计划,删除数据后使用索引的实际 IO 发生情况如图 9-95 所示,依然只有 1 次磁盘物理读。

```
Predicate Information (identified by operation id):
---------------------------------------------------

   1 - access("ID"=6000000)


统计信息
----------------------------------------------------------
         45  recursive calls
```

图 9-95　删除数据后使用索引的实际 IO 发生情况

```
        0  db block gets
       17  consistent gets
        1  physical reads
        0  redo size
      552  bytes sent via SQL*Net to client
      626  bytes received via SQL*Net from client
        2  SQL*Net roundtrips to/from client
        0  sorts (memory)
        0  sorts (disk)
        1  rows processed
```

图 9-95　删除数据后使用索引的实际 IO 发生情况（续）

而对于没有索引的列，删除数据后 SQL 为全表扫描，查询效率并没有改观，尽管没有查询到数据，但还是全表扫描，Cost 依然为 20845，如图 9-96 所示。

```
SQL> select RANDOM_STRING from xxgthousand where RANDOM_STRING='VMXIOB5085MB427S3B8P';

未选定行

已用时间：  00: 00: 03.19

执行计划
----------------------------------------------------------
Plan hash value: 3397830195

--------------------------------------------------------------------------------
| Id  | Operation          | Name         | Rows  | Bytes | Cost (%CPU)| Time     |
--------------------------------------------------------------------------------
|   0 | SELECT STATEMENT   |              |     1 |    21 | 20845   (1)| 00:00:01 |
|*  1 |  TABLE ACCESS FULL | XXGTHOUSAND  |     1 |    21 | 20845   (1)| 00:00:01 |
--------------------------------------------------------------------------------
```

图 9-96　删除数据后 SQL 为全表扫描

最终没有因为数据减少 50%而执行效率提升一倍。删除数据后无索引的实际 IO 发生情况如图 9-97 所示，产生了 76164 的磁盘读，和没有删除数据时一样。这个现象在 Oracle 数据库中叫作高水位线，数据不断存储就像水位上升一样，即使水位退了，那一条最高的水位线还留在那里。一旦发生全表扫描，Oracle 就从最高水位线的地方进行扫描，这也就是为什么删除数据无助于全表扫描的 SQL 性能。

```
Predicate Information (identified by operation id):
---------------------------------------------------

   1 - filter("RANDOM_STRING"='VMXIOB5085MB427S3B8P')


统计信息
----------------------------------------------------------
          0  recursive calls
          0  db block gets
     114164  consistent gets
      76164  physical reads
          0  redo size
        365  bytes sent via SQL*Net to client
        431  bytes received via SQL*Net from client
          1  SQL*Net roundtrips to/from client
          0  sorts (memory)
          0  sorts (disk)
          0  rows processed
```

图 9-97　删除数据后无索引的实际 IO 发生情况

而这个常识对于大部分开发人员来说还是没有概念，依然采用这种方式进行数据治理。即使遇到没有高水位线的数据库也会因为删除数据不释放空间而产生碎片，在碎片率较高的情况下，SQL 效率依然或多或少地受到影响。所以不能通过简单粗暴的 DELETE 进行数据归档的收尾工作。

通常数据归档都是归档一些容易占用空间的流水表，这类数据有一个特点是数据量较多，按照时间维度容易切分。所以通常按照年月等维度将数据定期移动（注意这里不是删除）到非生产环境的归档库中。这样做的好处如下。

- 节约在线系统的宝贵磁盘空间。
- 为在线系统瘦身，减少在线系统备份的时长。

由于待归档数据有清晰的时间周期，所以可以在建表之初就给这些属性的表按照其生命周期管理定义成为分区表。这里有些读者朋友可能会说，在工作中这些都是开发人员主导的，这里纠正一下这种错误的流程。数据库设计是不应该由开发人员来完成的，应该有 DBA 的参与，这在 DBA 的岗位定义和工作职责中讲过了。DBA 是设计师和架构师，不仅仅是设计数据库对象，还要把握需求，这样才能控制好数据的存储和使用。

笔者曾经在公安行业的海量数据存储场景中，每天新增几千万到 1 亿的数据，必然会涉及这个问题。在上线之初，甚至在招投标之前，就已经对业务中每天的车辆数据有了较为清晰的估算，结合每年机动车的增加等因素预估了存储的大小以及数据过期的清理策略。

9.3　数据库的复制（克隆）与高可用受控切换的最佳实践

数据库切换是危险度仅次于使用备份还原的工作项目。一般遇到数据库切换的场景，都是数据库遇到灾难的极端场景。数据库的主备高可用架构一般就是用来进行容灾切换的。

数据库的切换分为受控切换和不受控切换。本节讲述的是计划停机内的受控切换，一般情况不受控切换（自动切换）有一定的风险。在工作中尽可能保证 RDBMS 采用本地高可用模式，如 Oracle 的 RAC 和 MySQL 的 MGR 等确保数据一致性的架构，这样的架构对于应用程序透明，不存在切换（角色）的问题。

当数据库需要切换时说明"灾难"已经来临。因为但凡数据库可以启动是不需要切换的，数据库切换通常是主数据库无法启动，或者主数据库服务器所在的服务器（或阵列）因为种种不可抗原因导致数据库不可用，那么这种情况下数据库是要进行切换的。在此特别强调，能不切换数据库就不要切换，数据库切换存在数据不一致性的风险，而重启不存在这个风险，很多时候重启的时间和切换的时间相比差别不大。

本节将讲述 Oracle 的 PDB 数据库克隆、MySQL 的克隆、MySQL 的 MGR 架构以及受控切换、PostgreSQL 高可用架构的主从切换、Redis 架构高可用切换等最佳实践。

9.3.1　Oracle 的 PDB 数据库克隆

数据库的迁移一直是一个耗时的问题。如果对一致性有要求，需要停机进行逻辑备份导出。在 Oracle 数据库与 Oracle 数据库之间的迁移可以采用 OGG 进行数据同步，再进行切换，停机时间很短。不过自从数据库克隆技术出来以后，这一切变得容易很多，效率也大大提升。因为克隆数据库是对数据库文件的物理复制，不需要逐条导出和导入。而且 Oracle 的 PDB 克隆不仅仅可以在 CDB 内复制克隆，也可以借助 dblink 进行跨主机的数据库克隆。本实验介绍在 Oracle 中克隆 PDB 的方式，演示同时使用 dblink 配合达到远程复制的效果。

数据源来自 IP 地址尾数为 101 的服务器，源端数据库情况如图 9-98 所示，在 101 数据库上建立了 SOURCEPDB 的数据库。

```
[oracle@dev-dbteam-143101 ~]$ s

SQL*Plus: Release 19.0.0.0.0 - Production on 星期四 6月 15 09:06:56 2023
Version 19.3.0.0.0

Copyright (c) 1982, 2019, Oracle.  All rights reserved.


连接到:
Oracle Database 19c Enterprise Edition Release 19.0.0.0.0 - Production
Version 19.3.0.0.0

SQL> show pdbs

    CON_ID CON_NAME                       OPEN MODE  RESTRICTED
---------- ------------------------------ ---------- ----------
         2 PDB$SEED                       READ ONLY  NO
         3 XXG                            READ WRITE NO
         4 XXG2                           READ WRITE NO
         5 TU                             READ WRITE NO
         6 PDBTANG                        READ WRITE NO
         7 SOURCEPDB                      READ WRITE NO
```

图 9-98　源端数据库情况

在 SOURCEPDB 数据库中建表并且模拟写入数据，用于验证数据库克隆效果。建立验证数据如图 9-99 所示。

```
SQL> create table source(id int);

表已创建。

SQL> insert into source values (101);

已创建 1 行。

SQL> select * from source;

        ID
----------
       101
```

图 9-99 建立验证数据

在 IP 地址尾号为 103 的目标库上进行连接配置，远程连接配置 tnsnames，建立从 103 连接到 101 的数据库连接字符串，如图 9-100 所示。

```
[oracle@adg1 ~]$ vim /u01/app/oracle/product/19.3.0/db/network/admin/tnsnames.ora
source =
  (DESCRIPTION =
    (ADDRESS_LIST =
      (ADDRESS = (PROTOCOL = TCP)(HOST = 10.60.143.101)(PORT = 1521))
    )
    (CONNECT_DATA =
      (SERVER = DEDICATED)
      (SERVICE_NAME = sourcepdb)
    )
  )

target =
  (DESCRIPTION =
    (ADDRESS_LIST =
      (ADDRESS = (PROTOCOL = TCP)(HOST = adg1)(PORT = 1521))
    )
    (CONNECT_DATA =
      (SERVER = DEDICATED)
      (SERVICE_NAME = targetpdb)
    )
  )
```

图 9-100 远程连接配置 tnsnames

在远程数据库上为将要迁移过来的 PDB 建立新的 PDB 做路径准备，建立目标数据库的规划路径，如图 9-101 所示。注意这里建立目录使用操作系统的 oracle 用户，可以确保权限正确。

```
[oracle@adg1 ~]$ cd /u01/app/oracle/oradata/O19C/
[oracle@adg1 O19C]$ mkdir -p targetpdb
```

图 9-101 建立目标数据库的规划路径

登录 103 目标服务器数据库的 CDB，检查目标数据库状态，如图 9-102 所示，主从数据库的 PDB 内容相差很多。

```
[oracle@adg1 019C]$ s

SQL*Plus: Release 19.0.0.0.0 - Production on 星期三 6月 14 14:38:11 2023
Version 19.3.0.0.0

Copyright (c) 1982, 2019, Oracle.  All rights reserved.


连接到:
Oracle Database 19c Enterprise Edition Release 19.0.0.0.0 - Production
Version 19.3.0.0.0

SQL> show pdbs;

    CON_ID CON_NAME                       OPEN MODE  RESTRICTED
---------- ------------------------------ ---------- ----------
         2 PDB$SEED                       READ ONLY  NO
         3 ORCLPDB                        READ WRITE NO
         4 PDB1                           READ WRITE NO
         5 PDB2                           READ WRITE NO
         6 TANG                           READ WRITE NO
```

图 9-102　检查目标数据库状态

建立 dblink 如图 9-103 所示，建立了一个连接名为 source_to_target 并且用户名和密码均为 tang 的 dblink 连接。该连接使用的是 tnsnames 中的 source 连接字符串。

```
SQL> create database link source_to_target connect to tang identified by tang using 'source';

数据库链接已创建。
```

图 9-103　建立 dblink

连接建立完成后为远程数据库直连打通连接。然后就可以进行远程克隆建立数据库的动作了，此时可能会发生报错。克隆数据库可能缺失的权限，如图 9-104 所示。

```
SQL> create pluggable database targetpdb from sourcepdb@source_to_target;
create pluggable database targetpdb from sourcepdb@source_to_target
*
第 1 行出现错误:
ORA-17628: 远程 Oracle 服务器返回了 Oracle 错误 1031 ORA-01031:
权限不足
```

图 9-104　克隆数据库可能缺失的权限

遇到以上提示权限不足的时候，需要到源数据库上对连接账号授权。对源端克隆账号授权，如图 9-105 所示。

```
SQL> alter session set container=sourcepdb;

会话已更改。

SQL> grant create pluggable database to tang;

授权成功。
```

图 9-105　对源端克隆账号授权

授权完成后，再到目标库上进行一次克隆的动作。远程克隆迁移数据库完成，如图 9-106 所示。数据库从网络上被物理复制完毕。

```
SQL> create pluggable database targetpdb from sourcepdb@source_to_target;
```

插接式数据库已创建。

图 9-106　远程克隆迁移数据库完成

将克隆过来的数据库启动打开后，可以使用源库的账号登录到被克隆过来的数据库，看到数据已经成功复制过来了。克隆数据库数据验证，数据完成复制过来，如图 9-107 所示。

远程克隆对于迁移数据库来说大大降低了迁移的复杂度且最大限度地减少了迁移数据库的停机时间。

```
[oracle@adg1 O19C]$ sqlplus tang/tang@target

SQL*Plus: Release 19.0.0.0.0 - Production on 星期三 6月 14 15:21:51 2023
Version 19.3.0.0.0

Copyright (c) 1982, 2019, Oracle.  All rights reserved.

上次成功登录时间：星期三  6月   14 2023 14:57:53 +08:00

连接到:
Oracle Database 19c Enterprise Edition Release 19.0.0.0.0 - Production
Version 19.3.0.0.0

SQL> select * from source;

        ID
----------
       101
```

图 9-107　克隆数据库数据验证

9.3.2　MySQL 克隆的关键点

从 MySQL8 开始，MySQL 数据库支持和 Oracle 数据库类似的数据库克隆技术。由于克隆使用的是物理文件的复制，所以比原始的 mysqldump 或 mysqlpump 主库数据速度快几十倍甚至上千倍。使用 MySQL 克隆方法搭建 MySQL 的主从高可用可以使效率大大提升。但是在实际工作中总会发生一些意想不到的问题。本节讲述一次 MySQL 克隆主从后反复出现问题的案例。

首先介绍 MySQL 克隆的操作方法，常常用于搭建主从等需要将当前数据库克隆至另一台数据库上的场景。

在演示环境的源数据库（IP 末尾地址是 138）服务器上，使用 root 账号登录，安装克隆插件，如图 9-108 所示。

```
mysql> install plugin clone soname 'mysql_clone.so';
Query OK, 0 rows affected (0.20 sec)

mysql> show plugins;
.
.
.
.
.
| clone                    | ACTIVE  | CLONE  | mysql_clone.so | GPL  |
+--------------------------+---------+--------+----------------+------+
46 rows in set (0.00 sec)
```

图 9-108　安装克隆插件

通过检查插件，可以看到克隆插件已经安装完毕了，再建立一个用于克隆的账户，并且授权，如图 9-109 所示。

```
mysql> create user clone@'%' identified by '1';
Query OK, 0 rows affected (0.10 sec)

mysql> grant backup_admin on *.* to 'clone';
Query OK, 0 rows affected (0.02 sec)

mysql> grant replication slave,replication client on *.* to clone@'%';
Query OK, 0 rows affected (0.00 sec)

mysql> grant clone_admin on *.* to clone;
Query OK, 0 rows affected (0.01 sec)
```

图 9-109　建立克隆账户并且授权

检查一下源库中的 databases 情况，如图 9-110 所示。

```
mysql> show databases;
+--------------------+
| Database           |
+--------------------+
| information_schema |
| mysql              |
| performance_schema |
| sys                |
| t                  |
| x                  |
| xxg                |
+--------------------+
7 rows in set (0.01 sec)
```

图 9-110　检查源库数据库列表

然后登录空的从库（IP 末尾地址为 91）服务器上。与主库进行同样的安装插件的动作，先用 root 账户登录，使用同样的命令进行安装。

在 91 数据库上建立克隆的用户 clone 并赋予权限。在从库上建立克隆账户（也可直接使用 root 账户克隆，需给 root 授予 clone_admin 权限），如图 9-111 所示。

```
mysql> create user clone@'%' identified by '1';
Query OK, 0 rows affected (0.10 sec)


mysql> grant clone_admin on *.* to clone;
Query OK, 0 rows affected (0.01 sec)

mysql> grant super on *.* to clone@'%';
Query OK, 0 rows affected, 1 warning (0.01 sec)
```

图 9-111 在从库上建立克隆账户

在从库 91 上，使用 clone 账号登录进行克隆动作。在从库上设置需要克隆的主库地址并且执行克隆命令，如图 9-112 所示。

```
[root@mysql8 mysql]# mysql -uclone -p
Welcome to the MySQL monitor.  Commands end with ; or \g.
Your MySQL connection id is 20
Server version: 8.0.29 MySQL Community Server - GPL

Copyright (c) 2000, 2022, Oracle and/or its affiliates.

Oracle is a registered trademark of Oracle Corporation and/or its
affiliates. Other names may be trademarks of their respective
owners.

Type 'help;' or '\h' for help. Type '\c' to clear the current input statement.

mysql> set global clone_valid_donor_list='10.60.143.138:3306';
Query OK, 0 rows affected (0.00 sec)

mysql> clone instance from clone@'10.60.143.138':3306 identified by '1';
Query OK, 0 rows affected (41.84 sec)
```

图 9-112 在从库上设置需要克隆的主库地址并且执行克隆命令

当克隆执行完毕后，再次使用 SQL 操作数据库，此时数据库连接会话会被断开，因为克隆完成后数据库自动重启，所以再次进行操作时会自动重连。克隆成功后数据库重启，如图 9-113 所示。

```
mysql> show databases;
ERROR 2013 (HY000): Lost connection to MySQL server during query
No connection. Trying to reconnect...
Connection id:    8
Current database: *** NONE ***

+--------------------+
| Database           |
+--------------------+
| information_schema |
+--------------------+
1 row in set (0.03 sec)
```

图 9-113 克隆成功后数据库重启

在从库 91 上重新切换至 root 账户检查，主库成功被全部物理复制过来，可以看到源库上的 databases 已经成功克隆到目标库了，如图 9-114 所示。

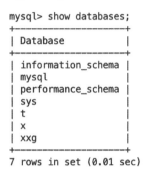

图 9-114　主库成功被全部物理复制过来

正因为是全部物理复制，那么主库上的所有对象都会被复制过来。本案例遇到过一次非常特殊的问题，在克隆数据库后，检查主从状态一致，并且模拟数据增删改查后依然是主从一致的。但是在运行 1～2 天后通过监控就发现主从已经中断，中断的原因是数据重复。再重新克隆建立主从后，依然是刚开始主从同步，但是过几天再次主从中断，中断的原因也是数据重复，重复涉及的表也一样，是一个报表。

经过大量排查以后检查到主库上有定时任务，而克隆数据库将这些定时任务全部照搬到从库。当到达一定的时间，主从数据库上定时任务同时执行生成报表，主库将这个数据传给从库时，从库的报表已经建立了，于是导致了数据冲突进而影响了数据库的主从复制。

综上所述，在使用物理克隆时需要考虑到主从上的存储过程、触发器等因素，在克隆完毕关闭相关对象或是在主库上设计数据库时就避免这些对象的生成。

9.3.3　MySQL 的 MGR 架构以及受控切换

MySQL 的高可用架构解决方案很多，不过官方出品的并不多。目前官方主推的是 MGR（MySQL Group Replication）。本节使用最简单的方式来演示 MGR 的搭建过程。

MySQL 至少需要三个节点，需要准备三个相同版本的数据库节点。演示环境为 IP 结尾

134、138、91 的三台服务机器。

首先在每个节点的配置文件上增加以下参数。MGR 的必需参数如图 9-115 所示，这里的端口不能和数据库使用同一个端口。

```
[root@mysql8 ~]# vim /etc/my.cnf

transaction_write_set_extraction=XXHASH64
loose-group_replication_group_name="aaaaaaaa-aaaa-aaaa-aaaa-aaaaaaaaaaaa"
loose-group_replication_start_on_boot=off
loose-group_replication_local_address= "10.60.143.91:3316"
loose-group_replication_group_seeds= "10.60.143.134:3316,10.60.143.138:3316,10.60.143.91:3316"
loose-group_replication_bootstrap_group=off
loose-group_replication_single_primary_mode=FALSE

[root@mysql8 ~]# service mysqld restart
Stopping mysqld:                                           [  OK  ]
Starting mysqld:                                           [  OK  ]
```

图 9-115 MGR 的必需参数

注意：以上配置需要在每个节点上实施，其中 loose-group_replication_group_seeds 参数为整个集群中每个节点的地址及端口号，loose-group_replication_local_address 参数为本机的节点和端口号。

配置完成后重启数据库，用管理员账户 root 登录数据库。建立连接账号及复制账号，并且安装克隆插件，如图 9-116 所示。

```
mysql> CREATE USER rpl_user@'%' IDENTIFIED BY '1';
Query OK, 0 rows affected (0.01 sec)

mysql> grant replication slave,replication client on *.* to rpl_user@'%';
Query OK, 0 rows affected (0.00 sec)

mysql> install plugin clone soname 'mysql_clone.so';
Query OK, 0 rows affected (0.20 sec)

mysql> create user clone_user@'%' identified by '1';
Query OK, 0 rows affected (0.01 sec)

mysql> grant backup_admin on *.* to clone_user;
Query OK, 0 rows affected (0.00 sec)

mysql> grant replication slave,replication client on *.* to clone_user;
Query OK, 0 rows affected (0.00 sec)

mysql> CHANGE MASTER TO MASTER_USER='rpl_user', MASTER_PASSWORD='1' FOR CHANNEL 'group_replication_recovery';
Query OK, 0 rows affected, 5 warnings (0.05 sec)
```

图 9-116 建立连接账号及复制账号

注意：以上演示环境为了方便，建立的账号密码为"1"，不建议在生产中如此实施。安装完插件后，注意通过 SHOWPLUGINS 命令检查 clone 插件及 group_replication 插件成功安装。

以上账号及插件的配置需要在每个节点上实施。

选取 134 数据库为集群的起始数据库，开始创建集群。开启 MGR 集群，如图 9-117 所示。

```
mysql> set global group_replication_bootstrap_group=ON;
Query OK, 0 rows affected (0.01 sec)

mysql> START group_replication;
Query OK, 0 rows affected (1.47 sec)

mysql> set global group_replication_bootstrap_group=OFF;
Query OK, 0 rows affected (0.00 sec)
```

图 9-117　开启 MGR 集群

成功开启组复制后，再来到其他的数据库节点（即 138 和 91）上进行克隆操作。克隆数据库如图 9-118 所示。指定被克隆的数据库地址和端口，执行克隆复制。

```
mysql> set global clone_valid_donor_list = '10.60.143.134:3306';
Query OK, 0 rows affected (0.00 sec)

mysql> clone instance from clone_user@'10.60.143.134':3306 identified by '1';
Query OK, 0 rows affected (23.03 sec)
```

图 9-118　克隆数据库

成功克隆数据库后，在 138 和 91 数据库上，可以将数据库加入到已经开启的 MGR 集群中去，如图 9-119 所示。START group_replication 是加入 MGR 的命令。后面的会话中断和尝试重连表示，克隆数据库成功以后数据库发生了重启。所以原会话被终止，最终返回 Query OK 说明加入 MGR 成功。

```
mysql> START group_replication;
ERROR 2013 (HY000): Lost connection to MySQL server during query
No connection. Trying to reconnect...
Connection id:    12
Current database: *** NONE ***

Query OK, 0 rows affected (1.79 sec)
```

图 9-119　加入 MGR

当所有节点都加入集群后，在任意节点都可以查询 MGR 的集群状态，如图 9-120 所示，看到状态都是 ONLINE（在本实验过程中遇到过集群连接问题，需要每个节点把所有 MEMBER_HOST 信息都写到自己的 /etc/hosts 中）。

```
mysql> SELECT * FROM performance_schema.replication_group_members;
| CHANNEL_NAME              | MEMBER_ID                            | MEMBER_HOST | MEMBER_PORT | MEMBER_STATE | MEMBER_ROLE | MEMBER_VERSION | MEMBER_COMMUNICATION_STACK |
| group_replication_applier | 685927e0-0c11-11ee-bf98-005056968d7c | mysql8      |        3306 | ONLINE       | PRIMARY     | 8.0.33         | XCom                       |
| group_replication_applier | bca44c90-8f26-11ed-a1b0-005056964c25 | base        |        3306 | ONLINE       | PRIMARY     | 8.0.33         | XCom                       |
| group_replication_applier | e3be5a41-d73b-11ec-9b9d-005056960765 | mysql8.0.20 |        3306 | ONLINE       | PRIMARY     | 8.0.33         | XCom                       |
3 rows in set (0.00 sec)
```

图 9-120　MGR 状态查询

此时多节点 MGR 已经搭建完毕了，可以在任意节点做写入操作。任意节点写入数据，在 134 数据库上写入，在 138 数据库和 91 数据库上都可以查询，如图 9-121 所示。

```
[root@mysql8 ~]# mysql -uroot -p1 -h 10.60.143.134 -e "insert into x.t134 values(134,134)"
mysql: [Warning] Using a password on the command line interface can be insecure.
[root@mysql8 ~]# mysql -uroot -p1 -h 10.60.143.138 -e "select * from x.t134"
mysql: [Warning] Using a password on the command line interface can be insecure.
+-----+-----+
| id  | a   |
+-----+-----+
|   0 |   0 |
|   1 |   1 |
|  91 |  91 |
| 134 | 134 |
+-----+-----+

[root@mysql8 ~]# mysql -uroot -p1 -h 10.60.143.91 -e "select * from x.t134"
mysql: [Warning] Using a password on the command line interface can be insecure.
+-----+-----+
| id  | a   |
+-----+-----+
|   0 |   0 |
|   1 |   1 |
|  91 |  91 |
| 134 | 134 |
+-----+-----+
```

图 9-121 任意节点写入数据

由于 MGR 的每个节点数据都是一致的，所以该架构下数据库不存在单节点故障后的切换问题，只是应用程序需要切换。如果想避免应用程序的切换需要 MySQL Router 的配合，应用程序连接 MySQL Router，MySQL Router 将应用的连接进行故障迁移。

9.3.4 PostgreSQL 的高可用切换

PostgreSQL 数据库的主从高可用在之前的章节已经讲述过了，由于在这种架构下和 Oracle 的 ADG 类似，从库是不可以写入的（不同于 MySQL 的多主 MGR 和 Oracle 的 RAC），那么当主库发生不可用时需要进行主从切换。本节讲述是最基本的手工切换方式。

在现在的模式下主库地址为 10.60.148.180，从库地址为 10.60.148.181。做切换前的准备工作。检查主库数据库角色和状态，148.180 当前角色是主库。select pg_is_in_recovery();的返回是"f"，如图 9-122 所示。f 是主库的标记，t 是从库的标记。同时执行 select client_addr,sync_state from pg_stat_replication;可以看到当前的从库是 10.60.148.181。

这个时候给主库中写入一条数据，检查一下从库是否接受词条数据的变更，以保证主从现在的通信正常以及数据同步正常。检查从库数据库角色和状态，148.181 当前角色是从库，如图 9-123 所示。select pg_is_in_recovery();的返回是"t"。同时执行 select client_addr,sync_state from pg_stat_replication;没有返回值，表示它不是任何数据库的主库。

```
[pg15@dev-dbteam-148180 pg_root]$ psql
psql (15.3)
Type "help" for help.

postgres=# \c xxg;
You are now connected to database "xxg" as user "postgres".

xxg=# insert into xuexiaogang values (4,4);
INSERT 0 1
xxg=# select * from xuexiaogang;
 id | n
----+---
  1 | 1
  2 | 2
  3 | 3
  4 | 4
(4 rows)

xxg=#  select pg_is_in_recovery();
 pg_is_in_recovery
-------------------
 f
(1 row)

xxg=# select client_addr,sync_state from pg_stat_replication;
  client_addr  | sync_state
---------------+------------
 10.60.148.181 | async
(1 row)
```

图 9-122　检查主库数据库角色和状态

```
[pg15@dev-dbteam-148181 pg_root]$ psql
psql (15.3)
Type "help" for help.
xxg=# select * from xuexiaogang;
 id | n
----+---
  1 | 1
  2 | 2
  3 | 3
  4 | 4
(4 rows)

xxg=# select client_addr,sync_state from pg_stat_replication;
 client_addr | sync_state
-------------+------------
(0 rows)

xxg=#  select pg_is_in_recovery();
 pg_is_in_recovery
-------------------
 t
(1 row)
```

图 9-123　检查从库数据库角色和状态

将主库关闭，模拟数据库不可用。手工关闭 PostgreSQL 主库，执行 **pg_ctl stop** 命令关闭主库，如图 9-124 所示。

```
[pg15@dev-dbteam-148180 pg_root]$ pg_ctl stop
waiting for server to shut down.... done
server stopped
[pg15@dev-dbteam-148180 pg_root]$
```

图 9-124 手工关闭 PostgreSQL 主库

此时从库还是只读状态，需要将从库的角色变成主库。PostgreSQL 当前版本的主从架构上，决定数据库角色的是这个 standby.signal 文件。若其存在于 PostgreSQL 主数据目录下，则这个数据库是从库，反之如果该 standby.signal 文件没有存在于 PostgreSQL 主数据目录下，说明该数据库是主库。

该文件仅仅是一个文件，没有实质性的内容。通过 pg_ctl promote 命令提升从库角色为主库。从库进行切换，在执行完 pg_ctl promote 命令后，主数据目录下的 standby.signal 文件没有了，如图 9-125 所示。

```
drwx------ 3 pg15 pg15  124 Jul  2 14:13 pg_wal
drwx------ 2 pg15 pg15   18 Jul  2 11:11 pg_xact
-rw------- 1 pg15 pg15  325 Jul  2 11:11 postgresql.auto.conf
-rw------- 1 pg15 pg15  30K Jul  2 11:11 postgresql.conf
-rw------- 1 pg15 pg15   29 Jul  2 11:21 postmaster.opts
-rw------- 1 pg15 pg15   79 Jul  2 11:21 postmaster.pid
-rw------- 1 pg15 pg15    0 Jul  2 11:11 standby.signal
[pg15@dev-dbteam-148181 pg_root]$ pg_ctl promote
waiting for server to promote.... done
server promoted
[pg15@dev-dbteam-148181 pg_root]$ ll
total 252K
-rw------- 1 pg15 pg15  225 Jul  2 11:11 backup_label.old
-rw------- 1 pg15 pg15 178K Jul  2 11:11 backup_manifest
drwx------ 6 pg15 pg15   46 Jul  2 11:11 base
-rw------- 1 pg15 pg15   44 Jul  2 11:21 current_logfiles
drwx------ 2 pg15 pg15 4.0K Jul  2 11:24 global
drwx------ 2 pg15 pg15  246 Jul  2 11:21 log
drwx------ 2 pg15 pg15    6 Jul  2 11:11 pg_commit_ts
drwx------ 2 pg15 pg15    6 Jul  2 11:11 pg_dynshmem
-rw------- 1 pg15 pg15 4.8K Jul  2 11:11 pg_hba.conf
-rw------- 1 pg15 pg15 1.6K Jul  2 11:11 pg_ident.conf
drwx------ 4 pg15 pg15   68 Jul  2 14:14 pg_logical
drwx------ 4 pg15 pg15   36 Jul  2 11:11 pg_multixact
drwx------ 2 pg15 pg15    6 Jul  2 11:11 pg_notify
drwx------ 2 pg15 pg15    6 Jul  2 11:11 pg_replslot
drwx------ 2 pg15 pg15    6 Jul  2 11:11 pg_serial
drwx------ 2 pg15 pg15    6 Jul  2 11:11 pg_snapshots
drwx------ 2 pg15 pg15    6 Jul  2 11:11 pg_stat
drwx------ 2 pg15 pg15    6 Jul  2 11:11 pg_stat_tmp
drwx------ 2 pg15 pg15   18 Jul  2 11:51 pg_subtrans
drwx------ 2 pg15 pg15    6 Jul  2 11:11 pg_tblspc
drwx------ 2 pg15 pg15    6 Jul  2 11:11 pg_twophase
-rw------- 1 pg15 pg15    3 Jul  2 11:11 PG_VERSION
drwx------ 3 pg15 pg15  188 Jul  2 14:14 pg_wal
drwx------ 2 pg15 pg15   18 Jul  2 11:11 pg_xact
-rw------- 1 pg15 pg15  325 Jul  2 11:11 postgresql.auto.conf
-rw------- 1 pg15 pg15  30K Jul  2 11:11 postgresql.conf
-rw------- 1 pg15 pg15   29 Jul  2 11:21 postmaster.opts
-rw------- 1 pg15 pg15   79 Jul  2 14:14 postmaster.pid
```

图 9-125 从库进行切换

此时登录 148.181，进行角色查询。原从库新主库状态如图 9-126 所示，select pg_is_in_recovery();返回的是 "f"，f 是主库的标记。

```
[pg15@dev-dbteam-148181 pg_root]$ psql
psql (15.3)
Type "help" for help.

postgres=# select pg_is_in_recovery();
 pg_is_in_recovery
-------------------
 f
(1 row)
```

图 9-126　原从库新主库状态

148.181 升为主库，那么 148.180 需要降为从库。原主库降为从库，在原主库 148.180 上，手工创建（无实质内容的空文件）standby.signal 文件，如图 9-127 所示。

```
-rw-------  1 pg15 pg15   88 Jul  1 17:54 postgresql.auto.conf
-rw-------  1 pg15 pg15  30K Jul  2 10:44 postgresql.conf
-rw-------  1 pg15 pg15   29 Jul  1 18:01 postmaster.opts
[pg15@dev-dbteam-148180 pg_root]$ touch standby.signal
[pg15@dev-dbteam-148180 pg_root]$ ll
total 64K
drwx------ 7 pg15 pg15   63 Jul  2 11:11 base
-rw------- 1 pg15 pg15   44 Jul  2 11:08 current_logfiles
drwx------ 2 pg15 pg15 4.0K Jul  2 10:38 global
drwx------ 2 pg15 pg15  166 Jul  2 00:00 log
drwx------ 2 pg15 pg15    6 Jul  1 17:54 pg_commit_ts
drwx------ 2 pg15 pg15    6 Jul  1 17:54 pg_dynshmem
-rw------- 1 pg15 pg15 4.8K Jul  2 11:08 pg_hba.conf
-rw------- 1 pg15 pg15 1.6K Jul  1 17:54 pg_ident.conf
drwx------ 4 pg15 pg15   68 Jul  2 14:13 pg_logical
drwx------ 4 pg15 pg15   36 Jul  1 17:54 pg_multixact
drwx------ 2 pg15 pg15    6 Jul  1 17:54 pg_notify
drwx------ 2 pg15 pg15    6 Jul  2 11:11 pg_replslot
drwx------ 2 pg15 pg15    6 Jul  1 17:54 pg_serial
drwx------ 2 pg15 pg15    6 Jul  1 17:54 pg_snapshots
drwx------ 2 pg15 pg15   25 Jul  2 14:13 pg_stat
drwx------ 2 pg15 pg15    6 Jul  1 17:54 pg_stat_tmp
drwx------ 2 pg15 pg15   18 Jul  1 17:54 pg_subtrans
drwx------ 2 pg15 pg15    6 Jul  1 17:54 pg_tblspc
drwx------ 2 pg15 pg15    6 Jul  1 17:54 pg_twophase
-rw------- 1 pg15 pg15    3 Jul  1 17:54 PG_VERSION
drwx------ 3 pg15 pg15  204 Jul  2 14:13 pg_wal
drwx------ 2 pg15 pg15   18 Jul  1 17:54 pg_xact
-rw------- 1 pg15 pg15   88 Jul  1 17:54 postgresql.auto.conf
-rw------- 1 pg15 pg15  30K Jul  2 10:44 postgresql.conf
-rw------- 1 pg15 pg15   29 Jul  1 18:01 postmaster.opts
-rw-rw-r-- 1 pg15 pg15    0 Jul  2 14:23 standby.signal
[pg15@dev-dbteam-148180 pg_root]$
```

图 9-127　原主库降为从库

现在主从的角色都完成了转换，但是连接关系还没有建立。需要手工指定连接关系，如图 9-128 所示，在数据目录下的 postgresql.auto.conf 配置文件中，增加一段连接字符串信息：

primary_conninfo = 'host=10.60.148.181 port=1922 user=postgres password=1'，其作用是让 148.180 知道新的主库地址是 148.181。

```
[pg15@dev-dbteam-148180 pg_root]$ echo "primary_conninfo = 'host=10.60.148.181 port=1922 user=postgres password=1'" >postgresql.auto.conf
[pg15@dev-dbteam-148180 pg_root]$ cat postgresql.auto.conf
primary_conninfo = 'host=10.60.148.181 port=1922 user=postgres password=1'
```

图 9-128　手工指定连接关系

设置完毕后，启动 148.180 的数据库，执行 pg_ctl start 命令启动新从库（原主库），并且验证 148.180 的角色为从库，如图 9-129 所示。

```
[pg15@dev-dbteam-148180 pg_root]$ pg_ctl start
waiting for server to start....2023-07-02 14:26:01.030 CST [23679] LOG:  redirecting log outpu
2023-07-02 14:26:01.030 CST [23679] HINT:  Future log output will appear in directory "log".
 done
server started
[pg15@dev-dbteam-148180 pg_root]$ psql
psql (15.3)
Type "help" for help.

postgres=# select pg_is_in_recovery();
 pg_is_in_recovery
-------------------
 t
(1 row)
```

图 9-129　启动新从库

而此时新的主库与新的从库的主从关系也可以从图 9-130 中看到。在 148.181 上运行 select client_addr,sync_state from pg_stat_replication;看到新的从库是 148.180。

```
[pg15@dev-dbteam-148181 pg_root]$ psql
psql (15.3)
Type "help" for help.

postgres=# select pg_is_in_recovery();
 pg_is_in_recovery
-------------------
 f
(1 row)

postgres=# select client_addr,sync_state from pg_stat_replication;
  client_addr  | sync_state
---------------+------------
 10.60.148.180 | async
(1 row)
```

图 9-130　新主从关系

此外还有一个判断主从关系和传输的方法，通过 ps 命令检查 wal 进程。如图 9-131 所示，通过进程检查主从角色和状态，执行 ps -ef|grep wal 命令可以看到在主库上 walsender 是发送者，而在从库上 walreceiver 是接收者。

最终写数据验证，在新主库 148.181 上写入一条 id 为 5 的数据，如图 9-132 所示。

数据验证成功，在新从库接收到数据，在 148.180 查到了新数据同步，至此主从切换成功，如图 9-133 所示。

```
[pg15@dev-dbteam-148181 pg_root]$ ps -ef|grep wal
pg15     31633 31559  0 14:14 ?        00:00:00 postgres: walwriter
pg15     31689 31559  0 14:34 ?        00:00:00 postgres: walsender postgres 10.60.148.180(58110) streaming 0/4000418
pg15     31798 31534  0 15:19 pts/0    00:00:00 grep --color=auto wal


[pg15@dev-dbteam-148180 pg_root]$ ps -ef|grep wal
pg15     23899 23679  0 14:34 ?        00:00:02 postgres: walreceiver streaming 0/4000418
pg15     23921 22869  0 15:19 pts/1    00:00:00 grep --color=auto wal
[pg15@dev-dbteam-148180 pg_root]$
```

图 9-131 通过进程检查主从角色和状态

```
[pg15@dev-dbteam-148181 pg_root]$ psql
psql (15.3)
Type "help" for help.

postgres=# \c xxg
You are now connected to database "xxg" as user "postgres".
xxg=# select * from xuexiaogang;
 id | n
----+---
  1 | 1
  2 | 2
  3 | 3
  4 | 4
(4 rows)

xxg=# insert into xuexiaogang values (5,5);
INSERT 0 1
xxg=# select * from xuexiaogang;
 id | n
----+---
  1 | 1
  2 | 2
  3 | 3
  4 | 4
  5 | 5
(5 rows)
```

图 9-132 新主库写入数据

```
[pg15@dev-dbteam-148180 pg_root]$ psql
psql (15.3)
Type "help" for help.

postgres=#  select pg_is_in_recovery();
 pg_is_in_recovery
-------------------
 t
(1 row)

postgres=# \c xxg;
You are now connected to database "xxg" as user "postgres".
xxg=# select * from xuexiaogang;
 id | n
----+---
  1 | 1
  2 | 2
  3 | 3
  4 | 4
  5 | 5
(5 rows)
```

图 9-133 在新从库接收到数据

PostgreSQL 的数据库切换比起 Oracle 来说步骤稍微多了一点，所以如果想完成自动化切换，可以参考开源的解决方案。这些方案有很多，在这里无法一一介绍，本节只是介绍了原生的手工切换方案。

9.3.5 Redis 的高可用切换

虽然哨兵系统很好地实现了主从切换的自动化，但对于客户端来说，使用起来仍然不够方便。因为主节点 IP 的自动切换导致客户端每次访问都需要先向哨兵系统询问当前的主节点 IP，才能访问。更好的解决方法是使用虚拟 IP 地址（Virtual IP Address，VIP）。

VIP 方案中，Redis 系统对外始终使用同一个 IP 地址。当 Redis 进行主从切换时，需要将 VIP 从之前的 Redis 节点移动到现在的新主 Redis 节点上。为实现 VIP 方案，需要使用第三方软件进行虚拟 IP 映射。本节选择了 keepalived 软件来实现。keepalived 安装如图 9-134 所示。在有 Yum 源的情况下，直接安装。没有网络环境的情况下，可以单独寻找 keepalived 的 rpm 安装包进行安装。在本次实验中使用两台机器，机器 IP 分别为 10.60.143.33 和 10.60.143.136。以下称 33 节点和 136 节点，keepalived 软件在两个节点都要安装。

```
yum -y install keepalived

sudo vim /etc/keepalived/keepalived.conf
```

图 9-134　keepalived 安装

打开配置文件，keepalived 主节点的配置如图 9-135 所示。

```
vrrp_script chk_redis {
        script "/etc/keepalived/scripts/redis_check.sh"
        interval 2
}
vrrp_instance VI_1 {
        state MASTER
        interface ens192
        virtual_router_id 51
        priority 101
        authentication {
                    auth_type PASS
                    auth_pass 1
        }
        track_script {
              chk_redis
        }
        virtual_ipaddress {
            10.60.143.131/24
        }
        notify_master /etc/keepalived/scripts/redis_master.sh
        notify_backup /etc/keepalived/scripts/redis_backup.sh
        notify_fault  /etc/keepalived/scripts/redis_fault.sh
        notify_stop   /etc/keepalived/scripts/redis_stop.sh
}
```

图 9-135　keepalived 主节点的配置

下面对配置文件进行一下解释，首先需要定义全局的配置选项 global_defs。该部分定义了一些全局变量。

1）vrrp_instance 部分定义了一个虚拟路由器实例。

2）state 表示当前服务器的状态（MASTER 表示主服务器，BACKUP 表示备用服务器）。

3）interface 表示网卡名称（使用 ifconfig 命令查看当前 IP 对应的网卡）。

4）virtual_router_id 表示虚拟路由器 ID（keepalived 的路由器 ID，可以任意设置两台 keepalive 的 ID 不同即可）。

5）priority 表示服务器的优先级（权重优先级，主库要比从库高，通常主库是 100，从库是 90）。

6）virtual_ipaddress 表示虚拟 IP 地址及子网掩码（这里的 10.60.143.131 是对外服务的虚拟 IP，通常这个 IP 会和主节点 IP 在同一台机器上，即主节点有双 IP。当主节点损坏时，这个 IP 就会到从节点上，即从节点有双 IP）。需确保这个 IP 是闲置的，无人使用。

7）track_script 表示要对哪些脚本进行健康检查。

8）authentication 表示认证信息（这里使用简单密码认证）。

在配置文件最上面的 vrrp_script 部分（容易忽视，这里再次截图说明），定义了一个脚本。Redis 检查脚本定义如图 9-136 所示，该脚本用于检查 Redis 的状态，可以根据实际需要编写脚本。

```
vrrp_script chk_redis {
        script "/etc/keepalived/scripts/redis_check.sh"
        interval 2
}
```

图 9-136　Redis 检查脚本定义

编写 Redis 健康检查脚本/etc/keepalived/check_redis.sh，用于检查 Redis 服务器的状态。一个简单的 Redis 检查脚本示例如图 9-137 所示，其目的是检查 Redis 实例的存活。如果返回值为 0，则表示 Redis 正常服务，否则表示 Redis 服务器出现故障。

```
[root@MySQL-M-connect scripts]# vi redis_check.sh
#!/bin/bash

ALIVE=`redis-cli PING`
if [ "$ALIVE" == "PONG" ]; then
  echo $ALIVE
  exit 0
else
  echo $ALIVE
  exit 1
fi
```

图 9-137　Redis 检查脚本内容

此外 keepalived 的配置文件中还提到了 4 个脚本。分别是 redis_master.sh、redis_backup.sh、redis_fault.sh、redis_stop.sh。主要是复制数据库的启停等工作。

编写 redis_master.sh 脚本，该脚本主从服务器相同。当服务器状态升级为 Master 时会自动调用该脚本，将 Redis 升级成为主节点，脚本内容如图 9-138 所示。

```
[root@MySQL-M-connect scripts]# vim /etc/keepalived/scripts/redis_master.sh
#!/bin/bash

REDISCLI="redis-cli"
LOGFILE="/var/log/keepalived-redis-state.log"

echo "[master]" >> $LOGFILE
date >> $LOGFILE
echo "Being master...." >> $LOGFILE 2>&1ls
echo "Run SLAVEOF NO ONE cmd ..." >> $LOGFILE
$REDISCLI SLAVEOF NO ONE >> $LOGFILE 2>&1
```

图 9-138　redis_master 脚本内容

编写 redis_backup.sh 脚本，主从服务器需要将 SLAVEOF 的 IP 换为对方。当服务器状态变为从服务器时会自动调用该脚本，将 Redis 降级为从节点，脚本内容如图 9-139 所示。

编写 redis_fault.sh 和 redis_stop.sh 脚本。这两个脚本主从服务器保持一致。错误日志脚本内容如图 9-140 所示，这两个脚本不是必须，许多环境中甚至没有这两个脚本也可以正常运行。

```
[root@MySQL-M-connect scripts]# vim /etc/keepalived/scripts/redis_backup.sh
#!/bin/bash

REDISCLI="redis-cli"
LOGFILE="/var/log/keepalived-redis-state.log"

echo "[backup]" >> $LOGFILE
date >> $LOGFILE
echo "Being slave...." >> $LOGFILE 2>&1

echo "Run SLAVEOF cmd ..." >> $LOGFILE
$REDISCLI SLAVEOF 10.60.143.33 6379 >> $LOGFILE  2>&1
```

图 9-139　redis_backup 脚本

```
[root@MySQL-M-connect scripts]# sudo vim /etc/keepalived/scripts/redis_fault.sh
#!/bin/bash

LOGFILE=/var/log/keepalived-redis-state.log

echo "[fault]" >> $LOGFILE
date >> $LOGFILE

[root@MySQL-M-connect scripts]# sudo vim /etc/keepalived/scripts/redis_stop.sh
#!/bin/bash

LOGFILE=/var/log/keepalived-redis-state.log

echo "[stop]" >> $LOGFILE
date >> $LOGFILE
```

图 9-140　错误日志脚本

接下来将 33 节点和 136 节点的 keepalived 服务分别启动并检查状态，此时 136 节点为主服务器。主节点启动 keepalived 如图 9-141 所示，看到 10.60.143.131 的虚拟 IP 已经分配给了 136 节点。

```
[root@MySQL-M-connect keepalived]# sudo systemctl start keepalived
[root@MySQL-M-connect keepalived]# sudo systemctl status keepalived
● keepalived.service - LVS and VRRP High Availability Monitor
   Loaded: loaded (/usr/lib/systemd/system/keepalived.service; disabled; vendor preset: disabled)
   Active: active (running) since 三 2023-07-19 14:55:42 CST; 1h 40min ago
  Process: 74276 ExecStart=/usr/sbin/keepalived $KEEPALIVED_OPTIONS (code=exited, status=0/SUCCESS)
 Main PID: 74277 (keepalived)
   CGroup: /system.slice/keepalived.service
           ├─74277 /usr/sbin/keepalived -D
           ├─74278 /usr/sbin/keepalived -D
           └─74279 /usr/sbin/keepalived -D

7月 19 14:58:27 MySQL-M-connect Keepalived_vrrp[74279]: Sending gratuitous ARP on ens192 for 10.60.143.131
7月 19 14:58:27 MySQL-M-connect Keepalived_vrrp[74279]: Sending gratuitous ARP on ens192 for 10.60.143.131
7月 19 14:58:27 MySQL-M-connect Keepalived_vrrp[74279]: Sending gratuitous ARP on ens192 for 10.60.143.131
7月 19 14:58:27 MySQL-M-connect Keepalived_vrrp[74279]: Opening script file /etc/keepalived/scripts/redis_master.sh
7月 19 14:58:32 MySQL-M-connect Keepalived_vrrp[74279]: Sending gratuitous ARP on ens192 for 10.60.143.131
7月 19 14:58:32 MySQL-M-connect Keepalived_vrrp[74279]: VRRP_Instance(VI_1) Sending/queueing gratuitous ARPs on ....131
```

图 9-141　主节点启动 keepalived

33 节点为从服务器。从节点启动 keepalived，如图 9-142 所示，启动过程调用 redis_backup.sh 脚本，可以看到输出日志倒数第四行显示该节点现在为 BACKUP 状态。

```
[root@MySQL8S keepalived]# sudo systemctl start keepalived
[root@MySQL8S keepalived]# sudo systemctl status keepalived
● keepalived.service - LVS and VRRP High Availability Monitor
   Loaded: loaded (/usr/lib/systemd/system/keepalived.service; disabled; vendor preset: disabled)
   Active: active (running) since 三 2023-07-19 16:31:30 CST; 3min 25s ago
  Process: 76551 ExecStart=/usr/sbin/keepalived $KEEPALIVED_OPTIONS (code=exited, status=0/SUCCESS)
 Main PID: 76552 (keepalived)
   CGroup: /system.slice/keepalived.service
           ├─76552 /usr/sbin/keepalived -D
           ├─76553 /usr/sbin/keepalived -D
           └─76554 /usr/sbin/keepalived -D

7月 19 16:31:30 MySQL8S Keepalived_vrrp[76554]: Registering gratuitous ARP shared channel
7月 19 16:31:30 MySQL8S Keepalived_vrrp[76554]: Opening file '/etc/keepalived/keepalived.conf'.
7月 19 16:31:30 MySQL8S Keepalived_vrrp[76554]: WARNING - default user 'keepalived_script' for script execution does not exist - please create.
7月 19 16:31:35 MySQL8S Keepalived_vrrp[76554]: SECURITY VIOLATION - scripts are being executed but script_security not enabled.
7月 19 16:31:35 MySQL8S Keepalived_vrrp[76554]: VRRP_Instance(VI_1) removing protocol VIPs.
7月 19 16:31:35 MySQL8S Keepalived_vrrp[76554]: Using LinkWatch kernel netlink reflector...
7月 19 16:31:35 MySQL8S Keepalived_vrrp[76554]: VRRP_Instance(VI_1) Entering BACKUP STATE
7月 19 16:31:35 MySQL8S Keepalived_vrrp[76554]: Opening script file /etc/keepalived/scripts/redis_backup.sh
7月 19 16:31:35 MySQL8S Keepalived_vrrp[76554]: VRRP sockpool: [ifindex(2), proto(112), unicast(0), fd(10,11)]
7月 19 16:31:35 MySQL8S Keepalived_vrrp[76554]: VRRP_Script(chk_redis) succeeded
```

图 9-142　从节点启动 keepalived

进而可以使用 ip addr 命令查看虚拟 IP 地址，主节点双 IP 地址如图 9-143 所示，此时可以看到 136 节点绑定了虚拟 IP 地址 131。

```
[root@MySQL-M-connect scripts]# ip addr
1: lo: <LOOPBACK,UP,LOWER_UP> mtu 65536 qdisc noqueue state UNKNOWN group default qlen 1000
    link/loopback 00:00:00:00:00:00 brd 00:00:00:00:00:00
    inet 127.0.0.1/8 scope host lo
       valid_lft forever preferred_lft forever
    inet6 ::1/128 scope host
       valid_lft forever preferred_lft forever
2: ens192: <BROADCAST,MULTICAST,UP,LOWER_UP> mtu 1500 qdisc mq state UP group default qlen 1000
    link/ether 00:50:56:96:dd:79 brd ff:ff:ff:ff:ff:ff
    inet 10.60.143.136/24 brd 10.60.143.255 scope global noprefixroute ens192
       valid_lft forever preferred_lft forever
    inet 10.60.143.131/24 scope global secondary ens192
       valid_lft forever preferred_lft forever
    inet6 fe80::250:56ff:fe96:dd79/64 scope link noprefixroute
       valid_lft forever preferred_lft forever
```

图 9-143　主节点双 IP 地址

下面测试一下 keepalived 的 VIP 功能。检查 Redis 状态，如图 9-144 所示，可以使用普通 IP 和虚拟 IP 分别访问 Redis 数据库。

```
[root@MySQL-M-connect keepalived]# redis-cli -h 10.60.143.136 -p 6379  info|grep role
role:master
[root@MySQL-M-connect keepalived]# redis-cli -h 10.60.143.33 -p 6379   info|grep role
role:slave
[root@MySQL-M-connect keepalived]# redis-cli -h 10.60.143.131 -p 6379  info|grep role
role:master
```

图 9-144　检查 Redis 状态

接下来模拟故障，关闭主节点 136 的 Redis 数据库。关闭主节点 Redis，如图 9-145 所示，主节点日志显示主节点已经出错。

而此时在 Redis 的从节点 33 上，已经可以看到 Redis 从节点自动接管，33 节点已经成为新的主节点，从节点 131 也同时拥有了虚拟 IP 地址 131，如图 9-146 所示。

```
[root@MySQL-M-connect keepalived]# ps -ef|grep "redis*"
root       54347  81810  0 14:38 pts/0    00:00:00 grep --color=auto redis*
redis      86405      1  0 13:08 ?        00:00:09 /usr/bin/redis-server 0.0.0.0:6379
[root@MySQL-M-connect keepalived]# kill -9 86405

查看主节点错误日志：
[root@MySQL-M-connect keepalived]# tailf /var/log/keepalived-redis-state.log
[fault]
Wed Jul 19 14:38:38 CST 2023
```

图 9-145　关闭主节点 Redis

```
[root@MySQL8S scripts]# tailf /var/log/keepalived-redis-state.log
[master]
Wed Jul 19 14:57:14 CST 2023
Being master....
Run SLAVEOF NO ONE cmd ...
OK


[root@MySQL8S keepalived]# redis-cli -h 10.60.143.33 info| grep role
role:master
[root@MySQL8S keepalived]# redis-cli -h 10.60.143.131 info| grep role
role:master
```

图 9-146　Redis 从节点自动接管

以上整个过程切换是自动的，此时使用通过虚拟 IP 地址仍然能正常连接 Redis 数据库。这种稳定的 IP 服务就是 VIP 系统的意义所在。

当重启原主节点 136 后，通过查看重启过程主从服务器日志内容。Redis 主节点抢占，可以看到重启后的原 136 节点重新抢占回主服务器身份，如图 9-147 所示。

通过以上 keepalived+Redis 主主的模式，Redis 可以自动进行健康检查，并在需要时进行主备切换，以实现高可用性。需要注意的是，在配置和使用 keepalived 时，需要根据实际情况进行调整。

```
[root@MySQL-M-connect keepalived]# sudo systemctl start redis

[root@MySQL-M-connect scripts]# tailf /var/log/keepalived-redis-state.log
[master]
Wed Jul 19 14:58:27 CST 2023
Being master....
Run SLAVEOF NO ONE cmd ...
OK

[root@MySQL8S scripts]# tailf /var/log/keepalived-redis-state.log
[backup]
Wed Jul 19 14:58:25 CST 2023
Being slave....
Run SLAVEOF cmd ...
OK

[root@MySQL8S scripts]# redis-cli -h 10.60.143.136 info| grep role
role:master
```

图 9-147 Redis 主节点抢占

在实际工作中 Redis 还有另外一个高可用称为 Redis-Cluster。此高可用模式至少需要三主三从，6 个实例（不一定是 6 台机器）。此高可用模式无须切换，不过当集群中半数主节点出现问题后，约等于集群损坏。

Redis-Cluster 目前不支持多 db 模式，在单实例和主从模式下有 db0、db1 这样使用习惯的应用程序，在 Redis-Cluster 中只有默认的 db。

9.4 三个 SQL 编写的最佳实践

同样一个需求或同样一个场景，SQL 的写法或实现方法可以有多种方式。这些方式可能是等效的，也可能是不等效的，这里的等效是结果相同并不代表效率相同。当然如果不等效自然是错误的。那么在等效的前提下，效率可能相差很多。一般来说，效率最优（也就是执行最快）的就是最佳实践。

9.4.1 关于 MyBatis 开发框架使用绑定变量的实践

Oracle 解析有软硬解析之分，而软硬解析和绑定变量有直接关系。过去一般知道 SELECT 语句中要注意绑定变量。随着工作环境的变化和场景的改变，发现了 INSERT 和 UPDATE 在绑定变量上的问题。其原因是在工作中经常发现应用程序批量写入时采用了框架的方法，而这种方法对数据库性能有很大的影响。

有的开发人员不知道批量提交，会逐条提交。这个在第 2 章中已经做过说明，从数据库的体系结构来说，单条提交的性能远低于批量提交。

而部分有经验的开发人员知道应该批量提交，但是批量的方法不对，造成了危害不亚于单条提交的写法。

其实批量操作的正确理解是多次 INSERTINTO 循环，一次性 Commit（提交），而不是把一批 INSERT 语句组合成为一个 SQL。这两者是有本质的区别的。

首先展示一下有些开发人员常使用的 MyBatis 批量写入的方式。最终的 MyBatis 的

oracleBatchInsert 批量写入方法如图 9-148 所示的，它是通过框架中在配置文件 Mapper.xml 中写入的 SQL 进行拼装来实现的。

```xml
<insert id="oracleBatchInsert" parameterType="list">
    insert into i (id, t)
    <foreach collection="list" item="item" index="index" separator="union all">
        select #{item}, sysdate from dual
    </foreach>
</insert>
```

图 9-148 MyBatis 的 oracleBatchInsert 批量写入

通过开发框架 MyBatis 组装后，在 Oracle 内运行。随后在 AWR 报告中得到的 SQL 如图 9-149 所示，大量绑定变量。

```
insert into i (id,  t)
   select :1 ,  sysdate from dual
union all
   select :2 ,  sysdate from dual
union all
   select :3 ,  sysdate from dual
union all
   select :4 ,  sysdate from dual
union all
   select :5 ,  sysdate from dual
union all
   select :6 ,  sysdate from dual
union all
   select :7 ,  sysdate from dual
union all
   select :8 ,  sysdate from dual
union all
   select :9 ,  sysdate from dual
union all
   select :10 ,  sysdate from dual
union all
   select :11 ,  sysdate from dual
union all
   select :12 ,  sysdate from dual
union all
   select :13 ,  sysdate from dual
union all
   .
   .
   .
```

图 9-149 AWR 报告中得到的 SQL

这种写法在早期的 Oracle 数据库版本上（12c），当变量大于 65535 时，会出现数据库关闭的严重影响。

可能是知道这个影响或担心数据库处理问题，开发人员通常会分批处理数据。这样虽然一次变量没有 6 万多个，但是 3 万、4 万还是比较常见的。这个取决于一次处理多少行，以及每行多少个字段。如果是 1000 行，每行 50 个字段，那么就是 5 万个变量。

出现这种大量的绑定变量的原因是：开发人员采用了 list 对象，根据前台应用传入的 list 对象大小进行组装，然后将组装完成的 SQL 语句一次性传入 Oracle 数据库执行，即将前面段落中提到的一批 INSERT 语句组合成为一个 SQL。这样会使得这个 SQL 又大又长，占据着数据库的内存。因为在 Oracle 中，每条 SQL 语句在执行之前都需要经过解析（Parse），为了避免这种硬解析，数据库会将执行过的 SQL 缓存起来。每次接收到 SQL 先在当前会话的 PGA 中查找是否存在匹配的缓存 Session Cursor，如果没有会去 SGA 中查找相关父游标。所以这些执行过的 SQL 会占据着数据库的内存资源，如果一个 SQL 很大，甚至出现多个版本，就会使得一个 SQL 的多版本占据着数据库上的大量内存，侵蚀着 SGA 和 PGA 资源，导致资源不足，引起数据库次生灾害故障。

下面请看另外一种写法，直接在程序中使用循环语句。MyBatis 框架的 oracleInsert 的单条提交方法如图 9-150 所示。在 Java 中，一个功能就是一个方法，oracleInsert 是一个方法，循环执行单条数据的插入，其对应的框架中的配置文件 Mapper.xml 实现如下。

```
<insert id="oracleInsert" parameterType="long">
    insert into i (id, t) values (#{id}, sysdate)
</insert>
```

图 9-150　MyBatis 框架的 oracleInsert 的单条提交

在数据库领域的常识中，单条写入的效率不如批量写入的方式。所以同样是单条循环写入，在 MyBatis 框架中也提供了一种优化的实现方式，即 ExecutorType.BATCH 的参数实现批量写入。

下面通过实验来说明三种代码带来的不一样的效果。Java 三种写入数据完整代码如图 9-151 所示，代码第 25～36 行是采用 list 方式将多个 INSERT 组装成一句的写法，这种写法以下简称 A 写法。代码第 37～50 行是采用 for 循环的方式执行的多次循环 INSERT 批量提交（但是没有用 MyBatis 框架的批量参数），这种写法以下简称 B 写法。代码第 51～64 行是采用 for 循环的方式执行的多次循环 INSERT 批量提交，且使用了 MyBatis 框架的批量参数，这种写法以下简称 C 写法。

```
package db;

import org.apache.ibatis.session.SqlSession;

import java.util.ArrayList;
import java.util.Date;
import java.util.List;
```

图 9-151　Java 三种写入数据完整代码

```java
public class mybatisInsert {

    public static void main(String[] args) {
        try{
            //0...10000的数值列表用于写入
            List<Integer> longList = new ArrayList<>();
            for (int i=0; i < 10000; i++){
                longList.add(i);
            }
            classicInsert(longList);
            loopInsert(longList);
            loopBatchInsert(longList);
        } catch (Exception e){
            e.printStackTrace();
        }
    }

    //目前使用Mybatis批量写入Oracle的普遍实现方式-Union All
    private static void classicInsert(List<Integer> longList){
        SqlSession sqlSession = MybatisUtils.getSqlSession();  //默认session
        mapperInterface mapperInterface = sqlSession.getMapper(mapperInterface.class);
        System.out.println("传统批量写入开始: ");
        Date startTime = new Date();
        mapperInterface.oracleBatchInsert(longList, type: "batch");
        sqlSession.commit();
        Date endTime = new Date();
        System.out.println("传统批量写入结束: " + (endTime.getTime()-startTime.getTime()) + "ms");
        sqlSession.close();
    }

    //使用for循环调用单条写入的批量写入方式
    private static void loopInsert(List<Integer> longList){
        SqlSession sqlSession = MybatisUtils.getSqlSession();  //默认session
        mapperInterface mapperInterface = sqlSession.getMapper(mapperInterface.class);
        System.out.println("循环单条写入开始: ");
        Date startTime = new Date();
        for (long i:longList){
            mapperInterface.oracleInsert(i, type: "loop");
        }
        sqlSession.commit();
        Date endTime = new Date();
        System.out.println("循环单条写入结束: " + (endTime.getTime()-startTime.getTime()) + "ms");
        sqlSession.close();
    }

    //使用for循环调用单条写入的批量写入方式，使用框架批量参数
    private static void loopBatchInsert(List<Integer> longList){
        SqlSession sqlSession = MybatisUtils.getBatchSqlSession();  //批量session
        mapperInterface mapperInterface = sqlSession.getMapper(mapperInterface.class);
```

图 9-151　Java 三种写入数据完整代码（续）

三种写入方法对应的 Mapper.xml 是一个。MyBatis 框架 Mapper 文件如图 9-152 所示，代码第 20～25 行是 list 方式将多个 INSERT 组装成一句的 Mapper 文件，对应的是 A 写法。代码第 27～29 行是 for 循环的方式执行的多次循环 INSERT 的 Mapper 文件，对应的是 B 写法和 C 写法。

```xml
<insert id="oracleBatchInsert">
    insert into i (id, t, a)
    <foreach collection="list" item="item" index="index" separator="union all">
        select #{item}, sysdate, #{type} from dual
    </foreach>
</insert>

<insert id="oracleInsert">
    insert into i (id, t, a) values (#{id}, sysdate, #{type})
</insert>
```

图 9-152　MyBatis 框架 Mapper 文件

在 MyBatis 的框架中还涉及 session，MyBatis 的 session 配置如图 9-153 所示，代码第 16 行指明数据库的连接配置文件，代码 24～29 行是关键。

```java
package db;

import org.apache.ibatis.io.Resources;
import org.apache.ibatis.session.ExecutorType;
import org.apache.ibatis.session.SqlSession;
import org.apache.ibatis.session.SqlSessionFactory;
import org.apache.ibatis.session.SqlSessionFactoryBuilder;

import java.io.InputStream;

public class MybatisUtils {
    private static SqlSessionFactory sqlSessionFactory;

    static {
        try {
            String resource = "mybatis-config.xml";
            InputStream inputStream = Resources.getResourceAsStream(resource);
            sqlSessionFactory = new SqlSessionFactoryBuilder().build(inputStream);
        }catch (Exception e){
            e.printStackTrace();
        }
    }

    public  static SqlSession getSqlSession(){
        return sqlSessionFactory.openSession();
    }
}
```

图 9-153　MyBatis 的 session 配置

```
public static SqlSession getBatchSqlSession(){
    return sqlSessionFactory.openSession(ExecutorType.BATCH);
}
```

图 9-153　MyBatis 的 session 配置（续）

整个 MyBatisUtils 类中有两个方法：getSqlSession 方法和 getBatchSqlSession 方法。前者用在 A 和 B 写法上，是常规建立会话的方式；后者用在 C 写法上，用于建立批量执行的会话。

为了排除干扰项，避免网络因素的影响，将程序打成 Jar 包，在数据库本地运行。三种写法相同场景的执行效率如图 9-154 所示，同是 1 万条数据的写入场景，A 写法耗时 117s，B 写法耗时约 7s，C 写法耗时 0.25s。

```
[root@oracle19c otcadmin]# java -jar insert.jar
传统批量写入开始：
传统批量写入结束：117736ms
循环单条写入开始：
循环单条写入结束：7727ms
循环单条批量写入开始：
循环单条批量写入结束：249ms
```

图 9-154　三种写法相同场景执行效率

这个数据说明性能差距巨大，A 写法性能最差，而在生产中使用的最多的还是 A 写法。

从另一个维度数据库数据写入的时间戳也能说明。数据库写入时间戳数据说明如图 9-155 所示。batch 是 A 写法的写入，即传统 union 的批量写入，由于是拼装成了一个语句执行，最终落地时间是相同的。loop 写 B 写法的写入，也就是循环单独写入数据，循环的方式可以看到第一条数据至最后一条数据落地时间之间有着 7s 的差距，证明是逐条落地的。batch_loop 是 C 写法的写入，使用了框架的批量循环。由于批量循环的方式在 1s 内完成，落地时间精度无法看出差别。对于 1 万条批量写入耗时 0.25s，这是数据库一个比较正常的性能。

```
SQL> select * from i where id in (0,9999);

       ID T                    A
--------- -------------------- --------------------
        0 2023:07:2113:24:04   batch
     9999 2023:07:2113:24:04   batch
        0 2023:07:2113:24:04   loop
     9999 2023:07:2113:24:12   loop
        0 2023:07:2113:24:12   batch_loop
     9999 2023:07:2113:24:12   batch_loop

已选择 6 行。
```

图 9-155　数据库写入时间戳数据说明

除了性能，更重要的是数据库解析，通过执行 SQL 的 AWR 报告可以得到更权威的数据。Oracle 的 AWR 如图 9-156 所示，图中第一行显示通过拼装执行的语句占用了大量的内存空间。

SQL ordered by Sharable Memory

- Only Statements with Sharable Memory greater than 1048576 are displayed

Sharable Mem (b)	Executions	% Total	SQL Id	SQL Module	PDB Name	SQL Text
32,410,614	1	4.84	a91kqgkvhk4gh	JDBC Thin Client	TANG	insert into i (id, t, a) selec...
1,632,760	285	0.24	4bugmu5z8szn5	GoldenGate	TANG	SELECT /*+ ordered push_pred(v...
1,511,078	1	0.23	8a252j0vgc2cp		TANG	SELECT * FROM (SELECT /*+ orde...
1,261,410	1	0.19	aas1jwv6uu93h		TANG	SELECT /*+ ordered OPT_PARAM(...
1,147,783	1	0.17	8x2ufxz8tcmaf	DBMS_SCHEDULER	TANG	/* SQL Analyze(1) */ select /*...

图 9-156　Oracle 的 AWR

对于事务繁忙的数据库来说，大量写入组装成超长 SQL 语句是一种相当"奢侈"的实现方式，并且在第一次执行时会消耗大量执行时间（重复运行时，如果语句在内存中可以提升执行效率，后续再执行会快些），而后两种循环执行的方式，对数据库的压力就显著减少，执行效率也能大幅度提升。当带上 ExecutorType.BATCH 的参数时，一万条数据的写入也只需耗时 200ms，数据量上升时，单条写入的平均耗时会更低。

希望更多的开发人员认识到这个问题，采用正确的方法，而不要采用性能低而且会造成数据库问题的方法。

9.4.2　使用 exists 的 SQL 语句的改写

数据库中运行的很多应用程序产生的 SQL 语句性能并不好。有时候通过改写 SQL 语句会有性能提升，即使性能没有显著提升，通过简化或是规范化的改写也可以对日后的维护和可读性有较大的帮助，当然前提一定是等效的改写。

在工作中经常会遇到使用 exists 关键字的场景，有时候是必要的，但是有时候其实可以避免。在有些公司的开发规范中会有尽可能避免使用 exists 的规定，因为许多场景中这种写法的效率并不高，所以对应的 SQL 查询语句效率也很低，在本节中进行一下比较看看性能有没有差别以及可读性如何。

首先准备样例表 A 和表 B，查询 A、B 表的数据详情，如图 9-157 所示，A 表为主表，B 表和 A 表是多对 1 的关系。A 表的 ID 与 B 表的 AID 相关联。

```
SQL> select * from a;

        ID NAME
---------- ----------
         1 A1
         2 A2
         3 A3

SQL> select * from b;

        ID        AID NAME
---------- ---------- ----------
         1          1 B1
         2          1 B2
         3          2 B3
         4          2 B4
         5          3 B5
         6          3 B6

6 rows selected
```

图 9-157　查询 A、B 表的数据详情

分别对比使用 exists 和非 exists 的等效写法，exists 的 SQL 写法对比如图 9-158 所示，在这种需求场景下，结果一致。不过非 exists 的写法可读性较好。

```
SQL> SELECT ID,NAME FROM A WHERE EXISTS(SELECT * FROM B WHERE A.ID = B.AID );   --exists写法

        ID NAME
---------- ----------
         1 A1
         2 A2
         3 A3


SQL> select distinct a.id,a.name from a,b where a.id=b.aid;      --非exists写法

        ID NAME
---------- ----------
         1 A1
         2 A2
         3 A3
```

图 9-158　exists 的 SQL 写法对比

使用 setautotraceon 进行 SQL 的消耗跟踪，exists 写法执行消耗如图 9-159 所示，Cost 为 6，consistent gets 为 11。

```
SQL> SELECT ID,NAME FROM A WHERE EXISTS(SELECT * FROM B WHERE A.ID = B.AID );   --exists写法执行消耗

        ID NAME
---------- ----------
         1 A1
         2 A2
         3 A3


执行计划
----------------------------------------------------------
Plan hash value: 3821265669

--------------------------------------------------------------------------
| Id  | Operation          | Name | Rows  | Bytes | Cost (%CPU)| Time     |
--------------------------------------------------------------------------
|   0 | SELECT STATEMENT   |      |     2 |    66 |     6   (0)| 00:00:01 |
|*  1 |  HASH JOIN SEMI    |      |     2 |    66 |     6   (0)| 00:00:01 |
|   2 |   TABLE ACCESS FULL| A    |     3 |    60 |     3   (0)| 00:00:01 |
|   3 |   TABLE ACCESS FULL| B    |     3 |    39 |     3   (0)| 00:00:01 |
--------------------------------------------------------------------------

Predicate Information (identified by operation id):
---------------------------------------------------

   1 - access("A"."ID"="B"."AID")

Note
-----
   - dynamic statistics used: dynamic sampling (level=2)


统计信息
----------------------------------------------------------
          0  recursive calls
```

图 9-159　exists 写法执行消耗

```
  0  db block gets
 11  consistent gets
  0  physical reads
  0  redo size
697  bytes sent via SQL*Net to client
433  bytes received via SQL*Net from client
  2  SQL*Net roundtrips to/from client
  0  sorts (memory)
  0  sorts (disk)
  3  rows processed
```

图 9-159 exists 写法执行消耗（续）

非 exists 写法执行消耗如图 9-160 所示，Cost 为 7，consistent gets 为 10。从这个上面比较，exists 和非 exists 写法几乎一致。非 exists 仅仅是可读性好一些。

```
SQL> select distinct a.id,a.name from a,b where a.id=b.aid;          --非exists写法执行消耗
        ID NAME
---------- ----------
         1 A1
         2 A2
         3 A3


执行计划
----------------------------------------------------------
Plan hash value: 3611140294

---------------------------------------------------------------------------
| Id  | Operation          | Name | Rows  | Bytes | Cost (%CPU)| Time     |
---------------------------------------------------------------------------
|   0 | SELECT STATEMENT   |      |     2 |    66 |     7  (15)| 00:00:01 |
|   1 |  HASH UNIQUE       |      |     2 |    66 |     7  (15)| 00:00:01 |
|*  2 |   HASH JOIN SEMI   |      |     2 |    66 |     6   (0)| 00:00:01 |
|   3 |    TABLE ACCESS FULL| A   |     3 |    60 |     3   (0)| 00:00:01 |
|   4 |    TABLE ACCESS FULL| B   |     3 |    39 |     3   (0)| 00:00:01 |
---------------------------------------------------------------------------

Predicate Information (identified by operation id):
---------------------------------------------------

   2 - access("A"."ID"="B"."AID")

Note
-----
   - dynamic statistics used: dynamic sampling (level=2)


统计信息
----------------------------------------------------------
  0  recursive calls
  0  db block gets
 10  consistent gets
  0  physical reads
  0  redo size
697  bytes sent via SQL*Net to client
415  bytes received via SQL*Net from client
  2  SQL*Net roundtrips to/from client
  0  sorts (memory)
  0  sorts (disk)
  3  rows processed
```

图 9-160 非 exists 写法执行消耗

既然小数据看不出来，在这个模型上加大数据量，看看和数据量有没有关系。如图 9-161 所示，建立 10 万级别的 C 表和 D 表，C 表和 D 表是一对多的关系。C 表的 ID 和 D 表的 AID 是 1：3 的比例。

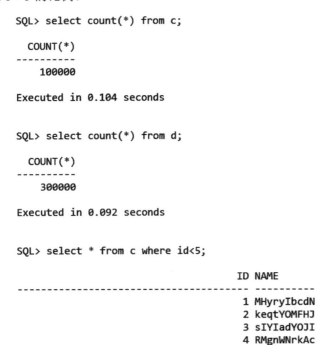

图 9-161　建立 10 万级别的 C 表和 D 表

在收集统计信息建立索引和直方图后，1 万行的 exists 查询消耗如图 9-162 所示，consistent gets 为 780。

```
SQL> SELECT ID,NAME FROM c WHERE  c.id<10001 and EXISTS(SELECT * FROM d WHERE c.id = d.aid );  --exists写法执行消耗
已选择 10000 行。

执行计划
----------------------------------------------------------
Plan hash value: 3543992594

--------------------------------------------------------------------------------
--------------
| Id  | Operation                            | Name  | Rows  | Bytes | Cost (%CP
U)| Time     |
--------------------------------------------------------------------------------
--------------
|   0 | SELECT STATEMENT                     |       |  9996 |  204K |   123   (
 1)| 00:00:01 |
|*  1 |  HASH JOIN SEMI                      |       |  9996 |  204K |   123   (
 1)| 00:00:01 |
|   2 |   TABLE ACCESS BY INDEX ROWID BATCHED| C     |  9996 |  156K |    53   (
 0)| 00:00:01 |
|*  3 |    INDEX RANGE SCAN                  | C_ID  |  9996 |       |    24   (
 0)| 00:00:01 |
|*  4 |   INDEX RANGE SCAN                   | D_AID | 30005 |  146K |    69   (
 0)| 00:00:01 |
--------------------------------------------------------------------------------
--------------

Predicate Information (identified by operation id):
---------------------------------------------------

   1 - access("C"."ID"="D"."AID")
   3 - access("C"."ID"<10001)
   4 - access("D"."AID"<10001)

Note
-----
   - this is an adaptive plan

统计信息
----------------------------------------------------------
          0  recursive calls
          0  db block gets
        780  consistent gets
          0  physical reads
          0  redo size
     289733  bytes sent via SQL*Net to client
       7996  bytes received via SQL*Net from client
        668  SQL*Net roundtrips to/from client
          0  sorts (memory)
          0  sorts (disk)
      10000  rows processed
```

图 9-162　1 万行的 exists 查询消耗

1 万行的非 exists 查询消耗如图 9-163 所示，consistent gets 为 115。两者的查询消耗对比大约是 1∶6.7 的关系。当然这些差距在实际执行的时候，消耗时间可能不是 1∶6.7 的关系，但是由于实际工作中没有这么理想的全是逻辑读，很多时候都是物理读。那么这个如果物理读是 1∶6.7 的关系，问题就严重了。

```
SQL> select distinct c.id,c.name from c,d where c.id=d.aid and c.id<10001;    --非exists写法执行消耗
已选择 10000 行。

执行计划
----------------------------------------------------------
Plan hash value: 3380385528

---------------------------------------------------------------------------------------
| Id  | Operation                            | Name  | Rows  | Bytes | Cost (%CPU)| Time     |
---------------------------------------------------------------------------------------
|   0 | SELECT STATEMENT                     |       |  9996 |  204K |   124   (2)| 00:00:01 |
|   1 |  HASH UNIQUE                         |       |  9996 |  204K |   124   (2)| 00:00:01 |
|*  2 |   HASH JOIN SEMI                     |       |  9996 |  204K |   123   (1)| 00:00:01 |
|   3 |    TABLE ACCESS BY INDEX ROWID BATCHED| C    |  9996 |  156K |    53   (0)| 00:00:01 |
|*  4 |     INDEX RANGE SCAN                 | C_ID  |  9996 |       |    24   (0)| 00:00:01 |
|*  5 |    INDEX RANGE SCAN                  | D_AID | 30005 |  146K |    69   (0)| 00:00:01 |
---------------------------------------------------------------------------------------

Predicate Information (identified by operation id):
---------------------------------------------------

   2 - access("C"."ID"="D"."AID")
   4 - access("C"."ID"<10001)
   5 - access("D"."AID"<10001)

Note
-----
   - this is an adaptive plan

统计信息
----------------------------------------------------------
          0  recursive calls
          0  db block gets
        115  consistent gets
          0  physical reads
          0  redo size
     289733  bytes sent via SQL*Net to client
       7756  bytes received via SQL*Net from client
        668  SQL*Net roundtrips to/from client
          0  sorts (memory)
          0  sorts (disk)
      10000  rows processed
```

图 9-163　1 万行的非 exists 查询消耗

继续加大返回行数，5 万行的 exists 查询消耗如图 9-164 所示，consistent gets 为 3979。

```
SQL> SELECT ID,NAME FROM c WHERE  c.id<50001 and EXISTS(SELECT * FROM d WHERE c.id = d.aid );   --exists写法执行消耗
已选择 50000 行。

执行计划
----------------------------------------------------------
Plan hash value: 1004759122

----------------------------------------------------------------------------
| Id  | Operation             | Name  | Rows  | Bytes | Cost (%CPU)| Time     |
----------------------------------------------------------------------------
|   0 | SELECT STATEMENT      |       | 49988 | 1025K |   287   (2)| 00:00:01 |
|*  1 |  HASH JOIN SEMI       |       | 49988 | 1025K |   287   (2)| 00:00:01 |
|*  2 |   TABLE ACCESS FULL   | C     | 49988 |  781K |   103   (1)| 00:00:01 |
|*  3 |   INDEX FAST FULL SCAN| D_AID |  150K |  732K |   184   (2)| 00:00:01 |
----------------------------------------------------------------------------

Predicate Information (identified by operation id):
---------------------------------------------------

   1 - access("C"."ID"="D"."AID")
   2 - filter("C"."ID"<50001)
   3 - filter("D"."AID"<50001)

Note
-----
   - this is an adaptive plan

统计信息
----------------------------------------------------------
          0  recursive calls
          0  db block gets
       3979  consistent gets
          0  physical reads
          0  redo size
    1486727  bytes sent via SQL*Net to client
      37333  bytes received via SQL*Net from client
       3335  SQL*Net roundtrips to/from client
          0  sorts (memory)
          0  sorts (disk)
      50000  rows processed
```

图 9-164　5 万行的 exists 查询消耗

5 万行的非 exists 查询消耗如图 9-165 所示，consistent gets 为 656，两者的查询消耗对比和之前的比例相差不大，成线性关系。本实验使用的是 2～3 列的简单表，而实际生产环境中都是特别复杂的大宽表，这个比例会被放大。再加之可能不是两表关联，而是多表关联，而且 C 表和 D 表不是 1∶2、1∶3 这种小比例，那么也会在一定程度上放大这个执行效率比例。

```
SQL> select distinct c.id,c.name from c,d where c.id=d.aid and c.id<50001;   --非exists写法执行消耗
已选择 50000 行。

执行计划
----------------------------------------------------------
Plan hash value: 1660526464

--------------------------------------------------------------------------------
| Id  | Operation            | Name  | Rows  | Bytes |TempSpc| Cost (%CPU)| Time     |
--------------------------------------------------------------------------------
```

图 9-165　5 万行的非 exists 查询消耗

```
|   0 | SELECT STATEMENT        |       | 49988 |  1025K|       |   613   (1)| 00:00:01 |
|   1 |  HASH UNIQUE            |       | 49988 |  1025K| 1384K |   613   (1)| 00:00:01 |
|*  2 |   HASH JOIN SEMI        |       | 49988 |  1025K|       |   287   (2)| 00:00:01 |
|*  3 |    TABLE ACCESS FULL    | C     | 49988 |   781K|       |   103   (1)| 00:00:01 |
|*  4 |    INDEX FAST FULL SCAN | D_AID |  150K |   732K|       |   184   (2)| 00:00:01 |
-----------------------------------------------------------------------------------------

Predicate Information (identified by operation id):
---------------------------------------------------

   2 - access("C"."ID"="D"."AID")
   3 - filter("C"."ID"<50001)
   4 - filter("D"."AID"<50001)

Note
-----
   - this is an adaptive plan


统计信息
----------------------------------------------------------
          0  recursive calls
          0  db block gets
        656  consistent gets
          0  physical reads
          0  redo size
    1486727  bytes sent via SQL*Net to client
      37093  bytes received via SQL*Net from client
       3335  SQL*Net roundtrips to/from client
          0  sorts (memory)
          0  sorts (disk)
      50000  rows processed
```

图 9-165　5 万行的非 exists 查询消耗（续）

所以 SQL 的改写不仅仅是增加了可读性，对于有的场景的性能能起到一定的提升。

9.4.3　设计上出发减少 SQL 的标量子查询

首先要说明一下什么是标量子查询。请看示例 SQL：SELECT A,B,C (SELECT D FROM T1 WHERE T1.ID=T2.ID AND ROWNUM=1) ,E FROM T2。类似示例 SQL 这样在 SELECT 之后、FROM 之前有一个子查询的就是标量子查询，其特点是返回单个值的子查询。

标量子查询大多数情况下是不建议的。主要原因之一是一个标量子查询就是多了一个表关联。在真实的开发环境中，一个 SQL 可能带了 7~8 个标量子查询，那么就会多关联 7~8 张表，无形中增加了运算量。原因之二是标量子查询还有可能因为关联表时，缺乏对应的索引导致部分表产生全表扫描。

除去性能问题，假设开发人员已经检查了所有的标量子查询的索引和执行计划，并没有发现问题。那么最后一个弊端就是可读性不高、后续维护成本增加。

产生标量子查询的原因只有两个：需求有问题或者设计有问题。

例如有两个关联表，A 表和 B 表。关联表对应关系如图 9-166 所示，A 与 B 是 1：2 的关系。这样清晰的数据展示可以方便读者直观地体会标量子查询的效果。

```
SQL> SELECT A.ID,A.NAME FROM A;

        ID NAME
---------- ----------
         1 A1
         2 A2
         3 A3

SQL> SELECT * FROM B;

        ID        AID NAME
---------- ---------- ----------
         1          1 B1
         2          1 B2
         3          2 B3
         4          2 B4
         5          3 B5
         6          3 B6
```

已选择 6 行。

图 9-166　关联表对应关系

因为标量子查询是返回单个值，所以有两种可能。

第一种为随机取数型，标量子查询随机取数如图 9-167 所示，为了展示 B 表的 NAME（B 表的 NAME 与 A 表的 NAME 明显不一样），采用了标量子查询的写法。

其特点是在一对多的关系下随机取了 B 表的第一条数据的 NAME 进行了返回。

```
SQL> SELECT A.ID,A.NAME,(SELECT NAME FROM B WHERE A.ID=B.AID AND ROWNUM=1) AS BL FROM A;

        ID NAME       BL
---------- ---------- ----------
         1 A1         B1
         2 A2         B3
         3 A3         B5
```

图 9-167　标量子查询随机取数

第二种为极值取数型，标量子查询极值取数如图 9-168 所示，也是为了展示 B 表的 NAME，只是用了 MAX 的方式（也可能会用 MIN 的方式）取了一行，采用了标量子查询的写法。

```
SQL> SELECT A.ID,A.NAME,(SELECT MAX(NAME) FROM B WHERE A.ID=B.AID) AS BL FROM A;

        ID NAME       BL
---------- ---------- ----------
         1 A1         B2
         2 A2         B4
         3 A3         B6
```

图 9-168　标量子查询极值取数

可以看到这两种返回的结果是不一样的。实际上，如果是上述的数据样本，其实是不适合用标量子查询的,因为在这种数据条件下,逻辑是错的。而在现实工作中可能由于开发人员失误,就是这样取数的。如果读者感兴趣可以到自己的环境中去抽查看看,是不是存在这种情况,这其实是业务逻辑发生错误的案例。

在现实生活中极有可能发生的情况是在 B 表上某列为相同值。使用标量子查询的数据样本如图 9-169 所示,这一列对于关联的数据来说都是一样的,取任意一行或者最大值、最小值都一样。

```
SQL> SELECT A.ID,A.NAME FROM A;

        ID NAME
---------- ----------
         1 A1
         2 A2
         3 A3


SQL> SELECT * FROM B;

        ID        AID NAME       CODE
---------- ---------- ---------- ----------
         1          1 B1         B
         2          1 B2         B
         3          2 B3         B
         4          2 B4         B
         5          3 B5         B
         6          3 B6         B
```

图 9-169 使用标量子查询的数据样本

那么 SQL 可能会写成如图 9-170 所示的真实标量子查询,这样写虽然达到了目的,但是多关联了一个表。其主要是表设计之初的问题。如果需求是在 A 表的这些列展示的时候需要 CODE 列,那么表设计的时候 CODE 就应该在 A 表上,当然 B 表上的 CODE 也可以适当冗余,那么这个查询就可以是单表查询了。也许有人会认为只是多关联了一个表而已,但当查询的 SQL 语句复杂时,查询性能就会下降。因为极有可能 A、B 表因为索引或者返回结果过大等因素发生全表扫描,那么另外一个表可能跟着全表扫描。关键是在实际工作中这种标量子查询随着开发的迭代不断往上加,最后就变成了无人能动的"祖传代码"。

```
SQL>
SQL> SELECT A.ID,A.NAME,(SELECT CODE FROM B WHERE A.ID=B.AID AND ROWNUM=1) AS BL FROM A;

        ID NAME       BL
---------- ---------- ----------
         1 A1         B
         2 A2         B
         3 A3         B

SQL> SELECT A.ID,A.NAME,(SELECT MAX(CODE) FROM B WHERE A.ID=B.AID) AS BL FROM A;

        ID NAME       BL
---------- ---------- ----------
         1 A1         B
         2 A2         B
         3 A3         B
```

图 9-170 真实标量子查询

还有另外一种标量子查询的场景。查询主表并且计算子表总数,如图 9-171 所示,就是在展示主表的字段时，一对多（可能是 2 也可能是几千）的具体数据量。

```
SQL> SELECT A.ID,A.NAME,(SELECT COUNT(*) FROM B WHERE A.ID=B.AID) AS C FROM A;

    ID NAME              C
---------- ---------- ----------
     1 A1                2
     2 A2                2
     3 A3                2
```

图 9-171 查询主表并且计算子表总数

同样也建议表设计之初就在 A 表上建立一个 C 列，每次 B 表增加或减少，在同一个事务中修改 A 表的 C 列。如果 A、B 两个表的对比关系是 1∶2，理论上性能差 2～3 倍。标量子查询的资源消耗如图 9-172 所示，查询 1 万条记录，有 2863 个逻辑读。看上去这个数值并不大，这是因为实验的表是只有几列的短小表，真实环境中的表都是几十个字段甚至上百个字段的表。

```
SQL> SELECT C.ID,C.NAME,(SELECT COUNT(*) FROM D WHERE C.ID=D.AID) AS C FROM C WHERE C.ID<10001;

已选择 10000 行。


执行计划
----------------------------------------------------------
Plan hash value: 690466705

--------------------------------------------------------------------------------
| Id  | Operation                             | Name | Rows  | Bytes | Cost (%CPU)| Time     |
--------------------------------------------------------------------------------
|   0 | SELECT STATEMENT                      |      | 9996  |  156K | 15051   (1)| 00:00:01 |
|   1 |  SORT AGGREGATE                       |      |     1 |     5 |            |          |
|*  2 |   INDEX RANGE SCAN                    | D_AID|     3 |    15 |     3   (0)| 00:00:01 |
|   3 |  TABLE ACCESS BY INDEX ROWID BATCHED  | C    | 9996  |  156K |    53   (0)| 00:00:01 |
|*  4 |   INDEX RANGE SCAN                    | C_ID | 9996  |       |    24   (0)| 00:00:01 |
--------------------------------------------------------------------------------

Predicate Information (identified by operation id):
---------------------------------------------------

   2 - access("D"."AID"=:B1)
   4 - access("C"."ID"<10001)
```

图 9-172 标量子查询的资源消耗

```
统计信息
----------------------------------------------------------
          0  recursive calls
          0  db block gets
       2863  consistent gets
          0  physical reads
          0  redo size
     327748  bytes sent via SQL*Net to client
       7777  bytes received via SQL*Net from client
        668  SQL*Net roundtrips to/from client
          0  sorts (memory)
          0  sorts (disk)
      10000  rows processed
```

图 9-172　标量子查询的资源消耗（续）

而如果把这个计算结果落地在 A 表上，去掉标量子查询的资源消耗如图 9-173 所示，在 FROM 之前用常量 3 表示对应关系的数据存在 A 表之上，不再需要关联 B 去实时计算 B 表的行数，在这个场景下的逻辑读是 1382。可见去掉标量子查询比使用标量子查询少了一半的消耗。也许有人可能觉得差别不大。但是如果 A：B 两个表的对比关系是 1：10000 的话，那么性能相差就非常大了。这和情况是真实存在的。在 2B 的场景中，一个合同可能有几万张发票，如果要看发票数量，其对比关系便是 1：10000 级别了。

```
SQL> SELECT C.ID,C.NAME, 3 FROM C WHERE C.ID<10001;
```

已选择 10000 行。

```
执行计划
----------------------------------------------------------
Plan hash value: 2702494104

--------------------------------------------------------------------------------------------
| Id  | Operation                           | Name | Rows  | Bytes | Cost (%CPU)| Time     |
--------------------------------------------------------------------------------------------
|   0 | SELECT STATEMENT                    |      |  9996 |  156K |    53   (0)| 00:00:01 |
|   1 |  TABLE ACCESS BY INDEX ROWID BATCHED| C    |  9996 |  156K |    53   (0)| 00:00:01 |
|*  2 |   INDEX RANGE SCAN                  | C_ID |  9996 |       |    24   (0)| 00:00:01 |
--------------------------------------------------------------------------------------------

Predicate Information (identified by operation id):
---------------------------------------------------
```

图 9-173　去掉标量子查询的资源消耗

```
    2 - access("C"."ID"<10001)

统计信息
----------------------------------------------------------
          0  recursive calls
          0  db block gets
       1382  consistent gets
          0  physical reads
          0  redo size
     327748  bytes sent via SQL*Net to client
       7733  bytes received via SQL*Net from client
        668  SQL*Net roundtrips to/from client
          0  sorts (memory)
          0  sorts (disk)
      10000  rows processed
```

图 9-173　去掉标量子查询的资源消耗（续）

以上都是从实际业务场景出发发现的问题，都是真实发生过的。

9.5　时序数据库使用的最佳实践

时序数据库场景相对单一，不需要对接大数据，因为其自身提供了海量数据分析场景。如果依靠时序数据库自身来完成分析，那么就需要在设计上下一番功夫，充分结合业务场景和规划来进行表的设计。

9.5.1　时序数据库的数据库表设计

本节将介绍有关时序数据库的对象设计。现实世界中的任何实体都可以看作对象，面向对象的设计本质上就是将现实世界的实体抽象成模型。以此为思路，钢铁行业供应链侧物流体系的抽象设计如图 9-174 所示，分为货物出厂、货物入库、货物在库、货物出库、货物交付 5 个环节。而在这 5 个环节之间形成了货物在途与货物在库两大场景。

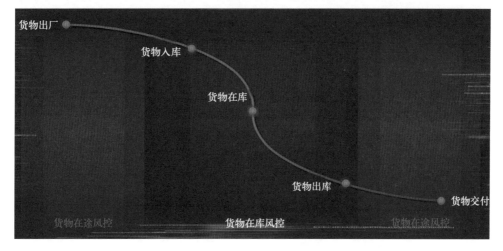

图 9-174　钢铁行业供应链侧物流体系的抽象设计

对于物流风控应用来说，货物在库是物流过程中耗时最长、投入物联技术最多的焦点场景，所涉及的典型物联系统有门禁系统、安防监控系统、视觉算法分析系统等。从技术路线来看，可分为传统物联感知技术与视觉算法分析技术，其中，物联感知技术可提供精准监管，但相对投运成本较高；而视觉算法分析技术则代表大数据驱动下的新型监管方式。

复杂的技术类型也意味复杂的数据类型，常见的包括非结构化数据（如安防监控系统中的视频图像）以及来自各类传感器的结构化数据（如用电量、温度和湿度等）。关于非结构化数据的处理，不在本节的讨论范围之内。下面以物联网门禁系统为例，重点分析时序数据库对于结构化数据对象设计的实践案例。

假设有一辆货车，每次进出仓库都会在仓库门禁系统中留下出入记录。这里每一条出入记录是典型的结构化数据，包含时间戳、车牌、开闸类型、进出场状态等字段，并被存储在时序数据库中。那么，同一辆货车每天进出不同的仓库，是否能将属于不同仓库的门禁系统的时序数据关联起来？

为了验证这一构想，可以进行以下的模拟实验。选取 TDengine 作为时序数据库系统，创建时序数据库模拟表，如图 9-175 所示，分表命名为 d1 和 d2。

```
taos> CREATE TABLE test.d1 (
    ->     ts TIMESTAMP,
    ->     direction INT,
    ->     velocity FLOAT,
    ->     latitude FLOAT,
    ->     longitude FLOAT,
    ->     plateno NCHAR(20)
    -> ) ;
Query OK, 0 of 0 row(s) in database (0.067965s)

taos> INSERT INTO d1 VALUES ('2022-11-01 08:01:00.000', 2, 1.00000, 1.00000, 1.00000, '沪AA27831');
Query OK, 1 of 1 row(s) in database (0.001530s)

taos> INSERT INTO d1 VALUES ('2022-11-01 08:02:00.000', 3, 2.00000, 2.00000, 2.00000, '沪AA27831');
Query OK, 1 of 1 row(s) in database (0.001246s)

taos> CREATE TABLE test.d2 (
    ->     ts TIMESTAMP,
    ->     direction INT,
    ->     velocity FLOAT,
    ->     latitude FLOAT,
    ->     longitude FLOAT,
    ->     plateno NCHAR(20)
    -> ) ;
Query OK, 0 of 0 row(s) in database (0.012299s)

taos> INSERT INTO d2 VALUES ('2022-11-01 08:01:00.000', 2, 1.00000, 1.00000, 1.00000, '沪AA27831');
Query OK, 1 of 1 row(s) in database (0.001965s)

taos> INSERT INTO d2 VALUES ('2022-11-01 08:02:00.000', 3, 2.00000, 2.00000, 2.00000, '沪AA27831');
Query OK, 1 of 1 row(s) in database (0.000987s)

taos> INSERT INTO d2 VALUES ('2022-11-01 08:03:00.000', 4, 3.00000, 3.00000, 3.00000, '沪A12345');
Query OK, 1 of 1 row(s) in database (0.001064s)

taos> INSERT INTO d2 VALUES ('2022-11-01 08:04:00.000', 5, 4.00000, 4.00000, 4.00000, '沪A12345');
Query OK, 1 of 1 row(s) in database (0.001067s)
```

图 9-175 创建时序数据库模拟表

这两个数据表分别对应两个物联门禁系统所产生的时序数据,时序数据库数据样例如图 9-176 所示。

```
taos> select * from d1;
           ts            |    direction    |     velocity     |     latitude     |    longitude     |      plateno      |
=============================================================================================================================
 2022-11-01 08:01:00.000 |              2  |         1.00000  |         1.00000  |         1.00000  | 沪AA27831         |
 2022-11-01 08:02:00.000 |              3  |         2.00000  |         2.00000  |         2.00000  | 沪AA27831         |
Query OK, 2 row(s) in set (0.084156s)

taos> select * from d2;
           ts            |    direction    |     velocity     |     latitude     |    longitude     |      plateno      |
=============================================================================================================================
 2022-11-01 08:01:00.000 |              2  |         1.00000  |         1.00000  |         1.00000  | 沪AA27831         |
 2022-11-01 08:02:00.000 |              3  |         2.00000  |         2.00000  |         2.00000  | 沪AA27831         |
 2022-11-01 08:03:00.000 |              4  |         3.00000  |         3.00000  |         3.00000  | 沪A12345          |
 2022-11-01 08:04:00.000 |              5  |         4.00000  |         4.00000  |         4.00000  | 沪A12345          |
Query OK, 4 row(s) in set (0.004214s)
```

图 9-176　时序数据库数据样例

先尝试使用车牌号码作为关联条件进行查询,但数据库提示该操作不受支持,如图 9-177 所示。实际上,这个结果并不出乎意料,因为时序数据库并非属于关系数据库管理系统(RDBMS)范畴,而更倾向于 NoSQL 类型,因此,无法直接进行关联操作。

```
taos> select * from d1,d2 where d1.plateno=d2.plateno and d1.plateno='沪AA27831';

DB error: invalid operation: not support ordinary column join (0.001798s)
```

图 9-177　时序数据库不支持关联查询

那么究竟该如何解决数据关联的问题?官方文档中介绍了一种名为超级表(Super Table)的解决方案,这里不妨来尝试一下(不同的时序数据库的做法不一样,本方法仅限于 TDengine)。

创建一个名为 t 的超级表,如图 9-178 所示,其包含了一个时间戳(TIMESTAMP)字段以及两个整型字段 x 和 y,同时使用 tags(标签)属性,用于存储采集设备的 id 和车牌号码。注意这里的 TAGS,目的是关联设备的 id 和车牌号码。同时创建子表 1,子表 1 中的用到了超级表的 TAGS 以及子表 2,子表 2 也用到了超级表的 TAGS。并通过使用超级表的 TAGS 属性,分别指定子表 1 的标签为('1', '沪 A12345'),子表 2 的标签为('2', '沪 A54321')。可见该数据库的产品在于提前设计,提前建模。

```
taos> create table t (ts TIMESTAMP,x int,y int) tags (id int,no nchar(100));

taos> CREATE TABLE t1 USING t TAGS ('1', '沪A12345');

Query OK, 0 of 0 row(s) in database (0.006030s)


taos> CREATE TABLE t2 USING t TAGS ('2', '沪A54321');

Query OK, 0 of 0 row(s) in database (0.017275s)
```

图 9-178　创建超级表

查看子表和超级表的表结构,如图 9-179 所示,从图中可以看到只有 t1 和 t2 两个表,但是请注意 stable_name 一列显示了它们的父表。

```
taos> show tables;
         table_name      |      created_time       | columns |     stable_name     |         uid         | tid  | vgId |
========================================================================================================================
 t1                      | 2023-07-17 09:20:36.426 |       3 | t                   |  2533274874300765   |    5 |    9 |
 d2                      | 2023-07-17 09:16:25.778 |       6 |                     |  2533274857523128   |    4 |    9 |
 t2                      | 2023-07-17 09:20:37.402 |       3 | t                   |  2533274891078295   |    6 |    9 |
 d1                      | 2023-07-17 09:12:20.762 |       6 |                     |  2533274840745330   |    3 |    9 |
Query OK, 6 row(s) in set (0.004890s)
```

图 9-179　查看子表和超级表的表结构

当分别向这两个子表插入一条数据后，可以直接在父表中进行查询。直接查询父表（超级表），实际上相当于在关系数据库中执行了"union all"操作，如图 9-180 所示。

```
taos> insert into t1 values ('2022-11-01T08:04:00.000+08:00',1,1);
Query OK, 1 of 1 row(s) in database (0.010983s)

taos> insert into t2 values ('2022-11-01T08:04:00.000+08:00',1,1);
Query OK, 1 of 1 row(s) in database (0.001937s)

taos> select * from t;
           ts            |      x       |      y       |      id      |        no         |
===========================================================================================
 2022-11-01 08:04:00.000 |            1 |            1 |            1 | 沪A12345          |
 2022-11-01 08:04:00.000 |            1 |            1 |            2 | 沪A54321          |
Query OK, 2 row(s) in set (0.039709s)
```

图 9-180　直接查询父表

最后，执行带有 WHERE 条件的 SQL 查询，时序数据库中变相关联如图 9-181 所示，实际上就相当于对这两个子表进行了关联操作。

```
taos> select * from t where no='沪A12345';
           ts            |      x       |      y       |      id      |        no         |
===========================================================================================
 2022-11-01 08:04:00.000 |            1 |            1 |            1 | 沪A12345          |
Query OK, 1 row(s) in set (0.018114s)
```

图 9-181　时序数据库中变相关联

因此，通过上述过程可以看出，时序数据库由于自身限制无法直接进行关联操作，但可以通过合理的设计（如如何设置标签以及设置多少个标签），来实现时序数据的关联分析。

物联网数据规模无疑是庞大的，如果使用传统的技术栈（成本暂且不提），处理这些数据将会非常烦琐。而时序数据库本身就是为解决时序数据的各种应用场景而设计的。这再一次提醒我们数据库选型和合理设计的重要性：没有最好的数据库，只有最合适的数据库。

9.5.2　时序数据库的数据分析

物流领域可提供多样性数据来源，包括代表信息流的业务数据和代表实物流的物联数据，而物联数据中又包含了异构的数据类型，即结构化数据和图片、视频等非结构化数据。本节将继续结合物联门禁系统，简单介绍仓储出入库环节数据分析的实践思路。

在出入库环节，除了门禁系统以外，监控系统在仓库的部署也十分广泛，因此可提供视频、图像作为货车出入记录的验证和补充。一般由监控设备得到的视频或图像会上传到对象存储服务器（如 S3），然后对目标轨迹、目标行为等进行算法推理分析，将图像信息转化为结构化数据，包括车辆类型、车牌号码、钢材类型等，同时存储在传统的关系数据库里。

此外，由物联门禁系统采集的时序数据存储在时序数据库中。理论上，监控系统和门禁系统的记录应一一对应，并可以根据时间戳或是车牌号码进行关联，然而在作业过程中，由于各种实际因素，如设备延迟、识别准确率等，往往难以百分之百确保。因此，对于风控应用场景，更关注从统计维度来反映仓库整体的运行情况。具体来说，可以是对以天为单位各类物联系统的数据进行汇总，例如，可以比较单位工作日内门禁系统和视觉系统分别记录的货车出入库数量，如果差值在一定误差范围内，就认为该项物联统计数据是真实有效的，确实反映了仓库的实际作业情况。

通过以上不同物联系统（监控系统与门禁系统）的综合应用，以及非结构化数据与时序数据的联动验证，进一步保障了物联统计数据的准确性。当我们能提供一定时间范围的物联统计数据并能保证其准确性后，便可以与上层系统中对应时间的业务数据相比较。风控应用中的时序数据联动分析如图 9-182 所示，若发现两者存在明显差异，可按照事先预设的规则形成风险事件，并推送到报警结果，从而完成风控应用中所需的从数据分析到事件预警的闭环。

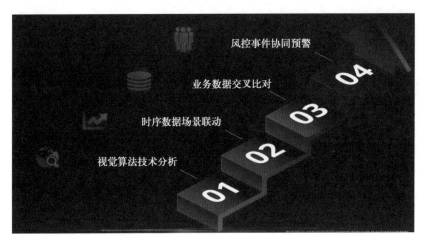

图 9-182　风控应用中的时序数据联动分析

时序数据库要不要和在线交易数据库对接？如果可以对接，那么势必能产生更大的价值。而本节介绍的是单一场景的时序数据库。如果要满足在线业务数据库和时序数据库的联合，那么就需要超融合数据库或者说是多模数据库才能达到。

附录

DBA 杂谈

DBA 是一个伴随着数据库产生和发展的群体，DBA 未来会如何发展？DBA 对相生相伴的数据库会有怎样的认知？DBA 生涯中有哪些感悟？本章就将分享这些引发广大 DBA 共鸣的话题。

1. 云计算时代，是否还需要 DBA

云计算带来了一个时代的改变。云计算厂商提供了 IaaS 和 PaaS，此外还有 SaaS 等服务模式，采用大量的开源和自研软件，加上自己的服务器管控以及共享的运维人员和专家，就构成了云服务，最典型的例子就是云数据库。云计算的本质是运维外包，还是一种具有规模效应的运维外包。它把运维工作都集中到了云计算厂商内部。一般上云的企业都通过公有云厂商以云租赁的方式购买 IaaS 和 PaaS 服务。这种情况下，DBA 的工作是不是被云厂商替代了？

其实无论用不用云，数据库都要面对以下几个问题：
- 不合理的需求。
- 不合理的设计。
- 不合理的实现。

这些问题都是需要人去解决的。如今云上的数据库会监控发现问题，能给的建议基本也是建立索引等。Oracle 数据库在 2018 年就实现了自治数据库，也就是自动化建立索引。在这个层面的 DBA 可以说已经或将要被替代了。

但是目前没有看到任何一个云数据库，给出了如何修改表结构、如何改 SQL 逻辑、如何修改表的关联，甚至去质疑业务逻辑和需求的。因为任何人在不了解业务逻辑的情况下是无法给出建议和意见的，而业务逻辑又不是简单地通过 AI 就可以掌握的。这些逻辑需要的是深入业务，甚至是深耕业务，才能知道如何优化，甚至是要改变业务才可以优化的。

根据笔者的实践经验，优化不仅仅是建立索引（当然这是必需的）。日常很多工作必须要触及实现、设计甚至需求才能优化。比如发现一个 SQL 有问题，问开发人员为什么要这样写？几乎 99% 的开发人员会说需求如此。找到需求方甚至是需求方的领导再去询问，得到的答复是我们的需求不是这样的。也就是说，需求与开发人员理解得不一样。只有通过改设计、改实现，使得需求变得合理，才能优化好数据库。

目前没有一个云数据库或云厂商可以针对上云的各家企业提供定制化的分析实际业务需求的服务。而这样的服务是当代 DBA 需要去做的。DBA 应该通过获取真实的需求，去引导需求方向着对数据库更加友好的方式提需求。然后指导开发人员怎么去实现，怎么去写代码。去修改表的设计以及对 SQL 逻辑的修改。

传统观点会说技术是为业务服务的，但这样的观点其实是阻碍了公司通过敏捷实践来获得收益。业务作为主导、技术作为辅助的观念在笔者看来需要一些改变——技术也是可以指导业务的。例如在 *A Seat at the Table* 一书中，作者 Mark Schwartz 讲述了 IT 领导者的作用是获取业务的结果，而不是交付给它的需求，而这反过来又要求我们的思考方式进行深刻的改变，重新思考管理、风险、团队结构以及 IT 部门与企业的其他部分之间的关系。IT 领导者不仅要对企业当前的 IT 能力集负责，还要对其 IT 资产的潜在价值负责——公司需要以一种敏捷的方式支持未来的需求。

云数据库的确解决了"部署、安装、备份、恢复"这些低价值工作。这些也就是传统的

DBA 的日常工作，也是 DBA 的基本功。在云时代这些简单重复性的工作会被公有云厂商完成，但是这并不代表 DBA 就要丢掉基本功，毕竟有这些基本功，处理问题时会比没有基本功的人要好一些。

综上所述，DBA 必须做出改变。不去尝试改写 SQL、不去设计数据库对象、不去控制需求的 DBA 不是合格的 DBA。因为不去做这些工作，那么数据库的稳定、高效就无从谈起。DBA 一定要进化成管控开发的 DBA，能够管理开发人员的代码质量。

2. DBA 是否要具备开发技能

DBA 确实需要具备一定的开发技能，但具体技能的掌握程度可能因职位和组织的不同而异。DBA 具备一定的开发技能。可以让 DBA 的职业之路走得更宽、更远。DBA 应该具备开发技能的原因如下：

1）在某些情况下，DBA 可能需要进行一些开发工作。例如，他们可能需要编写脚本或工具来实现数据库的自动化运维，或者使用一种或多种开发语言来协助开发团队。对于一些更复杂的数据处理任务，DBA 可能需要使用编程语言（如 Python、Java 等）来编写代码以处理数据。

2）若 DBA 具备一定的开发知识，可以站在开发人员的角度看待问题以及用开发人员的话术进行对话。从而在开发人员不知道怎么解决问题的时候给出开发层面的建议。

3）占应用程序的代码量一半的是 SQL，在代码中有大量复用的情况下，这个比例甚至可以达到 80%。可以说后端开发几乎就是以 SQL 为主体开发语言的，因为 Java 操作数据库依靠的还是 SQL。而 SQL 是 DBA 的基本技能之一。一个 DBA 应该知道如何编写高质量 SQL。

4）DBA 可以对开发人员进行画像，对后端程序代码（尤其是 SQL 代码）的质量进行管控从而提升数据库的性能。

5）若 DBA 具备开发技能，便可以更轻易地读一些数据库或者操作系统底层的代码对定位问题有非常大的帮助。

6）DBA 可以进行一些脚本或开源软件的二次开发及修改，使得其更加贴合自己使用，提升工作效率。

3. 盘点数据库的技术栈

数据库的技术栈有很多，具体如下。

1）关系数据库。以 Oracle、MySQL 等为代表。关系数据库主要以结构化数据为主，现在正在逐步融合半结构化的数据。对关系数据库需要了解其复杂的体系结构、锁、优化器以及高可用等，是所有数据库的基础，也是数据库应用最普遍的方向。

2）非关系数据库。如 Redis、MongoDB、列式数据库、图数据库。这些数据库属于半结构化数据库，不少人会因此把其归入中间件范畴。传统的关系型数据库在处理 Web 2.0 网站（特别是超大规模和高并发的 SNS 类型的 Web 2.0 纯动态网站）已经显得力不从心，出现了很多难以克服的问题，而非关系的数据库则由于其本身的特点得到了非常迅速的发展。

NoSQL 数据库的产生就是为了解决大规模数据集合多重数据种类带来的挑战,特别是大数据应用难题。

3)大数据方向。Hive、Impala、HBase 等 Hadoop 的组件都属于数据库的范畴,只是处理的方式方法不一,尤其是 Hive、Impala 基本使用的都是关系数据库的常用 SQL。

4)其他类型数据库:如搜索引擎、时序数据库、向量数据库、时空数据库等。

5)中间件方向。
- 有负责数据传输的 ETL 和 CDC。
- 有负责解耦的消息队列。
- 有负责对数据库读写的 JVM。

6)操作系统方向。数据库安装在操作系统之上,必须要和操作系统频繁交互。需要对 CPU、内存以及文件系统有深入的了解。
- 文件系统:了解操作系统中的文件系统是如何组织和管理数据的,以及文件系统的存储结构。
- 进程管理:了解操作系统如何管理和调度进程,以及进程之间的通信和同步方式。
- 内存管理:了解操作系统中的内存管理机制,包括虚拟内存和页面置换算法等内容。
- 文件和 I/O 管理:了解操作系统是如何管理文件和处理输入输出操作的。

7)网络方向。
- 网络知识:了解计算机网络的基本概念、协议以及网络架构,以便在数据库应用程序中进行网络连接和数据传输。
- 网络安全:了解网络安全的基本原理和概念,学习如何保护数据库系统的数据安全性。

8)硬件方向。需要了解各种存储对数据库的影响,选择合适数据库的硬件环境,并且与时俱进地更新硬件的技术进展。

9)开发方向。了解开发语言和工具,阅读开发编写的代码,找出其中的问题。

可能会由于认知原因没有将技术栈罗列完整,但是已经可以看出涉及的技术栈非常广泛。学习和从事数据库看似是一项工程,而实际上覆盖的面是超乎外人所想象的。在本书的第 7 章和第 8 章可以看到需要的不仅仅是开发技能,还有算法技能,在第 4 章可以看到还有逻辑和抽象建模技能等。

4. 怎么看待国产数据库的发展

国产数据库是最近几年大家谈论较多的话题之一。国产数据库分支图谱如图 A-1 所示(该图由徐戟老师提供,在此表示感谢),可以看到这个图谱枝繁叶茂。可能在不同人眼中有些分支应该合并或拆开,但这并不重要,因为还可能会出现新产品或改名字,所以只要不是大的分支有错误就没有关系。通过图谱可以看出,国产数据库主要,基于 MySQL 和 PostgreSQL 这两个生态的数据库占据大多数。

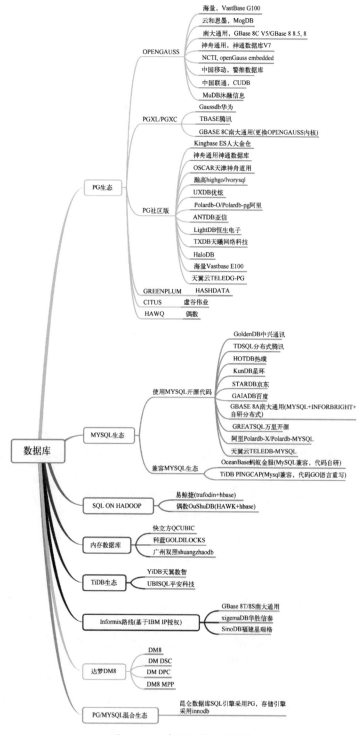

图 A-1 国产数据库分支图谱

国产数据库种类很多，有将近 300 种之多，因能力和精力有限只列出了一部分，会有新的数据库出现，也会有数据库消亡。对于国产数据库的发展，2023 年 1 月，TiDB、TDengine、TDSQL、OceanBase 数据库的掌门人一起相聚高端对话栏目《开谈》，共话中国开源数据库新格局。这次论谈中，大家预测在 2025 年的时候，可能还在健康运作的国产数据库会有 20 家左右。也就是说估计有 90%的国产数据库会消失。国产数据库会经历这一个时期的动荡，等到 2028 年左右最终会稳定下来。

5. 我们是否仍然需要 Hadoop

DBA 的工作是围绕着数据库展开的，这些年大数据是一个热词。大数据本身也是数据库技术的范畴。越来越多的公司要求 DBA 不仅要处理好关系数据库和非关系数据库，也要熟悉甚至是精通大数据技术。企业当中，Hadoop 是应用得较为广泛的、具有代表性的、基于分布式存储的大数据技术。

在第 6 章讲述了对于中小企业来说，使用 Hadoop 技术栈不是一个好的选择。技术也有时代性，在 Hadoop 刚刚诞生的时候，当时的硬件条件，无论是在存储容量上还是在 IO 吞吐量上，单机数据库或者说集中式数据库都容易触达当时硬件处理的瓶颈，在这种情况下确实需要分布式的解决方案。

其实分布式的思想早在 20 世纪就被提出了，只是当时由于技术的制约没有发展起来。主要是分布式对于网络的要求较高，以保证数据能够在服务器之间快速高效地传输，确保用户可以实时访问数据和应用程序，而不会遇到任何滞后或延迟。而在 21 世纪初，网络带宽是比较低的，随着网络中服务器数量的增加，传输的数据量也会增加，这意味着需要更大的网络带宽才能适应增加的流量。以上原因限制了分布式的发展，不过随着网络技术发展，网络带宽已不再是瓶颈，所以分布式也得以再次掀起热潮。

技术的进步不仅表现在网络层面。21 世纪初，一般的服务器内存的常规配置为 4~8GB，数据库的存储配上高端的阵列，存储空间可以达到几十 TB。现在的服务器物理机内存达到 1TB 已经是标配，已经可以做到把企业的所有数据全部加载到内存中处理。

大数据的分布式存储的主旨是用大量廉价存储、分散存储来弥补单一存储不足的问题。而这个难题被单盘的大容量所解决，单盘 SSD 的容量足以覆盖大多数企业的数据存储要求。

大数据的分布式计算主旨是缓解低效 IO，让所有服务器并行计算。通过 N 台机器的分散 IO 得到 N 倍的加速。举个例子，一个容器里有数量不详的红、黄、蓝、绿的豆子，让一个人统计这四种颜色的豆子各有多少个，这一个人就要从头数到尾，需要较长的时间。这个每次数一个豆子就是一次 IO 运算。而分布式计算就是找来 N 个孩子（孩子虽然数得慢，但是人多），每人拿一部分豆子，各自数各自的，每个孩子数一次豆子也就是一次 IO 运算。最终把每个孩子的结果汇总上来。由于是 N 个孩子一起，所以速度就提升了 N 倍。而现在随着使用非易失性存储器标准（NVMe）的技术，硬盘从 HDD 发展到 SSD，硬盘的存储容量和读写性能都得到了巨大提升。通过 NVMe 访问存储介质所花时间比以前的硬盘技术减少了 99.9%。也就是最初的那一个人数豆子的能力提升了 1000 倍，超过了找 1000 个孩子来数豆子的效率，那么就不需要再用 1000 个孩子这种劳动密集型的解决方案了。

之所以出现目前这种单机数据库热度再次超越 Hadoop 为代表的分布式解决方案,是因为硬件瓶颈的突破。数据库的瓶颈主要在于存储的 IO 吞吐量,而存储的吞吐量已经追赶上了数据增长的脚步。而分布式的协调、网络传输和架构使得开发和运维的成本远远高于单机。Hadoop 多年的实施和运维经验使得用户再次青睐单机(集中式)数据库的架构。

单机(集中式)数据库和大数据(分布式)都需要,只是大数据会合并到数据库中来,由集中式数据库或者分布式数据库直接完成,而不再需要单独的 Hadoop 来完成。越来越多高性价比的技术栈正在对 Hadoop 为代表的分布式大数据解决方案进行部分甚至全部的替换。

6. SQL 开发规范

几乎所有的 DBA 从使用和维护数据库的经验中得出的结论是:影响数据库稳定的主要的问题来自 SQL。这些 SQL 可能是人为执行的,但是更多的可能是应用程序执行的。通常来说,一个高质量(消耗数据库资源极少)的 SQL 是不会造成数据库故障的。造成数据库故障的通常是那些 IO 吞吐量大的 SQL。这里的 IO 吞吐量可能是逻辑 IO 也可能是物理 IO。一般来说,逻辑 IO 可能体现了系统的繁忙程度,可能是真的很忙,也可能是无效运作(比如因为表在内存中全表读)。而物理 IO 则非常容易造成性能问题甚至导致数据库宕机。在特殊情况下 SQL 影响的范围不仅仅是数据库的故障,还可能涉及中间件,造成连带问题。所以这里列出的开发规范几乎都是避免大 IO 吞吐量所定制的 SQL 规范。

1)对查询进行优化,尽量避免全表扫描,SQL 应该包含符合客观情况且有意义的 WHERE 条件。

说明:如果一个表每年有 100 万行数据,目前存储了 10 年的数据。WHERE 条件后面如果查询 8 年的数据,那么这个就是不客观或没有意义的。无论这个列是不是有索引,都极其有可能造成全表扫描。

2)尽量避免在 WHERE 子句中仅仅使用!=、<>或 or 操作符,应尽量避免在 WHERE 子句中对字段进行 null 值判断以及使用 or 来连接条件。

说明:仅仅使用不等于,无论这个列是不是有索引,都极有可能造成全表扫描。

3)谨慎使用 not in。

说明:仅仅使用 not in 约等于使用不等于,无论这个列是不是有索引,都极有可能造成全表扫描。

4)绝对禁止通配符第一位是%。

说明:第一位是通配符说明全表的任何数据都可能符合条件,一定会造成全表扫描。

5)禁止使用 select * from table for update。

说明:这种方式会造成本来不会锁的查询也开启了一个锁。

6)禁止在 WHERE 子句中的"="左边进行函数、算术运算或其他表达式运算。

说明:这种写法导致索引失效,一定会造成全表扫描。

7)避免复杂语句,循环嵌套不应该超过 3 层,关联表不要超过 3 张,避免不必要和无意义的排序。

说明:有的时候为了一个字段关联了一个表,而造成了非必要的多表计算。而按照非索引列排序则加剧了 CPU 的运算。当结果集很大时，内存无法完成排序会导致磁盘的排序。

8）禁止使用 NULL 和空格等不确定含义的字符作为业务判断逻辑。

说明：极有可能造成全表扫描。

9）WHERE 条件一定要包含索引的第一列。

说明：缺失前导列极有可能造成全表扫描。

10）对于分区表应该必选分区条件。

说明：如果没有选择分区条件，会造成跨所有分区的查询，其代价比不分区还要大。

11）对于大表的查询，一定要带分页。

说明：结果集很大，全部返回会导致性能较差。

12）对于大表不允许不带 WHERE 条件。

说明：没有 WHERE 条件会造成全表扫描，导致数据库 CPU 和 IO 高，甚至造成 JVM 的频繁 GC 甚至是 OOM。

13）不允许对 CLOB 和 BLOB 进行模糊查询。

说明：这种数据类型的模糊查询，直接导致全表扫描。

14）禁止使用框架的总页数计算，推荐使用分页，不显示总页数。

说明：除非每次查询结果集在 100 条以内，否则使用 MyBatis 的 PageHelper 会导致大量的 IO。

15）Update 必须带有条件，并且条件字段是带有索引的列。

说明：缺失索引或者无法用到索引，会导致全表扫描后再去更新，效率极低，容易造成锁。

16）多表关联禁止产生笛卡儿积。

说明：N 表关联时候必须有 N-1 个关联条件，否则容易产生笛卡儿积。而且关联列有索引，关联列缺失可能造成个别表全表扫描。

17）查询范围需要有上下限，不能只有一侧。

说明：没有明确的区间容易造成全表扫描，例如，时间范围只写了 1 小时以前，或者时间范围只写了 5 年以前。

18）谨慎控制 in 或 or 的个数。

说明：若 in 或 or 的值过多，容易造成性能低下。达到一定程度时，可能造成全表扫描。

19）or 涉及的字段必须都有索引。

说明：在 WHERE 条件后的 or 字段，如果任意一个字段缺失索引，则会导致全表扫描。

20）不允许拼接字段作为谓词充当过滤条件。

说明：拼接后的谓词造成了索引失效，发生全表扫描。

以上为 SQL 开发的规范，其实还有比 SQL 开发规范更加重要的，那就是数据库设计规范。因为在设计出问题的情况下，很多优化是收效甚微甚至是无能为力的。

以上列出的没有覆盖所有场景，在工作中总会发现以前没有的覆盖的场景，所以开发规范也是一个不断在修订的过程。但是以上的条目如果都能避免，或至少避免了常见的全表扫描，那么数据库的稳定性一定比之前有所提高。